Crossrail Project

Infrastructure design and construction

Crossrail

ice
Institution of Civil Engineers
publishing

Crossrail Project

Infrastructure design and construction

Edited by Mike Black, Christian Dodge
and Ursula Lawrence

Published by ICE Publishing, One Great George Street, Westminster, London SW1P 3AA.

Full details of ICE Publishing sales representatives and distributors can be found at:

www.icevirtuallibrary.com/info/printbooksales

Other titles by ICE Publishing:
Monitoring Underground Construction: A Best Practice Guide. British Tunnelling Society. ISBN 978-0-7277-4118-9

UK Specification for Ground Investigation, Second edition (Site Investigation in Construction Series).
Site Investigation Steering Group. ISBN 978-0-7277-3506-5

ICE Manual of Geotechnical Engineering. J. Burland, T. Chapman, H. Skinner and M. Brown (eds).
ISBN 978-0-7277-3652-9 (2 volumes)

www.icevirtuallibrary.com
A catalogue record for this book is available from the British Library

ISBN 978-0-7277-6078-4

Reprinted 2015

ICE Publishing is a division of Thomas Telford Ltd, a wholly-owned subsidiary of the Institution of Civil Engineers (ICE).

Commissioning Editor: Gavin Jamieson
Production Editor: Rebecca Taylor
Market Development Executive: Elizabeth Hobson

MIX
Paper from responsible sources
FSC www.fsc.org FSC® C013056

Printed and bound in Great Britain by
TJ International Ltd, Padstow, Cornwall

Contents

Foreword

This is the first in a series of volumes that will chart the technical progress of the Crossrail project.

The Crossrail Technical Papers Competition, which began in 2012, is a way of capturing some of these technical and innovative achievements. The competition has generated the papers that form the content for this publication. These papers record, in detail and with transparency, achievements and best practice made by the project teams during the design and delivery of this project.

The competition has been extremely successful in attracting high-quality contributions from authors across the programme. It is now an annual event in which personnel from the Crossrail delivery team, consultants, contractors, suppliers and third party stakeholders on the project can pass on their experiences and best practice to the rest of the project and the wider technical community.

This publication brings together the best of these papers from the first two years of the competition. There will be subsequent volumes covering the second half of the programme as Crossrail moves from the construction phase towards making the railway fully operational by 2019. It offers a unique opportunity to pass on invaluable learning to others planning, designing and delivering underground construction and urban metro projects throughout the world.

On behalf of Crossrail Ltd, we take great pride in the publication of this book which records the technical challenges and achievements of the teams delivering the Crossrail project. I hope you enjoy reading about this fascinating project and find its content both an interesting and informative read.

Chris Dulake
Crossrail Chief Engineer

An Introduction to the Crossrail Project

The concept of rail tunnels crossing central London to link Paddington and Liverpool Street stations dates back to the 19th century when the Regent's Canal Company was granted permission to build a railway that would link Paddington with London's docks. Although this was aborted, the idea reappeared in the 1940s and is alluded to in Sir Patrick Abercrombie's County of London Plan and Greater London Plan.

The idea has had many variations since then, but the principles of having twin-bore tunnels large enough to handle main line rolling stock, yet providing a metro-tube service without the need for passengers changing between main line stations and tube stations, remains the purpose to this day.

The construction of the Crossrail project began at North Dock in Canary Wharf in May 2009. Five years on, it is the biggest railway construction project in Europe and is one of the largest single infrastructure investments undertaken in the UK. It consists of 21 km of new twin-bore tunnels under central London and 10 new world-class stations constructed under the largest city in the European Union. The route will link existing Network Rail services from Reading and Heathrow in the west, and Shenfield and Abbey Wood in the east connecting new stations at Paddington, Bond Street, Tottenham Court Road, Farringdon, Liverpool Street, Whitechapel, Canary Wharf, Custom House, Woolwich and Abbey Wood.

The Crossrail tunnels are being constructed using eight Tunnel Boring Machines (TBMs) beneath the busy streets of London. These running tunnels, linking the stations, are constructed with precast reinforced concrete segments. The ten new ticket halls for the eight underground stations, five portals, plus the various ventilation and access points along the route, are constructed using secant and diaphragm walls, boxes and shafts. The station tunnels and caverns are being constructed using fibre reinforced sprayed concrete lining (SCL). In addition to this, there is a multitude of grouting shafts used to mitigate the effects of ground movement on London's existing infrastructure.

Currently, over 44 million hours have been worked, with the project employing over 10,000 people across more than 40 worksites throughout London and the South East. However, Crossrail is more than a civil engineering project. As the tunnelling and underground construction phase nears completion, the project is embarking on the next big challenge involving the mechanical, electrical and architectural disciplines related to rolling stock, rail systems and station fit outs.

Crossrail has relied on tried and tested technology and engineering principles throughout its design and construction programme while recognising its duty to embrace new technologies and innovation wherever practical.

As part of Crossrail's legacy, it is incumbent upon the organisation to share its experiences and best practices with the rest of the industry and to showcase the skills of the personnel involved and the successful delivery of each phase of works. This publication is the culmination of that experience and best practice celebrated by the Crossrail Technical Papers Competition, and is presented here.

Crossrail Project: Infrastructure design and construction
ISBN 978-0-7277-6078-4

doi: 10.1680/cpid.60784.001

Development of two different station designs – a comparison of Paddington and Farringdon Stations

James Paterson MA MSc DIC CEng FICE MIStructE CPEng FIEAust, Lead Civil & Structural Engineer for Farringdon Station, URS
Mark Raiss MA MSc DIC PhD CEng FICE FIStructE FCS, Crossrail Programme Director, URS.

ABSTRACT

Crossrail Central includes the construction of eight new underground stations and one new surface station linked together by 21km of new twin bore tunnels. This paper describes some key design features of two of the underground stations at Paddington and Farringdon, both of which have been designed by URS. Both stations are to be constructed on constrained sites in central London adjacent to historic buildings, and they are both key interchange stations which have to address many of the same design issues.

However there are major differences in their layout and forms of construction. The aim of this paper is to explore the reasons for the different solutions and to show that they are appropriate and cost-effective. It is hoped that this broad overview will be of assistance to designers when considering similar rail projects at feasibility stage. The paper also shows that, despite the fact that the two stations are more than 3 miles apart, they are inextricably linked by the planning and programming of the tunnelling operations.

INTRODUCTION

Crossrail Central includes 21km of new twin bore tunnels through central London and the construction of eight new underground stations, plus one new surface station. [*Fig 1*]
The purpose of this paper is to describe a number of key design features of the two underground stations at Paddington and at Farringdon. They have been selected as they are both to be constructed on constrained sites in central London adjacent to historic buildings. They both form key interchange stations and both have to address many of the same fundamental design issues. However there are major differences in layout and in the form & sequence of construction which are a direct consequence of their particular location and other constraints.

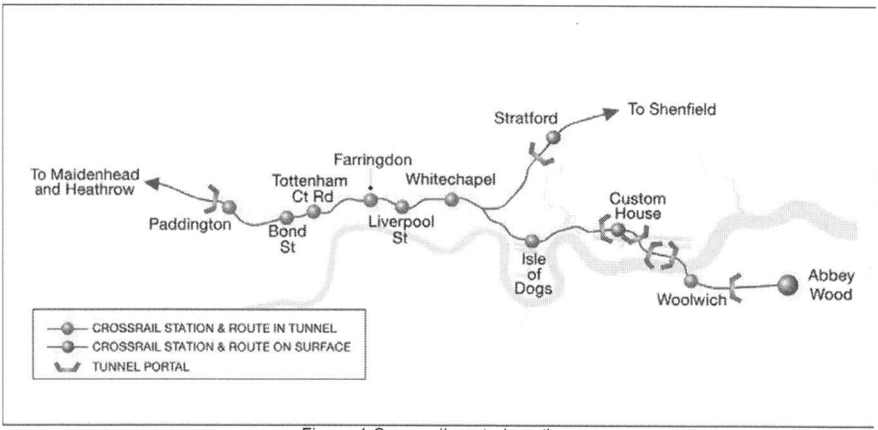

Figure 1 Crossrail central section

Also, while the two stations are more than 3 miles apart, they are connected in terms of the tunnelling works and the tunnelling construction programme, which have a significant impact on the design solutions.
This paper:

- Outlines some of the key design issues
- Describes the location and constraints of the two stations
- Explains how the two stations address particular key design issues
- Provides a table of key differences
- Shows how the solutions adopted are both appropriate and cost-effective.

This paper concentrates on the civil & structural engineering aspects. The authors hope that this broad summary will be of assistance to designers of similar projects. In preparing this paper, it has been assumed that the reader is familiar with the key features of the overall Crossrail Project.

Paddington Station is the subject of a separate paper - which covers the design in detail; this paper therefore only outlines certain features of the design of Paddington to allow comparison with Farringdon Station design.

KEY DESIGN ISSUES

Summary

There are very many different factors that need to be considered in the design process. However, for the purpose of this report, the relevant factors are taken to be:

- Adjacent existing structures / urban context
- Rail infrastructure connections
- Oversite developments
- Integration with tunnelling works
- Ground conditions
- Form of wall construction
- Sequence of construction
- Spaceproofing

The final design of both stations is driven by safety, buildability, physical constraints and the need to provide a station appropriate to its location and historical setting. The key design objective has been to provide the travelling public with an excellent passenger experience while providing Crossrail with a design that offers value for money and that can be built safely within the overall Crossrail programme.

While this report concentrates on civil & structural design issues, the authors wish to emphasise that all disciplines have worked collaboratively with each other, as well as with Crossrail, third party stakeholders and contractors, to develop solutions which respond to these constraints and drivers. The decision to co-locate the design teams with the Crossrail client has been important in fostering this spirit of co-operation.

Adjacent Existing Structures / Urban Context

Both stations are to be constructed in densely populated areas on restricted sites and adjacent to listed buildings. Limiting ground movements is a key to minimising the risk of damage to adjacent properties, utilities and infrastructure. This is particularly important for safety-critical items such as certain utilities, London Underground or Network Rail lines, as well as for historic structures.

Rail Infrastructure Connections

Both Paddington and Farringdon Crossrail Stations form key interchanges to London's existing rail infrastructure. [*Figure 2*]

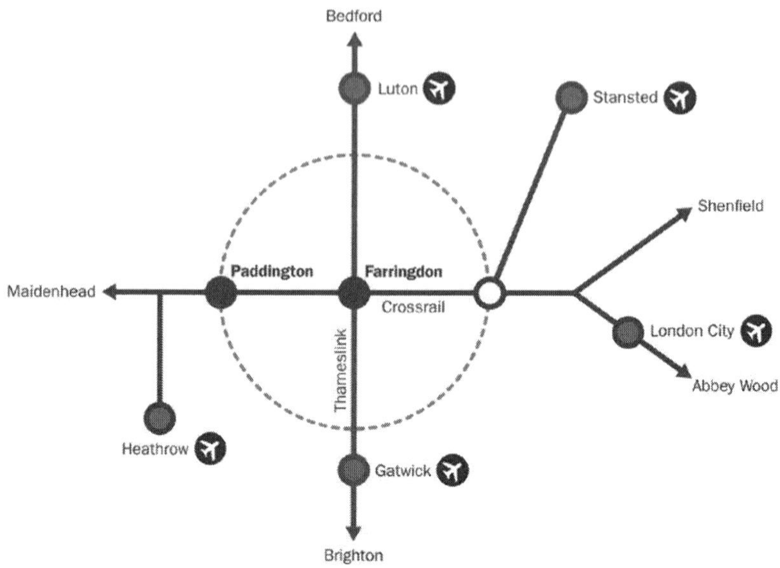

Figure 2 Key transportation links

Paddington Crossrail Station, the most westerly of the Central Stations, links in with Network Rail's terminus station at Paddington, as well as providing a direct link to London Underground's (LU's) Bakerloo Line.

Farringdon is located centrally in the heart of London. As well as providing links to the Circle, Hammersmith & City and Metropolitan LU lines, it also forms a major interchange with Thameslink – the only other cross-London mainline railway. On completion, Farringdon Station will provide direct rail links to four of London's five airports.

Oversite Developments (OSDs)

Part of the funding for Crossrail is derived from commercial OSDs, which also play an important role in re-establishing the built environment.

At Paddington Station there is no OSD directly above the station, while at Farringdon there are two OSDs, one above the West Ticket Hall (WTH) and one above the East Ticket Hall (ETH).

The significance is that Paddington Station represents a completed, final architectural statement, while at Farringdon both sites rely on the subsequent OSDs for their full integration into the urban settings.

Integration with Tunnelling Works

One of the key design considerations for any underground station is the way it integrates with the tunnelling works – both the permanent structural connections and the construction phase (including the dismantling & removal of the Tunnel Boring Machines (TBMs)).

Platforms are typically either

- formed within a cut-and-cover box [Paddington], or are
- constructed as mined tunnels using Sprayed Concrete Lining (SCL) construction techniques [Farringdon]

The choice of construction methodology depends on the alignment of the tunnels and the depth of construction. At Paddington, the depth from existing road to top of rail level is approximately 21m, while at Farringdon Station the depth varies from 27m to 32m.

Five twin-bore tunnels are to be constructed. Two of these twin-bore tunnels (X- and Y-Drives) affect the construction of Paddington and Farringdon Stations:

- The X-Drive (6.4 km long) commences at Royal Oak portal, just to the west of Paddington Station, proceeds in an easterly direction and terminates at Farringdon
- The Y-Drive (8.3 km long) commences at the Limmo shaft in east London and proceeds in a westerly direction also terminating at Farringdon

Consequently Farringdon Station is the termination point for two drives (ie 4 TBMs) and has to accommodate their arrival, decommissioning and removal. Paddington Station is near the start of the X-Drive, and consequently the sequence and method of construction of the station has to accommodate the passage of two TBMs through the station box relatively early in the station's construction sequence. [*Figure 3*]

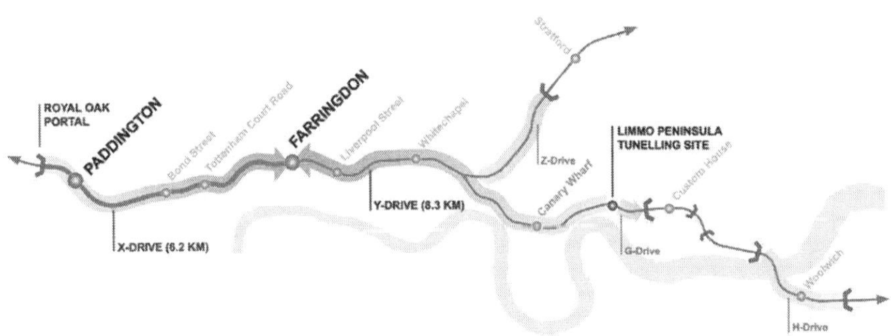

Figure 3 Crossrail central tunnelling drives

Ground Conditions

The ground conditions at the two stations are very different.

At Paddington, London Clay exists at shallow depth and extends well below the base of the excavation. These favourable ground conditions are beneficial to the tunnelling operation and can be exploited in the box construction methodology.

By contrast, at Farringdon, the ground is very complex and heavily faulted, with London Clay overlying Lambeth Group and Thanet Sands. Although, in terms of ground water levels, the site is under-drained at depth, there are perched water tables and water-bearing gravels which pose particular problems for the construction of mined SCL tunnels. [*Figure 4*] These ground conditions have had a major impact on the tunnelling strategy and a direct impact on the size, location and sequence of construction of the shafts.

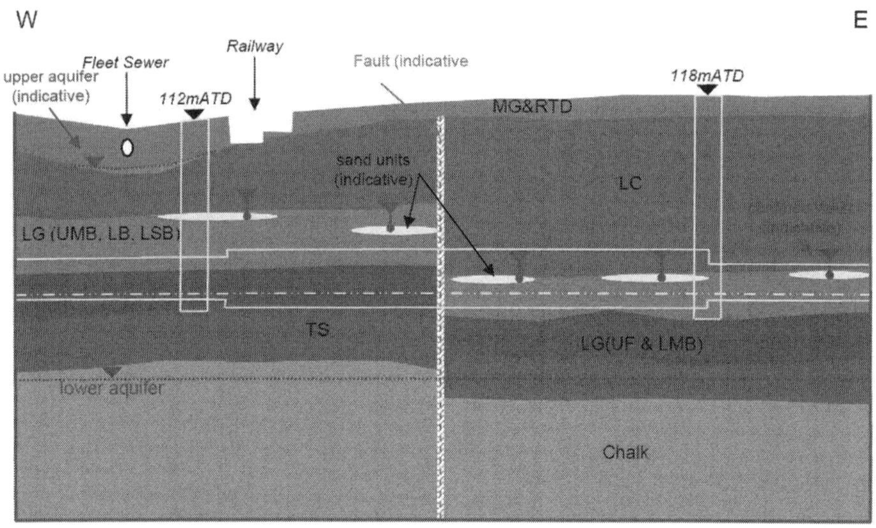

MG=Made Ground; RTD=River Terrace Deposits; LC=London Clay; LG=Lambeth Group; UMB=Upper Mottled Beds; LB=Laminated Beds; LSB=Lower Shelly Beds; Upnor= Upnor Formation; LMB=Lower Mottled Beds; TS=Thanet Sand

Figure 4 Farringdon diagrammatic geological section

Wall Construction

Deep basement construction is a feature of most Crossrail Central stations, whether in the form of a single cut-and-cover box or a series of discrete shafts & boxes.

While many methods of construction exist, perimeter walls are typically either

- diaphragm walled panel construction or

- secant or contiguous bored-pile walls.

While diaphragm-walled construction can offer considerable advantages, the size and type of plant means that it is not suitable for all sites.

Sequence of Construction

The sequence & method of construction for shafts and underground boxes is usually described as either “bottom-up” or “top-down”.

[These terms refer to the sequence in which the permanent works are cast within the perimeter walls, ie:

“Bottom-up” construction is when temporary propping (eg - steel frames or temporary anchors) is used to support the perimeter walls as excavation proceeds down to the bottom. The bottom levels of permanent works slabs are therefore constructed first and the other levels are constructed progressively upwards.

“Top-down” construction is when the upper level, permanent works slab is cast first, and the lower levels are constructed progressively as excavation proceeds. The slabs prop the perimeter walls as excavation proceeds and act as both permanent and temporary works props (ie the *permanent works* are constructed from the *top down*).] [*Figure 5*]

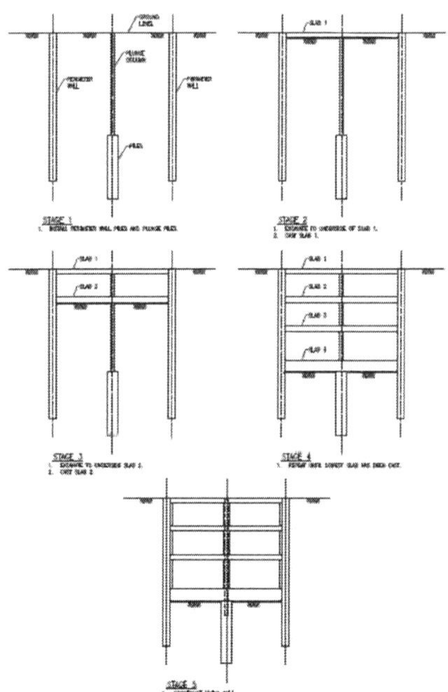

Figure 5 Top-down construction sequence

Either sequence of construction can be used with diaphragm walled or contiguous / secant bored pile wall construction.

Generally top-down construction is better at controlling ground movements. The top slab (which is cast first) can provide useful layout areas on congested sites and/or allow road access to be reinstated. While time to complete all of the permanent works can be similar for either method, the time to reach the base of the excavation using temporary steel props in bottom-up construction is usually quicker.

A feature of top-down construction is the use of plunge columns - steel columns plunged into bored piles to carry the weight of the slabs as excavation proceeds. Plunge columns can either be temporary (removed after construction of the permanent columns & walls), or permanent (cast into the permanent walls or columns).

Because the slabs are typically cast on the ground, there is a great advantage in simplifying the soffit shape and a flat soffit is generally preferable to a beam and slab arrangement. A clear route for horizontal loads in both the temporary and permanent condition is required.

Spaceproofing

The layout and configuration of a railway station has to respond to a complex range of requirements and constraints.

For deep underground stations, space-proofing is of particular importance. In addition to passenger flows, vertical circulation and associated space requirements, one of the key considerations is the space-proofing associated with tunnel ventilation, service risers, means of escape and fire-fighting access. Tunnel ventilation, in particular, requires risers with large cross-sectional areas and with a limited number of bends – with large fans requiring access for maintenance, repair and replacement.

Ultimately the purpose of the Crossrail project is to serve the travelling public. Key to that is to create routes with intuitive way-finding to improve the passenger experience for all users and to cater for growth in passenger demand. This has a big impact on the volumes and spaces created and hence on the structural layout.

STATION DESCRIPTIONS

The following descriptions should be read in conjunction with the following figures [*Figures 6, 7, 8 & 9*], which provide a direct visual comparison between the two stations.

Figure 6 Farringdon and Paddington plans at basement level

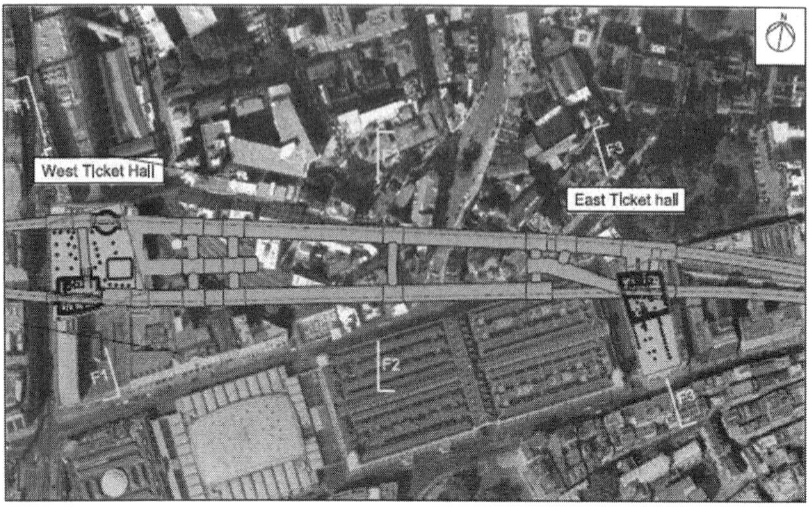

Farringdon Station
Platform Level (Scale 1:3000)

Paddington Station
Platform Level (Scale 1:3000)

Figure 7 Farringdon and Paddington plans at platform level

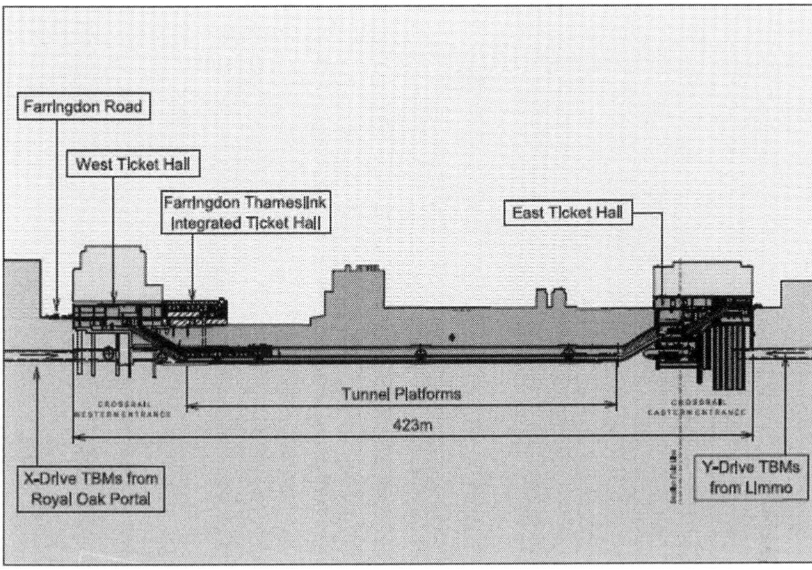

Farringdon - Longitudinal Section (Scale 1;3000)

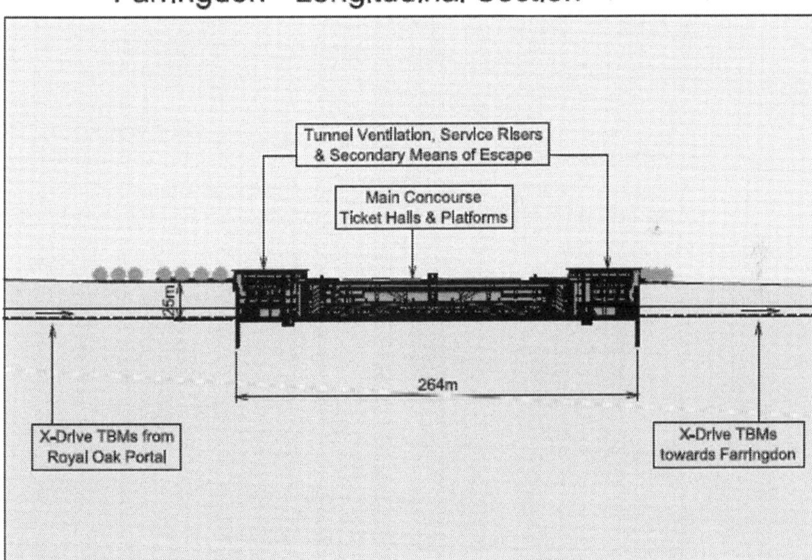

Paddington - Longitudinal Section (Scale 1:3000)

Figure 8 Farringdon and Paddington longitudinal sections

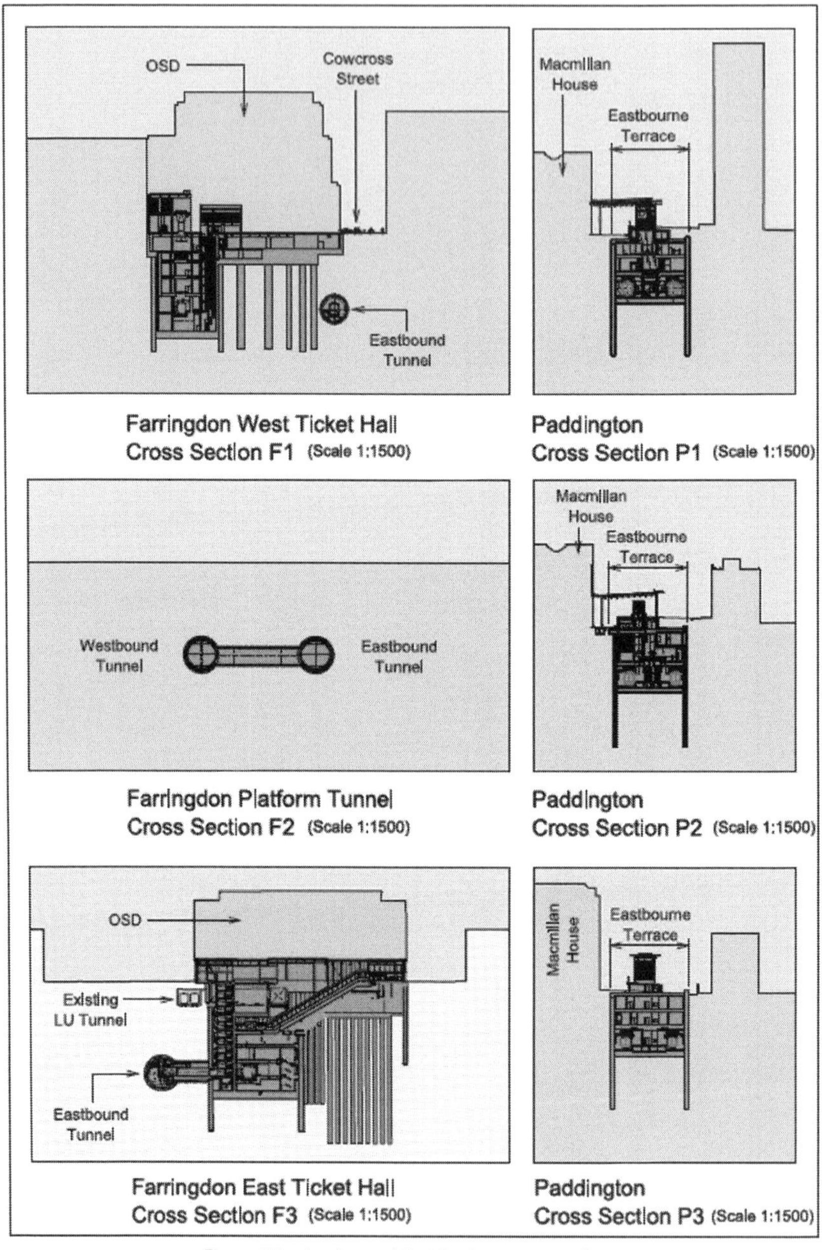

Figure 9 Farringdon and Paddington cross-sections

Paddington Station

Context

Paddington Crossrail Station is to be constructed under Eastbourne Terrace immediately adjacent to Paddington mainline station.

Paddington mainline station is the London terminus for frequent long-distance high-speed trains to the west of England as well as for local suburban trains, and is housed in Isambard Kingdom Brunel's historic shed which dates from 1854 and is Grade I listed. Macmillan House, a terrace adjacent to Eastbourne Terrace is also Grade I listed, while the Hilton Hotel building which fronts Paddington Station is Grade II*. [*Figure 10*]

Figure 10 Paddington historic photograph of Eastbourne Terrace

As noted above, there is no oversite development directly above the station (although there is the option for an oversite development above the "bomb gap" in Macmillan House in the future). The design of Paddington station therefore creates a single architectural entity, the design of which needs to sit sensitively alongside the existing Grade I building.

Although, like other Crossrail Central stations, Paddington is to be constructed on a constrained site, it does have the benefit that the entire station can be constructed beneath an existing road, Eastbourne Terrace. The complete station, including ticket hall, concourse, platforms etc can therefore be contained within a single rectangular box at a relatively shallow depth. [*Figures 11 & 12*]

Figure 11 Paddington typical cross-section

Figure 12 Paddington cut away section

(In contrast, at Farringdon, there is a split site; tunnel platforms are therefore required at considerable depth, as the platforms are located below existing buildings and infrastructure.)

Station Features

Some of the key features of the Paddington Station design are summarised to allow comparison with Farringdon Station - which is described in greater detail below.

There is a single island platform at track level for eastbound and westbound trains, with evacuation cores and ventilation & services risers at each end. The structure comprises a single elongated box running parallel to the existing mainline station (external dimensions 264m long x 25m wide). The large central void houses the two main escalator inclines. The concourse is protected by a glass canopy which replaces the existing Grade I listed canopy. [*Figure 13*]

Figure 13 Paddington concourse image

There is a central row of 1800 x 1250mm elliptically shaped columns at 18.33m centres. The roof slab is just below ground level and the general excavation depth is approximately 24m below Eastbourne Terrace.

Form of Wall Construction / Sequence of Construction

The box is to be constructed top-down using 1.0m and 1.2m thick diaphragm wall panels around the entire perimeter. At each end of the box, the diaphragm wall is stopped short to allow the subsequent tunnelling works. The total length of diaphragm wall is 575m, and the typical depths are 40m.

Top-down construction is used to limit movements of the adjacent Grade I listed buildings and to allow Eastbourne Terrace to be reinstated at the earliest opportunity.

The primary slabs (roof, intermediate, concourse and base) prop the diaphragm walls both during excavation and in the permanent case. One level of temporary steel propping is required at an intermediary level to control movements and reduce bending moments, and to limit stresses in the already constructed tunnel linings. These props are removed after casting the concourse slab. The 1750mm deep base slab acts as a raft foundation for the columns and walls and also resists the upward loads due to long term hydrostatic pressures and soil heave. The base slab is connected to a central line of tension piles and the diaphragm walls, which resist the upward loads.

Integration with TBM construction

The method and sequence of construction has been integrated with the planned tunnelling works. The strategy adopted for Paddington station is to construct critical sections of diaphragm wall in advance of the arrival of the X-Drive TBMs. This avoids the risks associated with constructing the diaphragm walls close to the tunnels while tunnelling operations are in progress.

Excavation and top-down construction commences after the TBM drives have passed through. Careful consideration has been given to the risks associated with the un-propped open tunnel bores during the excavation works and the impact these would have on the structural behaviour of the diaphragm wall panels. The temporary props at intermediary level are to be preloaded to allow continued use of the tunnels by others during the excavation until such time as the concourse slab is installed. The tunnels are closed before excavation can proceed below the concourse slab (with the tunnels segments being demolished within the station footprint) during the main station works.

In addition to the permanent openings required for vertical circulation and service risers, large temporary openings are formed within the slabs, improving construction access between levels and reducing loads on the temporary plunge columns. 51 No 1800mm diameter bored piles are provided along the centre of the box to act as tension piles against uplift due to heave and flotation. 49 of these piles also act as compression piles to support the temporary steel plunge columns.

Early designs were based on a partial closure of Eastbourne Terrace. However, following a careful optioneering study, full closure of the road has now been agreed for the first two years of construction. This results in a larger, safer worksite during the piling and during the construction of diaphragm wall, and greatly reduced construction times.

Farringdon Station

Context

Farringdon Crossrail Station is located in an area of London with a rich history.

Unlike Paddington Station, there is no opportunity to construct a cut-and-cover box as Farringdon is split between two sites (the West Ticket Hall [WTH] site and the East Ticket Hall [ETH] site) which are over 300m apart. [*Figure 14*] Tunnel platforms are required and a total of four discrete shafts, rather than a single box.

Figure 14 Farringdon – overall isometric view

Both sites were formerly occupied by buildings which have been demolished to allow construction of the ticket halls, shafts etc to take place. [*Figures 15 & 16*] At the WTH, there is a "collaborative" OSD (ie the developer is already established) while at the ETH there is a "non-collaborative" OSD where a scheme has been taken through to Planning Approval stage by the URS design team.

Figure 15 Farringdon western ticket hall – aerial view prior to demolition

Figure 16 Farringdon eastern ticket hall – aerial view prior to demolition

Both ticket hall structures have been designed to accommodate vertical loads and stability loads from the proposed development, as well as making provision for ground level entrances, retail frontages, lift cores, connections to utilities etc.

One of the key buildings in the area is the Grade II* Smithfield Market, the east elevation of which is adjacent to the ETH site. At the WTH, the recently constructed Thameslink Ticket Hall has been designed to accommodate the future Crossrail station construction and to form an integrated Crossrail / Thameslink ticket hall and interchange.

Station Features

Platform Construction

As noted above, the ground at Farringdon is complex and heavily faulted, and the water-bearing gravels create extremely challenging conditions for the construction of the platform tunnels.

It was established early on in the design process that the platforms should be formed by SCL enlargements to lined pilot tunnels constructed using TBMs. [*Figure 17*] The pilot tunnel allows water-bearing channels to be depressurised via wells drilled out from the pilot tunnel, greatly reducing the risk of face collapse during the subsequent SCL enlargement works.

Figure 17 Typical mined station

Impact of tunnelling strategy

The impact of the tunnelling strategy at Farringdon is also very challenging as the station is the termination point for four TBMs.

The original strategy had been to terminate the X-Drive at the WTH site and the Y-Drive at the ETH, and to separately procure a smaller diameter TBM to construct the pilot tunnels. However, following a detailed study, it was decided to take the X-Drive through to the ETH and to use the standard diameter X-Drive tunnels as the pilot bores. Consequently the X-Drive passes through the WTH and on to the ETH. The ETH is now the termination point for all four TBMs.

Description of Ticket Hall Sites

Both ticket hall sites required the demolition of existing structures and both are bounded by roads, railways and structures, with original ground level typically 4 to 6m below road level.

The two sites are roughly rectangular and their overall dimensions are approximately:

- WTH: 65m x 48m
- ETH: 70m x 32m

The WTH site is bounded by:

- Farringdon Road to the west - Red Route
- Cowcross Street to the north - heavily-used / pedestrianised
- Recently demolished building to the south
- Newly constructed Thameslink ticket hall to the east (some loads from WTH are carried by the structure of the Thameslink Integrated ticket hall)

The ETH site is bounded by:

- LU lines to the north
- Barbican LU station to the north-east
- listed Smithfield Market to the west
- narrow street to the east and south

The suspended Moorgate spur of the Thameslink line passes through the site - to be reinstated on completion.

Form of Wall Construction / Sequence of Construction

The four shafts at Farringdon are irregular in shape, and are of different sizes and depths.

Access to the piling platforms, several metres below road level, is difficult. That and the size of the sites and their proximity to the rail infrastructure meant that diaphragm walled construction was considered inappropriate - leading to the choice of secant bored-pile wall construction.

WTH – 3No Shafts

Large rectangular shaft on the westbound tunnel (25m x 21m):
Provision for tunnel ventilation, service risers and emergency means of escape. It is necessary to excavate the shaft quickly to allow the westbound X-Drive TBM to pass through. The shaft is immediately adjacent to Farringdon Road, which contains major services - making it particularly important to limit ground movements. The need to excavate the shaft rapidly has dictated the use of bottom-up construction - but ground movements have been kept within specified limits by using stiff steel temporary frames at close centres.

Relatively shallow rectangular shaft (15m x 11m):
One level of temporary props and built bottom-up. Its primary function is to facilitate the construction of an inclined SCL escalator barrel.

Circular shaft on the eastbound tunnel (13m diameter)
Provides access for the construction of an SCL stub tunnel. The stub tunnel is required to enable the existing under-reamed piles from the now-demolished 1960's building to be removed (these piles would otherwise have clashed with the passage of the incoming eastbound X-Drive TBM). It has circular concrete waling beams - subsequently cast integrally with the intermediary slabs. Rapid construction is required to allow for the construction of the stub tunnel prior to the arrival of the TBM. [*Figures 18 & 19*]

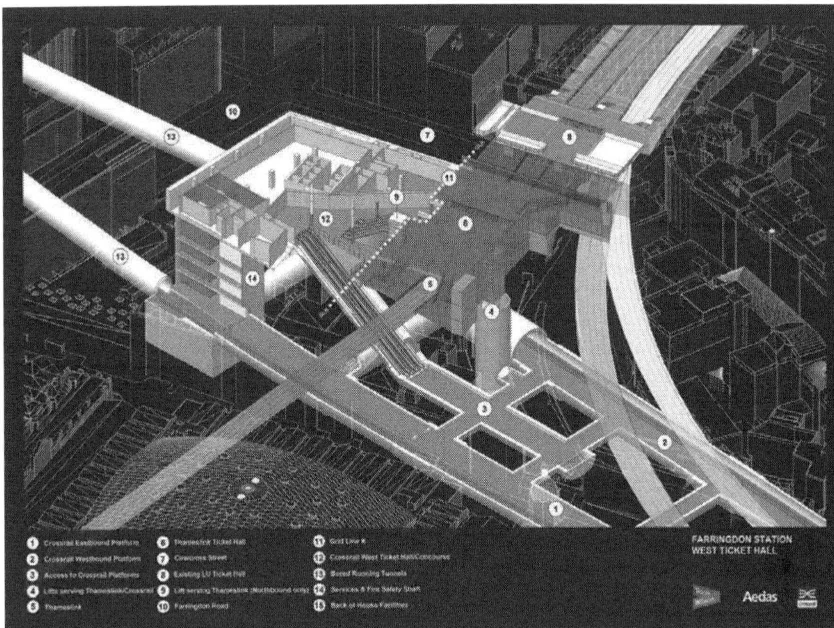

Figure 18 Farringdon western ticket hall isomentric view

Figure 19 Farringdon integrated western ticket hall

ETH – 1No Shaft

Trapezoidal Shaft (30m x 25m)
A two-storey basement over the south half of the site houses M&E plant, while the north half of the site is occupied by a large deep trapezoidal shaft, which extends 35m below road level.

Its plan size is dictated by the need to accommodate tunnel ventilation, service risers, plant rooms, vertical transportation (two escalators and an inclined lift) and emergency means of escape. Space-proofing requirements have meant that the shaft extends to the site boundary on three sides.

1.5m diameter secant bored piles are used in conjunction with top-down construction. Flat slabs prop the perimeter walls in the temporary and permanent condition – reducing wall displacements and hence controlling ground movements. These slabs are supported on the perimeter walls and internally on four steel columns, which are plunged into 1.5 m diameter piles. The steel plunge columns are subsequently cast into four 1.0 x 1.0m reinforced concrete columns, forming part of the permanent works.
[*Figure 20*]

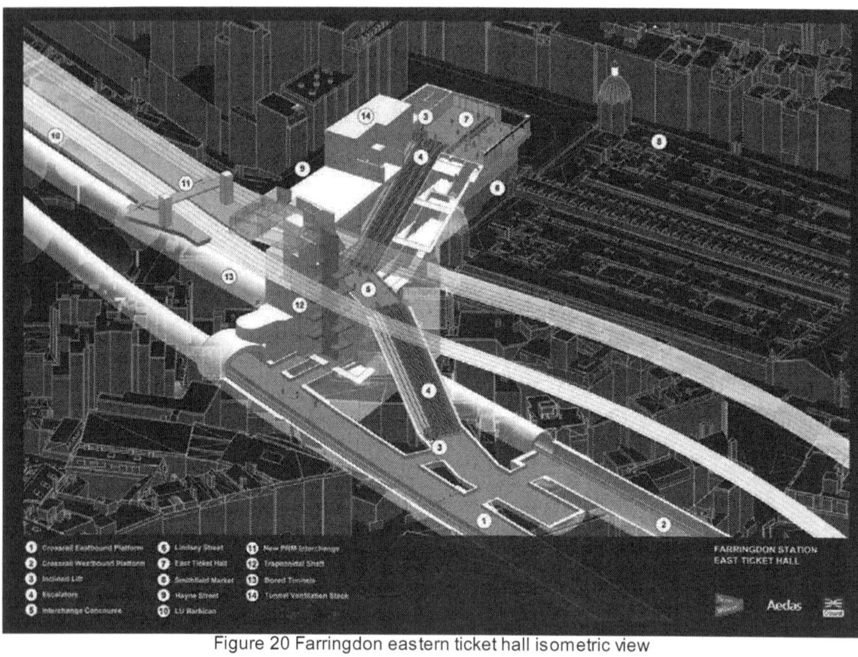

Figure 20 Farringdon eastern ticket hall isometric view

Integration with TBM Construction

The ETH trapezoidal shaft performs a vitally important role in terms of the tunnelling strategy as the termination point for four TBMs. Close collaboration was required with other consultants as well as with the two tunnelling contractors during the Optimal Contractor Involvement (OCI) phase.

X-Drive TBMs

- Large temporary voids are cast in the centre of the first three levels of permanent slab during the top-down construction. These voids allow the dismantled X-Drive TBM cutter heads to be craned out.
- A temporary SCL shaft is to be constructed within the trapezoidal shaft beneath the first three levels of slab.
- The westbound X-Drive TBM will bore through the west wall of the trapezoidal shaft and terminate within the temporary SCL shaft. The dismantled westbound X-Drive TBM cutter head will be taken out through the temporary voids.
- From the base of the temporary SCL shaft a temporary adit is to be mined out, passing through the north wall of the trapezoidal shaft to form a platform-sized reception chamber for the eastbound X-Drive, to the north of the trapezoidal shaft. The cutter head of this drive will also be removed through the ETH shaft.
- In both cases, the TBM train is taken back to the portal at Royal Oak, passing through Paddington Station.

Y-Drive TBMs

- The two Y-Drive TBMs arrive at the ETH approximately a year after the X-Drive, by which time the trapezoidal shaft will have been constructed
- The westbound Y-Drive TBM will therefore terminate in a platform sized reception chamber constructed immediately to the east of the trapezoidal shaft
- The eastbound Y-Drive TBM will terminate in the same platform-sized reception chamber as the eastbound X-Drive
- In both cases the whole of the TBM train, including cutter head, will return to the portal at Limmo, and will not be taken out through the ETH shaft. [*Figure 21*]

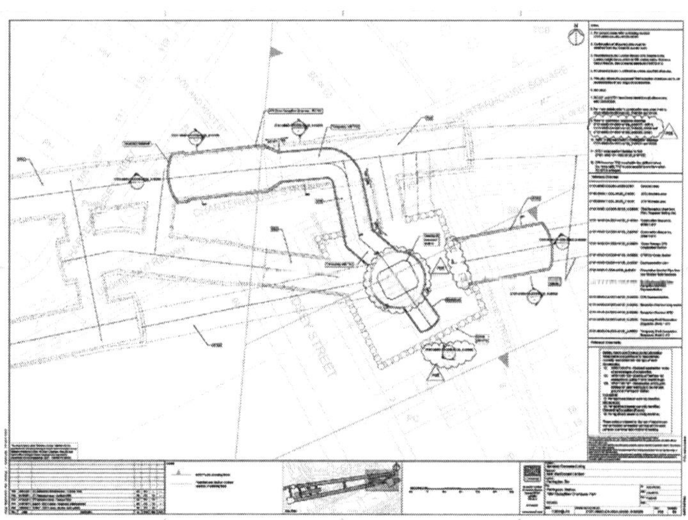

Figure 21 Farringdon eastern ticket hall temporary shaft and adits

The perimeter wall of the trapezoidal shaft is punctured by a total of six large openings:

- Two openings to accommodate the westbound running tunnel
- Three openings for tunnel ventilation and means of escape
- One opening for the inclined SCL escalator shaft.

Impact of OSDs

At both the WTH and ETH sites, the new ticket hall structures and their foundations are designed to support future oversite developments. Both ticket halls incorporate stub columns and a crash deck to allow construction at a future date. The Crossrail stations therefore do not form completed architectural entities but rely on the subsequent OSDs to re-establish the urban setting.

WTH OSD
The "collaborative" WTH OSD required close collaboration with the developer and their professional team to agree column & core locations. The developer's consultants specified the vertical and horizontal loads to be carried by the ticket hall structure and foundations.

The presence of existing piles at the WTH from the demolished building and the area sterilised by the future tunnels imposes severe restrictions on the location of new piles. The vertical loads from the OSD are carried by a combination of the 2.1m diameter foundation piles and the three shafts. Some piles are sleeved to avoid load transfer from the piles to the tunnels and/or to reduce the impact of negative skin friction.

A deep raft slab, together with a series of transfer structures, transfer superstructure loads to the foundations.

ETH OSD
The "non-collaborative" ETH OSD has required the preparation of a design taken to planning approval stage. The OSD and ticket hall loads are carried by a combination of foundation piles and the raft slab of the trapezoidal shaft. [*Figure 22*]

Figure 22 Farringdon eastern ticket hall hall image

DISCUSSION

While both stations:

- are on constrained inner city sites in highly developed areas of London,
- have links with major existing transportation infrastructure,
- are adjacent to historic listed structures,
- must integrate with the overall tunnelling strategy, and
- will provide a world-class railway,

it can be readily seen that the nature of the two sites has led to very different structural forms. The following table highlights some of the key differences:

Paddington	**Farringdon**
Entirely located beneath an existing road	Located beneath existing buildings and major infrastructure
Construction of ticket hall does not require demolition of major building structures (other than the taxi canopy)	Construction of both ticket halls requires demolition of numerous existing buildings
Single site	Two sites (300m apart)
Construction requires temporary closure of existing link road [Eastbourne Terrace]	Construction adjacent to red route [Farringdon Road] - which cannot be closed. Adjacent to numerous roads which serve Smithfield Market
Cut & cover platforms	Mined SCL platforms with pilot TBM
Relatively straightforward ground conditions. London Clay to depth	Very complex, poor, faulted ground. Water bearing gravels
Two TBMs pass through box, at an early stage in their journey. Station construction continues while tunnelling works progress eastward.	Four TBMs terminate at site. Reception chambers required for dismantling and extracting cutter heads etc.
Single Diaphragm-walled box - 1.0 & 1.2m thick panels. Total length = 575m	Four separate secant bored-pile walled shafts - 1.2 & 1.5m diameter piles. Total length = 400m (280 x 1.2m, 120 x 1.5m)
Top-down construction	Mixture of top-down and bottom up construction. Temporary SCL shafts and adits to facilitate cutter head removal.
Adjacent to Grade I listed building [Paddington Station & Macmillan House]	Adjacent to Grade II* listed building [Smithfield Market]
One local authority [City of Westminster]	Three local authorities involved [Camden & Islington Boroughs and City of London]
Adits linking to existing underground stations. Ground level connection to Paddington Mainline station.	Shared concourse integrated with Thameslink. Indirect link to Circle, Hammersmith & City and Metropolitan LU Lines at Farringdon, and direct link to LU lines at Barbican.
No OSD (other than potential for future OSD in the "bomb gap" in Macmillan House)	Two OSDs ("collaborative" and "non-collaborative"). Construction of associated crash decks and the provision of additional foundation and substructure capacity.

Ground movements are a function of wall deflection and these were controlled by a combination of the sequence of excavation and the stiffnesses of perimeter walls and of the horizontal props. Diaphragm walling is inherently stiffer than a secant bored pile wall of same overall thickness. This greater stiffness leads to smaller wall deflections and hence lower ground movements during excavation. However the size of plant required to construct the panels and to install the reinforcement can be very large. This, combined with need for bentonite storage, means that diaphragm walling is better suited to large sites with good access, and to wall layouts that are reasonably rectilinear.

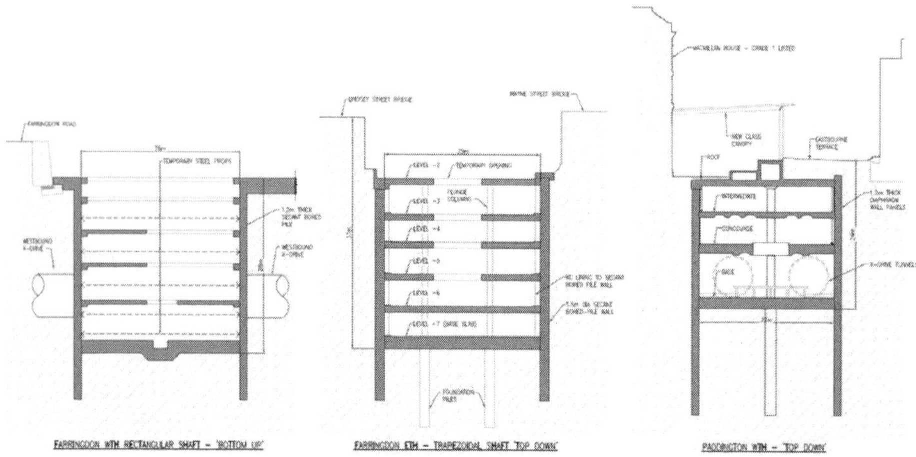

Figure 23 Paddington and Farringdon – comparison of shaft construction

In assessing the most appropriate form of wall construction at these two stations, it was considered that diaphragm walling techniques were best suited to the single regular box at Paddington. At Farringdon, however, the need to construct very deep, irregularly shaped shafts on sites with difficult access, close to rail infrastructure, meant that a secant bored pile solution was more appropriate. Top down construction was used except where ground movements were not critical or where programme requirements dictated the need to complete shaft excavation at the earliest opportunity (as is the case for the rectangular shaft at the WTH). In terms of the structural solutions, diagrams have been prepared to give a direct visual comparison of the resulting designs for the two major shafts at Farringdon Station and the main box at Paddington Station. [*Figure 23*] The key features of these three shafts / boxes are summarised in the table below:

Shaft / Box	Dims: Depth x Width	Construction Sequence	Wall type	Dimensions Moment of Inertia	Prop type
Farringdon WTH Rectangular Shaft	28m x 26m	Bottom-up	Secant bored-pile	1200mm Φ @ 1550 c/c 0.066 m^4/m	Temporary steel props
Farringdon ETH Trapezoidal Shaft	37m x 25m	Top-down	Secant bored-pile	1500mm Φ @ 1950 c/c 0.127 m^4/m	Flat slab
Paddington Main Box	25m x 25m	Top-down	Diaphragm wall	1200mm panels 0.144 m^4/m	Flat slab (with coffers)

CONCLUSION

Both stations share key issues. Both have to meet similar performance requirements, control ground movements and integrate with the tunnelling operations. However the particular constraints of the two sites have resulted in very different structural forms and different construction methodologies.

The underlying design philosophy and the key drivers of safety, economy and constructability remain the same [*Figures 24 & 25*], and the collaboration with other consultants, third party stakeholders and the contractor to develop these solutions to resolve design issues is vitally important for both stations.

Figure 24 Farringdon western ticket hall site progress photograph Sept 2012

Figure 25 Farringdon eastern ticket hall site progress photograph Sept 2012

Crossrail Project: Infrastructure design and construction
ISBN 978-0-7277-6078-4

doi: 10.1680/cpid.60784.029

Design of Paddington Station

Richard Scantlebury MEng MA(Cantab) CEng MICE , Senior Engineer, URS.
Gerard Brennan BE MSc DIC CEng MIStructE, Lead Structural Engineer for Paddington Station, URS.
Mark Raiss MA MSc DIC PhD CEng FICE FIStructE FCS, Crossrail Programme Director, URS.

Abstract

A flagship Crossrail structure, the new station at Paddington will be constructed at the heart of one of London's most significant existing transport hubs. The opportunity to connect Crossrail with international air travel, national mainline, and London transport both above and below ground, drives the need to develop an iconic piece of infrastructure worthy of such a location. This is a significant challenge given the position of the new station: below the existing taxi rank and busy Eastbourne Terrace, and within just 3 metres of Isambard Kingdom Brunel's 150 year old mainline station.

Since commencing design work on the station in 2009, URS have implemented fundamental changes to the original scheme. These will improve passenger experience, operational functionality, and reduce construction costs. The final scheme however is not without challenge: construction planning to minimise the impact on road traffic; complex links to adjacent underground stations; interfaces with the existing above ground buildings and mainline station. And of course construction of the main station box itself: 264 m long, 25 m wide, 25 m deep, framed with diaphragm walls and constructed top-down using sizeable plunge columns. A significant design and construction constraint is that the Crossrail running tunnels are driven through the end walls after diaphragm walling and remain live during the excavation of the box until the penultimate slab is installed.

This paper discusses the architectural and structural engineering challenges encountered, and solutions developed, during the detailed design of the station. In particular it focuses on the changes introduced since the preliminary design stage, reviewing the driving factors for structural form and discussing the final design solutions which were implemented. This in turn exposes the requirement for the structural engineer in such an interdisciplinary project to produce efficient solutions which respond to the various requirements of the integrated design team.

Introduction

Located to the South-West of the existing mainline station, underneath Eastbourne Terrace and Departures Road, the Crossrail station at Paddington will connect this new railway with an already bustling transport hub. Paddington connects four London Underground (LU) lines and a myriad of bus routes with trains to Heathrow and national mainline destinations. The existing infrastructure is rich in engineering and architectural heritage: Brunel's mainline station opened in 1854 and was shortly followed in 1863 by the western terminus of the Metropolitan Railway, the first underground railway in the world.

Building a new underground station for Crossrail in this location therefore presents significant challenges and opportunities both for design and construction. The final design has been driven by efficiency, buildability, physical constraints, functionality and the need to provide a station appropriate to this location and historical setting. This discussion considers in turn the constraints which drive so much of the final outcome, an overview of how the design has developed, and a description of the final structure both in terms of design and construction.

Constraints

Existing buildings and infrastructure

As with so many of the Crossrail stations in central London, existing structures provide stringent physical constraints which dictate limits for the new construction. Figures 1 and 2 show the location and surrounding buildings, which are described below:

- Mainline station to the north-east of the site, dates to 1854 and is Grade I listed. Originally the taxi rank servicing the station was located along Departures Road, however this has been relocated to the north of the station by an earlier phase of the project.
- The canopy above the taxi rank in Departures Road is listed by association, although it has been modified many times and is now deemed to have little historical value.
- London Underground (LU) stations for the Bakerloo line to the north-east of the mainline station, and District and Circle lines at the south-eastern end of Eastbourne Terrace.
- 'Macmillan House' between Departures Road and the mainline station. These terraced structures, owned by Network Rail, run alongside the original taxi-rank. Macmillan House includes the 'bomb-gap' area, a 50 m long portion of the structure which was destroyed during World War II and replaced with a low-rise office block.
- Modern office blocks to the south-west of the site, with thresholds at the Eastbourne Terrace street level some 3.5 m above Departures Road.
- 'Sub-station N', located below Macmillan House and providing power for the mainline station and other Network Rail assets.

- Hilton Hotel located at the eastern end of Departures Road.
- A capped brick sewer beneath Eastbourne Terrace, removed as part of the Crossrail works.
- The 'Osborne Tunnel' beneath the road at the western end of the site, used for servicing on-board catering for mainline trains.
- A utility corridor beneath the Eastbourne Terrace footpath, into which numerous services have been diverted by a prior contract.

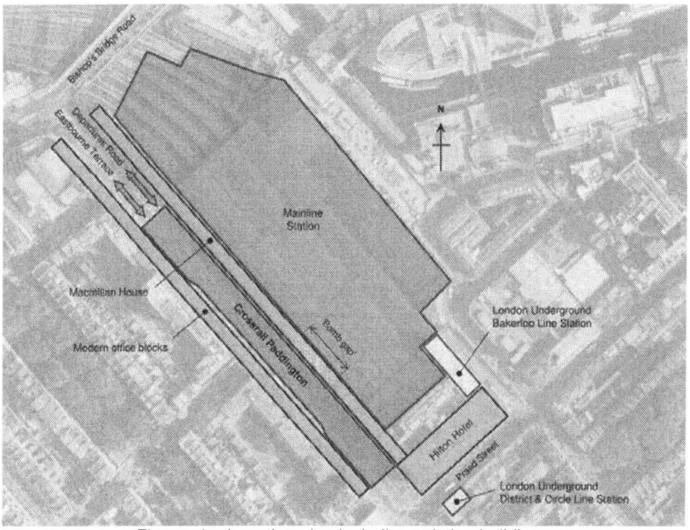

Figure: 1 – Location plan including existing buildings

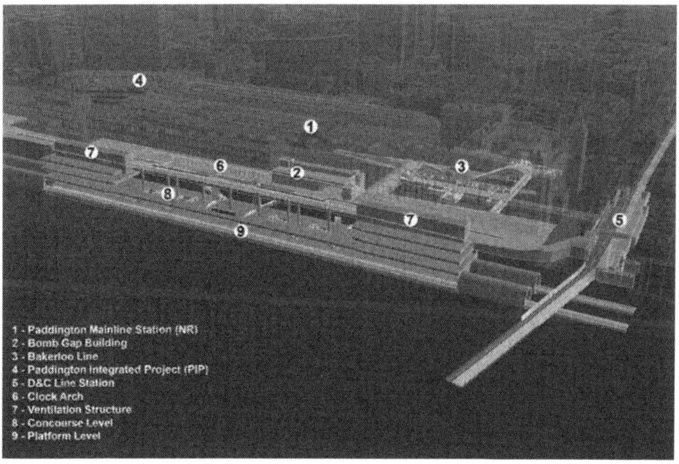

Figure 2: – Schematic graphic showing new structure in context

Geotechnical conditions

The soil profile comprises Made Ground (and some Langley Silt) above River Terrace Deposits over the London Clay.

Limited data shows the London Clay extends to about 60.5 m below Eastbourne Terrace (to elevation 66 mATD) and overlies a thin clayey silt and fine sand layer of the Harwich Formation above the Upper Mottled Beds and Laminated Beds of the Lambeth Group. The elevation of the surface of the London Clay falls from about 122.6 mATD at the west end of the Station box to about 118 mATD at the east. At the east end of the station box there is up to about 4.5 m of River Terrace Gravel above the London Clay.

The River Terrace Deposits become thinner toward the west and appear to be absent at the west end of the station box. Above the River Terrace Gravels the material is predominantly Made Ground. It also appears that the walls supporting Eastbourne Terrace are mainly retaining Made Ground.

As is typical for Central London there are two aquifers beneath the site; a shallow upper aquifer in the River Terrace Deposits and a lower deep aquifer in the Chalk, Thanet Sand and Upnor Formation. The two aquifers are separated by the low permeability London Clay and clay layers in the Lambeth Group.

Programme and interfaces

In addition to these physical constraints, a number of construction aspects, both local to the site and from the wider project, strongly influence the design:

- The proximity of Paddington to Royal Oak Portal from where the first tunnel boring machines (TBMs) were launched means that the tunnel drive passes through the station before perimeter wall installation is complete, and prior to any excavation. These tunnels are required to service the ongoing TBM drive and so must be kept live during all but the final stage of excavation at Paddington.

- Eastbourne Terrace is a heavily traffic part of London's road. It was decided to close the road completely during the early stages of the programme, but reopen two lanes of traffic after two years, to minimise disruption.

- Handover of the platform area to the system-wide contractor, for works on the Crossrail critical path.

Scheme Development

Preliminary scheme

The preliminary scheme design was taken to RIBA stage C and handed over to URS for further design in August of 2009. This scheme located the station box underneath Eastbourne Terrace, to be constructed using top-down methods.

The main station box itself was 264 m long, 25 m wide, 26 m deep, with four permanent slabs propping diaphragm wall panels both during construction and in the permanent case. In the central portion of the station, inclined concrete struts between the main slabs (in back of house areas) formed a 'structural truss', transferring vertical loads to the perimeter walls and removing the need for columns at platform level (figure 3). Long voids ran along the box centre line through each of the upper three slabs, creating a 'light well' from street level to platform, housing escalators and supplying daylight to the lowest areas of the station. The resultant anticipated view from platform level can be seen in figure 4.

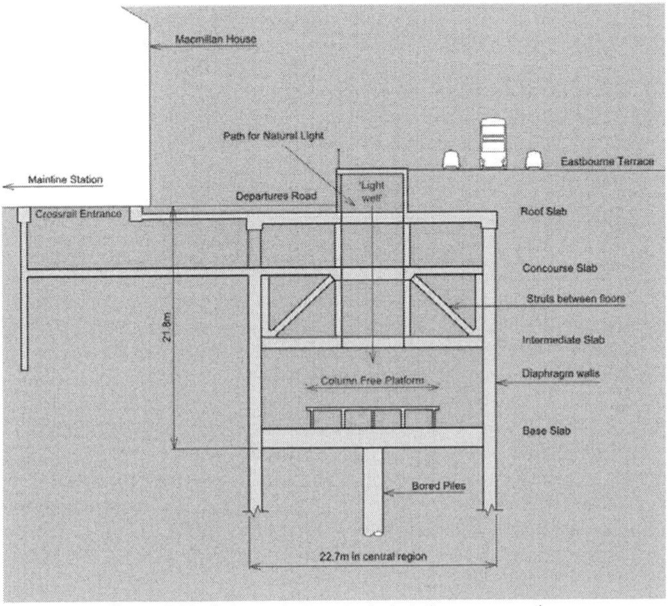

Figure 3: Preliminary design - typical station cross section

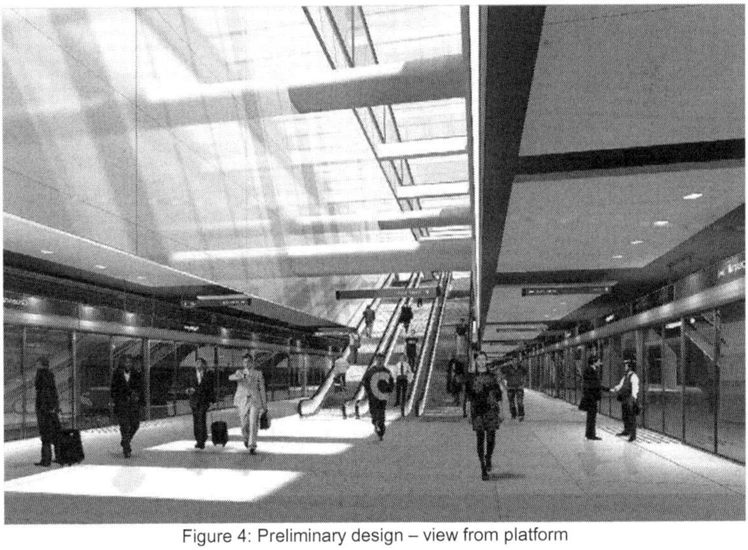

Figure 4: Preliminary design – view from platform

Street level access to the Crossrail station was via an entrance located within the mainline station concourse, behind the bomb-gap area. Passengers descended from the mainline concourse via a bank of escalators, before walking through a link tunnel beneath what is currently the post-war low-rise building, to gain access to the main station box. In order to construct this new entrance and link tunnel, the existing building at the bomb gap would be substantially modified, with large transfer structures introduced to support the structure above the new link. The tunnel would displace existing subterranean toilet blocks and sub-station 'N', both therefore relocated and demolished as part of the works.

Access to the LU Bakerloo line was proposed via a bored tunnel from the new concourse level which passed beneath the mainline station, 7 m below ground. A cut and cover link from the eastern end of the main box, underneath Praed St, would connect the new station to the District and Circle Lines.

Departures Road scheme

Whilst taking this preliminary scheme through the RIBA Stage D phase, URS, Weston Williamson and Crossrail worked together to develop an alternative design, known as the 'Departures Road Scheme' (DRS). It was recognised from the outset that this alternative could potentially deliver improvements over the preliminary design, and so both the original scheme and DRS were progressed simultaneously to RIBA stage D.

The most significant change in the DRS was to move the street level entrance out of the mainline station, locating it instead within Departures Road itself. This eliminated the need both for construction within the mainline station, and major rework of existing buildings in the bomb gap area, as the link tunnel was no longer required. In addition, a glazed lightweight canopy was proposed above the new Departures Road entrance. This provided cover above the Crossrail access and created an 'urban realm' between the mainline and proposed stations. Figures 5 and 6 summarise the revised form.

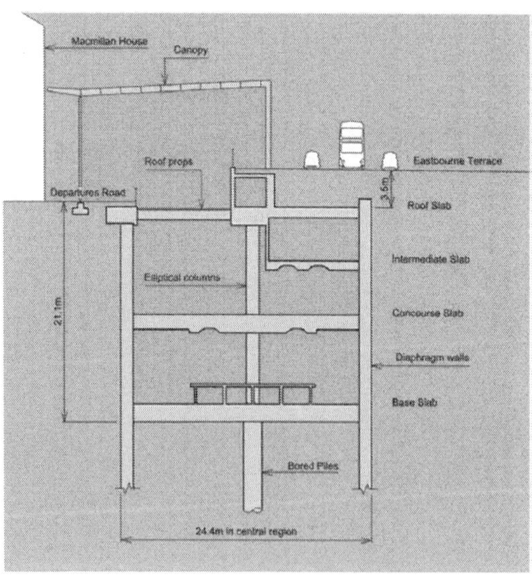

Figure 5: Final design – typical cross section

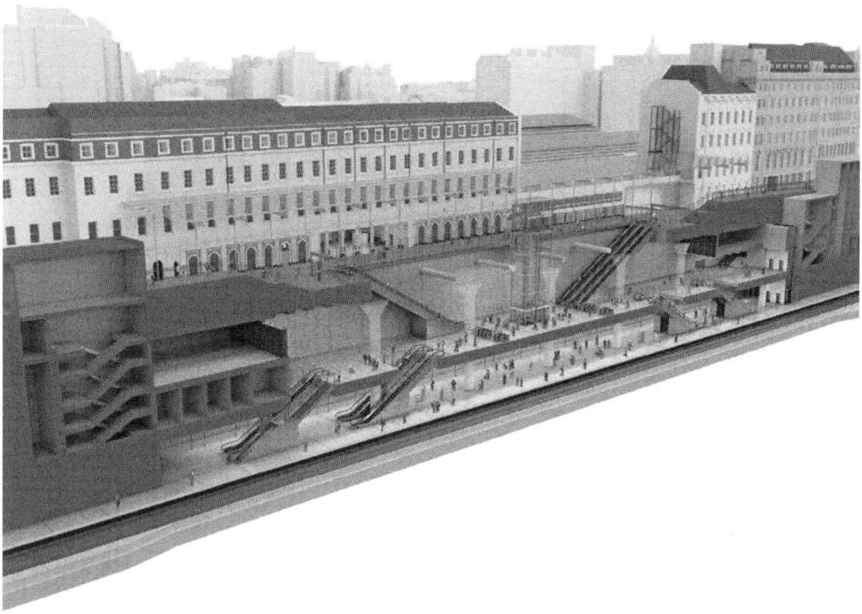

Figure 6: Final design axonometric view

By progressing both schemes to RIBA stage D, a fair comparison could be made between the two. It became clear through this process that the DRS provided significant benefits, including:

- Improved passenger comfort by reducing the time spent, and distance travelled, underground.
- Improved passenger experience as the Departures Road entrance and canopy create a grand 'atrium'. Passengers entering the station will enjoy direct line of sight from street level to the gate lines below, whilst those ascending from the platforms will emerge into a large, airy, naturally lit space. See figures 7 and 8.
- Columns are reintroduced in the central region of the station box, removing the inclined struts between slabs and therefore providing more space back of house for equipment and circulation.
- Less disruption to the mainline station during construction as no excavation is required for the entrance within the existing concourse.
- No link tunnel now required so demolition of the toilets and sub-station 'N' are no longer required. The major structural works in the bomb gap building are replaced by a refurbishment to accommodate the entrance from Departures Road to the man line station.
- Substantial cost savings as a result of all the above.

On this basis the DRS was identified as the preferred option and was progressed to detailed design and construction.

Figure 7: Final design - view of entrance and canopy structure

Figure 8: Final design - view from concourse

Further developments generally did not affect scope with the exception of the link tunnels to LU stations. Following a Crossrail value engineering review in 2011 the District and Circle line link was replaced by passive provision for a future tunnel. The Bakerloo line link was temporarily put on hold and ultimately lowered, as the initial proposed tunnel was located very close to the top of the clay layer and passed in close proximity to the Macmillan House and Brunel shed foundations.

Final Design

Passenger Experience

Despite the developments of the DRS, the overall station geometry and location are similar to the preliminary design stage. The main station box is in the same position, underneath Eastbourne Terrace and Departures Road. Overall plan dimensions are unchanged from the preliminary design, although the excavation depth is reduced by a about a metre. Existing street levels across the width of the station will be restored, with Eastbourne Terrace to the south up to 3.5 m higher than the urban realm at Departures Road.

Passengers descend into Crossrail from street level via a 65 m long void in the station roof, with banks of escalators at either end. These terminate at concourse level some 11 m below ground. Four circular concrete props traversing the roof slab penetration form the only structure between the glass canopy roof and the concourse-level gate lines (see figure 8). From concourse, escalators located along the box centre line provide transit to and from the platform level below.

The effect of this arrangement is to provide passengers entering the station with a pleasant, grand entrance and a sense of space whilst descending to concourse level. Similarly those exiting the platforms ascend into a large 'entrance hall' with unimpeded sight lines from concourse to the canopy roof 20 m above, and sky beyond. For passengers in both directions, the wayfinding afforded by the DRS is much clearer and so transit is less stressful.

Departures Road Canopy

The newly developed area in Departures Road (seen in figure 7) is covered by a lightweight steel-framed glass canopy spanning transversely from the change in street level to the mainline station, but remaining independent from the existing, listed buildings. This structure provides cover to the central stretch of the urban realm, and pedestrian route between Crossrail and mainline stations, whilst maintaining an open space flooded with natural light.

The structure itself is planar: 122 m long and 23.29 m wide, with a main span of 18.2 m (perpendicular to Departures Road). The canopy is a minimum of 9.75 m above the Departures Road level and 7.1 m above the Eastbourne Terrace pavement.

Primary structural elements are on a 6.11 m grid, derived from the MacMillan House arcade. The main beams are fabricated H-shaped hollow sections spanning between H columns founded on the upstand structure of the main box along Eastbourne Terrace, and slender circular columns set back 2 m from Macmillan House.

Longitudinal stability is provided through the base moment contribution of the columns adjacent to Eastbourne Terrace, and the roof acting as a plan vierendeel truss in order to distribute any horizontal loads. Secondary longitudinal framing members (parallel with Eastbourne Terrace) consisting of 550 mm deep by 30 mm thick glass fin beams are at 1.9 m centres and provide lateral torsional buckling resistance for the main frame members.

Structural Slabs

As with all top-down construction, the reinforced concrete floor slabs at each level serve to support vertical load and transfer horizontal soil propping forces throughout the full life of the structure, including construction. At Paddington four such slabs are provided:

- Roof slab below street level, 1000 mm thick.
- Intermediate slab, 900 mm thick.
- Concourse slab housing ticket offices and gate lines, up to 1750 mm thick.
- Base slab, 1750 mm thick.

At either end of the station box each of these slabs act to prop the walls and so resist horizontal forces. However in order to provide the desired entrance from street level, an alternative solution is required over the central 92 m of the structure, with a different propping arrangement.

At roof level the four props described previously provide the solution. The position of these props directly below the existing ground level allows for provision of a large waling beam along the northern edge of the entrance, below the street. This beam spans horizontally, transferring soil propping loads between the four circular props and thus to the southern side of the excavation, where the roof slab itself acts as a deep, internal waling beam. The one metre diameter concrete props are precast, to provide the high quality finish demanded by their prominent position.

The 'intermediate' slab, providing back of house areas for plant and staff, is bypassed by passengers travelling on the escalators between street level and concourse. In order to provide direct passage from street to concourse the intermediate slab is half-width in the central part of the station. In fact the slab stops short of the main central columns in this area; it is supported by the southern wall on one side and on the other by hangers anchored into the roof slab. The result, as can be seen from figure 8, is to give users an uninterrupted view of the main columns extending from concourse to roof level, mirroring the climb of the large entrance escalators. In this area therefore the intermediate slab does not act to prop the walls which span instead from concourse to roof.

Further down, the concourse slab provides transverse propping for the full length of the box. The resultant horizontal loads are significantly increased in line with the entrance, where the intermediate slab does not contribute. As a result the slab depth here is increased to 1750 mm, from 1000 mm in the rest of the station. Voids in the slab along the box centre-line accommodate stairs and escalators to and from platform. The remaining slab area either side acts as an internal deep waling beam, distributing the soil loads around the voids.

Soffits of the elevated slabs will generally be fair faced concrete. In areas accessible to the public, intermediate and concourse slab soffits are coffered, with a curved profile resulting in a 'spherical cap' of 420 mm depth removed from the slab, at 3.055 m centres (figure 9). These coffered profiles enable lighting to be recessed into the slab itself, increasing the proportion of reflected light and thus reducing glare. Structurally they serve to reduce the dead load of the slab whilst maintaining the overall depth and so produce an efficient section for bending. The coffered profile will be formed using precast panels, ensuring a high quality of finish for these exposed elements.

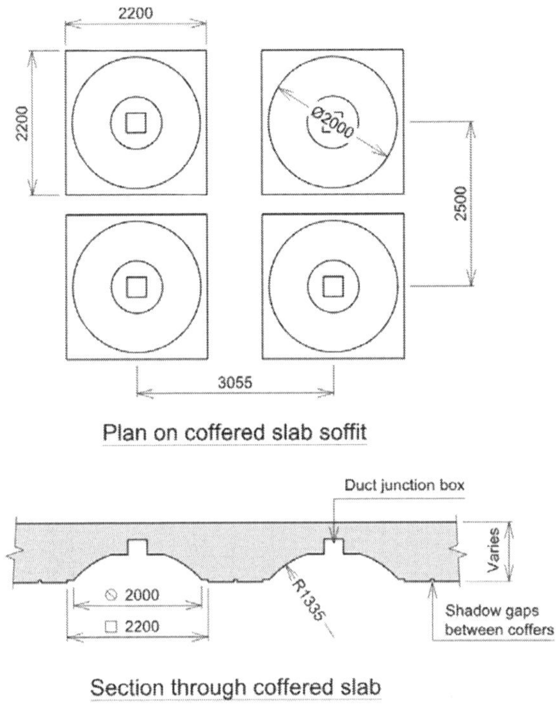

Figure 9: Coffer details at intermediate and concourse slabs

The base slab plays a relatively minor propping role as a result of the top-down sequence of construction, which loads the suspended levels to a much higher degree. Predominantly this deep slab therefore works to resist uplift forces (from water pressure and heave) by bending between diaphragm walls via the central row of piles. The slab itself is penetrated only by sumps at either end of the box, and the local excavation providing access to the Bakerloo line link.

This link, lowered by 12 m since the preliminary scheme, takes passengers down from platform level via escalators and stairs, to a tunnel passing underneath the track and through the diaphragm wall, before travelling beneath the mainline station in a 4900 mm diameter sprayed concrete passageway to the Bakerloo line platforms.
This lowered alternative provides a more direct route for passengers and reduces impact on the historic structures above.

Vertical Elements

In the central portion of the station, accessible to the public, a single row of elliptical columns is provided spaced at 18.33 m centres, with oversized flared column heads below the roof slab. The column section is 1250 mm wide throughout, and 1800 mm long increasing to 6200 mm at the head. By providing a solitary row of central columns on such a large grid, platform space is maintained and these elements possess a sense of grandeur which compliments the public areas through which they pass (see figure 8). The elliptical shape provides structural efficiency in bending and against buckling, whilst maintaining a softer aesthetic. The column heads, fundamentally required to provide punching shear resistance at roof level, are sized beyond the structural requirements in order to maintain a sense of proportion responding to the scale of the surrounding elements.

The reintroduction of columns as part of the DRS delivers a number of benefits. The inclined truss members between floors are removed, which increases accessible space for back of house areas and thus makes the most effective use of the slab footprint. The columns provide a simpler structural form, which makes for simpler detailing and reduced reinforcement congestion. Primarily however the multi-level entrance could not be accommodated with the structural layout from the preliminary scheme, and so once again the most obvious benefit is the improved passenger experience.

At the station ends, in back of house areas, the single row is replaced by pairs of columns with a reduced spacing, typically between seven and ten metres. This arrangement suits the requirements in these areas for numerous large slab penetrations accommodating plant and ventilation equipment. One metre square columns are provided in these regions, where visual impact is less of a concern.

Vertical loads around the perimeter of the structure are carried by diaphragm walls which also retain the ground in the temporary and permanent conditions. Diaphragm walls are stiffer and stronger than secant piles of the same thickness, and therefore better suited to this site with existing structures in close proximity. Typically the panels are 1.2 m thick, although at the box ends this reduces to 1.0 m to provide the necessary TBM clearances. The walls extend twenty metres below the base slab, working with the central bored piles to resist long term uplift forces and prevent flotation.

Above ground, the existing Eastbourne Terrace level will be restored during construction. Various alternatives were considered but clearly to minimise disruption to road alignments and thresholds of existing buildings the most favourable solution is to maintain the current ground profile. Two approaches to restoring this level are combined, minimising the impact on the roof slab, 3.5 m below the road:

- Where headroom requirements allow, concrete structures are provided on top of the roof slab, underneath the road. These provide usable space and access routes, whilst significantly reducing the volume of fill supported by the slab.
- Elsewhere lightweight fill is used to minimise loading on the slab. Standard fill is provided for the top metre, the rest made up using lightweight expanded clay aggregate.

The Eastbourne Terrace level will be restored after completion of the roof and prior to excavation, in order to open the road to traffic. Therefore these reductions in structural weight above ground yield material savings not only in the permanent slabs and columns but also in the size of plunge columns and piles required during construction.

Construction

Top down Methods

Two factors drove the decision to construct the main station box using top down methods. First, top down construction is stiffer, with smaller wall movements, minimising damage to the surrounding buildings and utility corridor. Second, the desire to reopen Eastbourne Terrace to traffic as early as possible required the roof slab to be constructed at the front end of the construction programme.

Diaphragm walls provide soil retention during the bulk excavation, propped by the permanent slabs. The walls are constructed in panels of 3.35 m width, excavated under bentonite to maintain stability during the two-day construction process. At its heaviest the reinforcement in these panels consists of two layers of 50 mm diameter bars, using couplers to eliminate laps and reduce congestion. Steel stop ends form construction joints between panels and retain concrete during the pour. Diaphragm walls are installed with a verticality tolerance of 1:200, using a hydraulic grab. Figures 10 and 11 show the plant in the context of the narrow site.

Figure 10: Photograph showing diaphragm wall plant on site

Figure 11: Construction site as mid 2012

During excavation vertical support to the slabs is provided in part by the diaphragm walls and in part by steel plunge columns. Forty-nine such columns support the slabs along the box centre line, typically spaced at 6.11 m longitudinally. All columns are 550 mm square H-sections, with web and flange plate thicknesses varying from 25 mm to 40 mm and 50 mm to 70 mm , respectively, dependant on column location. These steel columns, approximately 23 m long, are installed to a verticality tolerance of 1:400 and supported by 1800 mm diameter piles below the base slab level, founded 60 m below ground. Upon installation of each plunge column the surrounding pile bore is backfilled with pea gravel.

Design of the columns used finite element models and eigenvalue analysis to establish buckling loads for each element, at every stage of construction. Backfill around the column was conservatively assumed to provide no lateral restraint to the column. To limit the effective length during early stages therefore, two 500 mm thick insitu concrete discs are cast at designated levels during the bore backfill to provide temporary lateral restraint. This reduction in effective length enables smaller sections to be used and thus delivers material savings.

At eight locations, where plunge and permanent columns coincide, the steel section is retained and acts compositely with the insitu elliptical concrete section cast around it, to support the station in the permanent case. All other plunge columns are removed once the concrete elements are constructed, and load is transferred to the permanent works. The supporting piles however are tied to the base slab and provide long term resistance against heave and flotation. Thus the pile design in the short term is governed by downward forces but in the long term by uplift.

The intermediate slab is not constructed top down, but replaced with temporary steel struts and waling beams during the bulk dig and cast on formwork later in the programme. This accelerates excavation and protects the running tunnel lining as described in the next section.

Tunnelling interface

The tunnel drive passes through Paddington during construction of the diaphragm walls, so this interface is key to both the station and tunnelling contracts.

At the eastern end of the box a soft 'eye' is provided which the TBM can break through. This involved constructing full depth panels but casting the lower sections using an unreinforced cement/bentonite mix. However more recent analysis has shown that the clay at this depth is sufficiently cohesive to be self-supporting, until a permanent reinforced 'collar' is cast around the tunnel lining. At the west end therefore, which was constructed later, panels forming the return are stopped short of the tunnel crown, leaving uninterrupted clay below through which the TBM can pass.

The interaction between the station and tunnelling programmes creates a particular constraint for the construction sequence at Paddington. The TBM passes through the station box prior to excavation, but the running tunnels must be kept live in order to service the ongoing tunnel drive further along the line, for as long as possible. This is achieved using the following sequence of excavation:

1. Perimeter walls, plunge columns and roof slab completed.
2. Excavation to 3 m above crown of tunnels.
3. Install steel propping frame, with struts preloaded to 700 kN/m. This relieves the horizontal soil pressure on the tunnel lining, just as excavation has relieved the vertical pressure.
4. Excavate to concourse slab soffit, 500 mm above tunnel crown.
5. Construct concourse slab on blinding.
6. Construct intermediate slab on formwork and remove steel props.
7. Running tunnels handed over to Paddington contractor.
8. Excavate below concourse including demolition of tunnel lining within station length.

Detailed analysis, in conjunction with the running tunnel designers, has been undertaken in order to demonstrate that this sequence can be accommodated by the lining segments, without the need for internal propping (which would restrict movements up and down the tunnel).

Existing structures

A major benefit of the DRS is the reduced degree of construction required within the existing mainline station and Macmillan House footprint.

Some works of course are still required. 800 mm of the Macmillan House basement outer wall must be removed by wire-cutting, in order to make space for the diaphragm wall installation. In addition to this the entrance between Departures Road and the main line station will be refurbished to accommodate the predicted increase in passenger flows.

This however is significantly less than originally anticipated by the preliminary scheme, which required piling and excavation for an entrance and tunnel between the mainline and Crossrail Stations. In the DRS the subterranean toilet blocks, and substation N, remain in use, saving significant demolition and diversion costs. The additional transfer structures designed above the entrance, to allow removal of Macmillan House foundations, are also no longer needed.

The net result therefore is a much simpler interface with these structures and associated stakeholders, and far less disruption for the users of both Macmillan House and the mainline station.

Conclusion

Rarely are designers presented with the opportunity to design new structures in the heart of London, in locations with constraints and heritage as substantial as those at Paddington. As can be seen from this discussion however the primary drivers and constraints remain the same as for any major civil project. The design at Paddington is driven by passenger experience and the client's needs, by programme and value, and by the existing environment and infrastructure.

At each stage such considerations have dominated the design, with every discipline individually and collectively developing solutions which respond to these requirements. The final design, and in particular the changes implemented through the alternative scheme, deliver improved solutions for the client and passengers alike. A coordinated approach between stakeholders and design disciplines has produced an efficient, elegant solution which seeks to provide all users with an exceptional experience, adding value both during construction and for many decades to come.

Crossrail Project: Infrastructure design and construction
ISBN 978-0-7277-6078-4

doi: 10.1680/cpid.60784.045

The Value of Common Design – C100 Crossrail Architectural Components

Neill McClements, C100 Lead Architect, Partner, Grimshaw

Introduction

The C100 team have brought practical and commercial value to the Crossrail project by maximising the benefits of common design. Simple, forward-thinking and intuitive passenger environments have governed the design, which will be completed in 2018.

Atkins, Grimshaw and GIA Equation form the multi-disciplinary C100 consortium for the project. Taking passenger experience as the key driver, the consortium is designing the architectural components that will be used throughout the platform and tunnel environments to create an integrated line-wide identity. Working with the individual design teams for each station, C100 have prepared ten component packages, including wall and floor finishes, lifts, escalators, lighting and signage suitable for use in all the new stations.

This systematic approach creates a unified passenger experience, maximises value and efficiency yet allows each station to respond to its context with an individual identity at street level. The benefits of common design are manifold and are grounded by the following principles:

- Design efficiency
- Simplified testing and approvals
- Manufacturing and installation efficiency
- Simplified maintenance and future station upgrade
- Improved passenger experience

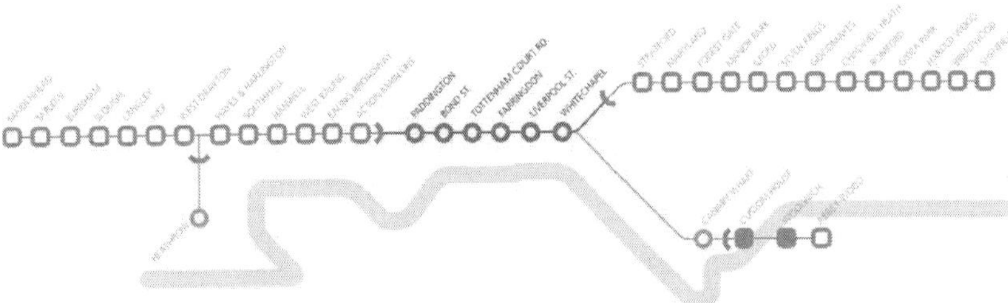

Fig. 1 The C100 common components to be applied to Crossrail central section stations

Fig.2 Artist impression of C100's design for the concourse tunnel

Design Efficiency

Efficiency is articulated across the design concept. The first task for the team was to engage collaboratively with the client and the different station design teams to develop an appropriate balance between common and individual design responses for the project. Through this collaborative dialogue a design concept emerged for the central stations which provided for a high degree of commonality in the below ground spaces while allowing for a stronger individual expression at street level.

Commonality, in particular, has been maximised in the tunnel spaces where the functional brief is similar across the different stations. In contrast the street level ticket halls are subject to unique local constraints and it was deemed appropriate that each station should have an individual design response to reflect this.

Once this design approach was established C100 worked closely with C121 tunnel engineering team to further explore opportunities for the standardisation of the different tunnel geometries. More than 80 different tunnel sections across the five tunnelled stations have been reduced to nine standard sections for the platforms, cross passages, concourses, and escalator tunnels.

In addition to this, the collaboration with C121 led also to the development of the distinctive curved transitions between the different standardised tunnels. These transitions exploit the physical properties of the proposed sprayed concrete engineering lining to widen the tunnel sections locally to junctions between tunnels.

This ultimately provides a number of functional benefits. For the passenger, this assists passenger flow and improves safety by eliminating blind spots when changes in direction occur. Design efficiency is further acknowledged during the construction stage; these smooth transitions assist with the application and waterproofing of the engineering lining at the junctions between tunnels.

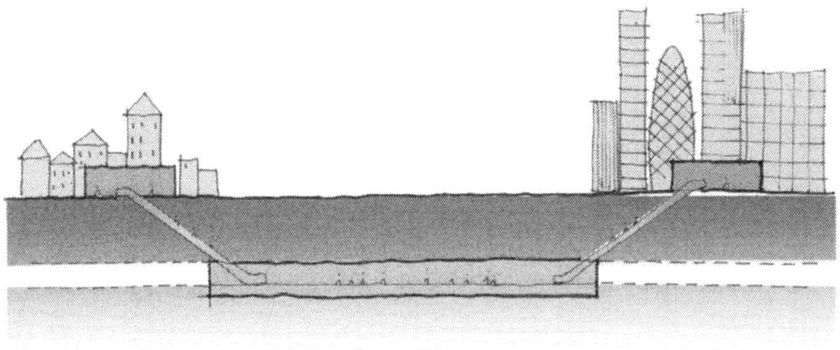

Fig.3 Diagram of common design in the underground levels and location specific design at street level

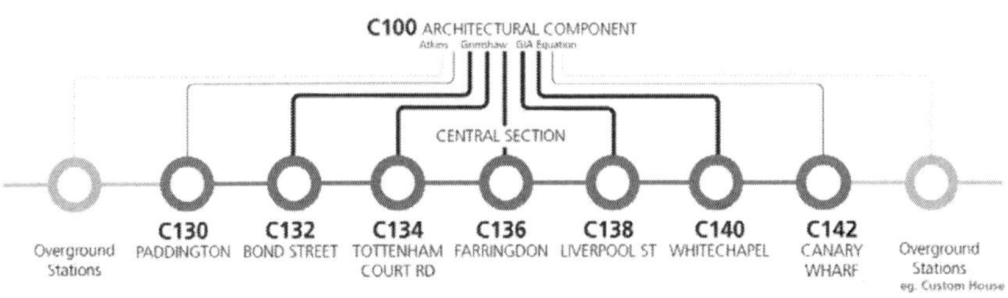

Fig.4 Common identity is strongest in the 5 tunnelled stations

It was decided that the rough finish to the sprayed concrete primary structural lining required a secondary cladding system in order to provide a high quality passenger environment which can be easily maintained. A lightweight fibre reinforced concrete panel system has been selected as a cost effective solution which allows the realisation of these distinctive curved transitions between tunnels. Central to this, has been the standardisation of the different tunnel sections which in turn results in standardised junctions between tunnels. The innovative standardisation of these complex double curvature geometries allows the re-use of moulds to exploit economies of scale across the stations. Similarly, the single curvature panels are set out based on standardised cladding grids across all stations driving down costs and simplifying future maintenance.

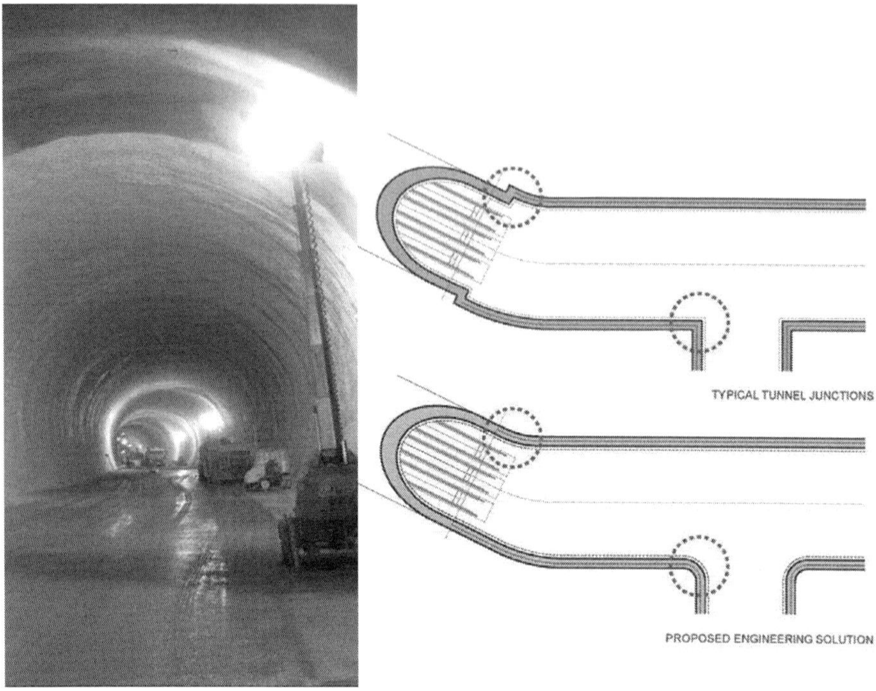

Fig.5 The use of sprayed concrete tunnel engineering (left) influences the curved tunnel transition design (right)

Fig.6 Curved tunnel transitions improve safety by reducing blind spots and vreating a distinctive signature for the Crossrail tunnel environment

The resulting kit of parts, rather like a giant plumbing kit, can then be applied and adapted by the different design teams to meet the specific layout arrangements at each station.

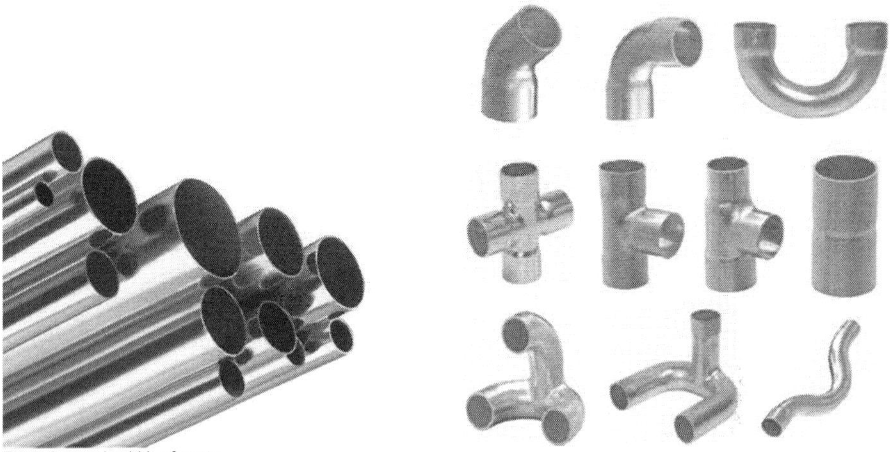

Fig.7 A standard kit of parts

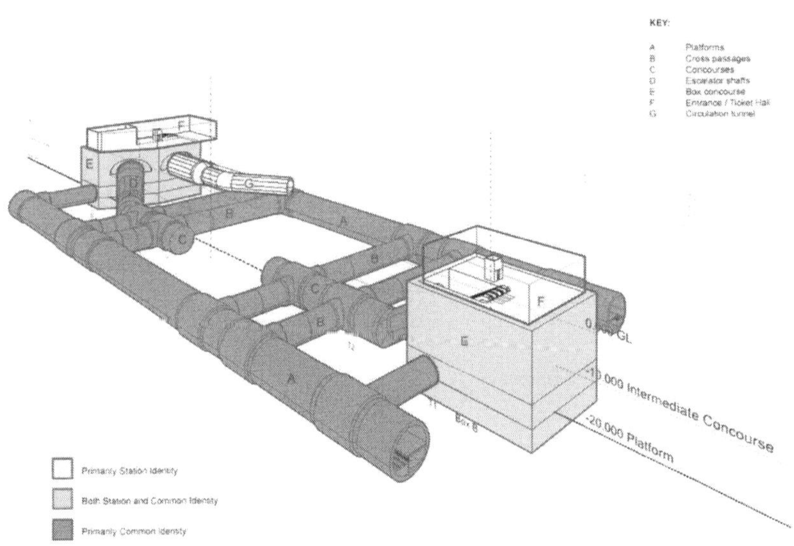

Fig.8 Diagram of standardised tunnels and junctions

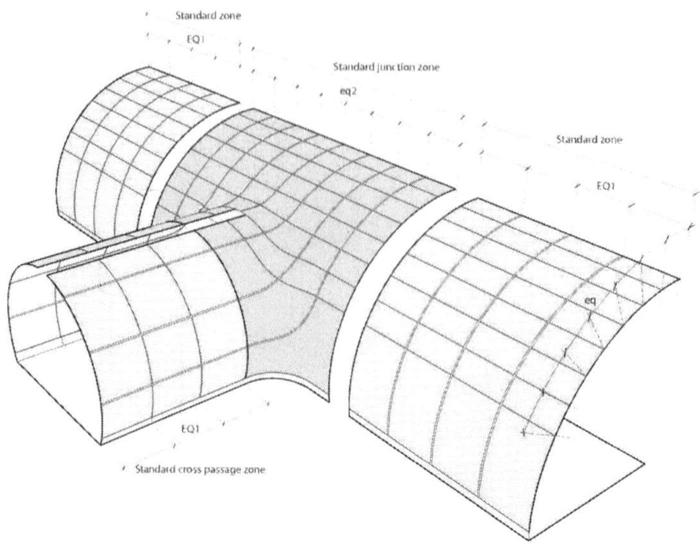

Fig.9 Standardisation of tunnel cladding system

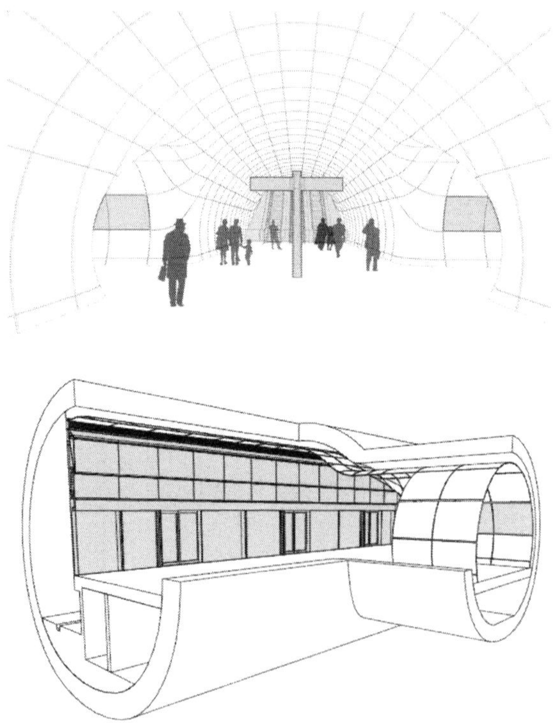

Fig.10 Zoning of technology elements (blue) e.g. Platform edge screen, totem, lighting and digital advertising

As well as developing common solutions for cladding and flooring within the tunnel environments C100 also developed, in collaboration with the station teams, a common approach to servicing the spaces, which maximised design efficiency. Servicing equipment has been zoned and focussed on a series of technology elements such as platform edge screens and services totems. These technology elements allow the de-cluttering of spaces and focus servicing equipment where it can be easily accessed. Like the tunnel finishes these have been standardised across stations to maximise economies of scale and drive further value into the project.

In addition to these major servicing elements, a common family of components such as signage, seating, gatelines and ticketing has been developed which will help to provide a coherent passenger environment across all spaces above and below ground. A standard approach to the design of these recurring finishes and components, along with the interface detailing with surrounding elements across a number of stations, significantly reduces the design resource required by the different station teams to complete the design. In light of this, Crossrail's design management resource requirement is also reduced.

This strategy therefore leaves station teams to focus on adapting the generic kit of common elements into their designs and provides sufficient time to focus on the unique aspects of each station location.

Simplified Testing and Approvals

A major infrastructure project such as Crossrail involves a number of key stakeholders including London underground and Rail for London who will be jointly responsible for the operation and maintenance of the infrastructure when it opens in 2018. It is therefore crucial that these key stakeholders are engaged in the design process and in the selection of finishes and materials to ensure that the chosen components are appropriate for their environments.

Thorough assurance procedures therefore need to be put in place to achieve progressive stakeholder sign off through the design process. By increasing commonality across the different stations the process is aided by reducing the number of different systems and materials that need to pass through this approvals process. A vital part of this strategy is the development of samples, mock-ups and prototypes as well as the reduction in the number of mock-ups and prototypes usually developed. C100 are currently developing a range of advanced mock-ups and prototypes of the common components which brings a number of benefits. They facilitate the early testing of the concept with end users to ensure that they are fit for purpose ahead of the completion of the design by contractors. In addition to this, the mock-ups and prototypes allows further design refinement with input from specialist suppliers as well as further stakeholder and input feedback. Initiating the production of the models provides vital constructability lessons to be tried and tested, which creates assurance before arrival on site.

Once the contractors are on-board a more extensive testing regime will be required which includes surveying the components in terms of fire performance; impact testing; acoustic testing; blast testing as well as materials and finishes testing. The line-wide procurement of the common components will subsequently reduce the amount of repeat testing required for each system resulting in significant cost reductions to the project.

Fig.11 Review of totem prototype

Fig.12 Fire Equipment cabinet mock-up

Fig.13 Fibre reinforced concrete tunnel cladding prototype

Fig.14 Review of curved wall signage

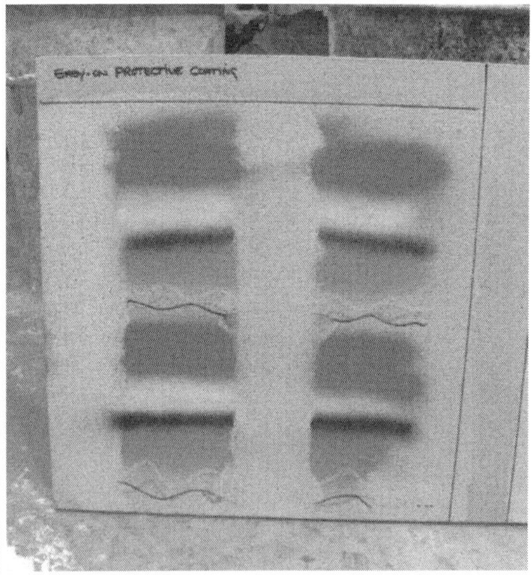

Fig.15 Advanced trialling of anti-graffiti coating

Manufacturing and Installation Efficiency

Fig.16 CAD/CAM fabrication of master mould for 3D curved cladding panel

Fig.17 Production mould

Common design provides a number of tangible benefits for manufacturing and installation. Taking the concrete cladding as an example, common design, when combined with line-wide procurement of this system, can lead to significant efficiencies and cost savings for the project. The raw materials can all be purchased from a single source ensuring consistent performance and visual characteristics while capitalising on economies of scale. Tooling set-up costs at the outset of production are also significantly reduced. In the case of the cladding this could be achieved through multiple re-use of the same cladding moulds across all five stations. Whilst the cost of producing a complex double curvature mould would be cost prohibitive if this was a one off, if the mould is used to produce multiple panels then this becomes a very cost effective cladding solution for the central stations. The line-wide procurement of the common elements will mean that the contractor designed panel profiles including structural stiffening ribs and secondary fixing structure will also being standardised.

This approach allows a consistent installation methodology to be developed for all stations which can be refined through the development of mock-ups and prototypes prior to commencing the installation on site. This helps to remove risk to the construction programme by allowing any installation problems to be identified early and designed before getting to site. Efficiency is also improved through repeat working as site teams become familiar with the installation methodology.
It is worth noting that the economies of scale offered by common design makes it beneficial for the installer to invest in specialist equipment which may be customised specifically for the

project. This investment will provide significant gains in terms of both productivity and the quality of the installed system.

In addition to this, site safety is also significantly improved through the increased predictability that a tried and tested installation methodology brings.

Fig.18 Finished 3D curved cladding panel

Fig.19 Trial installation

Simplified Maintenance and Station Upgrade

Fig.20 Specialist equipment for cladding installation / replacement

Lasting simplicity and coherence are inherent to the Crossrail ethos. There is a symmetry between the simplicity of the installation phase of the upgrade and the future access for maintenance work on the stations.

The implementation of standardised panel sizes, profiles and fixing systems means that access methodology remains consistent across all stations. The over-arching nature of common design is highlighted in the possibility of using the same bespoke handling equipment developed for the initial installation for future maintenance requirements. Maintenance of the stations is further simplified by the introduction of a single material for floor and cladding across all tunnel spaces, for which replacements and spares can be held centrally for urgent repairs should the need occur.

As the servicing elements themselves are also a part of a common family both their maintenance and future upgrade of stations is simplified. This includes elements such as PES wall, totems, service booms and lighting. These elements have been designed as modular 'plug and play' systems that can allow for the quick and easy maintenance and upgrade of services which accommodate the stations' journey into the future when new technologies emerge. As well as this, the components themselves utilise hard wearing self-finished materials which are easy to clean and are appropriate for regular handling by maintenance staff. This Common approach to servicing allows similar access & maintenance procedures to be followed across the different stations and could potentially result in one standard set of operation manuals. In light of this, maintenance staff skills will also be transferable across the five stations.

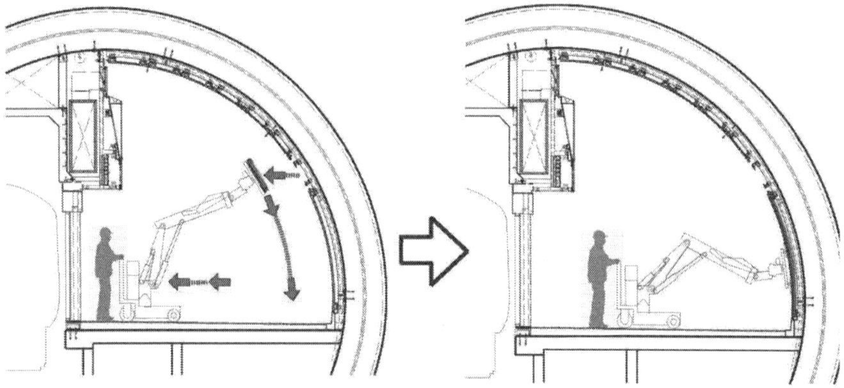

Fig.21 Cladding replacement strategy

Improved Passenger Experience

In addition to the tangible commercial and practical benefits described above, common design can play a key role in improving the overall passenger experience. The design puts people first, ensuring a highly efficient, safe and user friendly journey. Introducing commonality to similar elements across a number of stations helps to increase familiarisation and makes the environments more legible and intuitive for passengers to navigate through. These naturally intuitive environments are then supported by a common approach to the placement of key wayfinding signage which is located in a consistent manner across all stations. At a detailed level, the increased design development and refinement of products and finishes made possible through common design will create a higher quality passenger environment.

It is hoped that the completion of the Crossrail project will mark the arrival of a new world class transport environment. From a practical point of view securing raw materials from a common source and utilising the manufacturing processed will result in a consistent visual product across the different stations which will strengthen the perception that the stations are high quality environments. The repetitive use of elements that are unique to the project will also help to generate a unique and recognisable identity for Crossrail. For instance, the distinctive curvature of the flowing tunnel junctions will become an instantly recognisable motif of the Crossrail project. Similarly, the sleek aesthetic of the high technology elements such as the full height platform edge screens and services totems will help to reinforce this brand identity across both box and tunnelled stations. An additional layer of consistent finishes and components in the form of floor finish, lighting, signage, seating, speakers, and CCTV for example will further strengthen the unique Crossrail identity and provide a strong sense of coherency across all stages of the passenger journey from train to street.

Creating a timeless facility which is easy to use, manage and operate is a core value of the project. It is hoped that the Crossrail project will reinforce London's status as a growing world city and improve the public's view of London's transport system. The C100 team has used the process of common design to challenge and enhance traditional railway construction, making use of current best practice, tools and techniques to contribute to an innovative and affordable railway that will be fit for the 21st Century.

Crossrail
Crossrail Project: Infrastructure design and construction
ISBN 978-0-7277-6078-4

doi: 10.1680/cpid.60784.059

ice
Institution of Civil Engineers
publishing

The Importance of Construction Mock-ups and Trials

Stuart Moxon (IEng MCIOB MICE), Head of Compliance Testing, VINCI Construction UK Limited
Simon Atherton (B.Eng), Compliance Testing Engineer, VINCI Construction UK Limited

Abstract

The objective of this paper is to consider the importance of both full scale construction mock-ups and smaller trial samples to a large construction project such as Crossrail. The paper identifies the range of benefits and advantages to be gained by relatively small investment in mock-ups and trial samples and the enhanced value they can add to the final project.

The benefits of using mock-ups and trial samples have been identified in the following aspects of construction projects:

- Buildability
- Health and Safety
- Protection of the Environment
- Access Requirements
- Sustainability
- Aesthetics and Appearance
- Benchmarking
- Maintenance
- Publicity

Referring specifically to the Crossrail station mock-up further benefits have been achieved following completion of the model when it was widely publicised and a media event was held which included print and digital media. An item was also broadcast to millions of viewers in the London area, giving customers a vision of the future.

Visits to the model allow visualisation and assessment of the proposed final product in a unique way with a level of reality that cannot be provided by CAD, drawings or even 3D computer modelling.

To date over 500 people have visited the mock-up model including:

- Disability interest groups
- Designers and project teams
- Local authorities, government and councillors
- Potential sub-contractors and materials suppliers
- Key personnel from TfL
- Media

It is evident from the analysis of the benefits presented in this paper, that construction of a full-scale mock up, and testing on both the model and on the associated trial samples, can lead to reductions in construction costs, improvements in project programme management and delivery within planned timescales, reductions in unforeseen errors, increased safety during construction and can help reduce the impact on the environment. Ultimately, reviewing the mock-up of the finished product with the client before construction commences should ensure that there is greater transparency and all expectations are achieved.

Figure 1 – View of station platform

Introduction

Crossrail is a major new railway currently under construction in South East England. It includes the construction of 21 km of new tunnels under central London, connecting the Great Western Main Line near Paddington and the Great Eastern Main Line near Stratford, and is currently planned to open in 2018.

The proposed total budget is in the region of £15 billion with new Crossrail stations, costing in the region of £2 billion being constructed along the central route at Paddington, Tottenham Court Road, Bond Street, Farringdon, Liverpool Street, Whitechapel and Canary Wharf.

Technology Centre have constructed a full-size replication of a Crossrail station platform to provide a tangible preview of the future Crossrail station environment and passenger experience. It will aid future design decisions on aspects such as cladding type and

geometry, flooring, lighting, platform edge screen, CCTV, seating, ventilation, signage, space management, advertising and art.
This model has so far proved invaluable to the project in enabling people to visualise layouts, comment on finishes, experience the space and view lighting options. This is essential as it provides opportunity at an early stage of the project to comment on general impressions of the general layout and environment, to determine materials selection, to develop maintenance programmes, reduce risk, ensure sustainability and enable stakeholder buy-in.

The construction of this mock-up will be followed up with the construction of a number of smaller scale trial samples using real materials. These samples can be used as bench marks as well as aids for testing, training and process development.

Figure 2 – Totem

Figure 3 – Cladding & totem

When undertaking a construction project both the client and the main contractor will have a hierarchy of targets and constraints to consider. The primary driver for all commercial companies is profitability, however other important considerations are health and safety, Corporate Social Responsibility, meeting all the client's specifications and requirements and the company's reputation in the wider world.

When a large, costly or complicated construction project is planned it is often beneficial to consider smaller scale trials or mock-ups prior to the construction phase commencing. Some of the rewards of this approach can include:

- Honing construction techniques allowing the main project to be built in a shorter timescale at a reduced cost, with less wastage and a smaller impact on the environment and the neighbourhood.
- Assessing buildability, enabling obstacles to the construction process to be foreseen and designed out preventing delays and increased costs during the construction phase.
- Assessing safety hazards in a small controlled environment so they can be controlled or eliminated across the project, reducing injuries, costs and delays to the programme.
- Enable training of the workforce to reduce defects and improve efficiency.
- Ensuring the finished project will match the anticipation of all interested parties, reducing costs and extended programmes associated with modifications and late specification changes.

Buildability

The use of mock-ups as a preliminary step in the construction process is an invaluable tool for ensuring buildability. Difficulties encountered and mistakes made building the mock-up can then be avoided on the full-scale construction phase. Undertaking small scale trials will prove whether processes that appear practical on paper are equally so in reality or whether modifications are required. For example, can work areas be accessed to undertake works and if so what plant will be required to undertake the works.

Every construction company should be aiming for a target of zero defects, getting everything right the first time therefore removing the costs of "putting right" as well as reducing delays to the programme. One method of working toward this goal is the use of mock-ups and trials as a development and training aid. Problems can be identified and designed out through design improvements, better sequencing, better training and ensuring that the most appropriate construction methods and materials are employed.

Significant savings can be made by refining the construction process to ensure that the correct plant is made available when it is needed. In some cases a trial build followed by consultancy with the workforce can lead to small modifications in process or design that reduce the need for expensive plant or access equipment.

Modifications to the construction methods and procedures can be obtained by using mock-ups to assess the advantages of implementing off-site pre-fabricated components or alternatively reducing the size of components being brought to site. By assessing timescales and scheduling alternatives it may be possible to optimise delivery schedules during the construction phase, ensuring the required materials are always delivered in time without having materials arrive before they are required leading to storage problems where space is at a premium and potential damage to materials held in storage.

A major cost in construction can be remedial works or undertaking modifications. Trials and mock-ups enable improvements to the construction process to be achieved by identifying and illuminating any “mistakes” in a controlled environment before full-on construction begins. This leads to a reduction in the risk of errors being made, which can add to the construction cost and programme. Therefore greater control can be achieved during the actual project construction, with a safer, right first time method of work, with no need for reworks resulting from the time spend preparing the process of work on the mock-up.

Health and Safety

Health and safety is of major importance both in the construction phase of the project and after hand-over throughout the life of the facility. The use of a mock-up both allows safe methods of work to be developed where construction methods are innovative or a development is unique or ground-breaking and permits potential hazards to be identified and eradicated prior to the onset of the main construction phase of the project.

Construction phase health and safety goes hand in hand with buildability. Certain activities will have safety risks inherent in them such as working at height, hot works, transportation of heavy items and traffic movements both on-site and in the adjacent area with public access. The use of small scale trials on mock-ups can be used to establish the safest ways to undertake operations, identify unforeseen hazards which can then be guarded against and explore whether certain activities can be avoided.

There is generally considered to be a hierarchy of health and safety control measures with the primary option being the removal of the hazard. A good example of this would be working at height; if the works can be done at ground level then the hazard of working at height is removed. Experimentation with a mock-up can investigate if assembly is possible at ground level with a completed component then lifted into place. A 'dry run' can establish the feasibility of this and investigate whether other enabling works or alternative sequencing can facilitate this approach.

A major safety issue can be accessibility, how accessible is the work area, what access equipment is required, how much space is available to work in. Accessibility can be maximised by experimentation with a mock-up to optimise the sequencing and construction methods and to discover the most effective means of access.

Training can be undertaken utilising a full scale mock-up allowing the workforce to be instructed in, and practice using, safe working techniques in a safe and controlled environment. Mistakes can be made in a controlled environment with no repercussions in contrast to errors occurring during the construction phase, which are best avoided.

Protection of the Environment

The use of trials and mock-ups can be useful for reducing environmental impacts. By developing best practices the use of materials can be reduced, the amount of waste can be reduced and energy usage can be reduced.

Trials of equipment and working practices can be developed within a mock-up to determine the best methods of reducing noise, dust and other environmental impacts during the construction process. By developing means of optimising sequencing, the number deliveries can be reduced and storage of materials on site can be reduced, which will help to improve the environmental impact of construction works in and around the construction site.

Access Requirements

There are few buildings in the current age that are not required to be accessible to all regardless of age or mobility. There are legal requirements detailed in both the Disability Discrimination Act 2005 and more recently The Equality Act 2010 to ensure that access is readily available for everyone regardless of mobility or other disability. Part M of the Building Regulations details many access requirements which must be complied with. Any company having a regard to Corporate Social Responsibility will have a moral duty to ensure a policy of accessibility for all. This would normally be a desire of not only their boards of directors but also of shareholders and clients. Commercial operations including retail outlets such as department stores, shopping centres and other shops, transport terminals such as airports and stations, entertainment facilities such as cinemas, sports centres, theatres, museums and galleries will all have a vested interest in ensuring their potential customers are not excluded from their business.

Under the Equality Act 2010 a person is considered to have a disability if they have a physical or mental impairment and the impairment has a substantial and long-term adverse effect on their ability to perform normal day-to-day activities. To many people the stereotypical image of a disable person is someone in a wheelchair but the definition is clearly much wider. Disability could include the infirm such as those able to walk but with reduced mobility. A disability may be visual, the blind or partially sighted, or it could include hearing difficulties. In cases it may include those with learning difficulties and the elderly.

The challenge, therefore, to clients, planners, architects, designers and others concerned in the construction process is to design, build and deliver a facility which is accessible to all. There are many documents published giving guidance on access requirements but while some measures are straightforward such as ensuring a doorway is wide enough other issues are more complex and may not manifest themselves in a desk top study.

A full scale mock-up allows a range of persons with relevant expertise and interest in disability equality, access and service provision to appraise and evaluate the proposed arrangements allowing design modifications at an early stage should they be considered beneficial. A liaison with relevant interest groups such as the Disabled Experts Reference Group (part of the BSI Consumer & Public Interest Network) or local access groups can be highly beneficial.

Physical obstacles and restrictions are the most obvious; helping to determine early on if the access is wide enough, the access ramps are at a suitable inclination and are not too long without a landing, if adequate hand-holds are provided, are of the right kind and are at the correct location. In addition the mock-up will help to identify if there are physical obstructions which could be relocated or redesigned and if the floor is a suitable surface.

Visual aspects can also be determined earlier on with a mock-up model, including the correct lighting levels, planned access route acceptance and if the signage is clearly visible.

Personal perception of the future customer can also be determined early on within the full-scale mock-up. Confirmation of the whether or not the access appears straightforward and manageable or if the layout is confusing or daunting in any way.

Finally, a mock-up can identify any unforeseen pit-falls. If it is possible to go wrong, to get a wheelchair stuck, to take a wrong turn or if there are any trip hazards which could conceivably be encountered, then a mock-up of a typical platform and underground access section would give the added benefit of identifying these issues before full-scale construction takes place.

Sustainability

Construction mock-ups and trials are a very good example of sustainable solutions in practice through delivery of 'triple bottom line' benefits based on the widely recognised three pillars of sustainability: the Economical; Environmental and Societal factors.

Economical

The lessons learnt from other sectors such as the car industry and manufacturing sectors show that prototypes are an essential norm. Construction mock ups and trials reduce financial and performance risk, facilitate faster ways of working, understanding issues and scope. Therefore, such mock ups should be seen as an investment for project improvement and cost reduction rather than an additional cost.

Environmental

Construction Mock-ups and trials facilitate best practice resource efficiency and less waste through optimum design. Currently WRAP, a government funded organisation are strongly promoting resource efficiency research and best practice implementation. Best practice eventually finds its way to standard good practice, thus making more widespread reductions in environmental impact.

For repeat items such as installation of signs, lighting posts and tunnel sections, the reductions in CO_2 and environmental impact can be determined within a mock-up. These reductions and the resultant savings can be multiplied many times over. For technically complex one-off builds, the environmental savings can be seen by understanding potential environmental impacts due to rebuilds.

Mock-ups also give the added benefit of focusing attention on reducing environmental impact in many forms, including:

- Less base material
- Less waste
- Standardisation
- Optimum design
- Reducing the need to rebuild
- Consideration of whole life cycle 'cradle to grave'
- Understanding carbon impact and possibilities for reduction.

Societal

Construction Mock-ups and trials have proven to be an effective means of obtaining buy in of project stakeholders and investors in future projects.

Opinions of designers, experts and stakeholders can be gathered and acted upon before it gets built and becomes public. Getting this buy-in helps the construction move forward more positively and with more certainty.

Not just technical inputs but also the 'softer' side of aesthetics colour and form can be addressed. Getting it right first time, the avoidance of errors leads to client satisfaction and reassurance for stakeholders.

Using this sustainable solutions approach not only ensures an economically viable, environmentally sensitive and socially acceptable outcome, furthermore, the three factors help each other in such a way to deliver a 'triple bottom line' win.

Aesthetics and Appearance

The aesthetics of a building, facility or space can only ever be subjective but there will in most cases be a general consensus on what works and is pleasing to the eye and what is less appealing. Aesthetics is based on a multitude of elements including colour, texture, pattern and form. It will be affected by lighting, by layout, by the materials used and by colour schemes utilised. There can only ever be one way to assess aesthetics and that is to experience the layout first hand.

A full-scale mock-up allows a spectrum of people to walk through and gain an impression of a proposed finish and to comment on thoughts and feelings. Various options can be trialled to see whether a proposal will be acceptable and to see what colour schemes are attractive to prospective users of the space.

Aesthetic trials can be of economic benefit, it may be acceptable in some locations to use a cheaper finish; a painted or plastered concrete rather than expensive cladding may work in one location but not another. If the finish is to be bare concrete, can simple sprayed concrete be left on show or will it need floating off, painting or even covering completely. While the client will want a finished facility that is aesthetically attractive and pleasing, it may not be necessary to spend excessive amounts to achieve a satisfactory finish and different locations as well as different usages may dictate what is acceptable.

Benchmarking

The basic station design and layout developed at Technology Centre will be used as a model for six new underground stations with some design aspects possibly used as a standard for works undertaken at existing stations to be served by Crossrail. There is therefore a desire for consistency with the as-built stations matching all expectations. Trial samples constructed at Technology Centre can be referred back to as benchmark examples, furthermore construction of a trial sample demonstrates the quality which can be expected in the finished environment as well as enabling methods of achieving the desired standard to be developed.

Standards benchmarked can include:

- Aesthetic appearance - Colouring, Lighting, Signage, General impressions.
- Tactile aspects – Textures, Slip resistance, Abrasive surfaces.
- Quality of Workmanship.
- Ease of Cleaning.

Maintenance

Throughout the life of each station a programme of planned and reactive maintenance will be required. Maintenance will be undertaken by a wide range of contactors ranging from basic cleaners to electricians and specialist engineers. There are both costs and timescale issues associated with maintenance whether planned or reactive. The costs associated with maintenance can be minimised by reducing the level of maintenance required, increasing the ease of maintenance and reducing the duration of the maintenance period.

Maintenance can be expected to include the following activities:

- General Cleaning
- Specialist Cleaning
- Graffiti Removal

- Replacement of consumable such as lighting components
- Replacement of damaged items due to vandalism or accident
- Replacement of time expired items – general refurbishment of signs, furniture etc.
- Routine servicing and inspections – electrical installations etc.
- Changing advertising

Reactive maintenance can be reduced by ensuring components are as durable as possible and with high resistance to attack by vandals or accidental damage. Specialist cleaning and finishes which are relatively easy to clean and will not be permanently degraded by contact with various the substances they are likely to encounter throughout their life.

Ensuring safe access is readily available to undertake regular maintenance tasks will assist in keeping maintenance to simple routine operations without the need for extended timescales, increased manpower and corresponding increased costs. This may be achieved by small modifications to design to ensure that access is possible with standard equipment.

Test and Evaluation Schedule

The materials testing schedule for the Crossrail station mock-up and trial samples is varied and extensive. An overview of the testing is presented below, this is not necessarily exhaustive as additional testing can be undertaken, including bespoke testing to client's specifications.

- Impact Testing of a full size lower interface zone GFRC panel assessed against BS 8200:1995 "Code of practice for design of non-loadbearing external vertical enclosures of buildings" Appendix G: Hard and soft body impact tests.
- Impact Resistance testing to BS EN 12372:2006 Natural stone test methods. Determination of flexural strength under concentrated load of Terrazzo and Granite floor tiles.
- Abrasion Resistance testing to ASTM / C944M – 99 (2005) E1 – Standard Test Method for Abrasion Resistance of Concrete or Mortar Surfaces by Rotating – Cutting Method undertaken on polished and sandblasted finish GFRC flat test samples.
- Abrasion Resistance testing to BS EN 14157:2004 Natural stones. Determination of abrasion resistance of Terrazzo and Granite floor tiles.
- Cleanability (dust & dirt) assessment against document E3834 A3– Station Cleaning: Attachment 2 Walls, Ceilings and Soffit [Dust, hand and shoe mark removal] of GFRC panels, Terrazzo flooring and Granite flooring.
- Cleanability (common food and beverage spillages) assessment against document E3834 A3 – Station Cleaning: Attachment 2 Walls, Ceilings and Soffit [Cleaning of spillages of Cola, Coffee, Tomato ketchup sauce] of GFRC panels, Terrazzo flooring and Granite flooring.
- Cleanability (Chewing gum) assessment against document E3834 A3 – Station Cleaning: Attachment 2 Walls, Ceilings and Soffit of GFRC panels, Terrazzo flooring and Granite flooring.
- Cleanability (Body Fluid Spillages – blood stains) assessment against document E3834 A3 – Station Cleaning: Attachment 2 Walls, Ceilings and Soffit of GFRC panels,Terrazzo flooring and Granite flooring.
- Cleanability (Anti-Graffiti) assessment against document E3834 A3 – Station Cleaning: Attachment 2 Walls, Ceilings and Soffit and "Updated Anti-graffiti Testing Criteria" of polished and sandblasted finish GFRC test samples with surface sealant and anti-graffiti coating.
- Bomb Blast Testing of components
- Abrasion Resistance of Anti-scratch coatings used on PES wall passenger interface

zone, fire hydrant cabinet, balustrades, escalator side panels assessed against DIN 52348 – Testing of glass and plastics; abrasion test; sand trickling method.

- Visual appearance assessment of Anti-scratch film versus Permanent anti-scratch glass coating.
- Slip Resistance testing in accordance with BS 7976 of Terrazzo and Granite floor tiles.
- Spectrometer test of GFRC panels, Terrazzo and Granite floor tiles in accordance with BS8300:2009 A1:2010 Design of buildings and their approaches to meet the needs of disabled people – Code of practice.
- Spectrometer testing to confirm stainless steel handrail finish does not create reflective surface to cause hazard to partially sighted customers in accordance with BS8300:2009 A1:2010 Design of buildings and their approaches to meet the needs of disabled people – Code of practice.
- DDA – 'Warm to touch' assessment of handrail in accordance with BS 5395-1:2000: Stairs, ladders and walkways. Code of practice for the design, construction and maintenance of straight stairs and winders. A heat conductivity test to confirm alternative handrail finish is made of low conductive material to satisfy the "warm to touch requirement".
- General Design and Detailing Assessment of glass and stainless steel handrail. Assessment of robustness and ergonomics – Visual and tactile assessments on design and finishing details of glass and handrail elements to ensure elements shall be reasonably robust against removal by general hand held tools by the public and checking elements have no sharp edges, projections or finger traps.
- General Design and Detailing Assessment of glass and stainless steel handrail. Assessment of robustness and ergonomics – Visual and tactile assessments on design and finishing details of glass and handrail elements to ensure elements shall be reasonably robust against removal by general hand held tools by the public and checking elements have no sharp edges, projections or finger traps.
- Design Assessment – Confirmation that the current design allows for replacement of a single glass pane or railing section without extensive disturbance of the adjacent member.

Conclusions

Following completion, the model was widely publicised and a media event was held which included print and digital media, including presentation of information and views of the model on the Crossrail website: http://www.crossrail.co.uk/route/stations. Millions of viewers and future customers in the London area have also been given a vision of the future when a television news story was broadcast from the Crossrail mock-up.

To date over 500 people have visited the model including:

- Disability interest groups
- Designers and project teams
- Local authorities, government and councillors
- Potential sub-contractors and materials suppliers
- Key personnel from Transport for London (TfL).

Visits to the model have enabled the visualisation and assessment of the proposed final product in a unique way with a level of reality that cannot be provided by Computer Aided Design (CAD), drawings or even 3D computer modelling.

Testing undertaken both on the model and on trial samples can reduce costs and timescales, reduce errors, increase safety and help reduce the impact on the environment.
Ultimately allowing the client to see the finished product before construction begins should ensure that all expectations are met.

Crossrail Project: Infrastructure design and construction
ISBN 978-0-7277-6078-4

ICE Publishing: All rights reserved
doi: 10.1680/cpid.60784.071

New taxi facility on heritage structures at Paddington

Adrian Douglas, Principal Engineer Mott MacDonald
Param Hartop, Project Engineer, Crossrail Ltd
Chris Levy, Project Director Mott MacDonald

Abstract

As a preliminary part of the Crossrail works at Paddington Station, the existing deck spanning platforms and railway tracks is being transformed from a disused goods depot into a taxi and passenger interchange facility. Paddington Station is a Grade 1 listed heritage structure and all modifications to the existing fabric need to be sympathetic to the heritage environment.

The original deck was assessed for highway loading and additional loads from the new structures. A reinforced concrete framed building, constructed c.1920 using the Hennebique system had significant areas of delamination and corroded reinforcement. Surveys and evaluation of current condition generated a prioritised repair strategy. All work has been carried out in and around operational railway infrastructure, within a congested urban site, which has added complexity to both design and construction solutions.

Keywords: heritage; assessment; change of use; repair; strengthening; Hennebique; railway; interchange.

1. Introduction

This paper describes how as part of the Crossrail project the some of the various areas and levels of the site were changed to provide the new facility. The condition and use of the site before the works and the historic development of each area is described first, followed by the works carried out to create a unified facility. A general view of the new works areas is shown in Figure 1.

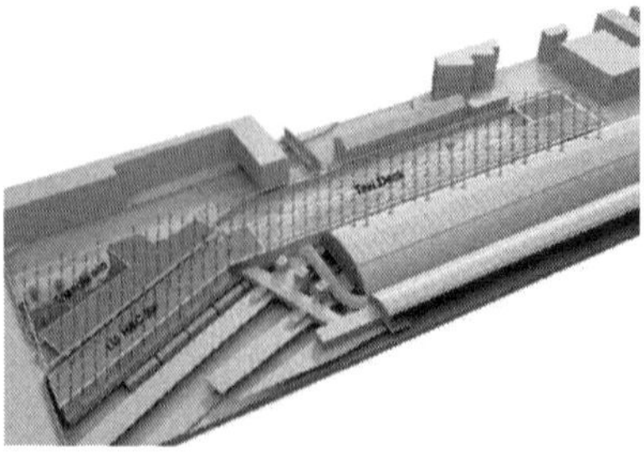

Fig 1: Arrangement of new works

Construction of the Crossrail station at Paddington required the relocation of the existing taxi rank from Eastbourne Terrace on the south side of the main line station. A suitable space existed on the unused parcels deck but with no clear means of access.

London Underground have an ongoing programme of providing step free access to platforms where practicable and relieving congestion. Significant congestion had been created at the Hammersmith and City line platform due to a recent reconfiguration hence the required upgrading. An integrated project meeting all of these demands was developed which provided:

- Taxi passenger interchange facility on the parcel and Hammersmith and City line deck;
- New concourse for the Hammersmith and City line station;
- New north entrance to the station complex on to a new plaza adjacent to the canal;
- New platform level concourse giving access to the taxi deck;
- Vertical circulation core between platform level and taxi.

This paper concentrates on the assessment and alterations to the Hammersmith and City line deck and repairs to the Mint building over the platform level concourse.

2. Pre-project and historical use of the site

2.1 General

Paddington Station is a major London Terminus for main line trains from the west of England and Heathrow airport. An aerial view of the site before the start of the works is shown in Figure 2. The station is served by three underground lines; the deep Bakerloo line and the shallow Hammersmith and City and Circle lines. The Hammersmith and City line runs along the north side of the station at the same level as the main line tracks. The surrounding area is mixed use with offices, flats and shops at street level. A hospital is adjacent to the station. The site slopes generally from north to south with the mainline track level in cutting 7m below. Bishop's Bridge Road bridge spans across the main line tracks at the west side of the site which is significant link in the local road network. A spur of the Regents Canal runs parallel to the tracks on the north side.

Paddington Station was first constructed on this site in 1854 [1]. The station was extensively re-modelled several times and included construction of a fourth span to the train shed and additional platforms between 1909 and 1916. The whole station complex is a Grade 1 listed heritage structure.

Crossrail Project: Infrastructure design and construction
ISBN 978-0-7277-6078-4

doi: 10.1680/cpid.60784.085

Transporting Tunnel Boring Machines

Chris Hickman (MEng), Graduate Civil Engineer, Bam Ferrovial Kier Joint Venture

Abstract

The world tunnelling market has seen significant growth over recent years with many tunnelling projects currently underway around the world. In the UK we are seeing resurgence with projects such as High Speed 2, Thames Tideway and potentially Crossrail 2 on the horizon. The future workload highlights the importance of recording the experience gained during the Crossrail project to help drive efficiency and innovation.

TBMs are rarely seen out in the open as they are often assembled and launched from portals or shafts with backshunts deep below ground. In some circumstances however, it is necessary to transport these huge machines whole. A task of this scale is a complex procedure and requires careful consideration and planning. The decision to transport a TBM will be influenced by many different factors, but may be considered more applicable to sites where the space available for assembly is at a premium, and projects where multiple launches are required along the route. A fast and efficient transit method therefore, is desirable to accelerate the programme of works and keep disruption to a minimum whilst maintaining high health and safety standards.

To provide a tangible example, this report draws on the experience of the engineers working on Crossrails Western Tunnels. This includes the transportation of two 1200 tonne, 140 m long TBMs from the assembly area, 600 m down to the launch portal. By presenting and analysing the various options available for transporting TBMs and considering the relative merits in terms of cost, time, risk, health and safety, operation and design, this report aims to provide guidance to parties considering undertaking similar operations. Conclusions are drawn highlighting the relative merits of the options available and the lessons learnt for the benefit of others.

Notations and Symbols:

TBM = Tunnel Boring Machine
PTFE – polytetrafluoroethylene
SPMT = self-propelled modular transporter t = tonnes (metric tons)
m = metres
km = kilometres h = hour

1 Introduction

As the world population grows our infrastructure needs increase but the space available for new development becomes scarcer. In our cities we look to push new infrastructure requirements below ground and we strive to provide faster and more efficient transport links. Examples of such projects in the UK include Crossrail and High Speed 2. To help deliver these projects the UK tunnelling industry is in resurgence and engineers are developing new technologies and construction methods to speed delivery, increase efficiency and reduce costs. Bearing in mind the increasing need for tunnels it is reasonable to assume that more TBM transits may be required in the UK before long. In developing transit methods it is always useful to draw on the experience of others who have carried out similar tasks.

TBM transits are rare but have been made possible in recent years by innovative thinking and technological advances in heavy lifting equipment. The task of moving a TBM whole is a complex procedure, carrying with it significant cost and risk implications, and therefore it is avoided in most circumstances. However, examples where a TBM transit may be considered include cases where space restrictions mean that the TBM cannot be assembled in launch position, transits through pre- formed caverns such as platform tunnels and on projects where a series of tunnels are required with relatively short gaps between tunnels.

This report aims to highlight the options available for transporting TBMs whole and highlight the advantages of each option. To provide a tangible example, the report draws on the experience of the engineers working on the Crossrail Western Running Tunnels project. Examples include the transportation of two 1200 tonne, 140 m long TBMs from the assembly area at Westbourne Park, 600 m down to the launch chamber at Royal Oak Portal and the options being developed for the TBM transit at Bond Street Station. The report also looks at options used on a number of other recent tunnelling projects across Europe including the Prague Metro Line Extension, Barcelona Line 9 and the Biel Bypass. It is hoped that this information can be used to help develop faster and more
efficient transit methods but also highlight issues that can be avoided at design stage.

2 Summary

2.1 TBM Transit Options

2.1.1 Continuous Cradles

The most common method of transporting a TBM whole is with the use of a continuous cradle. It is a continuous support structure installed for the entire length of the transit. The cradle structure is often formed of either a steel frame or a concrete invert slab profiled to suit the TBM with continuous rails cast into the surface.

An example of a continuous steel cradle is shown in Figure 1 below. In order to propel the TBM forward typically a steel beam will be fixed behind to the steel cradle. The propulsion rams in the shield can then be used to push against this beam and move forward. When the rams have reached their full extension the beam can be removed and moved forward and the process is repeated.

The backup gantries are ordinarily supported on inclined bogies running in the tunnel lining. However, in the example shown below no tunnel lining has been constructed and therefore the backup gantries are transferred onto a separate system of bogies running on tracks installed in the cradle.

Figure 1 - Prague Metro Line Extension - Cerveny Vrch Station – Showing a TBM transiting through a preformed station cavern on a continuous steel cradle [Stehlik, E, 2011]

In the case of continuous concrete cradles the TBM can be propelled forward by laying invert segments behind the TBM and shove against these. The gap between the concrete slab and the invert segments is packed and then later grouted in place to provide a stable structure to support the backup gantries. The arrangement has the advantage of being relatively simple and the number of invert segments can be selected to suit the inclination of the bogies and therefore, the backup train can continue uninterrupted. In addition to this advantage, if a temporary rail system is used to service tunnelling works the rail for the tunnel locos can be laid as it would be in the tunnel behind. This means that the advancement of the TBM and the supply of materials remain relatively free of disruption. Figures 2 and 3 below illustrate the arrangement in more detail.

Figure 2 - Prague Metro Line Extension - Petriny Station – Showing a continuous concrete invert slab with steel profiles cast into the surface [Stehlik, E, 2011]

Figure 3 - Prague Metro Line Extension - Petriny Station – Showing invert segments laid to propel the TBM forward and support the backup gantries [Stehlik, E, 2011]

Due to the significant amount of material required to construct a continuous cradle, TBM transits using these methods are generally restricted to transits along the tunnel route and are not more than a couple of hundred metres in length. As the transit length increases the cost of materials will increase and a point will be reached whereby alternative options are considered to provide savings. The solution is however, ideal for openings on the tunnel route such as platform tunnels where other considerations may offset the installation costs. The considerations include access for materials and equipment for transiting the TBMs as well as a need to minimise disruption to tunnelling activities and therefore, a fast and efficient transit method is required.

2.1.2 Mobile Cradles

In circumstances where a longer transit is required, alternative methods are generally considered to be more cost effective. Examples where TBMs have been transited out in the open are rare but have become more common in recent years. Arrangements generally involve a short cradle structure or frame welded to the TBM shield. The purpose of the frame is to provide a surface onto which hydraulic jacks can lift the shield section of the TBM. Propulsion is then provided by either push-pull jacks running in skid tracks or a SPMT. A separate arrangement is then used to support the TBM backup train. To provide more detail three examples of projects where TBMs have been transported
using a mobile cradle system are considered in the following section.

2.2 Case Studies

2.2.1 Barcelona Line 9, Pozo de Sagrera, Open Shaft Crossing

In 2007, engineers started looking at options to transit a 12.06 m diameter, 11.75 m long, 1500 tonne TBM shield and a further 1000 tonnes of backup gantries without the need to disassemble the machine from its backup. The transit length was 80 m though an open shaft with two curves with a radius of 285 and 300 m respectively.

Engineers adopted a standard skidding system incorporating hydraulic skid shoes with a stainless steel base running in steel skid tracks lined with PTFE pads to reduce friction. Four skidding shoes were employed for the task each with a vertical lift capacity of 500 tonnes. Two brackets were welded to the TBM shield with a beam below transferring the load of the machine onto the vertical jacks. Propulsion was provided by four hydraulic rams operating in the skid tracks. These operated with an extend and contract sequence intermittently locking to the skid track and moving the TBM (push-pull jacks). The horizontal thrust capacity of the rams was 64 tonnes and the pull capacity of 34 tonnes and a 620 mm stroke. All four jacks work together with two pushing and two pulling thus a total thrust force of 196 tonnes at any one time. The arrangement is illustrated in figure 4 below.

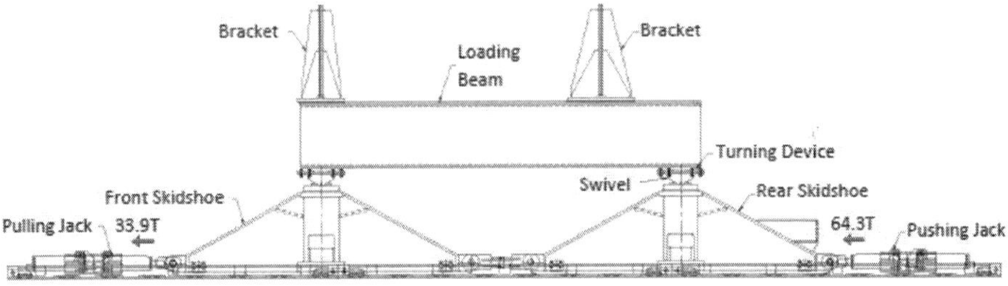

Figure 4 – Barcelona Line 9 - Detail showing TBM shield jacking arrangement [ALE, 2007]

In order to allow the TBM to pass through the curves in the transit route the skid tracks were laid out and fixed into place prior to the arrival of the TBM. To allow the rigid TBM Shield to pass through the tight curves each vertical jack was fitted with a swivel head arrangement to allow movement in both the longitudinal and transverse axis.

In order to support and transit the TBM backups engineers developed a special device to allow the backups to run on rails. The system comprised of legs extending from the underside of the backups to a special bogie car running on a pair of rails as shown in figure 5 below. The arrangement allowed the backup train to pass through the tight curves.

Figure 5 – Barcelona Line 9 - TBM shield traveling through shaft using skid shoes (right) and bogies for backup gantries running on rails (left) [ALE, 2007]

Once the TBM had reached the final position, the machine was lowered by vertical jacks onto concrete supports ready for re-launch. As the TBM advanced back into the newly constructed tunnel the bogie cars supporting the backup gantries were progressively removed. The whole transit operation was completed in eight days thus advancing at an average rate of 8 m/h.

2.2.2 Switzerland, Biel Bypass, Transit between Tunnels

As part of the construction of a new ring road in Switzerland several tunnels were required to allow the road to pass through the areas with steep and challenging topography. The spacing between each tunnel was very short and therefore, engineers chose to transport the TBM whole between tunnels rather than dismantling, transporting and re-assembling the TBM. The first of the two transits was considered to be the most complex and involved transiting the 12.56 m diameter, 110 m long, 2500 tonne machine 625 m between tunnels.

On route to the re-launch position an existing road bridge had to be lifted and the full length of the TBM had to pass in a 24 hour window to allow the bridge to re-open to reduce the impact on traffic. The TBM transit system therefore, had to be designed to allow the TBM to clear the bridge with time to carry out the bridge lifting and re-opening operations.

In a similar way to the system used in Barcelona, a lifting frame was welded to the side of the shield with four 500 tonne skid shoes lifting the TBM shield and providing a low friction surface on which to skid the TBM. However, unlike the case in Barcelona the transit route included a 2.5 % incline followed by a 3.7 % decline to the start of the next tunnel. Twelve 40 tonne strand jacks were used to haul the machine uphill and a further four stand jacks were used to provide restraint to the TBM on its route downhill. The moving procedure involved jacking the TBM up using the skid shoes, dragging it forward and setting it down again. Each cycle moved the TBM 4 m and an average speed of 7 m/h was achieved. The arrangement is shown in figure 6 below and illustrates the sheer scale of this operation.

Figure 6 - Biel Bypass – TBM shield supported on skid shoes transited using strand jacks with backup gantries running on concrete blocks [Thomas, I, 2011]

Unlike the bogie system used in Barcelona engineers chose to use the same inclined wheels which run in the tunnel lining. This involved developing a new system of pre-cast concrete blocks for the bogies to run on when they had left the tunnel. The concrete blocks came in sections and were lifted into place under the bridge (behind the shield and before the first backup). In order to minimise the number of blocks required only enough blocks to cover the length of the backup train were produced. As the TBM advanced, the last of the concrete bocks was removed and placed in front of the TBM. The system is illustrated in figure 7 below.

The restricting factor to the rate of movement was found to be the rate at which workers could carry the concrete blocks through the backup train and position them in place behind the TBM shield.

Despite this restriction successful transits were carried out in 2010 and 2011.

Figure 7- Biel Bypass – TBM laying concrete blocks for backup gantry support [Thomas, I, 2011]

2.2.3 Crossrail, Transit from Westbourne Park to Royal Oak Portal

In 2011 engineers working on the Crossrail Western Running Tunnels contract started looking into options to transit two TBMs whole from the assembly area at Westbourne Park (WBP) to the launch chamber 600 m away at Royal Oak Portal. Each TBM shield was 7.1 m in diameter, 12.5 m long and weighed 600 tonnes with the backup gantries adding a further 140 m to the length and an additional 500 tonnes. A number of obstacles were present on the transit route including curves with a radius of 200 m, multiple underground structures, a bridge, a 3.3 % decline, reduced headroom under permanent props in the launch chamber and a 1.7 m step. The transit method had to be selected to best tackle each of these obstacles whilst reducing disruption, risk and costs.

The existing footbridge located on the transit route was used by school children and commuters to cross the construction site, London Underground lines and Network Rail lines. Due to height restrictions the bridge had to be raised around 4 m to allow the shield to pass underneath. Regular use of the bridge by the public meant Westminster City Council (the local authority) was keen to minimise disruption to the public and therefore, the transit solution selected should allow the TBM to clear the bridge as quick as reasonably practical.

Two transit options were considered for transporting the TBMs; these included the skidding arrangement discussed in the examples from Barcelona and Switzerland and the use of SPMTs. The skidding options promised advance rates of 12 m/h for the straight sections reducing to 6 m/h for curves. The SPMT option offered a significantly faster rate of up to 6 km/hr. The cost of hiring SPMTs is higher than skidding equipment however, due to the increased speed and the reduced manpower requirements of this option, quotations from various heavy lifting specialists show that in this case the cost difference between each option was not significantly different. Due to the time pressures of clearing Hampden Street Footbridge and the small difference in cost between options SPMTs were chosen for the first section of the transit route. This allowed engineers to raise the bridge, transit the TBM through, lower the bridge and re-open the bridge within a 24 hour window.

In order to lift the TBM onto the SPMT for transport it was jacked up on a cradle structure using four 200 tonne climbing jacks to a height of 1.5 m. The SPMT was then maneuvered into position below the cradle before lowering the TBM into place. The arrangement used is shown in figure 8 below.

Figure 8 – Crossrail Western Running Tunnels – TBM shield supported on cradle structure transported on SPMT [Hickman, C, 2012]

In addition to time constraints, engineers also had to consider the capacity of the substructure and assets below to withstand the load of the TBM as it passed. The site information and ground investigation report highlighted a brick sewer and pumping chamber and a series of soft spots in the foundation soil. The design of the SPMTs was selected to both have capacity to support the TBM whilst also to sufficiently spreading the load onto the ground below. For this purpose a system incorporating two rows of twelve axles spread the load such that a ground bearing pressure of 8.7 tonnes/m^2 was exerted on the ground.

The backups were transited using fixed wheels running on a rail system similar to that used on the Barcelona Line 9 project. So that the SPMTs did not run over the rail, the rail was laid in 12 m sections behind the TBM shield and before the first backup gantry. Figure 9 below shows the backup train transport system as the TBM approaches the launch chamber.

Figure 9 – Crossrail Western Running Tunnels – TBM backups transported on a bogie and rail system [Hickman, C, 2012]

The system was successful on the straight sections of the transit route but issues arose as the TBM passed through the 200 m radii on the route. The problems were caused by lateral loads exerted on the fixed wheels as the TBM passed through these curves. Whilst the loads in question were small, a lack of suitable locations to install the wheels meant that several were welded to sections without stiffeners and bracing. The un-braced nature of the fixing locations meant that buckling of the supporting members occurred and alternative support methods had to be adopted in these locations. In hindsight, it would have been wise to remove the inclined bogies and replace them with the fixed wheels at these locations until the TBM was in launch position.

After passing the footbridge, the TBM descended down a 3.3 % gradient to the launch chamber. The structure of the launch chamber included a series of low level props. The lowest of the props was 7.45 m above the level of the baseslab and, as the TBM shield had a diameter of 7.1 m this left little room for the cradle structure and transport equipment. Therefore, in order to pass under the props the TBM had to be transferred onto skidding equipment for the final 100m of the transit.

Figure 10 - Crossrail Western Running Tunnels – TBM passing under first low level prop in launch chamber with only 20 mm clearance [Hickman, C, 2012]

The skidding equipment used included four skid shoes each with a 20 tonne push-pull unit running in PTFE lined skid tracks. To ensure that the TBM would not slide out of control, engineers checked the frictional characteristics of the PTFE pads to ensure the TBM would stop when required. In addition to this the skid tracks were fixed in place and the skid shoes intermittently locked into these to ensure that the TBM was restrained at all times. Operating on a push-pull cycle with a stroke of 600 mm the TBM advanced at a rate of 12 m/h towards launch position.

At the end of the launch chamber a 1.7 m high step had been included in the baseslab design. A steel bridge was installed so that the TBM could be skidded beyond the step. In order to get the TBM into launch position, the TBM was then jacked up using the climbing jacks strapped to the cradle structure. The steel bridge then had to be removed from below the TBM. The area below the TBM shield the bridge was designed so that all bracing could be removed from the outside to avoid the need for workers to access the area below the TBM. Chains were installed to the main bridge beams so that they could be dragged out from below the TBM using Tirfor winches.

Figure 11 - Crossrail Western Running Tunnels – TBM skidding onto steel bridge above launch position [Hickman, C, 2012]

2.2.4 Crossrail, TBM Transit through Bond Street Station

Advanced works were carried out at Bond Street Station to excavate the North West Shaft which will later form part of the Eastern Ticket Hall and construct a large proportion of the Eastbound Platform Tunnel. This meant that as the TBM arrived it would need to be transited trough the platform tunnel and then re-launch from the shaft at the other end. The layout of the platform tunnels was restricted by openings in the shaft structure and a curve in the permanent tunnel alignment. This meant the TBM had to negotiate a series of curves during transit. The alignment of the TBM is shown in figure 12 below.

Tight curves in the transit alignment can cause issues when transiting TBMs for two reasons. Firstly, the TBM shield may be considered as a rigid cylinder during transit and therefore, it will not easily pass through a curve. As the length of the shield increases and the radius of the curve reduces this issue is exasperated. Secondly, the backup train will follow the shield through gradual curves but snagging may occur when curves are tightened. If two consecutive curves are included in the route the problem is worsened as the backups will try to follow the shortest path and will consequently ride up resulting in snagging issues and thus slowing the advance of the TBM.

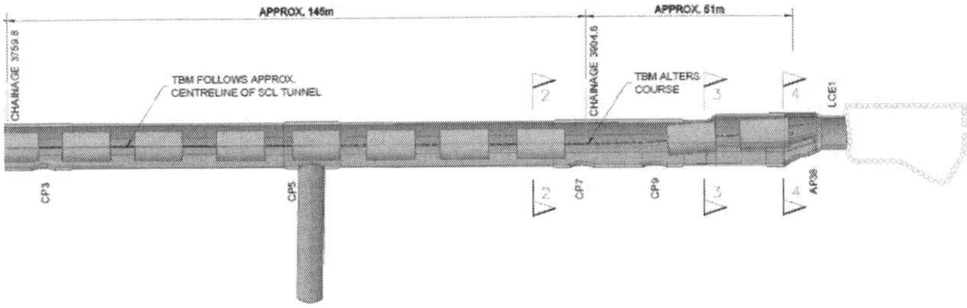

Figure 12 - Crossrail Western Running Tunnels – Plan showing TBM transit alignment through Bond Street Station [Freeman, J, 2012]

This problem was highlighted as miners attempted to negotiate a series of two consecutive 200 m radii at Paddington Station and one of the towing rams that connect the shield to the backup train failed. The cause of the failure could be attributed to a number of factors including casting defects, seizing of the bearing or excessive articulation of the joint but there is no doubt that the tunnel alignment was a contributing factor. The failed towing rams are shown in figure 13 below.

Figure 13 - Crossrail Western Running Tunnels – Photos showing fracture of towing rams from the TBM bridge assembly [Hickman, C, 2012]

An additional consideration for transiting TBMs below ground is the reception arrangement.

Guidance systems to control the TBM alignment has improved greatly over the years but some tolerance must be allowed for the location of the TBM arrival. There are several methods for accommodating this tolerance but in this case engineers chose to use a lean mix concrete invert with a transition section.

The lean mix concrete invert is cast prior to the arrival of the TBM and as the TBM arrives it cuts through the concrete invert to form a cradle at the correct height. Beyond the concrete invert, a steel or concrete cradle can be installed to suit the height and position of the TBM. In order to reduce the amount of work required after the TBM has arrived, the continuous cradle can be installed in advance and a transition arrangement used to adjust the level of the TBM. An example of the transition arrangement to be used at the Crossrail Bond Street Station is shown in figure 14 below.

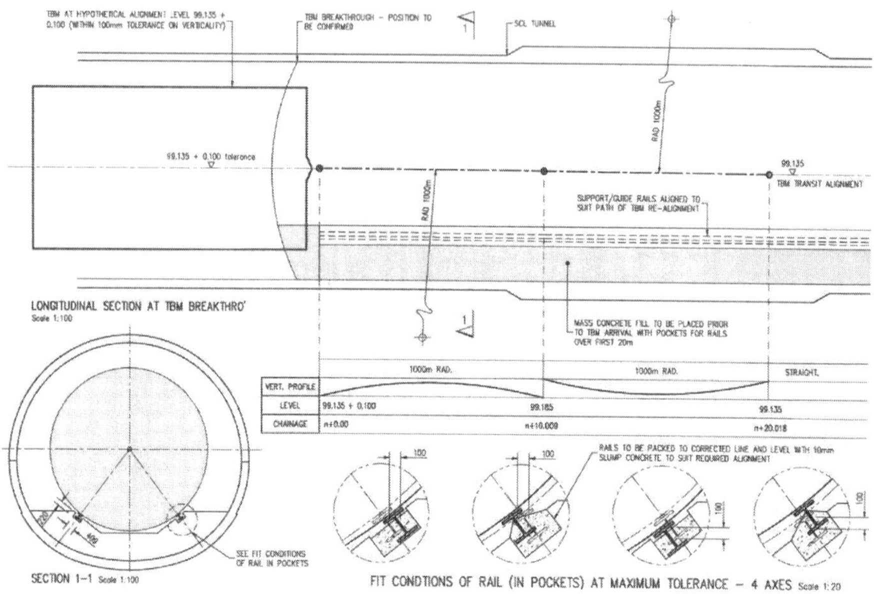

Figure 14 - Crossrail Western Running Tunnels – TBM reception and transition arrangement [Freeman, J, 2012]

2.2.5 Ireland, Dublin Port Tunnel

A single TBM was used to construct the two running tunnels for the Dublin Port Tunnels. The TBM was launched and constructed the first running tunnel and then had to be turned through 180° so that it could be re-launched for the return journey. In order to carry out this manoeuvre engineers chose to lift the 11.77 m diameter, 156 m long, 1600 tonne TBM using a combination of air bags and a crane. The TBM shield was detached from the backup train and lifted using airbags. Steel plates were laid across the transit area and were lubricated to reduce friction allowing the shield to be skidded into the new launch position. The backup gantries were then lifted one by one and joined to the back of the TBM.

The task of transporting a TBM whole is a complex procedure, the need for which should be identified early and considered carefully. If possible, the need for TBM transits and re-launches should be highlighted at design stage and considered in the design of launch chambers and stations. For example, removing the need for curves at stations will smooth and accelerate the transit and re- launch of a TBM.

Continuous cradle systems reduce disruption to tunnelling activities and are therefore ideal for use in cases where transits are required through shafts and caverns along the tunnel route. However, due to the high cost of this option in terms of labour and materials a cost benefit analysis may be required as openings increase in length.

Currently there are two main options for transporting TBMs over longer distances; these include skidding and SPMT systems. The decision on which transit option to use will be influenced by a number of factors such as the required advance rate, height restrictions and foundation considerations. Cost may not be an issue as quotations show that the difference between each option is not significant which is probably due to the increased time and manpower required for skidding systems.

In the authors opinion, of the systems that have been developed for transporting the TBM backup train, the bogie and rail system is the most cost effective and efficient method. However, the design of the system should carefully consider curves in the transit route.

The examples in this report provide a useful reference to parties considering similar operations. The issues highlighted and solutions discussed identify some of the issues that can be tacked early to reduce risks and delays.

Acknowledgements

The author would like to thank all those that have given their help in the preparation of this report. I would like to give special thanks for their support and guidance in writing this report:

Ivor Thomas, Tunnelling Manager, BFK Paul Glass, Technical Director, BFK

Andy Alder, Project Director, Crossrail C300

I would also like to thank Ermin Stehlik for the information regarding the TBM transits in Prague and John Freeman for the illustrations showing the TBM transit at Bond Street Station.

References and Bibliography

Dublin Port Tunnel, 2003. Dublin Port Tunnel Operations Information. [online] Available at: http://www.dublinporttunnel.ie/about/building/ [Accessed 07 September 2012].

Bern, K., Tunnel Boring Machine, Biel – Switzerland, 2011. VSL Job Reports. [online] Available at:
www.vsl.com/References/jobreport.php?getfile=JR2839 [Accessed 21 July 2012].

Stehlik, E., 2011. Figure 1 – Prague Metro Line Extension. [photograph] (Ermin Stehlik's own private collection).

Stehlik, E., 2011. Figure 2 – Prague Metro Line Extension. [photograph] (Ermin Stehlik's own private collection).

Stehlik, E., 2011. Figure 3 – Prague Metro Line Extension. [photograph] (Ermin Stehlik's own private collection).

ALE Heavylift Iberica, 2007. Figure 4 – Barcellona Line 9. [drawing] (BFK Technical Library).

ALE Heavylift Iberica, 2007. Figure 5 – Barcellona Line 9. [photograph] (BFK Technical Library).

Thomas, I., 2011. Figure 6 – Biel Bypass. [photograph] (Ivor Thomas's own private collection).
Thomas, I., 2011. Figure 7 – Biel Bypass. [photograph] (Ivor Thomas's own private collection).

Hickman, C., 2012. Figure 8 – Crossrail Western Running Tunnels. [photograph] (Chris Hickman's own private collection).

Hickman, C., 2012. Figure 9 – Crossrail Western Running Tunnels. [photograph] (Chris Hickman's own private collection).

Hickman, C., 2012. Figure 10 – Crossrail Western Running Tunnels. [photograph] (Chris Hickman's own private collection).

Hickman, C., 2012. Figure 11 – Crossrail Western Running Tunnels. [photograph] (Chris Hickman's own private collection).

Freeman, J., 2012. Figure 12 – Crossrail Western Running Tunnels. [drawing] (Kier Construction - Engineering design sketch).

Hickman, C., 2012. Figure 13 – Crossrail Western Running Tunnels. [photograph] (Chris Hickman's own private collection).

Freeman, J., 2012. Figure 14 – Crossrail Western Running Tunnels. [drawing] (Kier Construction - Engineering design sketch).

Crossrail Project: Infrastructure design and construction
ISBN 978-0-7277-6078-4

doi: 10.1680/cpid.60784.101

The Design of Thermal Tunnel Energy Segments for Crossrail

D P Nicholson BSc MSc DIC CEng MICE, Director, Ove Arup and Partners Ltd
Q Chen BSc MSc FGS, Senior Hydrogeologist, Ove Arup and Partners Ltd
M de Silva BSc PhD MCIWEM MIEEM CSci, Sustainability Manager, Crossrail Ltd
A Winter BSc MSc CEng MICE, Bored Tunnel Engineer, Crossrail Ltd
R Winterling MSc Dipl.-Ing, Technical Manager, Rehau AG+Co, Germany

Abstract

In recent years the drive to reduce carbon emissions and reliance on oil and gas has led to the increase in the use of renewable energy heating systems. The ground source heat pump systems are an example of these renewable systems. Recently tunnels have been considered as a way of connecting building heat pump systems to the energy stored in the soil. On Crossrail it is also recognised that heat is also given off by the trains in particular during braking, stopping at platforms and acceleration away from stations. A complex ventilation system including shaft, fans and Under Platform Extraction (UPE), has been designed to manage the rising temperature in tunnels and stations. This conventional approach results in high energy consumption for running the fans and neglects the possibility to use the extracted heat above ground in building, whilst reducing the energy consumption as well as CO2 emissions.

The Crossrail Tunnel Energy Segment (TES) design seeks to provide an alternative solution to cool the tunnel and the surrounding ground, and transfer this heat to adjacent buildings. Heat pumps are used in the buildings to step up the temperature for hot water supply throughout the year and space heating during winter.

This paper describes the technical challenges associated with:

- Assessing the available tunnel heat and linking this heat to adjacent buildings.
- Modeling the heat transfer from the tunnel heat to the pipe network in the segments.
- Design and construction of the segment pipework and box-outs
- Header pipe hydraulic design
- Access to the tunnels from shafts and boreholes.
- Integration of TES heating with adjacent buildings.
- Durability and operational considerations of TES including thermal effects and fire
- Operational and commercial benefits to cool the tunnel via TES

The paper also outlines the design for an 860m long demonstration tunnel on Crossrail between Tottenham Court Road and Fisher Street shaft.

The TES modeling of the Crossrail system, which is not dissimilar to other metro-style/frequency urban underground services, indicates that this is a viable technology for the future "smart" cooling of tunnels. It is "smart" because it extracts heat in a form that can be utilised in buildings, with a potential revenue stream and because it can be used to reduce the reliance on UPE and tunnel ventilation systems, thus saving on electrical energy. These two benefits provide significant carbon and financial savings at a time where there is a focus on carbon reduction and when fuel costs are anticipated to increase between 50-100% decade on decade in real terms.

Installation of a TES system should be considered in any future metro system, and the development of this concept on Crossrail provides a vital building block in understanding the benefits of the system so that it can be fully implemented on future projects.

I. Introduction

The shallow ground has a large potential of low enthalpy energy that can be used for heating and cooling purposes. With the advances in ground source heat pump technology it becomes increasingly feasible to capitalise on this geothermal energy present in the ground. The biggest advantage of the geothermal energy is that the soil temperature reaches a constant at a depth of approximately 10 to 15 m below surface. For example, in central London the soil temperature at 14 m below surface remains undisturbed at 14 °C. Outside London, the soil temperature is slightly lower.

Recent development has seen the use of foundation piles, diaphragm walls, and base slabs as ground heat exchangers (Adam and Markiewicz, 2009; Brandl et al, 2006; Fry, 2009). For example, thermal piles and walls involve attaching polymer absorber pipes to the reinforcement cages and this approach has been applied to Crossrail stations. Other application examples involving sprayed concrete linings include the tunnels at Stuttgart Metro U6 (Schneider, Moormann, 2010), Lainzer Tunnel (Franzius et al, 2011) and the metro stations (Brandl et al, 2010). Heat-exchange pipes have been placed along the lengths of the Channel Tunnel to extract heat from inside the tunnel (Personal communications).

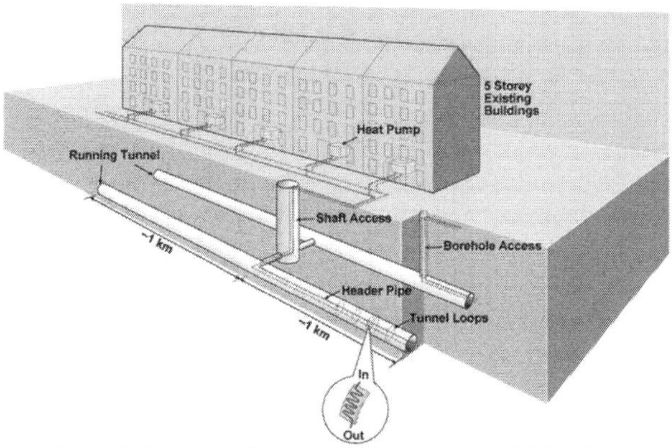

Figure 1: Schematic of the tunnel energy segment (TES) system

The paper by (Franzius and Pralle, 2011) discusses embedding plastic absorber pipes in the precast concrete tunnel segments of a conventional metro to collector heat from the adjacent ground. However, at Crossrail the primarily heat source is the waste heat rejected from trains and the secondary source is the surrounding ground. The heat is then transferred via header pipes to the ground surface, and then connected to heat pumps to supply the heat to buildings. Figure 1 shows the conceptual design of the Tunnel Energy Segment (TES) system.

II. Assessment of the heat inside tunnel

Operation of the London underground network over the past fifty years has gradually warmed the surrounding ground, which can no longer effectively absorb the waste heat from trains. The consequence is that today's temperature in many underground lines remains high throughout the year, with temperatures of more than 30 °C in some parts of the network (Botelle et al, 2010). Increasing air temperature presents a big challenge to the underground train operators. In future climate change will exacerbate the current situation, with mean temperatures predicated to increase by 2 to 6 °C higher than normal within the next 40 years (Defra, 2012).

To start with, an assessment of the available heat energy inside tunnels was carried out. Train heat is emitted from the motors, air conditioning systems and brakes. This waste heat is injected inside the tunnel which warms the air and surrounding soil mass.

As seen in Table 1, at peak time there are 24 trains each hour on one direction, meaning one train per 2.5 minutes. The length of Crossrail tunnel is about 21 km long. Assuming the average train speed is 60 km/hr, a train will take 21 minutes to complete the journey. Hence, the number of trains in the tunnel at peak time is 21/2.5 = 8.4 trains.

Assuming one train motor generates on average of 1 MW net heat, this is focused on the tunnel length where the trains leaves the station. Its air-conditioning unit generates about 0.1 MW, the total heat output from 8.4 trains is 1.1 x 8.4 = 9.2 MW. Crossrail comprises about 42km of tunnels with a surface area at intrados is about 410,000 m^2. Therefore, the peak heat input in terms of tunnel lining area is 9.2 x 10^6 / 410,000 = 22 W per square metre of tunnel surface area. On a weekly basis, the average number of trains is 14 trains per hour. The average heat output is thus 22 x 14/24 = 13 W/m^2. In addition, the ground behind the tunnel linings would also serve as a heat source similar to a closed-loop ground-source heat pump design. The train's heat output also increases dramatically during braking as the train approaches the platform and accelerates away from the station. Therefore these sections have higher heat output and are the locations where TES should be focused. For comparison closed loop boreholes typically transfer 35 W/m run and small diameter foundation piles about 60 W per m run.

Based on the above simplified calculation, the base heat output from the tunnel was about 7 W/m^2 base and the peak output was about 30 W/m^2 peak. On this basis a 500 m length of tunnel would be able to support about 100 family apartments with a total heat demand of 1,200 MWh/yr which would be met by a combination of a 400 kW gas boiler and a 400 kW heat pump. The full network of Crossrail tunnels (2 No x 21 km tunnels) would comprise 84 such lengths and would consequently have the capacity to provide heat for 8,400 families (30,000 people) or equivalent building heating.

Table 1 Crossrail designed train frequency during normal operation (SP2)

From	To	Hours	Train Freq	No of Trains
Monday – Friday				
00:00	00:30	0.50	12	6
00:30	05:45	5.25	0	0
05:45	06:15	0.50	12	6
06:15	07:00	0.75	16	12
07:00	07:45	0.75	20	15
07:45	09:15	1.50	24	36
09:15	10:00	0.75	20	15
10:00	16:00	6.00	18	108
16:00	16:45	0.75	20	15
16:45	18:15	1.50	24	36
18:15	19:00	0.75	20	15
19:00	22:00	3.00	16	48
22:00	00:00	2.00	12	24
			Total No.	336
Saturday				
00:00	00:30	0.50	12	6
00:30	05:45	5.25	0	0

05:45	09:00	3.25	12	39
09:00	21:00	12.00	18	216
21:00	00:00	3.00	12	36
			Total No.	297
Sunday				
00:00	07:00	7.00	0	0
07:00	12:00	5.00	12	60
12:00	21:00	9.00	18	162
21:00	23:45	2.75	12	33
23:45	00:00	0.25	0	0
			Total No.	255

III. Design of Tunnel Energy Segments system

1. Previous experience

There are a number of existing case studies on retrieving heat energy from tunnels. The Vienna LT22 testing plant was the first geothermally activated sprayed concrete lining tunnel in Austria. The absorber pipes were attached to a geosynthetic and placed between the primary and secondary tunnel lining (Adam and Markiewicz, 2009) followed by an installation in Germany at Stuttgart Metro U6, (Schneider and Moormann, 2010). More recently, the thermal activation of tunnel structures has been applied to TBM Tunnels. Züblin and REHAU have championed this development with a field test in the Katzenberg high-speed rail tunnel in Germany followed by the Jenbach twin track high-speed rail tunnel in Austria incorporated a 54 m long demonstration section with thermally activated segmental tunnel lining equipped with heat exchanger pipes. These pipes were connected to a heat pump in the utility centre building to provide hot water and space heating (Franzius and Pralle, 2011). It should be noted that the above schemes are taking heat from the ground mass rather than the tunnel air.

2. Crossrail experience

The Crossrail TES work built on this previous experience to help solve the very demanding technical and management challenges faced by the design team.

- Firstly, the TES design was started after the tunnel design had been completed and tendered. The TES design needed to cater for the absorber pipe requirements, whilst still meeting all the existing design requirements for structural, fire, H&V etc. For example, a design method had to be developed to enable the existing tunnel H&V design to be revised to incorporate the TES. The introduction of box out pockets at the ends of the segments to form pipe connection needed to take account of the caulking groove featured in the segment structural design.

- Secondly, discussions were held with the tunnel contractors to develop ways of incorporating the absorber pipes in segment moulds and ways of forming tunnel ring connections so that the TES did not delay construction or introduce high costs.

- Thirdly, TES needed to locate and manage a network of building who could use the heat.

- Fourthly the Crossrail Board needed to have design risk assessments and budget costs so that they could make informative decisions.

All these challenges need to be resolved to a demanding time frame. The multi-disciplinary inputs included tunneling, building services, structures, material, fire, geotechnics and hydrogeology, energy, hydraulics, costing, etc.

3. Thermal energy segments

In order to extract heat from the tunnel, the concept design is to embed 20 mm internal diameter absorber pipes into conventional concrete segments before casting. The segments are fabricated using a standard segmental lining manufacturing process. The pipe lengths of 10 ~ 20 m are placed into each segment in a meandering fashion as shown in Figure 2. For the reinforced segments, the pipe can be tied to the cage or inserted on a separate support mesh. On Crossrail reinforced segments are used on the floating track sections. For fibre reinforced concrete segments, the pipes are softened in a hot water bath and then attached to a light cage which is then positioned inside the segment mould. To minimise fire damage, the pipes are placed at least 200 mm from the intrados. To meet minimum pipe bending radii a 300 mm pipe spacing has been adopted using the PE-Xa polymer pipe specification. Figure 2 shows a typical pipe arrangement for a segment with a reinforcement cage before concrete is poured.

Figure 2: Absorber pipes assembled to a reinforcement segment before casting. Left – plane view, right – cross-section view (photos taken from Jenbach tunnel by Rehau Ltd)

In order to form the connections between segments, box-outs are specially designed at the ends of segments. These pockets are about 200 mm long, 100 mm wide and 70 mm deep to allow a permanent mechanical coupling of the adjacent pipe-ends across the box-out, see Figure 3. The Crossrail box-outs have reverse tapers to retain the mortar filler.

The impact on structural integrity and fire safety induced by these changes to standard segment design are discussed later.

The pipe connections are formed immediately after the segment ring is erected at the tunnel face. The majority of work being carried out from the erection train towed behind the tunnel boring machine. The connection system was initially developed for thermally activated building structures (TABS)s in highrise buildings for pipe connections buried in the concrete structure. The system allows permanent mechanical connections (Everloc) to be rapidly made using PE-Xa pipe in a small space to reduce the installation time and have a robust and reliable connection. Once formed, these connections cannot be undone. The permanent couplings have been subjected to pressure tests and the results show that the polymer pipe fails before the joint, and this provides confidence in the integrity of the couplings.

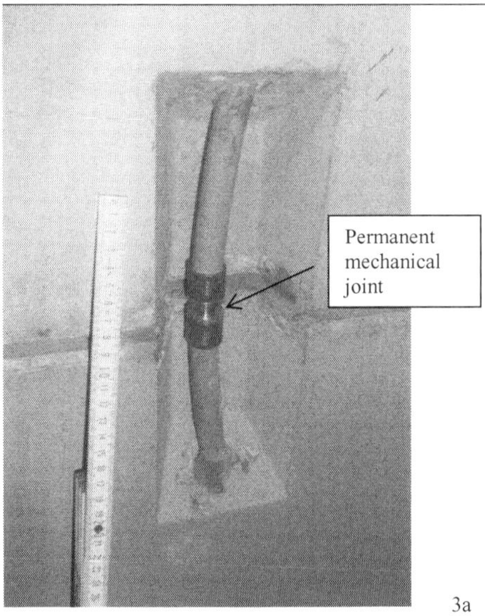

3a

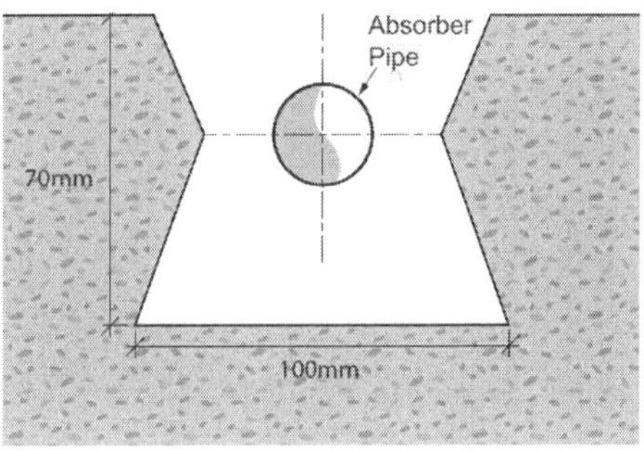

3b

Figure 1: Box-out for pipes' radial joint at the end of segments. 3a – tapered box-out used in Jenbach tunnel; 3b – reversed tapered box-out design for TES

To form the complete circuitry, the absorber pipes need to be joined up to from a continuous ring of six segments. About 5 neighboring rings are connected to form a circuit, see Figure 4. Each circuit is connected to flow and return header pipes. Control valves are provided to isolate each circuit in the event of a local failure and to assist with de-airing. The ring to ring connection positions cannot be predetermined because each ring is slightly tapered. The orientation of the taper is selected during tunnel construction to steer the tunnel as the work progresses. Hence the ring to ring pipe connections are formed from short lengths of plastic polymer mounted on the tunnel surface, see Figure 5.

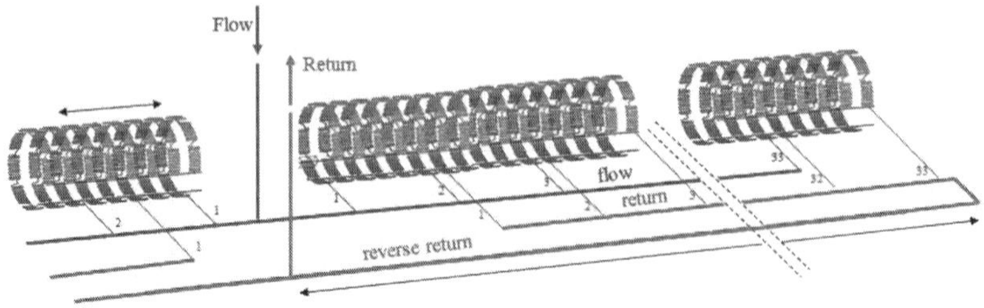

Figure 2: Schematic of pipework connections with three header pipes configuration.

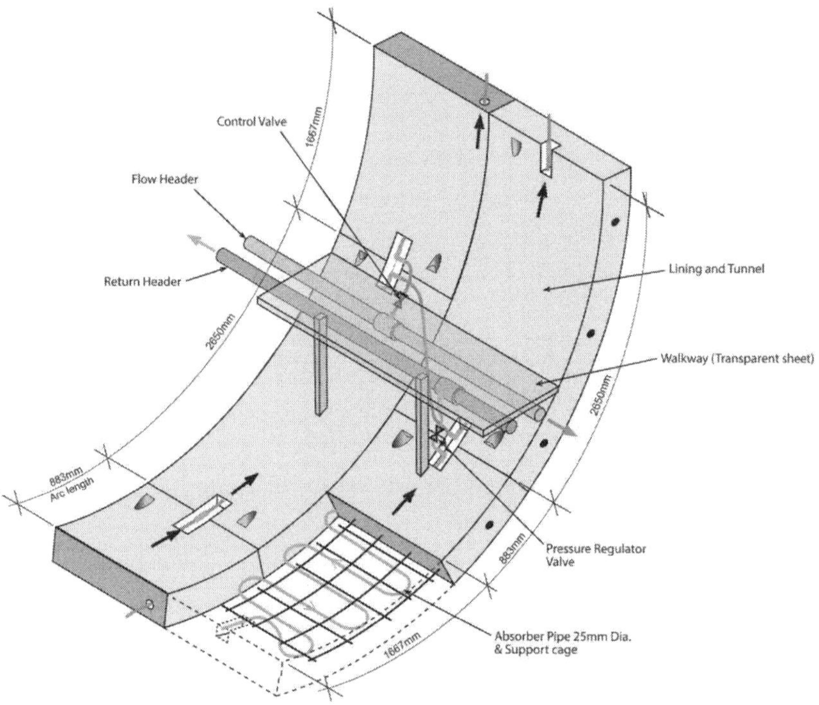

Figure 3: Schematic of TES design.

One complete circuit comprises about 11 m of pipe inside a standard segment and there are six segments plus a keystone in one ring. There are five rings in one circuit and 31 circuits to cover a tunnel length of about 250 m length. Tunnel header pipes would be connected to the surface though shafts and stations. Alternatively dedicated access boreholes could be provided at preferred surface locations for the header pipes, see Figure 1.

It has been planned that heat from the tunnels would be connected to district heating networks supplying heat energy to groups of buildings along the alignment of the tunnels. The income from the sale of heat to the district heating Energy Supply Companies (ESCos) provides a revenue stream to the tunnel operator which finances the installation and operation of the TES.

4. TES system design: Pipe work hydraulics and plant room.

The three header pipe "reverse return" arrangement was selected in preference to the two pipe arrangement because of the advantage of self-balancing. A procedure for removing trapped air was developed. Pressure losses were calculated for normal operation, and flushing of individual circuits and header pipes to remove trapped air bubbles.

Flow rate through the circuit varied depending on the heat extraction rate. The flow rates of 0.06 l/s achieved 10 W/m^2 heat extraction rate and 0.12 l/s achieved 30 W/m^2. The pressure losses for all fittings are calculated in accordance with CIBSE Guide C (CIBSE, 2007) except for the long radius U-bends that make up the meander of the pipes within the segments. The calculations for normal operations included a 50 kPa pressure drop for fittings on the circulation pump, and an additional pressure drop of 100 kPa for the heat exchanger. During flushing of air bubbles, the heat exchanger would be by-passed, so only 50 kPa pressure drop has been included in the calculations. The system is balanced using a reverse-return arrangement with valves on each circuit and an additional valve at the connections between the flow and return pipes. A schematic of the assumed layout is shown in Figure 5. A design margin of 15% is applied on the calculated pressure drop across the pipe network. The flow in the circuits needs to maintain turbulent state to ensure efficient heat exchange.

For normal operation header pipe lengths of 250 m were planned. Increasing the length to 500 m resulted in the system's maximum pressure loss rising from 7.5 bar to 9.1 bar for 90 mm internal diameter header pipes.

To ensure the smooth running of the system, it needs to be flushed to remove any trapped air bubbles from the circuits. First, each circuit will be flushed individually. This procedure involves closing all valves, except one, leading to the circuits, and supply water to the system to get a minimum water flow velocity of 1.0 m/s in the loops, based on the recommendations of MIS 3005 Standard. Secondly, air is to be removed from the system by flushed the header pipe. The maximum pressure head loss is 8.3 bar for a 250 m long section, and 12.6 bar for a 500m section.

The header pipe length of between 250 m and 400 m on each side of an access point (e.g. shaft or borehole) was adopted to ensure the hydraulic performance could be maintained.

IV. Potential market for the heat

The London Heat Map and building inventory along the Crossrail alignment were used to establish adjacent buildings that could use the heat from the TES. These buildings were mapped in GIS. Two selection criteria were considered: Firstly, the buildings should have a sufficiently large heat demand to justify connection to the system. The minimum threshold of heat demand was based on the available heat from the tunnel, which is classified by annual heat demand of 600 MWh/yr, 1,200 MWh/yr, 1,800 MWh/yr and 2,400 MWh/yr. Secondly, by building type, the buildings should have a heat load profile which was complemented the efficient heat pump operation, i.e. relatively constant heating base load. These criteria were used to divide the buildings into tier 1 – most suitable buildings, tier 2 - moderate suitability and tier 3 -low suitability buildings.

Tier 1 - Hotels, Large Residential, Hospitals
Tier 2 - Schools, Higher Education, Public libraries, Museums
Tier 3 - Offices, Leisure Centres, Retail

As a result of this investigation 365 buildings with a heat load greater than 600 MWh/yr were identified within 100 m of the Crossrail Tunnel alignment. These buildings comprised 34 Nos Tier 1, 4 Nos Tier 2, and 327 Nos Tier 3 buildings. Some of these are shown in Figure 6.

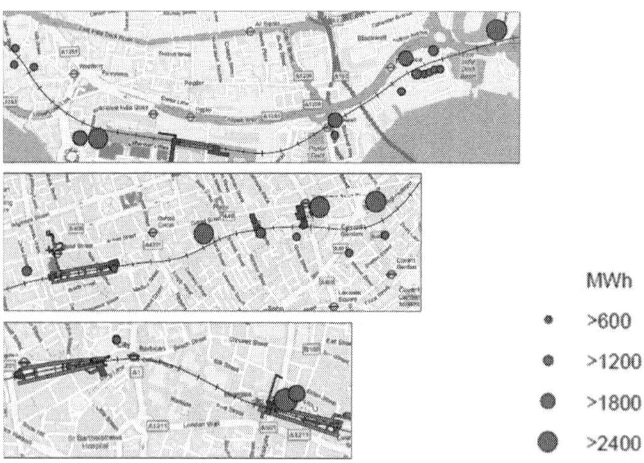

Figure 4: GIS study of existing energy users along Crossrail route: Tier 1 buildings at three locations – Canary Wharf, Bond Street, and Liverpool Street are shown as above.

The technically feasible surface connection points based on shaft and tunnel access routes are shown on Figure 7. A 250 m diameter circle is shown for the standard connection header pipe length.

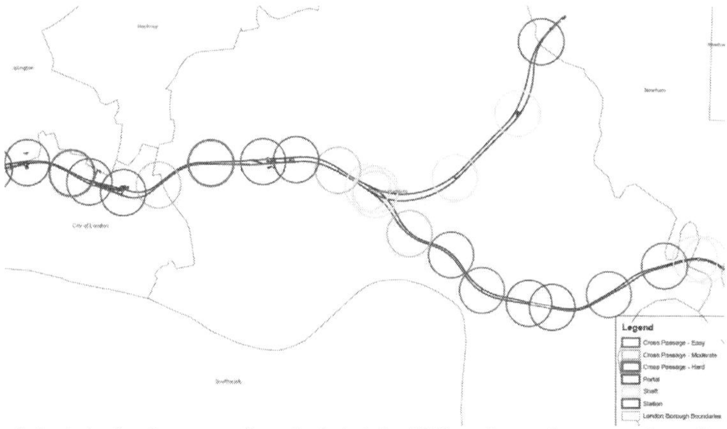

Figure 5: Analysis of surface connections. Each circle is of 250m radius, surface connection options are considered for shafts, cross-passages and directional drilled boreholes.

In some locations it was necessary to increase the capacity of the header pipes where the circles did not intersect. This approach enabled the energy segments to be connected to suitable buildings along the majority of the Crossrail tunnel.

V. Modelling of TES heat transfer

A finite-element numerical model was developed using numerical code LS-DYNA to simulate the thermal transfer behaviour of the TES, see Figure 8. The model is a 3-dimensional slice through the tunnel. In practice, the water in the pipe will be cooler near the inlet and warmer near the outlet; this model represents an average temperature along the length of the pipe. The analysis was carried out for 3 years without heat extraction in order to achieve a steady state condition within the soil. Heat was then extracted from the tunnel for a further 7 years to allow the system to change dynamically with varying tunnel temperature.

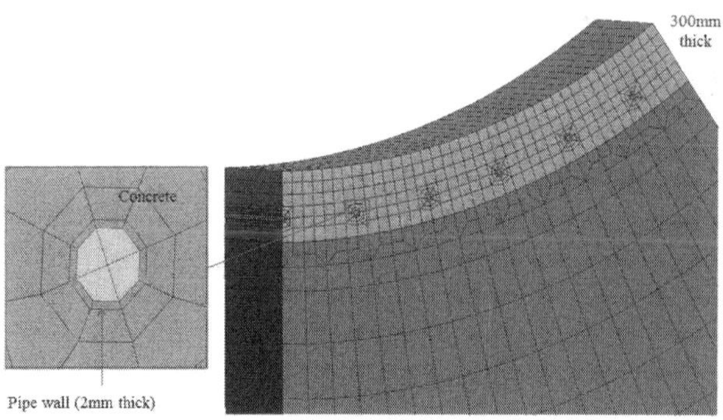

Figure 6: Set-up of the numerical thermal model in LS-DYNA to simulate heat transfer to absorber pipes

The thermal parameters used are summarised in Table 2. Furthermore, the boundary condition assumed an ambient soil temperature of 14.8°C at a distance of 100 m from the tunnel. Seasonal soil temperature variation was considered insignificant. In addition, the tunnel air temperature varies from 36°C in summer to 17°C in winter.

Table 2 Thermal parameters used in thermal analysis

Material	Density (kg/m^3)	Conductivity (W/m.K)	Heat Capacity (J/kg.K)	Young's Modulus (kPa)
Soil	2,000	1.8	1,000	150,000
Concrete	2,500	1.33	750	31 x 10^6
Absorber pipe wall	1,000	0.087	1,000	100
Heat transfer coefficient from tunnel air to tunnel surface	5 W/m²K			

Sensitivity analysis was carried out for heat extraction rates ranging from 5 to 35 W/m^2. The resulting fluid temperature inside pipes over time is shown in Figure 9. The system reaches quasi-steady state within the first year operation. To avoid freezing the pipes it was recommended that the continuous heat extraction rate be limited by 30 W/m^2 in winter.

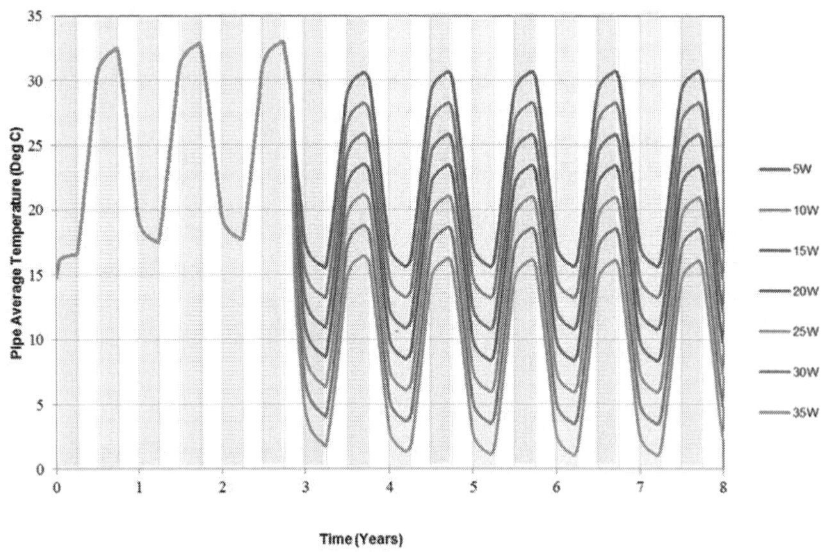

Figure 7: Predicated fluid temperature inside pipes over time for various extraction rates. Intital three years with no heat extraction.

VI. Tunnel cooling study

To understand the cooling effect of the TES inside the tunnel, a detailed 1-D ventilation model developed for the Crossrail tunnels was modified to include the TES, see Figure 10. The new model was validated by comparing the results with the LS DYNA analytical results.

Thermal simulations were carried out for the design full load train service pattern SP2 (240 m long trains, 24 TPH peak hour service frequency), with TES operating at a continuous predefined heat exchange rate.

The ventilation model predicts a reduction in peak tunnel air temperatures of 4 °C (Dry Bulb, DB) with the pipework installed along the full length of the running tunnel when the flow rate was set to 7.44 l/s in order to achieve 10 W/m^2 extraction rate, see Figure 11. Greater temperature reductions were achieved in areas such as tunnel east of Stepney Green junction. It was found that with the TES pipework installed in the proximity of the stations, the peak tunnel air temperature would be reduced up to 4 °C DB when the flow rate was set higher to 14.88 l/s, ie 20 W/m^2 extraction rate. This is due to focusing on the parts of the tunnel where braking and acceleration occurs.

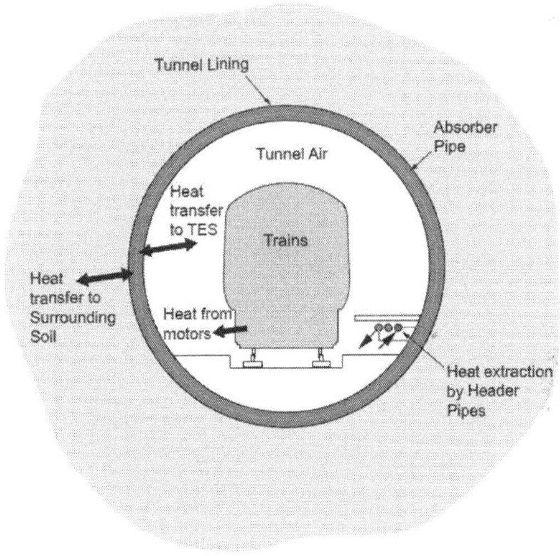

Figure 8: Set-up of the ventilation model showing heat transfer to tunnel and surrounding soil.

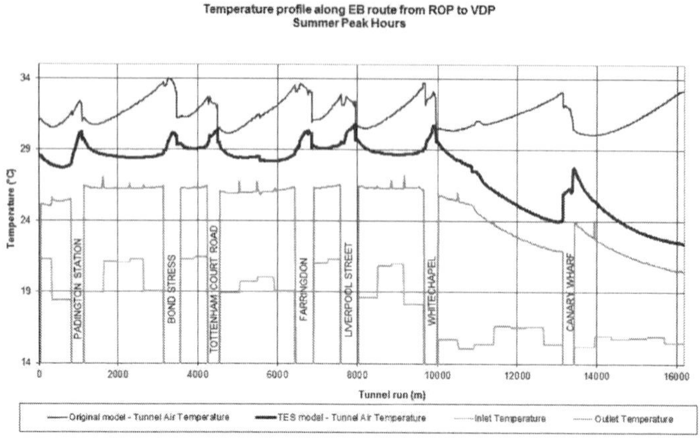

Figure 9: Predicated tunnel air temperature along Eastbound route from Royal Oak Portal to Canary Wharf Victoria Dock Portal during summer peak hours, with TES installed throughout the tunnel route. The absorber pipe inlet and outlet temperatures are also shown.

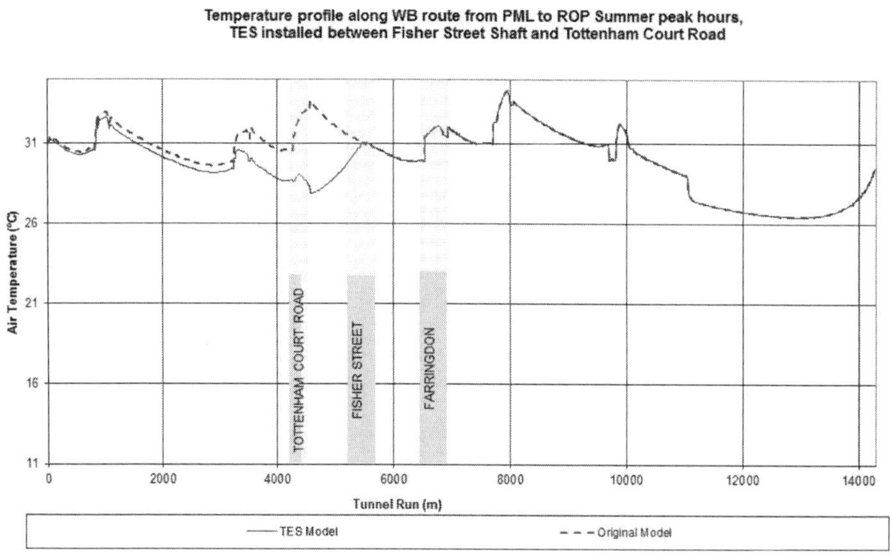

Figure 10: Predicated tunnel air temperature along Westbound route from Pudding Mill Lane Portal to Royal Oak Portal during summer peak hours, with TES installed between Fisher Street Shaft and Tottenham Court Road.

It should be noted that whenever the train heat inputs to the tunnels are lower than this full service pattern (for example in the early years of operation), lower heat extraction would be required to control water temperatures. This would have an impact on output from the heat pumps.

When the Crossrail system is operating, the bored tunnels are expected to be warmer than the surrounding ground due to waste heat emitted by the trains. Under the current design some of the heat is removed by air exchange through the draught relief ducts, and some by conduction through the tunnel walls to surrounding ground. TES will offer significant benefits as it will prevent a heat-build-up in the ground adjacent to the tunnel walls, and reduce the tunnel air temperature. This will lead to savings in the tunnel ventilation system operational cost and future proof the tunnel design for future increase in temperature due to climate change or higher train frequencies.

VII. Durability and operational considerations

1. Pipe durability

The PE-Xa grade pipe, comprising high-pressure cross-linked polyethylene, or equivalent, has been specified because:

- Tight pipe bending radii can be achieved.
- Robustness against notches, scratches and puncture loads during segment production and installation.
- Allows for segment joint extension at the box-outs.
- Permanent mechanical connections can be formed.
- Durability requirements in line with the tunnel design life (eg 120 years) at the pipe operating pressure and temperatures in accordance with EN 16892.

2. Thermal effects

In order to understand the thermal effect on the tunnel structure, an additional finite element model was developed using LS-DYNA to represent the running tunnel with symmetry boundary conditions on the two vertical planes at right angles to the tunnel axis. The model is a 3-dimensional 1.6m slice through the tunnel with the absorber pipe, see Figure 13.

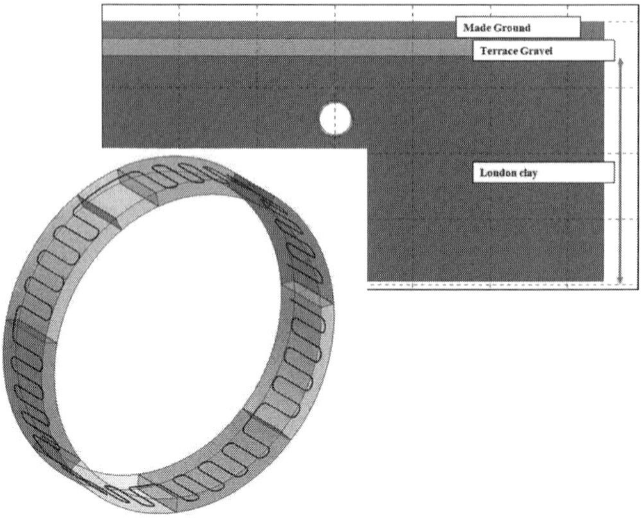

Figure 11: Set-up of the thermal-hydraulic-mechanical FE model for Crossrail tunnel linings in LS-DYNA. The black lines represent the absorber pipes.

The model using the elastic Mohr-Coulomb parameters, with the material properties summarised in Table 3. These soils were considered drained at all stages during the analysis. Their thermal properties used are given in Table 2. Thermal expansion coefficient of the soil skeleton and the pore fluid were assumed to be 3.0 x 10^{-5} / °C.

Table 3 Geotechnical parameters used in the LS-DYNA analysis

Stratigraphy	Initial stress conditions			Stiffness parameters			Strength parameters			
Stratum	γ (kN/m^3)	Ko (top)	Ko (base)	E'v (top)	E'v (base)	Poisson's Ratio	c' (kPa)	φ_p' (°)	c_u (kPa) (top)	c_u (kPa) (base)
Made Ground	18	0.5	0.5	10	10	0.2	0	25	-	-
River Terrace Deposits	18	0.5	0.5	25	25	0.25	0	40	-	-
London Clay	20	1.2	1.2	24	29	0.2	5	25	104	137
Lambeth Group	20	1	1	70	70	0.2	10	28	350	350
Thanet Sand	21	1	1	250	250	0.2	0	36	-	-

The thermal analyses were carried out for heat extraction rates from 0 (no heat extraction) to 30 W/m^2. Under natural conditions (ie no heat extraction), the maximum computed tunnel diameter change is +1.1 mm in the summer and 1.6 mm in the winter. There is negligible deformation at crown and invert of the tunnel. At heat extraction of 30 W/m^2 the tunnel shrinks at the axis level and expands at the crown during summer, and the tunnel shrinks towards the centre during winter. The changes in tunnel diameter are summarised in Table 4, 30 W/m^2 extraction results in a further 1 mm deformation compared to the case of no heat extraction.

Table 4 Thermal effect on tunnel dimensions

Simulation stage	Tunnel diameter change (mm)	
	Crown-Invert	Axis Level
Summer (no extraction)	+0.1	-1.1
Winter (no extraction)	-0.5	-1.6
Summer (30 w/m^2 heat extraction)	+0.3	-1.9
Winter (30 w/m^2 heat extraction)	-0.6	-2.6

Hoop stress in the tunnel lining without thermal effects is 4 MPa at a depth of 16 m. Hand calculated hoop stress in the tunnel lining, assuming a depth of 16 m and soil density of 20 kN/m^3, is approximately the same. As expected, the maximum hoop stress occurs at the convex to convex segmental joints. When no heat is extracted, the hoop stress (average through the segment thickness) shows a seasonal variation, being maximum in summer when it is 7% greater than the as-constructed condition. This rises by a further 2% when 30W/m^2 heat extraction takes place. In winter, the average hoop stress is negligibly different from the as-constructed condition, whether heat is extracted or not. Although the heat extraction induces relatively large temperature gradient across the thickness of the segments, the hoop stresses at both surfaces are predicted to remain compressive.

The computed tensile stress for no heat extraction and 30 W/m^2 during summer and winter is shown in Figure 14. For the ground conditions and geometry adopted in the analysis, the computed maximum tensile (bursting) stress occurs at the convex to convex joints. It is likely that the tensile bursting stress will be proportional to the contact force at the joints, which in turn is proportional to the average hoop stress. Therefore it is assumed that the bursting stress will rise by 7% compared to the condition of no heat extraction, and by a further 2% if 30 W/m^2 is extracted.

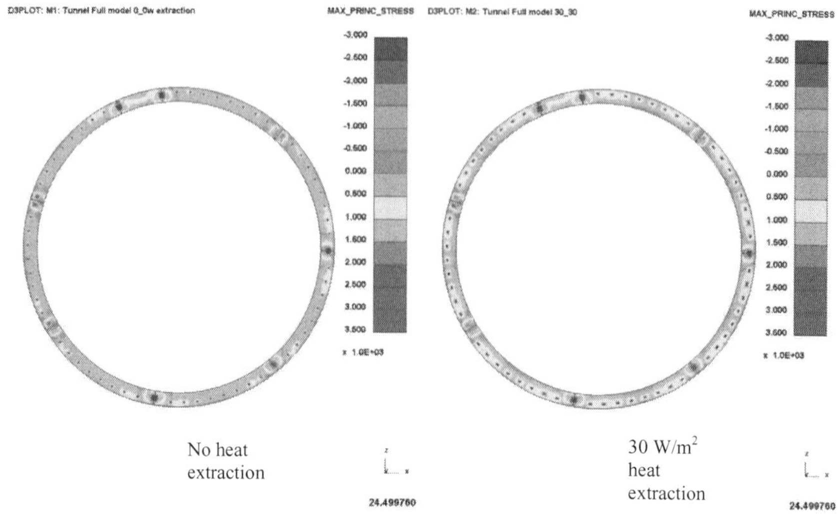

Figure 12: Comparison of maximum principal stress distribution in segments without and with heat extraction during summer

Tensile stress is also predicted around the pipes when heat is extracted, it reaches a maximum of 3.5 MPa when 30W/m^2 is extracted, which is below the design tensile stress limit of 5.0 MPa.

3. Fire safety

The piping is made of cross-linked polyethylene (PE-Xa) which is a form of polymer comprising molecular chains made up of hydrogen and carbon atoms. Though the cross-linking process additional links are built between neighouring carbon atoms. The Crossrail Fire Safety Strategy for the Central Tunnels states that materials within the tunnels should be chosen based on the London Underground standard 1-085 Fire safety performance of materials. Section 5.6.1 of the Crossrail Tunnel Fire Safety Strategy states that:

"*Any materials which do not comply with this standard will be risk-assessed to determine whether they are acceptable*".

The London Underground 1-085 standard has fire safety requirements for materials with regard to flammability, smoke emission, toxic fume emission. The proposed TES inclusion of exposed cross-linked polyethylene pipe with the tunnel is not fully compliant with the Standard in relation to flammability and potentially smoke emission. The Crossrail Fire Safety Strategy for the Central Tunnels allows a risk assessment to be carried out in lieu of strict compliance with the 1-085 Standard.

The embedded absorber piping is connected to a set of 3 header pipes, which are made of PE-Xa and 90 mm in diameter. To mitigate fire risk the design is based on burying the header pipes in mass concrete.

VIII. Operational and commercial benefits

It is important to assess the fundamental financial viability of the TES system. The elements to consider with respect to costs are initial capital tunneling cost, surface connection cost, cost incurred to connect the system to buildings, and operational cost.

The TES brings the following benefits to the tunnel operators:

1. Operational cost saving

Excess heat can cause operational difficulties. For example, metro tunnels may see a large heat input from trains, whereas power cables may raise the internal temperature of service tunnels. Therefore tunnel cooling often needs to be provided. The most cost-effective solution to achieve cooling would be natural ventilation, however this is rarely sufficient and conventionally fans are used to provide forced ventilation. Cooling the tunnel by means of the TES system will lead to saving in the operational cost of the ventilation fans. The heat capacity of the fluid in the absorber pipes embedded in the tunnel segments is much larger than air and hence allows for more efficient cooling than forced ventilation.

2. Capital cost saving

The physical size of the heat pumps may be significantly smaller than the air ventilation fans and associated air ducts that would normally be required for cooling the tunnel, hence allowing a more compact design and resulting in capital cost savings.

In some instances, using TES instead of forced air ventilation may even allow to reduce the tunnel diameter, where it is governed by the required air flow area. In a similar way, intermediate ventilation shafts may be downsized or removed because of the reduced air flow. In the future if these opportunities are exploited, significant capital cost savings may be achieved. At Crossrail the TES design options could not be developed fast enough to enable reduction in air ventilation and UPE usage to be incorporated into the design.

4. Carbon Saving

The provision of heat energy to buildings adjacent to tunnels reduces their heat requirement, for example the need for those buildings to heat air or water using conventional sources (gas, electricity, heating oil). On new construction work the TES heat could be used at the Planning stage as part of the Renewable Heat Incentive or the Part L renewable heating component. Accordingly, the use of TES provides a reduction in carbon output for the building being supplied with this heat energy. For example, 10 W/m^2 tunnel heat extraction provides enough heat to supply continuous summer low temperature demand of 250 kW, and approximately 170 m^3/day domestic hot water load preheat (approximately equivalent to a 500 bed luxury hotel), achieved carbon savings is about 42%.

5. Heat revenue

Selling the heat harvested from the tunnels to nearby properties will generate some direct income, in addition income may be possible through the renewable heat incentive and/or the selling of carbon offsets.

With the energy price rising and supply uncertainty, the TES may offer an alternative solution. Compared with geothermal piles and vertical borehole loops, the unit energy price is lower for TES system.

IX. Case Study: Fisher Street to Tottenham Court Road

A case study design of 860 m long demonstration TES tunnel between Tottenham Court Road station and Fisher St Shaft was develop to a design for construction to show that all the technical issues could be addressed and costs could be quantified. In order to make full use of the tunnel length available the chosen tunnel was divided into two sections of about 430 m with two sets of header pipes coming in and out of the Fisher St Shaft, as shown in Figure 15.

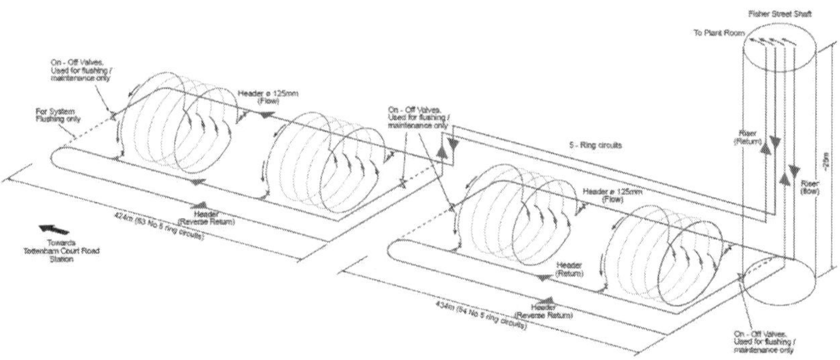

Figure 13: Schematic design of the TCR-FIS demonstration tunnel

The pressure drop at extraction rate of 30 W/m^2 is calculated as 7.5 Bar during normal operation, and 10.3 Bar during flushing. The effect of the TES on the tunnel temperatures is illustrated in Figure 12. Due to the space limitations the plant room layout was modified to fit to an area of 4 m by 10 m. The plant room is to accommodate the primary circuit i.e. circulation and standby pumps, the secondary and tertiary circuit are located at the end-users premises.

In order to reduce the fire load, the headers are designed to be buried in mass concrete beneath the evacuation walkway. As a result, the fire load introduced by the TES over a 20 m length of tunnel duration 15 minutes is reduced to 375 kW, and this is 4.2% of the designed train fire load of 8.8 MW over 20m long carriage. Pure polyethylene complies with the LU material requirements in relation to toxic fume emissions.

The instrumentation and monitoring specifications were also developed both inside the tunnel and the plant room. The cost of the TES is relatively low compared to the tunnel construction budget.

The following technical issues have been addressed for the development of the demonstrator TES section:-

- An assessment of the local heat available from the trains and the soil.
- The development of the demonstrator section ventilation model which included TES.
- The TES and header pipe layouts to extract heat at Fisher Street shaft.
- The pipe layouts for the steel cage reinforced segments for the floating track slabs.
- The connection details between segments and rings and header pipes.
- The space proofing and plant room arrangements.
- The stress analysis for the thermal effects on tunnel linings
- The assessment of fire risks.
- The construction processes for the segment moulds and their tunnel erection.
- The assessment of surface buildings within 100m from the Crossrail route for heat use.
- The development of drawings, specifications, quantities and budget costs

X. Summary and Conclusion

This paper provides an overview of a tunnel energy segment (TES) system for Crossrail where heat is generated by the trains in the tunnels all year round. The system uses embedded closed-loop water filled pipework in tunnel segments to extract this heat. In this way it both cools the tunnels and provides heat for adjacent buildings. The water temperature supplied to the heat pumps is relatively high enabling the system to operate at higher coefficient of performance (COP) compared with conventional ground source heating systems.

A major challenge for the TES project was the coordination of the multidiscipline team to deliver the design and costing studies for the TES project.

The TES is a viable technology for the future "smart" cooling of tunnels; "smart" because it extracts heat in a form that can be utilised and with a potential revenue stream and because it can be used to reduce the reliance on UPE and tunnel ventilation systems, thus saving on electrical energy. These two benefits provide significant carbon and financial savings at a time where there is a focus on carbon reduction and when fuel costs are anticipated to increase between 50 ~100% decade on decade in real terms. The development of this technology on Crossrail provides a vital building block in understanding the benefits of the system such that it can be implemented at scale on future projects.

It is considered that the technical and commercial problems associated with the TES have been sufficiently addressed to warrant a demonstration tunnel to prove the system for use in future tunnels. Thinking one step ahead to future proofing during new metro design and construction is important.

The project receives the 2011 "Technical Innovation of the Year" by International Tunnelling Awards and the 2011 "Innovative use of Equipment" by Tunnel & Tunnelling International Awards.

XI. References

1. Adam, D. & Markiewicz, R. (2009). Energy from earth-coupled structures, foundations, tunnels and sewers, Géotechnique 59, No. 3, 229–236
2. Botelle, M., Payne, K., Redhead, B. (2010), Squeezing the heat out of London's Tube, Proceedings of the ICE - Civil Engineering, Volume 163, Issue 3, 114 –122
3. Brandl, H. (2006). "Energy foundation and other thermo-active ground structures," Géotechnique, Vol. 56. No. 2 81-122.
4. Brandl, H., Adam, D., Markiewicz, R., Utenberger, W. and Hofinger, H. (2010). "Massivabsorbertechnologie zur Erdwarmenutzung bei der Wiener U-Bahnlinie U2," Osterr. Ingenieur-und Architekten-Zeitschrift, Vol. 155, No 7-9 and 10-12, 1-7.
5. CIBSE (2007) Guide C: Reference Data, 227p.
6. Fry, V.A. (2009) Lessons from London: regulation of open-loop ground source heat pumps in central London. Quarterly Journal of Engineering Geology and Hydrogeology 42 325-34.
7. Franzius, J. N. and Pralle, N. (2011). "Turning segmental tunnels into sources of renewable energy." Proceeding of ICE, Civil Engineering Vol. 164, 35-40.
8. Schneider, M. and Moormann, C. (2010). "GeoTU6 – a geothermal research project for tunnels", Tunnel 02/2010, pp 14-21.

Crossrail Project: Infrastructure design and construction
ISBN 978-0-7277-6078-4

doi: 10.1680/cpid.60784.123

Design of Sprayed Concrete Linings in Soft Ground – A Crossrail Perspective

Jiang Su BEng MPhil, Mott MacDonald
Alun Thomas BA PhD, Ramboll (formerly Mott MacDonald)

Abstract

Sprayed concrete lined (SCL) tunnelling has seen rapid development over the last twenty years in the UK. The purpose of this paper is to give an update on the latest design philosophies and methodologies, and to discuss the future development trends for SCL tunnelling in soft ground.

The paper will start with brief overview of past soft ground SCL tunnel design; especially the temporary sprayed concrete lining. After that, the paper will introduce the recent and most significant shift in design philosophy; the transition from temporary sprayed concrete lining to permanent sprayed concrete lining, with its implications for construction safety, cost and environmental impact. Some recent developments in the numerical modelling techniques to facilitate the design of the permanent sprayed concrete lining will be introduced. In conclusion, the paper will discuss the possible future trends of SCL tunnel design, in particular the "composite shell" lining, with its potential benefits for construction programme, cost and environmental impact, followed by additional thoughts on possible improvements in the numerical modelling of soft ground SCL tunnels.

1 Introduction

Many people in the past have questioned the method so this paper starts with a definition of the term: An SCL tunnel is a tunnel with a Sprayed Concrete Lining. The Sprayed Concrete Lining interacts with the ground and, through the choice of excavation sequence, this interaction can be optimized to minimize the loads on the ground support. Monitoring is required for "performance verification", to ensure that the tunnel is performing within the limits required by the design (given natural variations, for example, in advance rates and ground conditions), to determine when certain contingency measures are needed (such as temporary inverts).

The use of Sprayed Concrete Lining is an established method of soft ground tunnelling using sprayed concrete to support the excavation both temporarily and permanently. Compared with other tunnelling methods, such as using a Tunnel Boring Machine (TBM), the biggest advantage of the SCL technique is it allows tunnels and junctions to be built at varying cross sections and sizes with relative ease. As well as this, the SCL technique has low mobilisation costs, making it suitable for short lengths of tunnel construction.

2 SCL designed as temporary structures

Until recently, SCL tunnels, consisting of a layer of temporary sprayed primary lining, a layer of sheet waterproofing membrane and another layer of permanent cast secondary lining, dominated the market; mainly because of concerns over safety, watertightness and durability. For the purpose of this paper, this is called sacrificial primary SCL. The reason why the SCL was only treated as a temporary structure was due to two aspects: limitations of sprayed concrete technology and design method limitations, which will be discussed separately in the following sections.

2.1 Limitations of SCL technology

Twenty years ago, the early-age strength development was the top priority for sprayed concrete working as a ground support measure while many other properties had to be compromised, leading to SCL being treated only as a temporary structure. The main concerns at that time from the tunnelling industry about the SCL included:

- Low final strength in the long-term due to accelerator dosage (Kusterle 1997)
- Long-term durability issues due to "shadows" behind steel reinforcement (Thomas 2008)
- Poor quality control due to use of the dry-mix process and manual spraying(Austin & Robins 1995)

2.2 Limitations of design methodology

The lack of reliable design methods was another reason why SCL was perceived only as a temporary structure, especially in soft ground strata, such as London Clay. Tunnelling used to be an experience-based industry where empirical methods were widely adopted, such as using the Gaussian curve to predict the ground surface settlement profile (Peck 1969) or the Q-system (Barton*et al*, 1974). However, the empirical methods for soft ground were mostly based on the assessment of previous local practices and lacked universal acceptance (ICE 2004).

Closed-form analytical methods were also developed for tunnelling design. However, most of them are not suitable for SCL tunnels as the complex soil-structure interaction caused by the multi-step SCL construction sequence could not be properly modelled. The famous closed-form Curtis-Muir Wood formula (Muir-Wood 1975, Curtis 1976), originally developed to confirm the design of segmental linings, could only be used as a check for SCL tunnels because this method assumes that the whole lining is installed immediately after the tunnel is excavated, homogeneous ground conditions and that the tunnel is perfectly circular, neither of which are generally true for SCL tunnels(ICE 2004). The convergence-confinement method (Panet and Guénot 1982), commonly used to predict the displacement field due to the SCL tunnelling, cannot predict the bending moments within the SCL linings and suffers from similar limitations to other closed-form solutions. For the beam-spring model, where the SCL lining is modelled as a series of beams connected to each other and to the ground by radial and tangential springs that simulate the ground support interaction, choosing the correct value of spring stiffness is considered to be very difficult. Besides, the loads to be applied to the lining are evaluated separately from another system, disconnecting the interaction between soil and structures (O'Rourke 1984).

The numerical modelling was not considered as a viable design tool for SCL tunnels in soft ground until 1990s, mainly due to insufficient computing capacity. The Heathrow Express Trial Tunnel, the first soft ground SCL tunnel constructed in London in 1992, was designed by using finite difference software package FLAC2D (Deane & Bassett 1995). This project represented the most advanced design at that time in the UK. The main features are:

- A conservative approach using mean and the worst case parameters for London Clay was adopted due to the absence of prior local projects

- A small strain stiffness non-linear isotropic model was used to model the pre-peak behaviour of London Clay
- A strain softening stress-strain relationship was adopted for its post-peak behaviour
- The Hypothetical Modulus of Elasticity (HME) approach was used to model the early-age sprayed concrete and the 3D effects tunnel face effect. Despite some attempts to give it a theoretical basis, the HME is effectively an empirical correction factor (Thomas 2008) –a very useful one but empirical nonetheless.

Figure 1 Temporary SCLlining of Heathrow Express crossover cavern

The success of the Heathrow Express Trial Tunnel gave the industry significant confidence in building SCL tunnels in soft ground in London. However, the Heathrow Express project still posed a much bigger challenge to the designers as it comprised three large-diameter parallel SCL tunnels, in a close proximity under the terminal building (Van der Berg *et al* 2003). There were also cross-passage connections between these tunnels. There were four major challenges:

- Little prior experience of additional loading to the SCL tunnels caused by compensation grouting (Linney & Essler 1994)
- Difficulty in estimating potential building damage due to SCL tunnelling induced ground movement
- Lack of knowledge on the extra loading on the adjacent SCL tunnels caused by the construction of new SCL tunnels
- Scarce data available to estimate the stress concentrations around the junctions due to the construction of the cross-passages

To ensure safety, designers put a large quantity of reinforcement in the SCL tunnels. For a 9m diameter platform tunnel, the temporary primary lining consisted of 300mm of sprayed concrete lining, reinforced by two layers of 8mm diameter mesh and full-section lattice girders, leading to a reinforcement ratio at 0.23%. This quantity of reinforcement made the spraying very difficult, leading to concerns of quality control and long-term durability of the lining.

It could be argued that the perception of SCL as only temporary structures was driven not only by insufficient development of the sprayed concrete technology, but also by the lack of efficient design methods. As an aside, it should be noted that, despite the limitations of the design, the causes for the collapse at Heathrow stemmed from failures in the construction management, particularly quality control and interpretation of monitoring data (HSE 2000).

2.3 Other issues related to temporary SCL

Until the mid 1990's in the UK, most spraying was carried out using the dry mix method which required tunnel workers to stand very close to the tunnel face to spray the concrete, resulting in significant health and safety issues. Dry mix creates much more dust than wet mix spraying. In addition, the temporary SCL caused serious concerns of material waste, cost inefficiency and negative environment impact.

3 SCL designed as permanent structures

In order to improve the performance of SCL tunnels, constant efforts have been made by different parts of the tunnelling industry, such as material suppliers, machinery manufacturers, designers and contractors. Several advances have made the use of SCL as a permanent structure possible:

- The shift in producing SCL from the dry-mix process to a wet-mix process substantially improved the quality control and productivity (Austin & Robins 1995)
- The adoption of alkali-free accelerators enabled rapid setting, improved early and final strengths, reduced environmental impact and enhanced safety for tunnel workers (Kusterle 1997)
- The use of fibre reinforcement instead of mesh reinforcement eliminated the concern of "shadows", shortened construction programme and saved overall cost (Thomas 2008)
- The shift from hand spraying to robotic spraying speeded up progress and produced less material waste (Franzén 1992). It also complies with increasingly strict H&S laws in the UK.
- The use of real-time surveying with total stations instead of lattice girders to ensure the correct profile of the tunnel excavation and control the lining thickness has accelerated construction and also removed a key durability concern. This has been a huge leap forward. A recent project using the total station for profile control was the Heathrow Terminal 5 project, which together with the inclined tunnel face was marketed as the "Laser Shell" method developed by Morgan Est (Jones *et al.* 2008)
- The improvement in the quality of sprayed concrete has enabled the mature sprayed concrete to be treated as normal cast in-situ concrete, with the same long-term strength development, low permeability (in the order of $10^{-12}$and 10^{-14}m/s) and durability performance (Annett *et al.* 1997)

Figure 2 Permanent sprayed concrete primary lining at Limmo for Crossrail

Permanent sprayed concrete has been widely accepted in certain sectors –most notably hydropower projects –and certain countries (e.g. Norway) for many years but it has only recently gained this acceptance more widely in the world and in soft ground applications. For soft ground SCL tunnelling in the UK, for the purpose of this paper, the latest design option is called Double Shell SCL, which consists of a layer of permanent sprayed concrete primary lining, a layer of spray applied waterproofing membrane and a layer of sprayed or cast secondary lining, with no adhesive and shear bond assumed at the sprayed concrete-membrane interfaces. This was the design assumption due to the lack of evidence on the existence of adhesive and shear bond and thus no bond is assumed across the interface. Steel fibres are used as the main reinforcement, and no steel bars and meshes are used except at the tunnel junctions. Lattice girders are eliminated and the tunnel profile is controlled by the total station. This design option has been adopted on several important projects, such as A3 Hindhead and Crossrail.

3.1 Design requirements

In the latest design option, the permanent sprayed primary lining is designed to take the full short-term soil, water and any other pressures, such as surface surcharge and compensation grouting, that occur before the installation of the secondary lining. In some ground conditions there is a time-dependent ground loading as well as long term changes in water pressure. London clay is one such case. The soft ground is assumed to be impermeable within two years in an undrained short-term analysis. This assumption is consistent with the work of Wongsaroj et al (2007) who found that permeability for the clay should be in the region of 10^{-10} to 10^{-11}m/s to replicate the observed behaviour in the JLE Green Park field. It also coincides with Addenbrook's (1996) back-analysis of the running tunnels at the same field that the ground took between 10-20 years for most of the movements due to water flow to cease.

In the case of London clay, the long-term increase in soil pressure and surface surcharge are shared in proportion to the thickness between the two linings and the long-term water pressure is assumed to apply solely to the secondary lining in a drained long-term analysis. This is determined by the assumptions of cracking in the primary lining and of no adhesive and shear bond assumed across the sprayed concrete-membrane interface as mentioned above. Ground water is assumed to be able to travel from the extrados to the intrados of the primary lining through cracks and the water pressure directly applies to the sprayed waterproofing membrane and hence to the secondary lining. Depending on the time and budget, either complicated coupled consolidation analysis or simply fully permeable analysis could be carried out as the drained long-term analysis in a numerical model.

An initial layer of primary lining, normally between 25-75mm, is designed as sacrificial and disregarded in the long-term design case. This is mainly due to the quality control consideration that the quality of this layer of sprayed concrete will not be as good as the following layers. Another consideration of this sacrificial layer is to resist the chemicals in the groundwater, particularly sulphates, which would cause corrosion to the steel fibres, leading to a reduction in the effective primary lining thickness in the long-term. The thickness of the sacrificial layer is determined mainly by the type and density of chemicals in the groundwater. Arguably 75 mm is a quite conservative assumption.

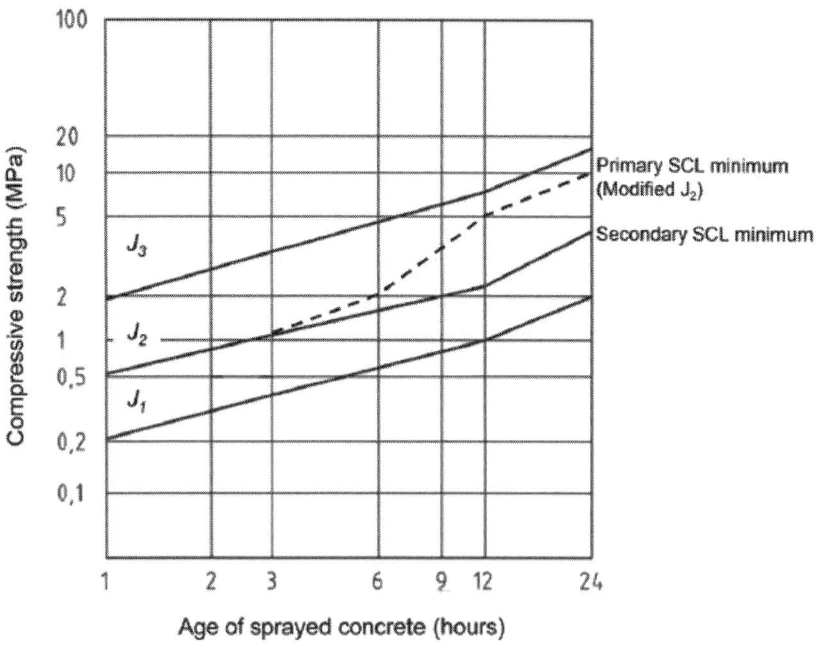

Figure 3 Early age strength development for sprayed concrete –modified J2 curve (based on EN 14487-1:2005)

A moderate strength of sprayed concrete is specified in the design and used in the construction. The lower limit of characteristic cylinder compressive strength for both linings is28N/mm2at 28 days and 32N/mm^2 at 90 days respectively, and the concrete should exceed a modified J2 curve (EN 14487-1:2005) within the first 24 hours. The specification of the lower limit of compressive strength of sprayed concrete is to guarantee that the lining would fulfil its functional requirements. Although structural synthetic fibres are more economic and better in resisting corrosion (Bernard 2009), steel fibres have been chosen as the reinforcement mainly due to the better performance in resisting long-term creep in tension (Bernard 2004). As the understanding of macro-synthetic fibres develops confidence in their performance under creep is likely to increase. A 50mm fire proofing sprayed concrete layer mixed with micro-synthetic fibres is designed inside of the secondary lining to reduce the spalling of concrete in a fire situation (Shuttleworth 2001).

3.2 Design methods

Since numerical modelling is now widely used in the design of SCL tunnels, it can be considered that the way to improve the design depends on the accuracy of the constitutive models. There are three main areas where numerical modelling results could be improved: the constitutive model for the ground, the constitutive model for the sprayed concrete, and calibration of modelling approach.

Industry and academia have spent significant efforts in collecting field and laboratory test data to develop constitutive models for ground strata such as London Clay (Gasparre 2005, Jardine *et al*1984, 1986). It could be argued that the constitutive models for London Clay have developed to a point where doubling the time and resource input may only improve the results very slightly. Hence, it relies on the designers' judgment to balance the accuracy of the results with the input resource to the modelling work. It should also be borne in mind that numerical modelling is just an approximation of reality, and its accuracy will be affected significantly by the selection of ground model and uncertainties of the ground (Thomas2008).

For Crossrail SCL tunnel design, A* model (Eadington & O'Brien 2011), a small strain stiffness non-linear elastic model, has been adopted. The model recognises important pre-failure behaviour of London Clay at the strains appropriate to tunnelling (Atkinson 2000, Addenbrook 1996, Addenbrook *et al* 1997), especially for cases where several tunnels are close together. The Tresca failure criterion with an associated plastic flow rule and the Mohr-Coulomb failure criterion with an unassociated plastic flow rule have been adopted to model the undrained and drained ground behaviour at failure and beyond respectively. The concept of a strain dependent stiffness model and failure criteria are both well-known in industry and academia and backed by significant historical data. Besides, the parameters used in these models are relatively easy to obtain, and the computing time is modest compared with that using more sophisticated models.

In contrast to the universally accepted design methods and criteria for the cast in-situ concrete, variations in SCL mix design and workmanship in different countries for different projects can affect the strength and stiffness development significantly. Tests on sprayed concrete at both early ages and long-term have been carried out by various researchers, leading to various stiffness/strength development equations (Thomas 2008). Therefore, where possible it is best to base the constitutive models on testing results from a similar, recently completed SCL project. For Crossrail SCL tunnel design, age-dependent strength curve for sprayed concrete were developed based on the modified J2 curve up to 24 hours, which was derived from the previous SCL projects, and Chang & Stille approach (Chang & Stille 1993) thereafter. The stiffness of sprayed concrete at each age was also derived from the Chang & Stille's stress-stiffness equation. The data of sprayed concrete stiffness and strength from Heathrow T5 project and A3 Hindhead were used to check the inputs. Eurocode 2 (BSI 2004) defines the compressive strain in normal concrete at peak stress (ε_c) as 0.2% and ultimate compressive strain (ε_{cu}) as 0.35%. There is currently no universally accepted stress-strain relationship for the concrete in the tension area as the tensile strength is usually very small hence not taken into account in the structural analysis. Plain concrete is a brittle material and will only exhibit a certain degree of ductility in tension when bar reinforcement or structural fibre is added. Therefore, tensile stress-strain relationships from different standards and methods are reviewed and selected for the design.

In addition, adding plastic hinges into the lining could be another way to improve efficiency in the design. Bending moment redistribution is a well-established theory in concrete structure design, fitting SCL tunnels especially well due to the structural redundancy. The combination of using an age-dependent linear elastic model and plastic hinges may substantially reduce the peak bending moment in the lining, leading to more evenly distributed bending moment and more efficient lining design at the cost of a slightly increased ground movement. This overcomes the limitations of using linear elastic elements to simulate a material which is nonlinear elasto-plastic and which creeps. The method corresponds with the experience in the field that SCL tunnels can easily redistribute stresses within themselves and the ground, provided that the ground does not become overstressed. This is a safe design method since the numerical model will predict collapse if the ground load continues to press on the area with plastic hinges, instead of redistributing.

3.3 Calibration

With clearly defined constitutive models for the ground and sprayed concrete, a 3D model with the correct construction sequence would be straightforward for modellers and the results should be fairly accurate. It is widely acknowledged that the 3D modelling is far better than the 2D modelling for SCL tunnels (Thomas 2003, Jones 2007) as the former can model the complicated construction sequence, and three-dimensional stress redistribution around the tunnel face explicitly. For Crossrail SCL tunnel design, 3D modelling has been extensively used for calibration works, compensation grouting and tunnel junction studies, as well as for tunnels in faulted ground and in irregular shapes. The 3D model and the construction of the Whitechapel junction enlargement are shown below in Figure 4 (a) and (b) respectively. However, it must be recognised that, compared with the 2D modelling, the 3D modelling is very time consuming and has been used only if it can be seen to improve the accuracy of results significantly.

Figure 4 3D Model (left) and the construction (right) of the Whitechapel junction enlargement using permanent SCL

In Crossrail, most SCL tunnels are still modelled by 2D analysis. To date, more than 100 2D primary and secondary lining sections have been analysed. Extensive calibration works, including the Jubilee Line Extension (JLE) London Bridge Station, Heathrow T5 and Kings Cross Station, were carried out at the beginning of design in order to verify the constitutive models and input parameters for ground and sprayed concrete. After that, a valid 2D modelling methodology was formulated and adopted for routine 2D analysis. For calibration of the numerical model, the most relevant data is tunnel convergence and lining stress directly obtained from an in-tunnel monitoring system. Tunnel designers are primarily concerned with the lining loads while ground movements are a secondary concern, when compared to the stability of the tunnel. An ideal calibration work is carried out as follows:

- Firstly, obtain field data for a completed SCL tunnel project, including ground surface and subsurface settlement curves, lining stresses and lining deformations, from a reliable source, such as a company's own database or published papers. It should be noted there is a scarcity of such data and it is rare to find cases in the same ground
- Build a3D model for these SCL tunnels explicitly modelling the construction sequence and using the real sprayed concrete stiffness and strength
- Compare the predicted lining stress, lining deformation, ground surface and subsurface settlement curves with the field data
- It is widely acknowledged that predicting the tunnelling induced ground surface settlement trough is still a very difficult task even now (Potts &Zdravkovic2001, Addenbrook 1997). It is satisfactory if the model predicts a surface settlement profile in reasonable agreement with the measured and a volume loss similar to the calibrated data, when the primary focus is the lining design
- For SCL tunnels constructed in London Clay, the small strain stiffness non-linearity of the ground should be considered not only for predicting ground deformation, but also for assessing the strain of the ground where adjacent tunnels are to be built
- If consistency between the field data and numerical modelling prediction is unachievable for ground deformation, lining stress and convergence at the same time, a priority should be given to the set of input parameters which predict the closest lining performance to the field data
- Match the 2D modelling results with the 3D modelling results with different ground pressure relaxation percentage and using the real sprayed concrete stiffness and strength
- Ideally, the 2D relaxation percentage should enable the predicted results, including both ground deformation and lining performance, consistent with the field data
- For lining performance, it is not only necessary to check whether the predicted maximum lining stress is consistent with the field data, but also whether the lining deformation pattern is consistent with the field observation
- Different ground relaxation percentage may be developed for different construction methods, such as pilot/enlargement or top heading/bench/invert

A typical 3D/2D numerical modelling calibration procedure is shown in Figure 5.

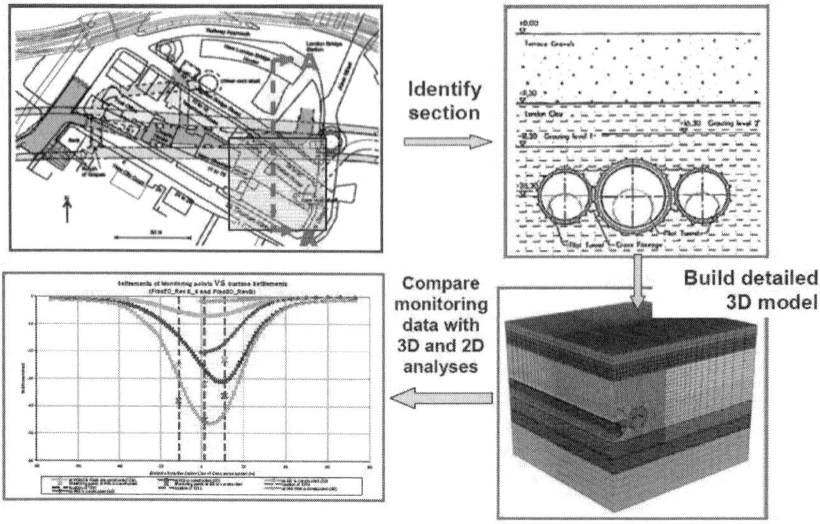

Figure 5 Typical 3D/2D numerical modelling calibration procedure

All of the above was successfully done for the Crossrail project which represents a major achievement and an important factor in giving the project team the confidence to design and build such large and complex structures in the sensitive environment of central London.

4 Further development of SCL

Huge progress has been made over the last twenty years, during which SCL has moved from being only a temporary structure to a permanent structure without any compromise in structural capacity, watertightness, durability and safety to tunnel workers, while, at the same time, achieving shorter construction periods and savings in the materials used. The next step is to discuss whether SCL can be improved further, with even thinner linings and shorter construction periods.

4.1 Further development of SCL technology

One promising design option, which stands out and attracts most attention at the moment, is the Composite Shell SCL. Composite Shell SCL, in the sense of tunnel construction, is the same as the Double Shell SCL, made of a layer of permanent sprayed concrete primary lining, a layer of spray applied waterproofing membrane and a layer of permanent sprayed concrete secondary lining. The major difference between them is that the former assumes a certain degree of adhesive and shear bond exists across the sprayed concrete-membrane interfaces while the latter does not. This difference makes the two lining design options behave in totally different ways under the soil and water pressures. For the Double Shell SCL, because there are no adhesive and shear bonds assumed at the interfaces, water could permeate through cracks in the primary lining and seep along the interface, applying pressure to the extrados of the waterproofing membrane. This has been the conservative design assumption, even though the test data from the manufacturers has shown a fully bonded behaviour (Holter & Nermoen 2011). The water pressure will be supported by the secondary lining alone in the long-term, becoming the main design case for the secondary lining thickness. By contrast, for the Composite Shell SCL, water permeating through cracks will be prevented from seeping along the interface, and most water will not reach the waterproofing membrane in its 120 years design life, providing that the SCL satisfies the maximum permeability requirements. (Jones 2012) The water pressure will be supported by both the primary and the secondary linings in the long-term, leading to a substantial thickness reduction in the secondary lining. In the case of the Composite Shell SCL, the thickness of the secondary lining could be as little as 60 mm (Holter& Nermoen 2011).The loading conditions for the Double Shell SCL and the Composite Shell SCL at the short and long-term are shown in Figure6.

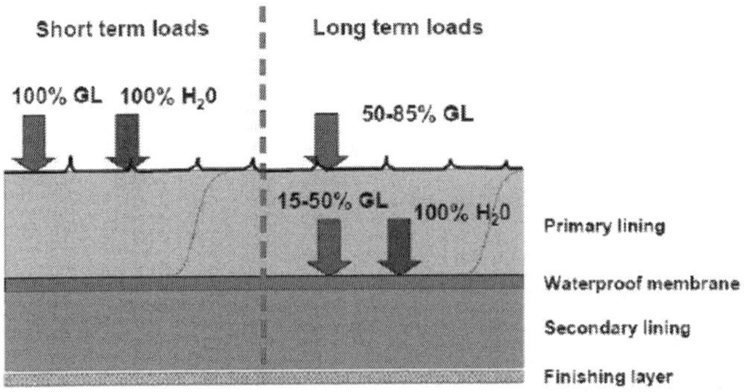

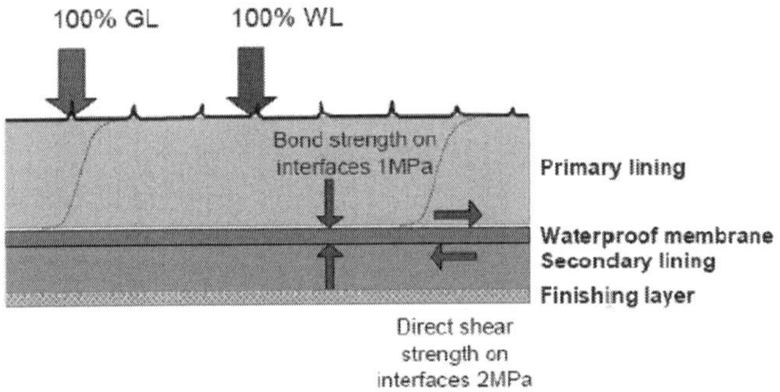

Figure 6 Typical loading conditions for Double Shell SCL (top) and Composite Shell SCL (bottom)

A research programme, in collaboration with Mott MacDonald, Tam-Normet and University of Southampton and carried out by the author, is in progress to investigate the behaviour of Composite Shell SCL tunnels. As part of the testing programme, a series of compression, tension and direct shear tests have been carried out on samples cut from panels, built up from a sprayed primary layer, spray applied waterproofing membrane(TamSeal 800)and sprayed secondary layer. The aim of the tests has been to understand the fundamental properties of sprayed concrete-membrane bonding systems, especially the impact of the interface roughness and membrane thickness on strength and stiffness. The testing has reported that, for TamSeal 800 sprayed waterproofing membrane, the adhesive bond strength is around 1MPa and the shear strength is above 2MPa at the sprayed concrete-membrane interface. This test result has made the composite action feasible at the sprayed concrete-membrane interface. With further testing and numerical modelling analysis, a design guideline shall be developed for the Composite Shell SCL tunnels in the near future.

4.2 Further development in design methods

One possible area comes from the further understanding of the creep effect of early age sprayed concrete. It is understood that the creep effect is stress-dependent (Thomas 2008) and little information is currently available regarding this issue. An understanding of this issue could lead to more accurate modelling of the SCL at early age. The field of materials science continues to develop. In the future it is likely that macro-synthetic fibres will be widely used and that fully bonded composite action will be assumed in design using spray applied waterproofing membranes. The latter could significantly reduce the overall lining thickness. Last but not the least, more back-analysis should be carried out so refinements can be made to the constitutive models for the ground and sprayed concrete as well as the 2D relaxation methodology. In order to do so, it is very important to determine the location and the techniques to install monitoring systems so that more reliable data will be available for back analysis. The Crossrail project represents a significant opportunity for the tunnelling industry to learn more and apply the benefits to future projects such as HS2 and Crossrail 2.

5 Conclusion

The last twenty years has seen huge progress for sprayed concrete linings, including the shift from being a temporary structure to a permanent structure. Once the composite action across the sprayed concrete-membrane interface become a consensus in the tunnelling industry, the SCL design philosophy will experience another revolution, which sees the Composite Shell SCL replacing the Sacrificial Primary SCL and the Double Shell SCL, switching the main burden of the permanent structure from the secondary lining to the primary lining. The secondary lining will then act mostly as a supporting layer for the purposes of fire resistance and internal fixings.

6 Acknowledgements

The author would like to acknowledge the contribution of many colleagues at Mott MacDonald in the development of the design approaches, especially to Tony Deane, Chris Pound, Ross Dimmock and Bethan Haig (both are now of Normet UK Ltd).

References

Addenbrook, T.I., Potts, D.M., & Puzrin, A.M. (1997). The influence of pre-failure soil stiffness on the numerical analysis of tunnel construction. Geotechinique 47, No.3, 693-712

Addenbrook,T.L. (1996).Numerical analysis of tunnelling in stiff clay, PhD thesis, Imperial College, University of London

Annett, M. F., Earnshaw, G. & Leggett, M. (1997). Permanent sprayed concrete linings at Heathrow Airport. Proc. Tunnelling '97. 517-534, London: The Institute of Mining & Metallurgy.

Atkinson, J.H. (2000). Non-linear soil stiffness in routine design. Geotechnique 50, No. 5, 487-508.

Austin, S & Robins, P. (1995). Sprayed Concrete: Properties, design and application, Bristol: Whittles Publishing.

Barton, N., Lien, R. & Lunde, J. (1974).Engineering classification of rock masses for the design of tunnel support. Rock Mechanics. 6: 4: 189-236.

Bernard, E.S. (2004). Creep of cracked fibre-reinforced shotcrete panels, Shotcrete: More Engineering Developments, Bernard (ed.), pp 47-58, Taylor & Francis, London.

Bernard, E.S. (2009).Design of fibre reinforced shotcrete linings with micro-synthetic fibres. Proc. Shotcrete for underground support XI.2009.

Chang, Y. and Stille, H. (1993). Influence of early age properties of shotcrete on tunnel construction sequence. Shotcrete for Underground Support VI, American Society of Civil Engineers, Reston, pp. 110-117

Curtis, D.J. (1976). Discussion on the circular tunnel in elastic ground, Geotechnique, vol. 26, No.1, pp 231-237.

Deane, A.P & Bassett, R.H. (1995). The Heathrow Express Trial Tunnel, Proceedings of the ICE -Geotechnical Engineering, Volume 113, Issue 3, pages 144 –156

Eadington, J & O'Brien, T (2011), Stiffness parameters for a deep tunnel –Developing a robust parameter selection framework. Proc. 15th European Conference on Soil Mechanics and Geotechnical Engineering. Athens, Greece.

EN 14487-1: 2005 (2005). "Sprayed Concrete —Part 1: Definitions, Specifications and Conformity", British Standards Institution, London.

Eurocode 2 (2004).EN 1992-1-1:2004 "Design of concrete structures —Part 1-1: General rules and rules for buildings", British Standards Institution, London.

Franzén, T. (1992). Shotcrete for underground support: a state-of-the-art report with focus on steel-fibre reinforcement. Tunnelling and Underground Space Technology, 7. No.4 383-391

Gasparre A (2005) -"Advanced laboratory characterisation of London Clay". PhD Thesis. University of London.

HSE (2000). The collapse of NATM tunnels at Heathrow Airport. Health & Safety Executive Books, HMSO, Norwich.

Holter, K.G & Nermoen, B. (2011). Permanent waterproof tunnel lining based on sprayed concrete and sprayed-applied double –bonded membrane. Proc. World Tunnelling Congress 2011.

ICE (2004). Tunnel Lining Design Guide, The Institution of Civil Engineer, 184pp. London: Thomas Telford Publishing

Jardine, R.J., Symes, M.J.P.R. , Burland, J.B.(1984).The measurement of soil stiffness in the triaxial apparatus, Geotechnique, 1984, Vol:34, Pages:323-340

Jardine, R.J., Potts, D.M. , Fourie, A.B., et al,(1986).Studies of the influence of non linear stress-strain characteristics in soil-structure interaction, Geotechnique, 1986, Vol:36, Pages:377-396

Jones, B. D. (2007). Stress in Sprayed Concrete Tunnel Junctions. Ph.D. thesis. University of Southampton.

Jones, B.D., Thomas, A.H., Hsu, Y.S.& Hilar, M.(2008). Evaluation of innovative sprayed-concrete-lined tunnelling. Proc Inst. Civ. Engrs Geotech. Engng 161, 137-149

Jones, B.D. (2012). Waterproofing sprayed concrete tunnels. Tunnelling Journal, 2012, June/July, Page:35-36

Kusterle, W.A. (1997). New ecologically desirable sprayed concrete. Proc. Tunnelling '97. pp 263-274, London: The Institution of Mining & Metallurgy.

Linney, L.F. & Essler, R.D. (1994), Compensation grouting trial works at Redcross Way, London. Grouting in the Ground. pp 313-326

Muir-Wood, A.M. (1975). The circular tunnel in elastic ground. Geotechnique. Vol.25, No.1. pp 115-127.

O'Rourke, T. D. (1984). Guidelines for Tunnel Lining Design. Technical Committee 011 Tunnel Lining Design of the Underground Technology Research Council of the ASCE Technical Council.

Peck, R.B. (1969). Deep excavations and tunneling in soft ground. Proc. 7th International Conference Soil Mechanics and Foundation Engineering, Mexico City, State of the Art Volume, pp. 225-190

Panet, M. & Guénot, A. (1982). Analysis of convergence behind face of a tunnel. Proc. Tunnelling'82, pp. 197–203. IMM, London.

Potts, D.M. & Zdravkovic, L. (2001). Finite element analysis in geotechnical engineering, vol. 2: Application. London: Thomas Telford.

Shuttleworth, P. (2001). Fire protection of concrete tunnel linings, Proc. 3rd Int. Conf. on Tunnel Fires and Escape From Tunnels, Washington DC, USA, 9-11 October, pp. 157-165.

Thomas, A. H. (2003). Numerical Modelling of Sprayed Concrete Lined (SCL) tunnels. Ph.D. thesis, University of Southampton.

Thomas, A.H. (2008). Sprayed Concrete Lined Tunnels. 264pp. Abingdon: Taylor and Francis Publishing.

Van der Berg, J.P., Clayton, C.R.I. and Powell, D.B. (2003).Displacements ahead of an advancing NATM tunnel in the London clay.Geotechnique, 53, (9), 767-784.

Wongsaroj, J., Soga, K. & Mair, R.J. (2007). Modelling of long term ground response to tunnelling under St James's Park, London. Geotechnique 57, No. 1, 75–90.

Crossrail Project: Infrastructure design and construction
ISBN 978-0-7277-6078-4

doi: 10.1680/cpid.60784.137

Crossrail Sprayed Concrete Linings

Andrew Pickett BEng CEng MICE, Design Team Leader C121 Mott MacDonald

Abstract

This paper looks at the sprayed concrete lined (SCL) tunnel design by Mott MacDonald for the Crossrail project. The SCL design consists of five new underground stations, shafts and two Crossovers which will be the largest SCL tunnels to be built in London. It also discusses how the design team was set up and the approach and methodology behind the design following Crossrail's values.

It will also give an overview of the project design, the challenges faced and the main project design interfaces and how they were managed. It also presents how the Mott MacDonald design has been coordinated with the architecture design to meet the Architect aspirations for the overall Crossrail aesthetic.

The paper also generally discusses details of design including the varying geology and will set out the methodology and ethos behind the design and how Mott MacDonald developed their design techniques to meet the challenges of the project while giving Crossrail best value over the design period of outline to tender, through the Optimised Contractor Involvement stage and to Detailed design.

Introduction

In June 2009 Mott MacDonald undertook the framework design package C121 for the design of all Sprayed Concrete Lined (SCL) tunnels for the Crossrail project. Since then the design has gone through RIBA D to RIBA F and Issue for Construction design stages.

This paper introduces the scope of the work, how C121 set out the basis of design, the challenges faced and the main project design interfaces and how they were managed. It will also present how the Mott MacDonald design has been coordinated with the architecture design to meet the Architect aspirations for the overall Crossrail aesthetic.

The paper will also generally discuss details of design including the varying geology and will set out the methodology and ethos behind the design and how Mott MacDonald developed their design techniques to meet the challenges of the project while giving Crossrail best value over the design period of outline to tender, through the Optimised Contractor Involvement stage and to Detailed design.

C121 Structures

The C121 design package consisted of significant structures on the central section of the Crossrail project. C121 were responsible for designing the following structures:

- All station tunnels/ adits at:
    - Bond Street
    - Tottenham Court Road
    - Farringdon
    - Liverpool Street Station
    - Whitechapel
- All intermediate shafts, crossovers and cross passages
    - Cross passages for running tunnel packages
    - Stepney Green Crossover
    - Whitechapel Crossover
    - Eleanor Street
    - Mile End
    - Limmo

Figure 1 shows a typical station layout (with station interfaces) and Figure 2 shows a cross section of Stepney Green Crossover.

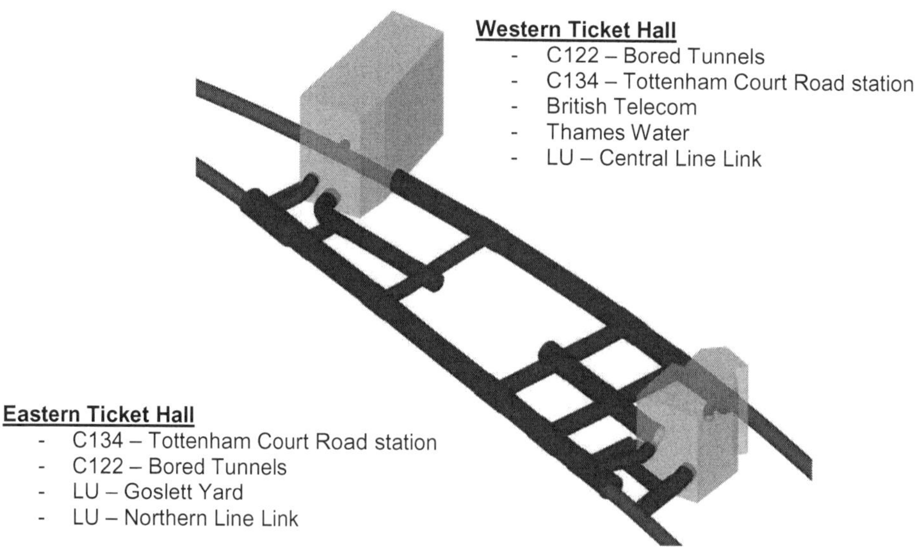

Figure 1 – Tottenham Court Road Station and design interfaces

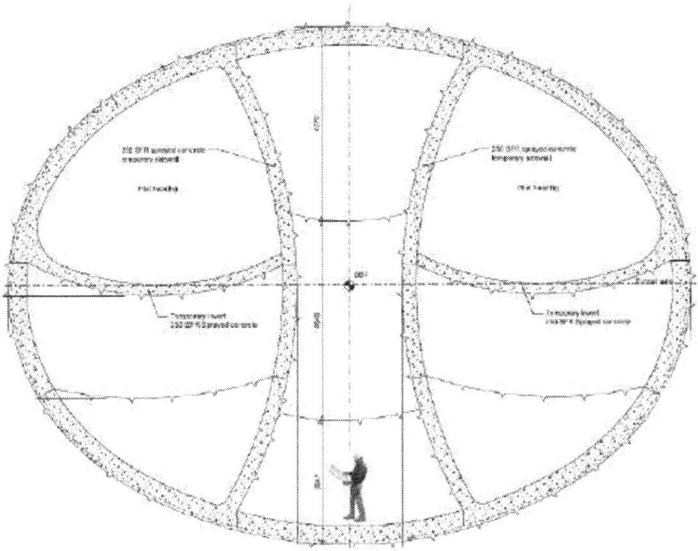

Figure 2 – Stepney Green Crossover Cavern

The platform tunnels at each station are typically 240m in length and the Stepney Green Crossover cavern will be the largest tunnel constructed in London in SCL. So the challenges at the beginning of the contract were clear, and required logical approach to the design and organisation of the C121 design team.

C121 Team

Figure 3 shows the organisation chart for the C121 design team and Figure 4 shows a typical set of interfaces for each station package.

The team had to have a robust set up to be able to meet the design challenges of designing all the structures in a consistent format. It was based around a core design delivery teams for discreet packages, which were supported consistently from specific teams such as, Geotechnical, Environment, Construction planning and Environment. It was also supported by Design challenge teams or peer assists that would come in and take a holistic review of the Crossrail SCL structures. This facilitated an efficient consistent design delivery and enabled consistent interfaces across other design packages as shown in Figure 4.

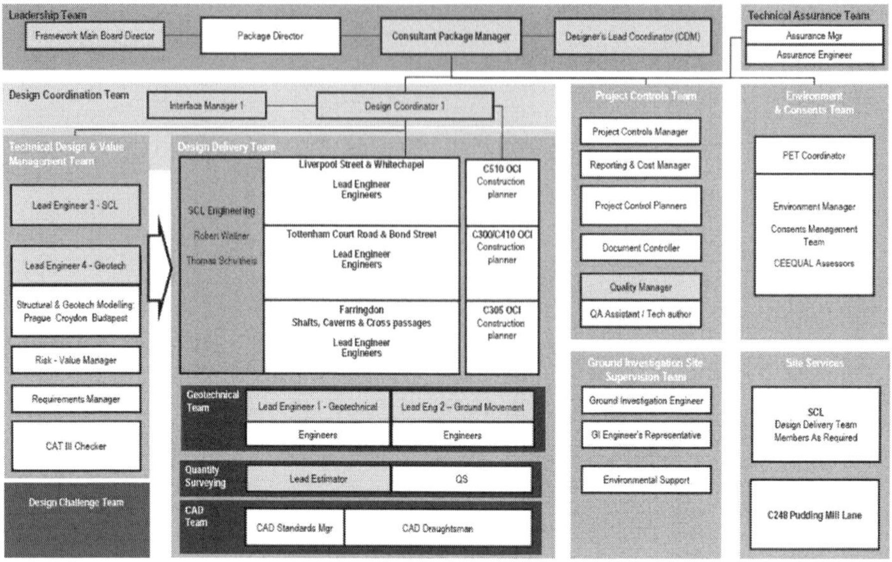

Figure 3 – C121 Organisation chart

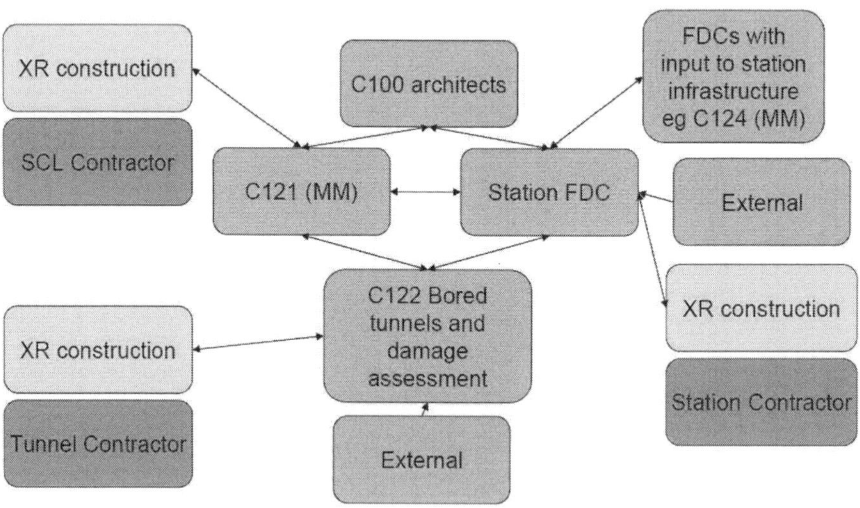

Figure 4 – Typical interfaces for one Station package

Basis of Design

Design Criteria

The Sprayed Concrete Linings (SCL) for the Crossrail Project are to be designed taking account of previous experience of the design and construction in soft ground of such tunnels. Due to being in London and the high profile of the project the consequences are high. The risks associated with SCL construction are mitigated through the use of precedent proven technologies. The tunnels are to be sized to comply with the project aims and they must be capable of being built safely and cost effectively within the programme requirements.

Safety considerations include the wellbeing of those constructing the tunnels, those people in adjacent structures (including existing tunnels) and the general public that may be affected by tunnel construction.

The priorities identified for achieving these aims were:

- Mechanise construction methods with the primary aim to keeping men away from the face, maximising safety with benefits for productivity and e.g. robotic spraying.

- Reducing men working at height, near to the face and using hand held equipment. The design minimises the number of joints in the permanent works and where possible uses fibre reinforcement thereby avoiding risks associated with traditional reinforcement installation.

- Incorporating issues of practicality for construction logistics and CDM risks for the construction and maintenance stages.

- Robust designs that are compliant with Crossrail Design Standards and current Best Practice and have an operational design life of 120 years with low levels of maintenance.

The following design techniques have been used:

- Bench marked the design against previous best practice in SCL projects and proven technologies.

- Carried out Comparative Risk Assessment to demonstrate the design risks are ALARP by assessing the safety critical activities and comparing the relative risks of different options.

- Used the current best practice methods to numerically model the ground: structure behaviour and establish the required lining thicknesses and structural requirements. Calibration of the analysis has been carried out against previous projects in London. Sensitivity analysis were carried out to establish the effects of variation in the main ground parameters selected.

- Carried out ongoing Peer reviews to allow for independent scrutiny of options.

Design Drivers

The design drivers for SCL structures are as follows:

- Permanent sprayed primary linings incorporating fibre reinforcement. The external 75mm of the lining considered to be sacrificial in the long term.

- Utilising the primary sprayed concrete as a permanent lining at junctions where it incorporates standard steel reinforcement.

- Excavation face sizes based on experience and stability calculations. Where large tunnels are required, and the geometry is appropriate, the use of pilot tunnel construction sequences rather than sidewall drift construction to provide suitably sized headings.

- The use of modern electronic surveying and setting out techniques which combined with the use of pilot tunnels, eliminates the requirement for lattice girders for profile control in tunnels.

- A full design based on moderately conservative ground parameters. A sensitivity analysis to confirm that the design using these parameters is robust.

- Junction opening designs which allow later erection of the permanent structural reinforcement within a completed sprayed concrete shell, rather than during the advance of the tunnel.

- Spray applied waterproof membranes (except Farringdon and tunnels in the Lambeth Group)

- Sprayed secondary linings, where there is a sprayed waterproof membrane.

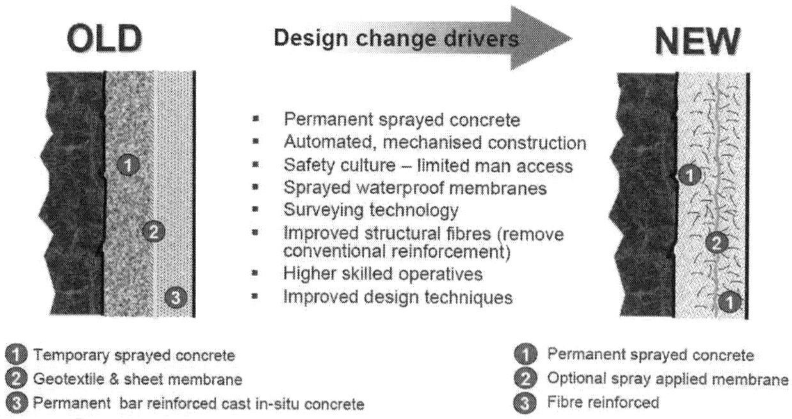

Figure 5 – Rationale for permanent primary and secondary SCL

The Design

The lining system for the SCL tunnels on Crossrail consists of a fibre reinforced primary lining, a waterproofing membrane and a fibre reinforced secondary lining. The lining system is currently a double shell lining with both linings considered part of the permanent load bearing structure. A spray applied waterproofing membrane is used in all tunnels except those found in the Lambeth Group soils, such as Farringdon Station. The linings are currently designed to share the loading, and the primary lining is considered load bearing throughout the design life of the structures. The linings are not currently designed as acting compositely where shear transfer and full bonding performance between at the interface between primary lining, waterproofing membrane and secondary lining would be required.

Primary Lining

The permanent primary lining has been designed to take the full short-term applied ground load and any other loads, such as from compensation grouting and surface surcharges, expected in the two years prior to secondary lining installation. Any additional long-term loads, such as those applied through ongoing consolidation of the London Clay, subsequent to the installation of the secondary lining is shared between the two linings, according to the results of numerical modelling.

The primary lining will be a sprayed concrete lining containing structural fibre reinforcement. The structural fibres are proposed in order to increase the ductility of the concrete and provide toughness and post crack resistance in the long term.

The excavation profile and thickness of the primary lining for each advance will be controlled using target-less remote laser scanning equipment, set up at a safe distance from the excavation face. The profile control system will be subject to testing and approval processes prior to application in the Crossrail tunnels.

Lattice girders have not been included due to the availability of modern setting/surveying techniques for shape control.

This allows for the removal of lattice girders for all tunnels apart from turnout tunnels employing sidewall drift construction sequences.

Secondary Lining

The secondary lining is designed for:

- Long term water pressure,
- Internal loads, such as mechanical and electrical equipment,
- Long term load conditions
- Temperature and shrinkage,

The proportion of consolidation loading applied to the secondary lining has been calculated using numerical modelling methods. The proportion of load transferred between the two linings is a function of the interface between them, i.e. the chosen waterproofing method. A spray applied membrane which bonds to both structural linings allows a greater degree of sharing between concentric linings. A sheet membrane, which allows slip and leaves a small space between the linings will not transfer load to the inner lining to the same degree.

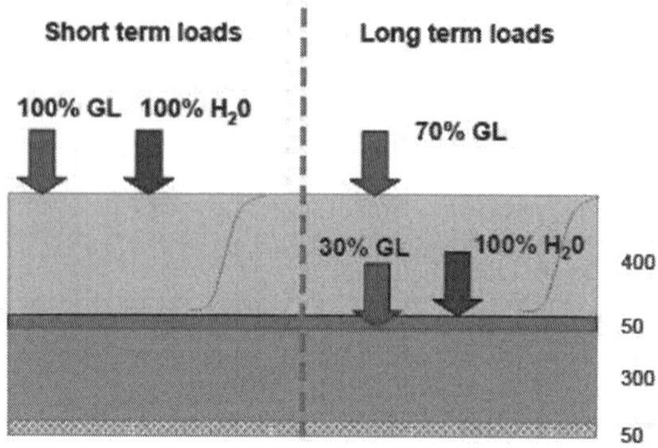

Figure 6 – Load sharing between primary and secondary SCL linings

Numerical modelling methods allow for load sharing between the tunnel lining and the ground and stress relaxation due to tunnel excavation using a sequential SCL approach. As a result, in the long term it is possible that the calculated radial pressure acting on the linings is less than the total overburden load. Values used for Crossrail modelling have been calibrated against tunnels built previously in London Clay, i.e. Heathrow, Kings Cross and London Bridge and further calibrated against 3D models of the construction sequence proposed at this stage of the design.

Secondary linings are designed to carry sufficient residual capacity to resist ground loading after a EUREKA time/temperature fire curve, as defined in the Technical Specification for Interoperability – Safety in Railway Tunnels TSI-SRT and will contain micro-synthetic fibres in order to limit explosive spalling and maintain structural integrity. It has been shown in extensive fire testing for projects such as Heathrow Terminal 5, A3 Hindhead, and CTRL that the inclusion of micro synthetic fibres in high strength, low permeability concrete mixes (such as proposed for the Crossrail linings) significantly reduces the risk of explosive spalling when exposed to severe hydrocarbon fires.

Waterproofing Systems

Waterproofing membranes will be installed between the primary and secondary linings of all SCL tunnels, both public and non-public. The waterproofing system proposed is to provide an “un-drained” lining; the groundwater will be prevented from entering the tunnel by the waterproofing membrane and not be managed in the tunnel drainage system.

Waterproofing of sprayed concrete lined tunnels has traditionally been achieved through installation of polymer sheet membranes between the primary sprayed concrete layer and the secondary cast in-situ concrete layer. However recent projects such as Hindhead have demonstrated that the developments in sprayed waterproof membranes technology have shown sprayed waterproof membranes to be as good as cast sheet membranes in low permeability geology, and therefore, to facilitate their economical use.

Much of the Crossrail SCL works are located in London Clay, or other equally low permeability strata. In many ways this makes the tunnels ideal for watertight concrete design. The advantages of a spray applied waterproofing membrane are:

- Facilitates the use of spray applied secondary linings as spraying onto sheet membranes is recognised as a difficult operation with potential risk of defects.
- The safest of the waterproofing and secondary lining options considered in the Comparative Risk Assessment, notably in terms of reducing work at height.
- A tunnel that is resistant to water ingress.
- Lowest cost option for waterproofing.
- Shortest construction programme.
- Positive precedent projects in UK and internationally

In the case where, due to the potentially higher water permeability of the ground (for example sand channels and laminated beds in the Lambeth Group), there is a greater risk of a substrate with active water ingress a sheet membrane has been chosen as a more suitable method for waterproofing. The most significant example of this is the proposed Farringdon Station.

Design process

The design process for C121 SCL Design is summarised in Figure 7. The process described takes into account the outline design processes provided in the BTS-ICE Tunnel Lining Design Guide (2004), and the ICE Design and Practice Guide – Sprayed Concrete Linings for Tunnels in Soft Ground (1996).

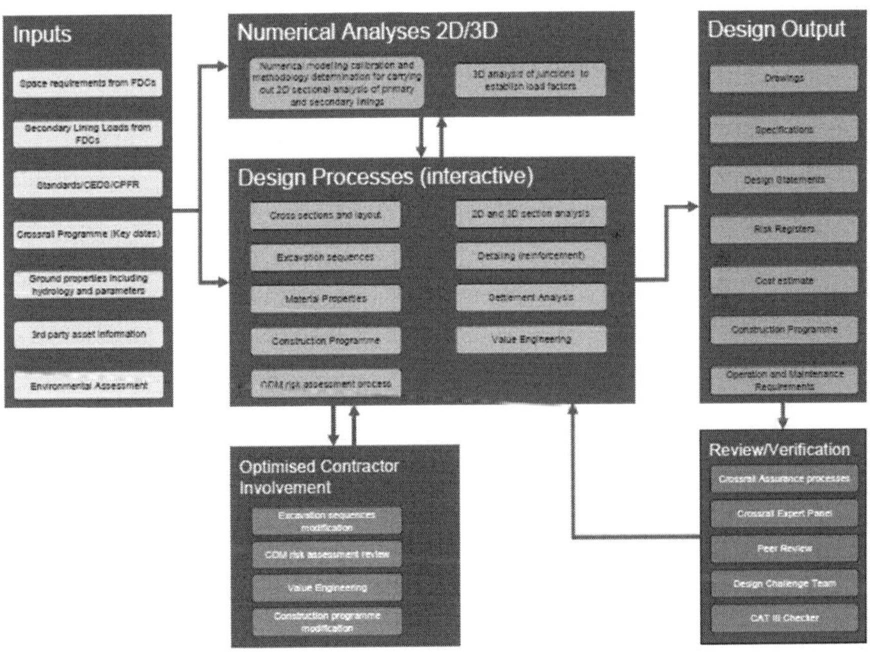

Figure 7 – Summary design process for C121 SCL Design

Expert Design Review/ Peer Review

In addition to Crossrail's Technical Assurance procedures, for example IDRs and Constructability Reviews and Category III Design Checks, Mott MacDonald employ the additional valuable process of Peer Review on all their major projects. The Peer Review comprises recognised design and construction experts not directly involved with the project to provide an independent assessment of the design, ensuring safety, and offering improvements where appropriate.

Challenge Team

Mott MacDonald formed a team of industry experts to periodically be called upon to challenge our design philosophy and approach, but also to provide additional views and ideas that assist the SCL design development. This approach has proved highly valuable on major projects, and to date has provided expert guidance on support on the Crossrail project.

Innovation and Value management

There have been several innovative solutions that have been put forward for the SCL design, these have included those already covered within the design which have improved safety from removing men from the face and reducing working at height while giving added programme benefits, such as:

- Sprayed waterproof membranes and secondary linings
- Elimination of lattice girders
- The use of steel fibre reinforcement instead of conventional reinforcement

We have also provided innovation and design solutions which have programme and wider architectural benefits, three examples are as follows:

- Junction construction methodology
- Standardised cross sections
- Curved junctions

Junctions and Openings

The design of SCL junctions is driven by the following:

- Designed to permit no disruption to the main tunnel construction programme. Therefore this allows the primary lining to be installed without reinforcement at junctions, but enlarged locally to accommodate later installation of rebar. (Traditional design approach was to install steel reinforcement, or thickening beams "as you go" round-by-round which is difficult and can cause durability risks.

- The primary shell will be sprayed using structural fibre reinforced sprayed concrete.

- The design approach provides a safe environment for creating the opening through an already stressed main tunnel shell.

- The design is to limit the amount of ground settlement above junctions.

The design sequence for tunnel junctions is as follows:

- Construct the main tunnel with an enlargement at the opening location. During this construction no reinforcement is installed.

- Later in the programme, remove backfill and install reinforcement in the entire area around the opening and spray a thickening patch of primary lining in junction zone. This lining including its reinforcement is designed to carry all the loads up to the installation of the secondary lining, and a proportion of the ground load in the long term.

- Break-out from the main tunnel and construct the primary lining of the smaller tunnel, once thickening patch has acquired 28 day design strength.

- At a later stage, install the waterproofing system, erect reinforcement around the opening for the secondary lining, and cast or spray the secondary lining.

Installing reinforcement into the primary lining while constructing the main tunnel was rejected based on the need for men to approach the face.

The chosen method allows for the construction of the main bore to be carried out with minimal changes from a standard excavation (slight increase in profile only) and install long bars of reinforcement in the required location once the lining is at a suitable strength to safely work nearby, all without impeding the productivity of the face, and in a much more controlled environment.

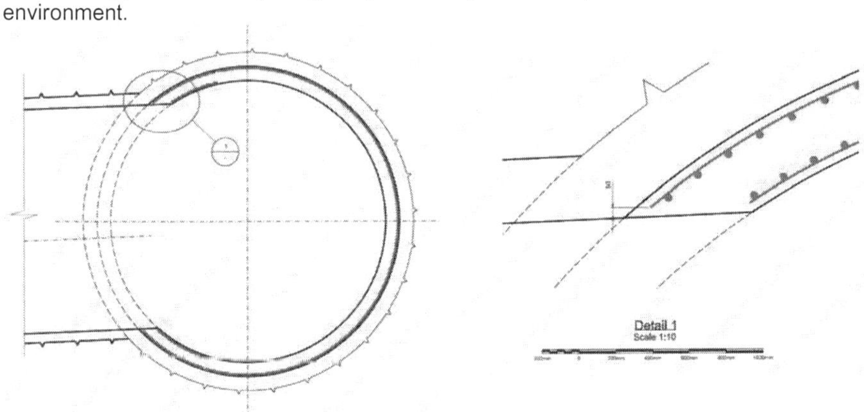

Figure 8 – Typical reinforced junction detail

Standardised cross sections

When C121 started RIBA D, there were in excess of 100 different cross sections. C121's aspiration was to reduce the number to maximise a set of standard cross sections. By standardising the amount of cross sections there were a number of benefits as follows:

- Reducing the number of different tunnel profiles and methods required for their construction.
- Repetition makes tasks easier to remember and therefore safer and more repeatable.
- Avoiding setting out errors and confusion.
- Reducing costs of fixings and cladding through consistent solutions system wide
- Facilitating the cladding design and a platform for architectural innovation, such as curved junctions.

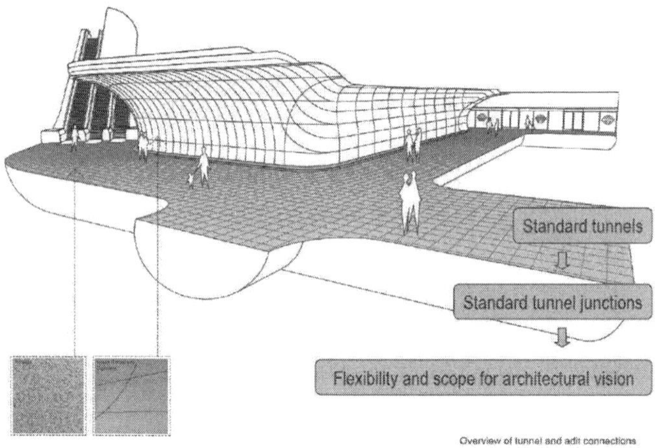

Figure 9 – Standardisation

Curved junctions

Following the standardisation of the tunnel cross sections C121 worked with the architects to provide SCL solutions for improved aesthetic and functionality (curved junctions). SCL secondary lings lend themselves to smooth transitions and rounded junctions, however on conventional design they have always been shown as right angles. C121 in collaboration with the architects maximised the reality of how SCL is formed to provide a platform for the architects to innovate and present ideas for curved cladding thereby to ease wayfinding, promote passenger flow, avoiding blind spots for improved personal safety and presenting a pleasing modern aesthetic across the Crossrail stations. An example of a platform tunnel with a curved junction is shown in Figure 10.

Figure 10 – Curved junctions

Value management

While a permanent SCL primary lining has a higher material cost than a conventional temporary lining by combining the following design steps there can be significant cost saving in programme. The following design steps combine to have a programme saving when compared to the conventionally reinforced temporary lining with lattice girders for shape control, sheet waterproof membrane and cast in place secondary lining, as follows:

- Steel fibre reinforced primary linings – reduces programme time for erection and fixing of conventional reinforcement.
- Junction innovation as described in 'Junctions and openings' reduces programme and allows no stoppage when tunnelling long tunnel drives (such as platform tunnels)
- Improved survey/setting out techniques – removal of lattice girders and therefore reduction in the erection and placement of lattice girders during primary lining installation
- Sprayed waterproof membrane – reduces programme compared with installation of sheet waterproof membrane, and allows bonding of a sprayed secondary lining;

- Sprayed secondary lining is quicker to install in comparison to placement of shutters for a cast secondary lining.
- Standardised cross sections increases familiarity for efficiency in plant and setting out.

These innovations have led to typical saving of 5-8 months in programme for a typical station compared to conventional SCL tunnels, as shown in Figure 11.

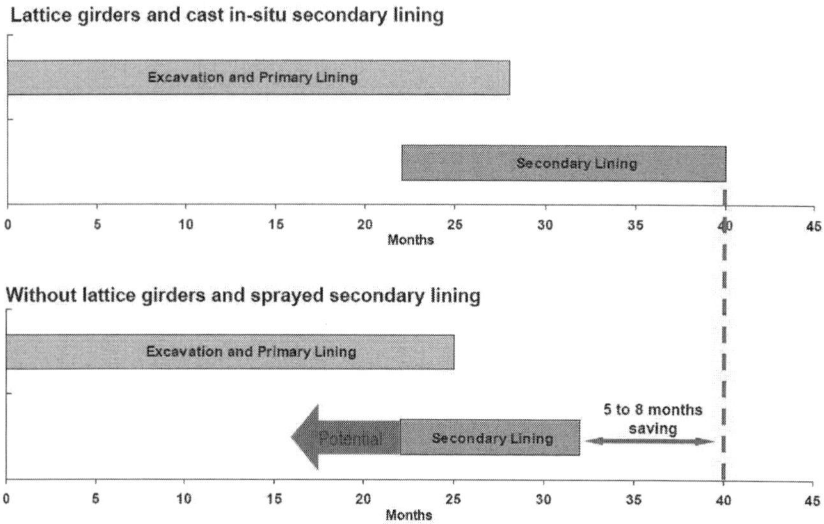

Figure 11 – Programme savings

Construction

The construction excavation sequences for the tunnels have also been standardised for differing tunnel sizes. The sequences have been chosen based on industry best practice and through a Comparative Risk Assessment CRA process to determine to meet CDM requirements as designers, examples of this were CRA carried out comparing pilot to sidewall drift techniques for tunnels up to 13m in excavated diameter. A summary of the excavation sequences are shown in Table 1.

Table 1: Construction sequences proposed for Crossrail Project

Diameter/ span	Description	Typical profile	Applicable Crossrail Tunnels	Precedent
Up to 5m	Full face		Service Tunnels Temporary access/ construction tunnels	Heathrow Express Rail Link Heathrow T5 Channel Tunnel Rail Link Ph 2 CTRL Phase 2 Wayside & Corsica Street Vent Shafts Thames Water Tunnels – Hampton (London Clay), Honor Oak Park (Chalk)
5 to 9m	Top heading, bench, invert (May include temporary invert strut in top heading)	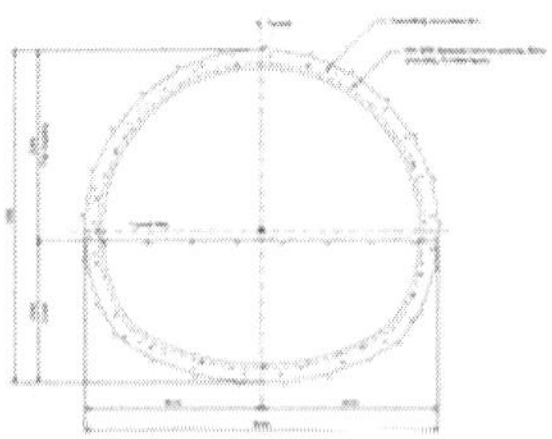	Running tunnel cross passages Station cross Passages Ventilation Tunnels Escalator tunnels	Channel Tunnel (Chalk Marl) Heathrow Baggage Tunnel, Heathrow Express Rail Link, Heathrow T5 (London Clay) Channel Tunnel Ph I – North Downs Tunnel (Chalk) Channel Tunnel Rail Link Ph2 - Corsica Street Shaft – vent tunnel enlargements (Lambeth Group – silty sands) A3 Hindhead Tunnel (Sands/weak sandstone)
9 to 13m	Pilot-enlargement	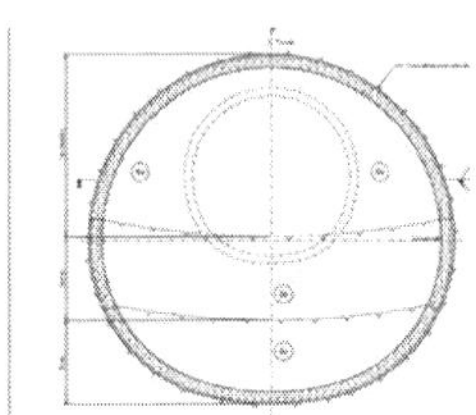	Platform Tunnels Concourse Tunnels	Jubilee Line Extension C102 – Waterloo Station 9.5m OD Platform Tunnels, 12.5m OD Concourse Tunnels Jubilee Line Extension C104 – London Bridge Station 9m OD Platform Tunnels, 12m OD Concourse Tunnels Kings Cross Station Modernisation 7.5m OD PLA & NLA concourse, 9.8m OD NLA Escalator

Diameter/ span	Description	Typical profile	Applicable Crossrail Tunnels	Precedent
12 to 14m	Single sidewall drift	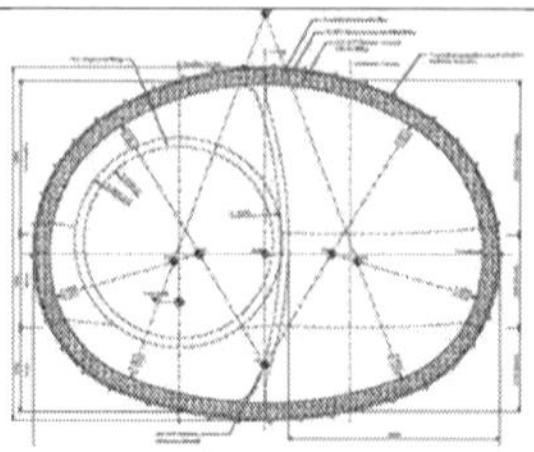 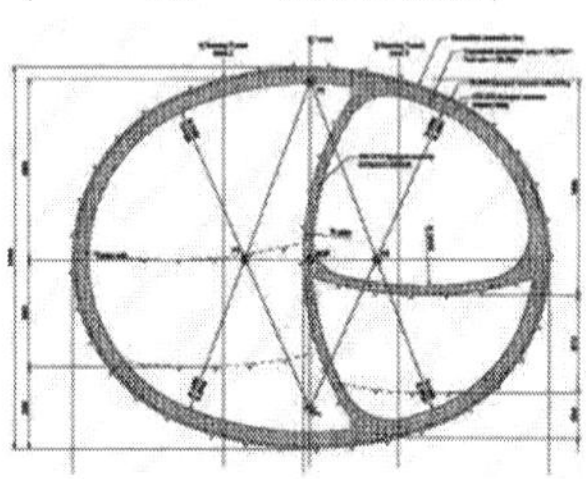	Fisher Street and Stepney Green tunnels	Jubilee Line C102 – Turnout tunnels (London clay) Jubilee Line C104 – Turnout tunnels (London clay)
14 to 18m	Double sidewall drift	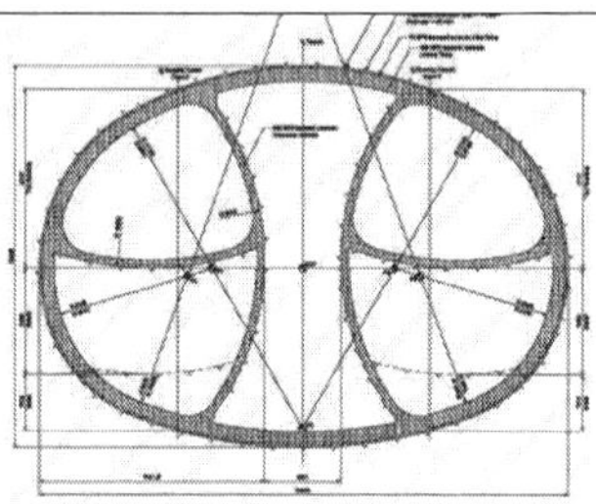	Stepney Green	Heathrow Express Rail Link (London clay) – Turnout tunnels and Crossover cavern Channel Tunnel Crossover Cavern (Chalk marl)

Optimised Contractor Involvement

Due to its size and programme the Crossrail contracts did not lend themselves to traditional Early Contractor involvement, i.e. contractors being involved in the design prior to the Invitation to tender (ITT) documentation issues, therefore the Employer took the opportunity to involve the Contractor immediately after the Contract was let and during the detailed design stage of the project in the form of Optimised Contractor Involvement (OCI).

The key OCI objectives set out were as follows:

- Obtain the Contractor's acknowledgement that the Employer's design is buildable
- Identify opportunities for the Employer's design to be modified to allow for safer or more cost effective construction;
- Coordinate the basis and detailed design for construction temporary works with the Employer's designers;
- Promote a better understanding by all parties of the contract scope
- Promote a better understanding by all parties of the potential risks and allow targeted elimination and mitigation of risk

Initially this was achieved through targeted workshops followed by separate meetings to concentrate on specific issues. The advantages of the OCI process were:

- A team approach to the final solution
- Increased opportunity for innovation
- Better integration of construction methods
- Timely procurement of materials
- Fewer variations
- Collaborative approach to sustainable issues
- Reduction of construction risk

Conclusion

In conclusion there were big challenges faced by Mott MacDonald on commencement of the Crossrail design, through concentrating primarily on safe design by reducing the requirement for tunnel operatives at the tunnel and designing best practice with enhanced robustment. Mott MacDonald have achieved delivering a Crossrail design to programme and best value which has been benchmarked on recent SCL design in the UK such as the A3 Hindhead tunnel which successfully opening in Summer 2011.

Acknowledgements

The author would like to acknowledge the contributions of many colleagues at Mott MacDonald in the development of the design approaches and Ross Dimmock at TAM International for their contributions in developing some of these concepts.

Crossrail
Crossrail Project: Infrastructure design and construction
ISBN 978-0-7277-6078-4

doi: 10.1680/cpid.60784.155

ice
Institution of Civil Engineers
publishing

Farringdon station SCL Design – Reducing risk at the heart of Crossrail

Andrew Davis BSc MSc FGS C.Geol, Geotechnical Team Leader C121 Mott MacDonald
Philip Duarte MEng CEng MICE, Design Team Leader C121 Mott MacDonald

Abstract

The ground conditions expected at Farringdon Station are amongst the most challenging on Crossrail, exacerbated further by four TBMs (C300 and C305 drives – also known as "X" and "Y" drives respectively) all terminating in reception chambers required to facilitate their extraction. The Sprayed Concrete Lined (SCL) tunnels are expected to be within the Lambeth Group, but crucially subject to intermittent, water bearing sand channels that could unexpectedly occur anywhere in the vicinity of the tunnels. This uncertainty, verified through various ground investigations, has dominated the design with five years' worth of detailed design carried out to ensure the construction risk is ALARP. Moreover, Farringdon is situated within a highly faulted area and these risks have also been managed in the design. Extensive work has demonstrated how to safely construct the platform tunnels at Farringdon station from within which other tunnelling works (such as cross passages and ventilation adits) can subsequently be managed. The SCL design process was closely scrutinised through the use of CDM registers, comparative risk registers, peer review challenge teams and client expert panels; adding to the robustness of the proposed design.

C121 worked closely with the client (CRL) to reduce the risks associated with excavating tunnels in this highly variable, challenging ground using ALARP principles. Particular focus was given to the geotechnical challenges underpinning the design as well as the potential different tunnelling options for the station works. Farringdon station is unique and the work carried out by C121 (Contract 121 – Sprayed Concrete Lining design for all Crossrail structures) shows the specific approach taken and how C121 deviated from the standard design adopted for the majority of SCL tunnels on Crossrail to produce a tailored solution for a very challenging station.

Tunnelling in Farringdon

Farringdon Station consists of approximately 1 km of tunnels to be constructed using a "Sprayed Concrete Lining" (*SCL*). The tunnel layout comprises a network of subterranean platform tunnels, connecting cross passages, concourse tunnels, escalator inclines and ventilation adits (see figure 1). The tunnel excavation diameters vary from approximately 6m to 11m and will be constructed at depths of up to 35m below ground surface beneath a congested area of central London that forms a major transport hub and employment centre for thousands of commuters. Impacts from tunnelling on these activities could be serious and potentially affect two London Underground lines, Thameslink services and Smithfield Market.

Farringdon station is situated in the middle of the central tunnelled section of the Crossrail alignment and is one of the final construction activities where the running tunnel TBM's meet for dismantling. Logistically, this represents a challenge to ensure that the construction of this station meets key programme milestones and therefore not delay the completion of the network. In isolation, this is already a key constraint; however Farringdon station has many unique challenges not faced elsewhere on the project with the main differentiator being the expected ground conditions (Figures 2 and 3). This was fundamental to understanding how to manage the engineering risks to safely build the tunnels, risks which needed to be shared and understood by all involved in the design, including the Client.

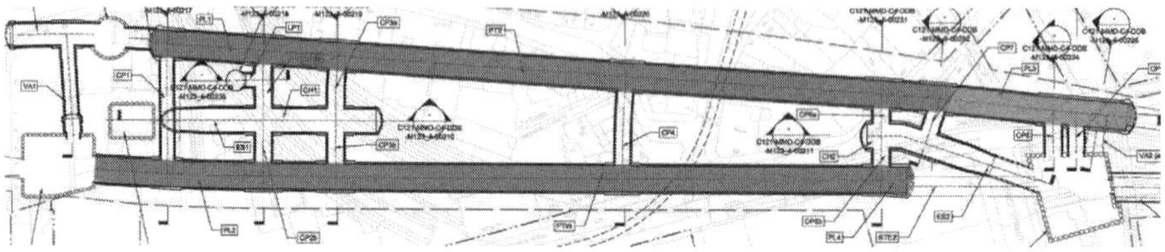

Figure 1 – Tunnels subject to further investigation

Sprayed Concrete Lining

SCL tunnel excavation is undertaken through spraying concrete directly onto the exposed excavated ground, forming an immediate and intimate bond with the strata. It interacts actively with the excavated ground surface instead of passively requiring the ground to move before it begins to resist that movement. In addition, the lining can easily follow complex profiles and layouts, in-fill over-break, and forms a strong, durable and watertight shell. As such it is an ideal method for the construction of complex station layouts. If required, initial ground stability can be enhanced through dewatering, the installation of canopy tubes and spiles, grouting, sub-division of the face and/or use of compressed air (thought the latter is fast becoming obsolete in the UK due to reduced working efficiency and H&S concerns).

Typically, in soft ground, excavation is carried out in different headings (for example a top heading, bench and invert) with SCL support applied as required. On the Crossrail project support is provided through an immediate SCL sealing layer of approx. 75mm (or thicker if a longer construction activity is anticipated – such as breaking out of a box) applied to any freshly excavated ground followed by a primary lining sprayed to the required design thickness. A waterproofing membrane is then applied (sheet or sprayed) with a final secondary lining, which can be sprayed or cast in situ. Depending on the size of the tunnel, a pilot tunnel may be required (a smaller tunnel within the larger tunnel) to facilitate excavation, control face stability and reduce risks. Due to the risks at Farringdon pilot tunnels were widely adopted but the ones which caused the greatest concern, and the focus of this paper, were the large platform tunnels, representing approx. 650 m of tunnelling

SCL is more dependent on the individuals carrying out the tunnelling and therefore has a greater reliance on human judgement than other tunnelling methods. For this reason, a substantial amount of work was carried out looking at ways to reduce this variable as well as provide a comparatively safer way of forming the pilot tunnels for the platform tunnels, in this highly variable stratum.

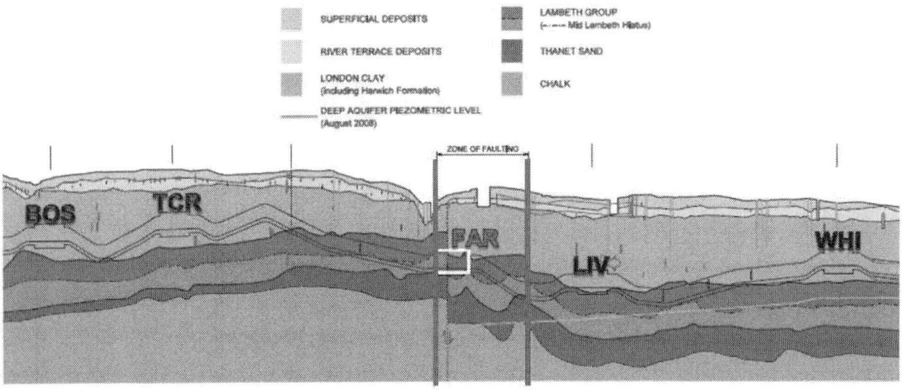

Figure 2 – Farringdon station Location and route geology

Geology

The geology at Farringdon Station is typical for that of east central London; comprising of a sequence of strata from Thanet Sand, the complex Lambeth Group strata and the lower A2 & A3 units of the London Clay. The only stratigraphical units that appear to be absent are the Upper Shelly Beds of the Lambeth Group and the Harwich Formation. A summary is provided in Table 1 and a longitudinal section as Figure 3. The spatial variation of superficial deposits are controlled by the historical route of the Fleet River at the western end of the station which has eroded down though the River Terrace Deposits and into the London Clay forming a buried valley to some 10m below current ground level infilled with alluvial and Made Ground deposits.

The SCL tunnels are located within all of the strata from Thanet Sand through to London Clay, the majority of the excavations being within the Lambeth Group. The shallow escalator incline at the western end has a crown level in close proximity to possible superficial deposits associated with the Fleet River.

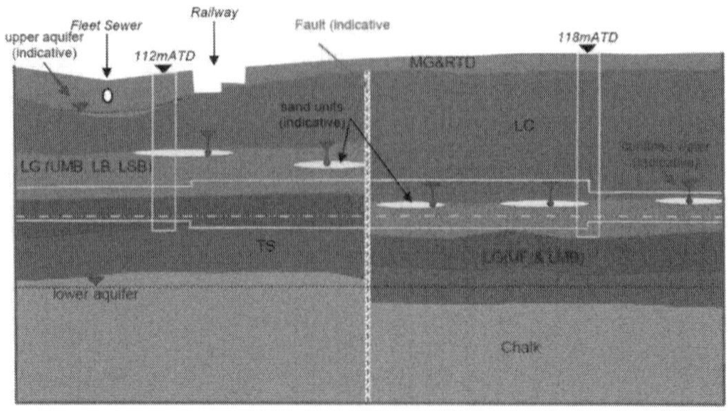

Figure 3 – Indicative geology anticipated in vicinity of Farringdon station

Strata		Thickness (m)	Typical Description
Made Ground (MG)		2 to 4.5	Variable
Alluvium (ALL)		1.5	Soft to firm dark grey and grey brown sandy slightly gravely organic silty CLAY.
River Terrace Deposits (RTD)		1 to 2	Medium dense orange brown fine to coarse SAND/GRAVEL and rare cobbles of flint.
London Clay (LC)		5 to 23	Firm to very stiff fissured CLAY with silt partings and laminations. Increase in silt partings and becoming slightly sandy in the A2 unit.
Lambeth Group (LG)	Upper Mottled Beds (UMB)	6 to 7	Very stiff to hard (friable microfissured) very closely to extremely closely fissured light grey to blue grey mottled light brown and red sandy to very sandy silty CLAY.
	Laminated Beds (LMB)	1 to 2.5	Stiff to very stiff thinly laminated fissured grey to dark grey, sandy to slightly sandy, silty to very silty CLAY, with extremely to very closely spaced thin laminations of light grey slightly sandy silt.
	Lower Shelly Clay (LSC)	<1	Stiff to very stiff fissured thinly laminated to thinly bedded dark grey and grey, silty to locally very silty CLAY with occasional bands of light grey sandy silt. Frequent to shells and shell fragments.
	Lower Mottled Beds (LMB)	2 to 8	Stiff to very stiff multi-coloured extremely closely fissured silty to very silty, slightly to very sandy CLAY.
	Upnor Formation (UF)	2 to 5	Very stiff dark grey-green, brown very sandy silty CLAY, with occasional fine to coarse rounded flint gravel. Very dense black rounded medium and coarse flint GRAVEL Very dense brown or green to dark green speckled clayey to very clayey silty fine to coarse SAND, with a little to much fine and medium rounded flint gravel.
Thanet Sand (TS)		3 to 11	Very dense grey green silty fine and medium SAND.

Table 1 Summary of strata through which SCL tunnelling will occur at Farringdon.

Sand Channels

Within the Lambeth Group there occur relict sand channel deposits related to the paeleo-environment present during the geological period of deposition, representing a series of meandering channels in an estuarine type environment. These channels vary in frequency, width, depth, length and infill material and, as a result, are spatially unpredictable and therefore problematic to identify by Ground Investigation (GI) with any degree of certainty. There are examples of large extensive channels, such as below Bond Street station, where identification and extrapolation is more certain. The size of channels identified at Farringdon however tend to be thinner, typically to a maximum thickness of 3m, less extensive and encountered variably throughout the Lambeth Group. These sand channels were the subject of various internal and collaborative workshops on how best to mitigate their presence in the tunnel alignment.

Faulting

Whilst faulting associated with major geological basement structural features was known to occur below London, the extensive GI programmes undertaken by Crossrail over twenty years has identified them to be more frequent than previously thought. The presence of faulting at Farringdon was postulated during early design phases due to the presence of the Fleet River alignment and the untypical change in strata levels across the station footprint. Subsequent phases of investigation however indicated that faulting was more widespread, with one major fault and many minor faults. Due to the concern at the effect of faulting and strata dislocation on tunnel excavation behaviour a separate independent review was undertaken by the British Geological Survey (BGS), combining the Crossrail GI database with that of the BGS to produce a detailed three dimensional model of the area. That review indicated the existence of at least four main faults, and four subsidiary faults, with throws typically of up to 4m (see Figure 4). As noted earlier the major fault, the "Smithfield Fault", has a throw of 8m and directly juxtaposes Upper and Lower Lambeth Group strata within the tunnel face. Encountering these faults represents a high risk activity not present in any of the other stations in severity or persistence. Figure 4 summarises the proposed prescriptive
mitigation measures following much consideration on the most cost effective risk option to proceed. C121 employed the use of comparative risk assessments, peer reviews and expert panels to satisfy both CRL and the CAT 3 checker that an ALARP design was achieved.

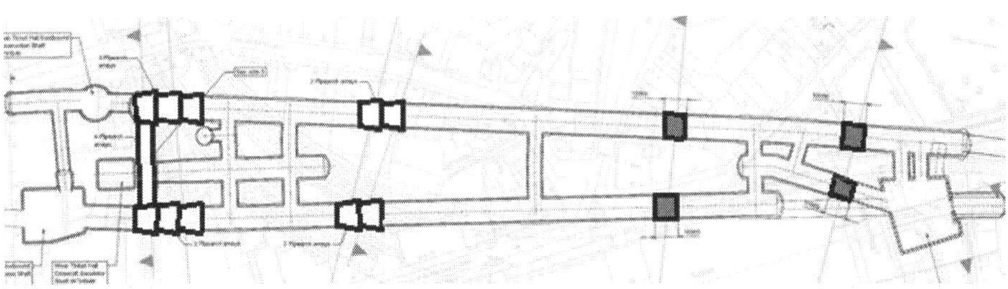

Figure 4 - Faulting in the vicinity of Farringdon Station

Notes:
Light shading – Grouted pipe arch
Dark shading – Grouted spiles

Groundwater

The groundwater profile at Farringdon is typical for the London area comprising of an upper aquifer located in the superficial deposits and a lower aquifer within the Chalk, Thanet Sand and Upnor Formation. The historical effects of groundwater abstraction from the Lower Aquifer has resulted in an underdrained profile with pore pressures within the Thanet Sand and Upnor Formation being much lower than those above in the London Clay and upper Lambeth Group. Due to the dislocation of strata across the main Smithfield Fault the effects of underdrainage are different to the west and east. The current profiles are shown in Figure 5 and indicate maximum pore pressures at platform tunnel level of 100kPa to the west of the fault and 150kPa to the east.

Ground Related Risks

Although the geology and groundwater conditions at Farringdon are not unusual for central London it is the requirement to undertake the station construction using the open face SCL technique that introduces greater ground risk. The need to control groundwater ingress and pressures where granular materials occur in an open face is paramount if stability is to be maintained prior to application of the sprayed concrete.

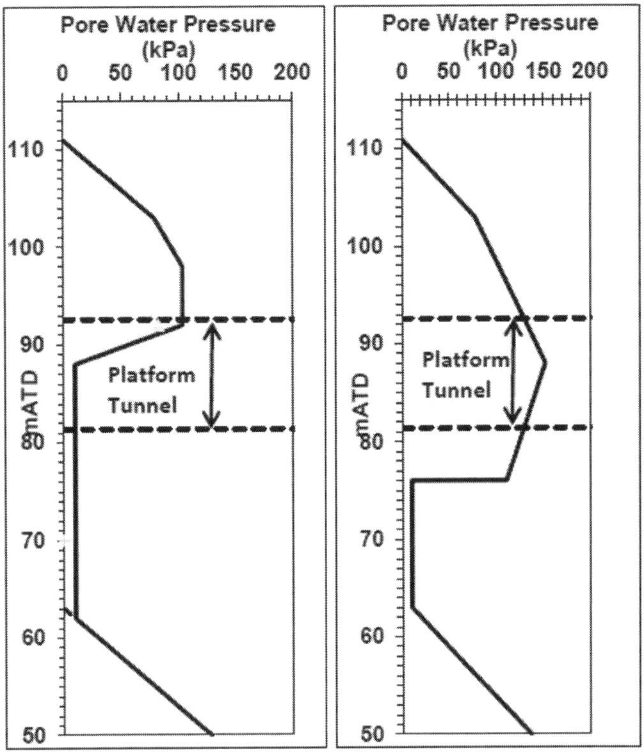

Figure 5 Pore pressure distributions west & east of the Smithfield Fault

If the groundwater pressures at the face of the excavation are not reduced adequately they would cause instability of any sand layer resulting in a tendency for the sand to run into the excavation. If the granular strata comprise a significant proportion of the face then direct face instability can occur. In situations where minor sand/silt strata occur (such as laminae) the wash in of granular material may cause local undermining of the face and subsequent indirect instability. Any such loss of material from the face could trigger a fall of ground from the face, overbreak and promote greater than anticipated settlement. In addition it can result in poor working conditions which are unacceptable on health and safety grounds.

The extent of the risk and the form of the mitigation will depend on a number of variables:

- The size of the sand units that controls the quantity of potential stored water;
- Their permeability which controls the rate of water inflow;
- Their ability to recharge which controls the time for which inflow may occur,
- The magnitude of the pore water pressures present which again affects the amount of inflow, and
- The location of the sand channel in relation to the tunnel excavation profile

The majority of previous SCL works undertaken in London have been undertaken within the relatively uniform and dry London Clay. Similarly for Crossrail the majority of SCL to be undertaken at other locations will be within the London Clay where the potential for water bearing granular materials only occurs at depth below invert level. The associated groundwater risk in these conditions is low and related to the potential for invert heave. At Farringdon however the Lambeth Group contains granular water bearing materials that, because of the faulted strata dislocation, may occur at any point around the tunnel periphery and in the tunnel face. As such the risk of water induced face instability is a particular site specific issue. The greater risk of groundwater induced instability is highlighted by a simple comparison of the number of wells allocated during tender design which is summarised in Table 2.

Station	Geology	Well Allowance	
		Surface Sites	In-tunnel
Bond Street	London Clay	None	140
Tottenham Court Road	London Clay	None	120
Farringdon	Lambeth Group	50	3250
Liverpool Street	London Clay & Upper Lambeth Group	28	850
Whitechapel	London Clay	None	400

Table 2 Summary of SCL station well quantities at Contract tender

Faced with the prospect of encountering these highly variable ground conditions, C121 produced a design which aimed to address the main risk of encountering the isolated pockets of potentially water bearing sand lenses. Many options were considered but the chosen method for providing a robust solution to the problem was extensive use of in-tunnel and surface depressurisation targeting the intermittent variable sand lenses and underpins the rest of the design at Farringdon.

As noted earlier Farringdon also harbours a further complication: widespread faulting. The major Smithfield fault essentially dissects the station in half which means the west of the station has different geology to the east, the former representing more pessimistic tunnelling conditions – this is recognised in the prescriptive pre-support shown on Figure 4. Coupled with the already challenging ground conditions, the faulting adds a further dimension to the complexity for open face tunnelling. In particular faulting can result in localised increased fissuring, presence of persistent polished surfaces (colloquially termed “greasybacks”), softening and potential pathway for groundwater ingress. Therefore, in addition to the depressurisation specified, further ground treatment was proposed to allow safe construction through the faulted areas, which in this case amounted to proposed use of pipe arches west of Smithfield fault and grouted spiles to the east, in the crown of the tunnels. The difference in methodology being related to the different strata expected to be present at crown level either side of the fault.

Depressurisation

In order to address the critical groundwater risk an extensive investigation and depressurisation scheme was developed combining both surface and in-tunnel works. Farringdon Station is located in a congested area and the potential to maximise use of surface well sites is extremely constrained and cannot solely be used to address the risk. It is not therefore possible to confidently identify and depressurise all water bearing materials by surface works alone. To adequately reduce the level of risk there is a need to undertake SCL excavation with a degree of confidence that unexpected granular materials are unlikely to be encountered. In order that this risk can be maintained ALARP there is a requirement that in-tunnel investigation works are comprehensive (see indicative Figure 5). In order to achieve that level of acceptable risk the in-tunnel investigation needs to consider the following:

- Different investigation hole layouts related to different tunnel dimensions and excavation sequences.
- Provision of adequate investigated ground coverage ahead, and around the tunnel profile.
- Adoption of a method of investigation that can identify changes in strata that would typically be missed by conventional open holed probing techniques.

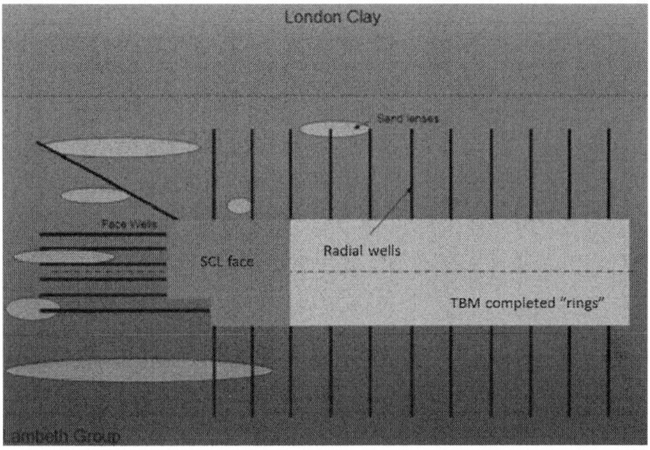

Figure 5 – Indicative schematic showing depressurisation for an SCL / TBM tunnel and sand lenses

The key difference between a closed face TBM pilot tunnel and a SCL pilot is the need to investigate ahead of the advancing face to ensure no "surprises" are encountered in the advancing face (Figure 5).

Given the impersistent channelised nature of the sand units, the drilling is considered to be of an investigatory nature and the findings of the investigation drilling will determine the subsequent risk and course of action. It is not expected that all holes will encounter groundwater bearing sand that requires a subsequent well installation. Where potential risk is identified by initial investigation then wells will be installed and the pore pressure reduced to predetermined appropriate levels. The assessment, direction and implementation of these mitigation items would be controlled by the construction Shift Review Group (SRG) who has direct day to day experience of the construction performance and understanding of impacts on the Works.

Tunnel Construction Options Selection

Numerous studies were commissioned to interrogate an alternative design methodology for the platform tunnels. It was widely agreed that a pilot tunnel was required for the platform tunnels, with systematic depressurisation, but the method for constructing these pilots was challenged extensively. The conclusion of all these studies was that a "newly" purchased TBM, forming a segmentally lined pilot tunnel for the platform tunnels, was the preferred solution and therefore all other methods and their respective risk profiles were compared to this base case. The options initially considered as part of this process are summarised in Table 3. Alternatives 5 to 8 were rapidly assessed as being inappropriate and discounted, the remaining base case and alternatives 1 to 4 were then carried forward for more detailed comparative assessment.

The enlargement of the platform tunnels from the pilot was also considered for each option. However, for the purposes of comparison, safety of the enlargement was deemed to be equal for each option. There would generally be two different methods considered, one for a segmentally lined pilot tunnel and one for enlarging from an SCL tunnel. Neither presented a great challenge over the other, but it was agreed that the most critical component of enlargement would be the depressurisation. As long as the pilot tunnel construction method allowed sufficient depressurisation to take place for the platform tunnel to be constructed, the enlargement method was not considered to be a deciding factor.

Peer Review

C121 operated a system of internal peer reviews using a group of experienced senior tunnelling professionals that were independent of the project. Regular "challenge" meetings were undertaken for the C121 option proposals to be subject to open questioning. This process was invaluable in obtaining a range of informed options that were considered to address the known risks and be constructible. In addition Crossrail appointed an external independent panel of tunnelling experts that reviewed all aspects of tunnelling on Crossrail, the proposals for Farringdon being subject to frequent rigorous examination. These two levels of independent review provide additional confidence to the Client and external Stakeholders that risks are being managed in accordance with industry knowledge and precedent experience.

Health & Safety Review

In order that the assessment approach did not become dominated by the technical, commercial and programme aspects of the different options it was critical that the health and safety risks were not sidelined. Construction Design Management (CDM) risk assessments were used to understand the key risks for the chosen option and subject to monthly reviews by the Client CDM-C. This was the final rigorous assessment of the chosen solution, before which all options were compared comparatively to identify the lowest risk option. Both CRL and C121 combined very effectively to make safety the top consideration for Farringdon, despite the fact that there were clear delivery programme and cost concerns (since Farringdon is where the lines connect).

Option	Description
Base Case	Pilot tunnels constructed using a "new" Earth Pressure Balance (EPB) tunnelling machine. Vacuum wells installed radially from the pilot tunnels. Pilot then enlarged by SCL to form the platform tunnels.
Alternative 1	Pilot tunnels constructed using an open faced shield with depressurisation wells drilled ahead of the open face. Vacuum wells also installed radially from the pilot tunnels then the pilot tunnel enlarged by SCL to form the platform tunnels.
Alternative 2	One pilot tunnel constructed using an open faced shield and the second constructed using SCL with wells drilled ahead of the open face in both. Vacuum wells also installed radially from the pilot tunnels then the pilot tunnel enlarged by SCL to form the platform tunnels.
Alternative 3	Pilot tunnels constructed using SCL with wells drilled ahead of the open face. Vacuum wells also installed radially from the pilot tunnels then the pilot tunnel enlarged to form the platform tunnels.

Alternative 4	Extending the X drive TBMs through Farringdon, so that pilot tunnels are constructed by the Running Tunnel TBMs. Vacuum wells installed radially from the pilot tunnels then the pilot tunnel enlarged by SCL to form the platform tunnels.
Alternative 5	Pilot tunnels constructed using a 'Bessac' machine which supports the face with compressed air. Vacuum wells installed radially from the pilot tunnels then the pilot tunnel enlarged by SCL to form the platform tunnels. **Discounted because:** The water in the Lambeth Group is generally within discrete sand pockets within the low permeability clay matrix. Many of these pockets are not connected to other permeable areas. Compressed air may not be effective in this type of ground. Rather than creating a high pressure area which acts against the water pressure, with local pockets of water bearing permeable material there is a risk that the air will displace the water within the sand lens and push the water into the face, resulting in an unstable excavated face. This results due to the compressed air not providing a variable pressure at the face to match the earth pressure (as an EPBM does). Only one manufacturer produces machines of this type, 'Bessac' in France, and there is limited experience of their use, particularly in the UK.
Alternative 6	Constructing the platform tunnels 'full size' using a closed face TBM. No need for depressurisation wells. **Discounted because:** The west ticket hall shafts would not accommodate installation of a closed face machine able to build the platform tunnels (approx..12.5m dia). Shaft enlargement could achieve this but at increased cost. Initial reviews showed that there insufficient space at the worksites for this. The TBM segmental lining would require openings for cross passages and ventilation tunnels. The design requires that these are of significant width, typically 5m. Construction of openings of this size would require major structural elements to be installed and take considerable time.
Alternative 7	Improving the ground for construction of open face pilot tunnels using horizontal jet grouting. **Discounted because:** Horizontal jet grouting could be used to stabilise the ground for construction of the platform tunnels, but given the uncertainties about the location of the sand channels it would need to be installed for the full length of both platform tunnels. This would require access for drill rigs and grouting equipment and would generate a significant volume of fluid cementitious spoil that would need to be removed from the tunnel. Except at the shafts at the west end of the station, access for installation is extremely difficult and there is limited UK experience.
Alternative 8	Improving the ground ahead of open face pilot tunnels using ground freezing. **Discounted because:** Ground freezing has major installation issues (impacting on programme and cost). Although only a low percentage of the ground needs to be frozen, it would be necessary to install the pipes along the full length of the tunnels. Risk that unfrozen windows of ground may occur over the extended length.

Table 3 Summary of construction options considered in risk analyses

Comparative Risk Assessments

A useful tool employed for assessing risk is a Comparative Risk Assessment (CRA) which allows multiple options to be compared relative to each other. The fundamental principle is that a peer review(s) are undertaken whereby each option under consideration (in this particular situation, the base case compared to options 1 to 4) are rated numerically against one another. These options were originally compared at a high level on cost, programme and risk grounds to see if there was a standout "winner", but it was clear that a more rigorous assessment was needed due to the nuance of each option presented. Therefore a more rigorous quantitative risk assessment was subsequently undertaken which for comparative purposes apportioned risk based costs leading to the following simple relationship and the decision tree shown in Figure 8:

Expected total cost (£) = Base cost $_{(1)}$ + Risk cost $_{(2)}$

(1) Base cost includes the definite cost items associated with each option
(2) Risk cost includes the costs associated with potential risks of face loss associated with each option

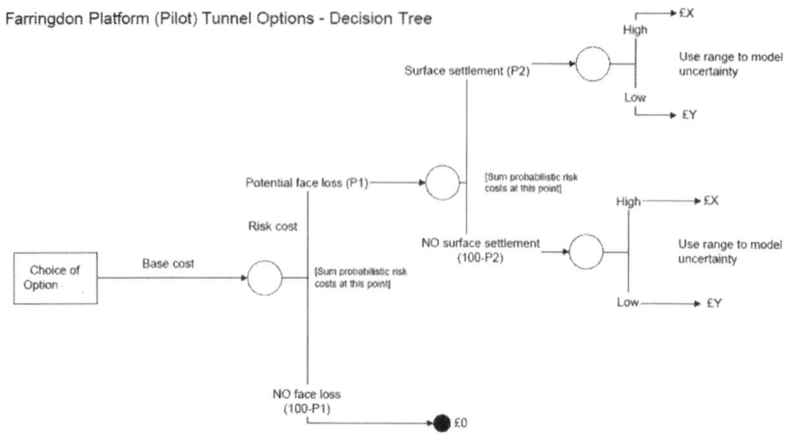

Figure 8 – Decision tree for QRA

The estimated costs and probabilities of potential face loss (P1) and subsequent surface settlement (P2) were then analysed using a "monte-carlo" simulation to calculate the expected total risk based cost for each option. In essence this analysis is based on health and safety as it is relating to face loss and settlement, which affect risk to personnel in tunnel and third parties at the surface.

The base costs for each option were estimated to include all direct costs (for plant, equipment, labour and materials) associated with construction the pilot and related platform tunnels. Risk cost includes direct and indirect (schedule) costs associated with a loss of face risk event. The latter focussed on the risk of face loss (during construction of the pilot tunnel) and related ground settlement. This was seen as the most variable and significant risk for alternative options which impacted the construction cost. This risk was also seen as having significant consequences to the safety of construction workers and third parties.

Otherwise, all options were considered to have similar risk profiles and meet the same safety requirements including ALARP residual risk exposure.

Tunnel Instability Risk Cost Estimates

The potential for tunnel instability and the resultant costs can be divided into two broad types:

1) Additional construction costs associated with a minor loss of face material (up to a team agreed value of $50m^3$)

2) Third Party Liability (TPL) costs associated with a surface settlement due to major loss of face material (greater than $50m^3$).

Additional Construction Cost – Tunnel Face Loss

The principle mechanism for significant face loss was considered to be through ground water ingress from gravel/sand pockets in the Lambeth Group, or poor design/workmanship (e.g. dewatering scheme). Loss of ground is seen by insurers to be the highest risk due to higher reconstruction costs and frequency of occurrence (source; International Association of Engineering Insurers Paper Tunnel Boring Machine Applications in Soft Ground Conditions 2001).

Additional direct construction costs associated with tunnel face loss were estimated to include the direct cost of ground stabilisation, temporary measures, surveys, repairs, debris removal etc. Indirect construction costs associated with the works being halted were also estimated, the project taking longer to complete whilst the problem is resolved. The indirect costs were based on an increase in the preliminary costs associated with the additional time to complete the works.

It should be noted that mechanical failure or poor operation of the plant or equipment (other than dewatering) was not considered as a differentiating factor causing tunnel face loss.

Third Party Liability (TPL) Cost – Surface Settlement

TPL costs associated with surface settlement are one of the biggest issues when tunnelling in urban environment and at Farringdon this is particularly the case due to close proximity of utilities, operational railways and third party properties.

The potential costs associated with TPL loss can be grouped into two areas:

1) Losses associated with direct material damage or bodily injury of third parties, and

2) Consequential financial loss e.g. reduced revenue or profit associated with business interruption. Note that these losses are not typically covered by insurers and will be dependent on exclusions and level of cover required for a major event. It was estimated that the majority of TPL losses are up to £50m for all construction options considered.

A tenfold multiplier to the direct TPL cost was agreed with the expert team to represent the consequential third party financial losses.

Results of Cost Model Analysis

A "monte-carlo" simulation was performed to produce a distribution of expected total cost for each option using the derived cost and probability ranges. A resulting cumulative distribution sets out the confidence level in the expected outturn cost. The graphics in Figure 9 show the analysis results. The actual values are not particularly important, but the comparative values for each are.

The distribution shows that the SCL pilot option (red line) is the most cost effective option should a low level of confidence in the expected total cost be acceptable, that being only at a level of 35%. Above 35% confidence, use of the existing running tunnel EPBM's (black line) was the most cost effective solution.

At approximately the 97% confidence level the existing and new TBM options (green line) have a similar cost/risk profile. A decision to adopt the use of the existing TBM's however needed to be balanced against any disadvantages of the later start of SCL tunnel works due to the later arrival of the running tunnel EPBM's resulting in overall project programme delay (not quantified comparatively in this analysis). That overriding project level risk was considered too high and therefore the seemingly more expensive option, based on mean total cost, of providing a bespoke EPBM was adopted.

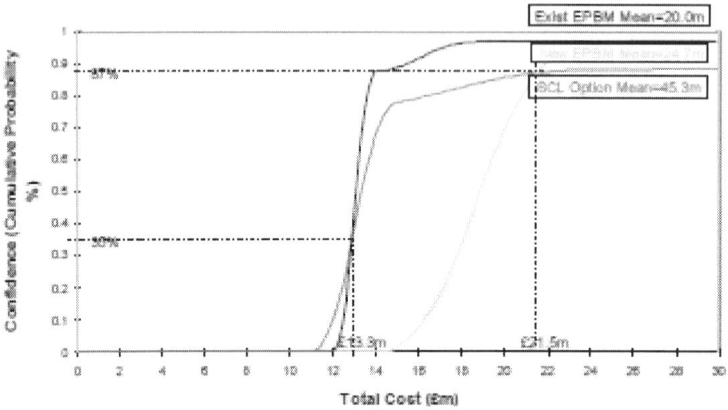

Figure 9– Confidence level of total cost

The average expected total cost over the total cumulative range for each option is summarised in Table 4.

Option	Cost
Additional EPBM	£24.7m
Existing EPBM	£20.0m*
SCL	£45.3m.

Table 4 – Summary of average expected total cost

* Note that this price does not take into account associated programme delays

Summary

Farringdon Station has a number of ground related and project delivery constraints not seen anywhere else on the route which in combination make the SCL works a significant risk. The combination of faulting and water bearing intermittent sands are unique for Crossrail SCL locations and therefore the work carried out at Farringdon is bespoke and required a collaborative approach with all involved (e.g. Client, Stakeholders, designers etc). The major risks were considered and managed in a number of ways to cover health & safety, cost, programme and technical issues. To collate the often competing risks, rigorous assessments were undertaken which were subject to stringent internal and external review. Specifically, for the platform tunnel construction, a qualitative risk assessment tool was employed using a "monte carlo" statistical analysis to compare the range of potential outturn costs based on risks cost for different options. The outcome resulted in a seemingly expensive option being adopted but that which has a very high level of confidence in the outturn cost providing Crossrail with greater programme and cost certainty at a global project level.

At the time of writing this paper, OCI and VE options are being discussed with the contractor and therefore the employers design, presented herein, may be subject to change. However the tools employed illustrate how high risk elements of design can be discretised and analysed to find the lowest risk option amongst a selection of potential solutions. The success of this approach was dependent on a willing client and collaborative environment where the common goal was to reduce risk at the heart of Crossrail. By having these key components in place, it was possible to develop a base line design which can be considered ALARP – which in central London, with thousands of people potentially affected and millions of pounds at stake, was exactly what Farringdon required.

Crossrail Project: Infrastructure design and construction
ISBN 978-0-7277-6078-4

doi: 10.1680/cpid.60784.171

Crossrail UXO Risk Assessment – Pre-empting best practice

Dr Phil Smith BEng MSc ACGI PhD DIC, Senior Geotechnical Engineer, Geotechnical Consulting Group
Dr Ursula Lawrence PhD Eur.Geol C.Geol FGS C.Sci, Geotechnical Engineer, Crossrail Limited.
Sarah Terry MSc C.Geol FGS, Formerly Geotechnical Engineer, Crossrail Limited
Simon Cooke BSc MA MBA MIExplE MAPM MIoR, Managing Director, 6 Alpha Associates Ltd.

Abstract

Site investigation works undertaken as part of the Crossrail project over the last ten years have involved extensive intrusive works into the ground in areas of London which were known to have been heavily bombed during World War Two. This bombing created a legacy of unexploded ordnance risk to such works. This paper explains how a procedure was developed to enable the risk to the ground investigation works to be assessed and for appropriate and effective measures to be identified where necessary to mitigate this risk. It is shown that the developed procedure is consistent with that now recommended by the CIRIA guide on unexploded ordnance, and provides a practical illustration of how an appropriate risk assessment procedure can avoid extensive and unnecessary mitigation works on site, while ensuring that when mitigation measures are undertaken, it is where and when they are truly needed to enable work to progress safely.

Introduction

The publication in October 2009 of the CIRIA report C681 *Unexploded ordnance (UXO) A guide for the construction industry* (CIRIA 2009) has for the first time established an industry wide best practice for the assessment of unexploded ordnance (UXO) risk to construction activities in the United Kingdom. This guide shows how potential UXO risk can be assessed and managed through a staged approach of preliminary risk assessment, detailed risk assessment, risk mitigation and implementation (see Figure 1).

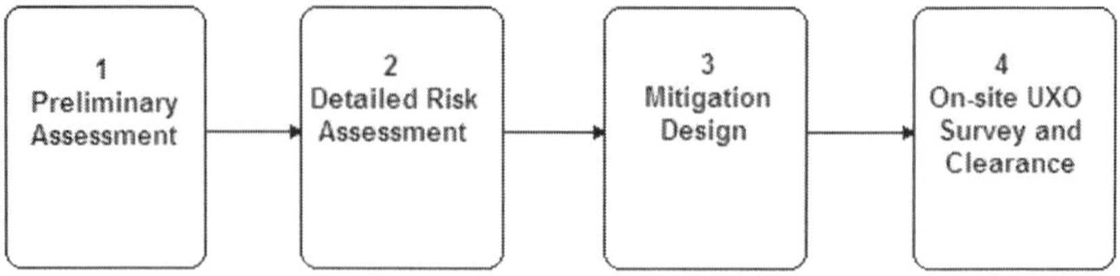

Figure 1. UXO Risk Assessment Procedure from CIRIA C681, showing the four main steps of the CIRIA UXO risk assessment procedure

Since much of the UXO that forms the hazard, thus generating the risk, dates from World War Two (WWII), the requirement to assess UXO risk in some recognised, consistent and professional manner has long been understood, not only by developers but also by designers and principal contractors.

This paper sets out how a procedure for UXO risk assessment was developed and implemented to support ongoing ground investigations being undertaken on behalf of Crossrail Limited for the Crossrail project prior to the CIRIA guide being commissioned, and how this procedure contains a staged assessment compatible with what the CIRIA guide has now established as recommended practice in the UK.

This paper provides a case study in how UXO risk can be assessed and mitigated through a logical process, commencing with data collection and analysis, followed by a review of available risk mitigation techniques, which identifies any limitations that may apply, enabling sensible and effective mitigation measures to be put in place where required. It shows the importance of considering any particular risk not in isolation, but in association with all relevant risks to the site activity, such that appropriate risk mitigation measures are applied where necessary, while ensuring that unnecessary or ineffective measures are not undertaken.

The Crossrail Scheme

Crossrail is a proposed rail link connecting Maidenhead in Berkshire through Central London to Shenfield in Essex and Abbey Wood in Kent. Most of the route across London is to be within twin tunnels, each of which are approximately 6.5m internal diameter. The tunnelled alignment runs for 21km from Royal Oak Portal which is situated west of Paddington, to Pudding Mill Lane Portal at the River Lea (near Stratford) and Plumstead Portal to the east of Woolwich, see Figure 2. As such, the route passes through those areas of London most heavily bombed during WWII, including the City and docklands, as shown in Figure 3.

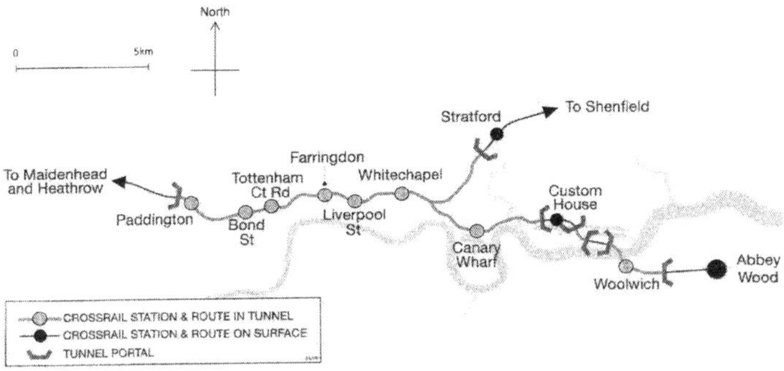

Figure 2. Central section Crossrail route, showing the Crossrail alignment across London from Royal Oak Portal in the west to Pudding Mill Lane and Plumstead Portals in the east.

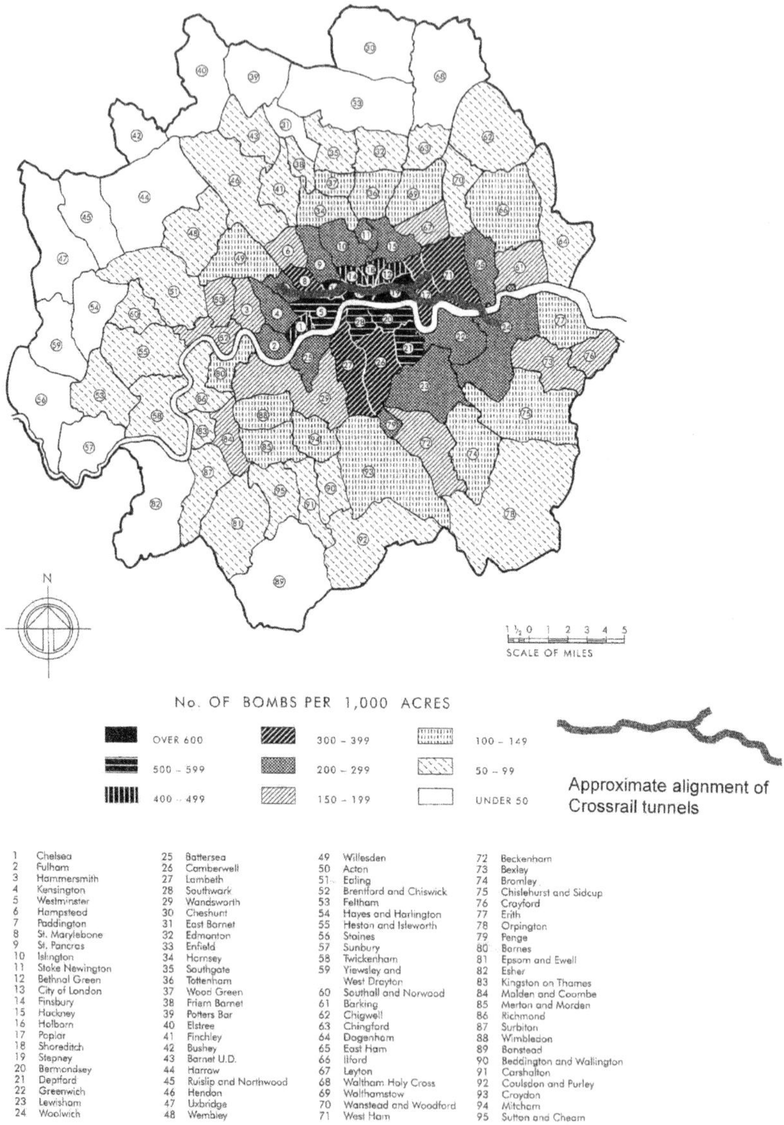

Figure 3. London WW2 Bomb Density plot (After LMA 2005, redrawn from PRO HO193/53), showing the numbers of bombs per 1000 acres to fall on London during WW2, with the Crossrail central (tunnelled section) alignment shown superimposed.

Crossrail ground investigations and the need for the UXO risk assessment procedure

Crossrail was originally proposed in the early 1990's, and initial ground investigations were undertaken at that time across West London, between Paddington and Liverpool Street inclusive. However, no specific measures to address UXO risk to the ground investigations are understood to have been applied at this time. Due to the prevailing economic conditions, Crossrail was then suspended until 2000, when works re-commenced.

Revisions to the scheme alignment saw the route extending further east than had previously been proposed, to approximately the current alignment, and ground investigation works re-commenced in 2001 to obtain geotechnical and geological data, with the initial focus on the eastern section of the route not covered by the earlier investigation works. The first of these new investigations were undertaken through the eastern central portion of the proposed alignment, and were again undertaken without specific UXO risk assessments or mitigation.

However, in 2004, a package of ground investigations was planned through the Silvertown area, between West India Dock on the Isle of Dogs and North Woolwich.
Given the intensive bombing that occurred in this area during WWII, it was clear that some assessment of the UXO risk, and potentially mitigation measures to reduce it, would be required.

The initial response was to commission a desk study report on the UXO risk from one of the established UXO contractors. However, while the specific locations of the proposed site investigation works were provided, the resulting desk study report proved of limited value, since it did little other than confirm that each of the proposed investigation locations lay within an area of high bombing density during WWII, and was accordingly in a 'high risk' area, in relative terms.

As a result of this, it was decided that all intrusive ground investigations undertaken on this package of works were to be preceded by intrusive Cone Penetration Test (CPT) -based magnetometer surveys (mag-cone), to prove the absence of UXO. However, during the course of those works, this approach was found to be untenable, firstly because the use of mag-cone screening added considerably to the direct cost of the works, and to their duration. Second, in many locations, the combination of shallow obstructions within the ground, such as buried concrete and/or brick in the Made Ground, and/or underground services/utilities, required multiple attempts to penetrate into the ground with the mag-cone. Third, the results from the mag-cone were often inconclusive, because of significant, background electromagnetic 'noise' generated by ferrous metal objects within the ground.

It also became clear that by making many more penetrations into the ground than was strictly required for conventional ground investigation, while the UXO risk was reduced slightly, the risk of a utility strike was much increased, and that therefore the overall risk faced by the ground investigation works in some cases was considered to have actually increased by those UXO mitigation measures that were, or might have been, undertaken.

Accordingly, following the end of works review meeting, Crossrail Limited (Cross London Rail Links Limited as was), as client, and Geotechnical Consulting Group (GCG), as Engineer, jointly developed a UXO risk assessment procedure, for application to all intrusive ground investigation works to be undertaken on behalf of the Crossrail project. The procedure, once developed, was reviewed and amended by 6 Alpha Associates, an external UXO consultancy, before being fully implemented.

Developing the Risk Assessment procedure: False Starts

The initial attempt to develop an UXO risk assessment procedure adopted a purely quantitative risk assessment procedure. However, it soon became clear that for UXO assessment work, this approach was unsuitable. The most simple approach of Risk = Probability multiplied by Consequence, does not work well, at least not in a practical and meaningful way, especially for UXO assessment where the probability of encountering and initiating UXO is very low (tending to zero), but the consequence is very severe, including potential multiple deaths (consequence tends to one).
Therefore, attempts were made to produce a procedure that gave much better discrimination: it was considered that it may be possible to determine the approximate statistical chance of encountering UXO if bombing density records were consulted. However, this approach was also considered deeply flawed. One reason for this was that historic bombing density records tend to be recorded and organised along WWII local government administrative boundaries, as illustrated in Figure 3, where the bombing density is related to the boroughs of London in WWII. Clearly though, WWII bombing was not undertaken to a precise or uniform density across a county or borough, with a resultant reduction, or perhaps increase in bomb density at the local authority boundary. Rather, bombing within an administrative boundary is likely to be concentrated around a particular area target. Therefore, the strict employment of average bomb density records for a given area would tend to over-estimate the probability of UXO in some areas, and under-estimate in others (perhaps where the UXO risk is likely to be higher and thus the real risk may have been masked).

Moreover, it was recognised that while the records of bombing of targets in the UK are reasonably good, they are, by the nature of the way in which they were observed and collected (ie in extremely hazardous and uncertain circumstances during or between bombing raids) understandably incomplete. Additionally, there are numerous other 'secondary' sources of UXO that should also be considered, e.g. hidden caches of Army/Home Guard munitions (both conventional and improvised varieties), high explosive pipe mines or cratering charges laid for Home Defence purposes as well as locations where either munitions were manufactured, stockpiled or transported, and may have been discarded or deliberately dumped. None of these factors would have been identified from the bombing density records.

Accordingly, whilst the bombing density information and records were borne in mind, the UXO risk assessment process adopted a qualitative approach.

The Adopted Risk Assessment Procedure

The philosophy behind the adopted qualitative risk assessment procedure was to look at historic land use in conjunction with bombing records for a particular site, to identify where there was a credible risk of UXO remaining on site and therefore of potentially being encountered. This risk was then reviewed relative to the site conditions and specific ground investigation technique(s), as well as the other known or potential underground hazards on the site, in order to determine what practical UXO risk mitigation measures could be undertaken that would be effective in reducing the holistic risk to the proposed works: for example, undertaking additional intrusive works to mitigate a low UXO risk may significantly increase the already high risk of striking uncharted services, thus actually increasing the overall ground investigation risk. The stages of the qualitative risk assessment procedure are illustrated in Figure 4 and are described in detail below.

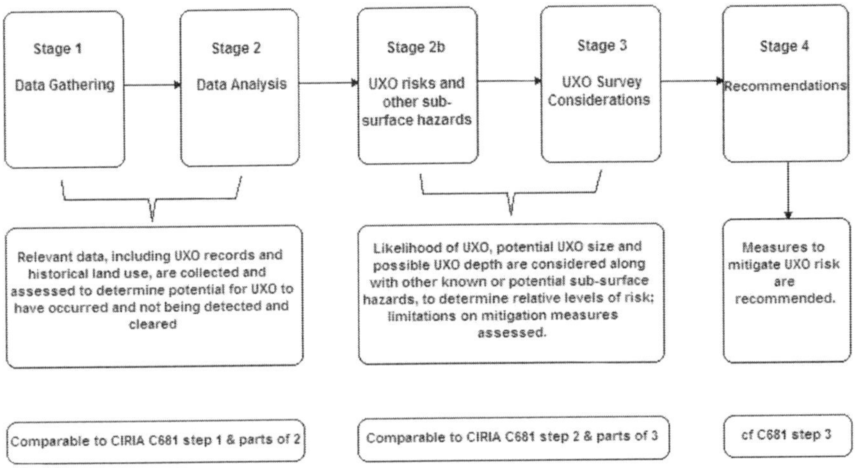

Figure 4. Crossrail qualitative UXO risk assessment procedure, showing the five main stages of the Crossrail UXO risk assessment procedure, and how these relate to the steps of the CIRIA guide C681.

Known UXO records

Stage 1 of the procedure started with the collection and assessment of the known bomb records. Details of average bombing densities were gathered (which were obtained from one of the established UXO consultants, but they are also widely published), along with bomb damage maps (LMA, 2005) while UXO records of so called abandoned bombs (i.e. UXO reported, but not cleared or discredited) for the London area were obtained from a number of sources, most notably from a list published in response to a parliamentary question (since reproduced in CIRIA C681), and from direct requests made to the archives of 33 Engineer Regiment (EOD), Royal Engineers. The Royal Engineers' records also provide details of those UXO that were located and cleared both during and post WWII, thus improving the understanding of the severity of bombing in any area under review.

For the whole Crossrail route, a UXO desk study had been commissioned from one of the established UXO consultants who employed many of the same primary data sources, supplemented by their own records of munition production facilities, military airfields, anti-aircraft artillery battery locations, etc.

Review historical land use

During stage 2 of the procedure, the study of known or likely bomb impacts was supplemented by reviewing the historic Ordnance Survey (O/S) mapping sheets for the area under review (at 1:2500 and 1:1250 scales, where they were obtainable). The purpose of the O/S map reviews was firstly to look for signs of bomb damage, by comparing land use prior to WWII to that shortly after it. For example, large areas of cleared land where formerly there was terraced housing provided an obvious indication of possible bomb damage. Notations of ruin, or the replacement of terraced properties by small detached properties (most likely post-war prefabricated housing), provided a similar indication of potential bomb damage, and therefore the likelihood of encountering potential unexploded bombs (UXBs).

The map review also provided an indication of land use during the war, giving an indication of likely targets in the area and enabling the potential for UXO to have remained undetected, or been detected but not cleared. Where land was undeveloped and not overlooked (i.e. by casual observers or neighbours who would have perhaps noticed an UXB), the potential for a UXB to fall unnoticed was deemed a higher probability.

Conversely, it was assumed in the methodology that those areas that may have been bombed which were continuously occupied and/or involved in processes important to the war effort, such as railways or factories, would have had fire watchers and other regular inspections, so it is unlikely that UXO would have fallen in these locations undetected, nor would the authorities have left known UXO in such locations.

Finally, the map review was extended to include all available mapping to present day, to enable an assessment of any post-war redevelopment of the site. Extensive, deep ground works in particular would provide a significant opportunity for UXO to have been detected, if present, while reductions or increases in ground level may reduce the volume of soil within which the risk of UXO need be considered.

This process enabled a qualitative assessment of UXB risk for the relatively small areas that constituted individual ground investigation locations, being routinely applied to areas of a few tens of square metres. The resulting assessment provided an indication of the relative UXO risks, rather than an absolute number, and the requirements to mitigate those risks were then dependent on a further stage of assessment, stage 2b, which involved the following process.

Assessment of potential for mitigation

The geology of the site was reviewed, to determine the maximum likely bomb penetration depth for the site, and also to assess the likely maximum horizontal movement of an air-dropped bomb through the ground (bombs dropped by aircraft have a forward velocity as well as a vertical velocity, so tend to move downwards and laterally through the ground, if they do not detonate on impact). Unexploded bombs also commonly conform to a so called "j-curve" underground, often coming to rest at a depth that is shallower than their theoretical (or actual) maximum; see Figure 5. For more details of the sub-surface behaviour of bombs, see for example Smith and Hetherington (1994). Typically, the maximum horizontal distance between the point of entry of a bomb into the ground and its final position is taken as one third its maximum vertical penetration into the ground. The maximum penetration of a bomb into the soil is highly dependent on a range of factors including geology, the presence of surface structures, bomb type (including shape and mass), height and speed at which it was released, etc. However, maximum penetration is generally held to be no greater than about 20m (in most situations, very much less). The maximum horizontal distance a bomb is likely to move through soil is therefore typically less than 7m.

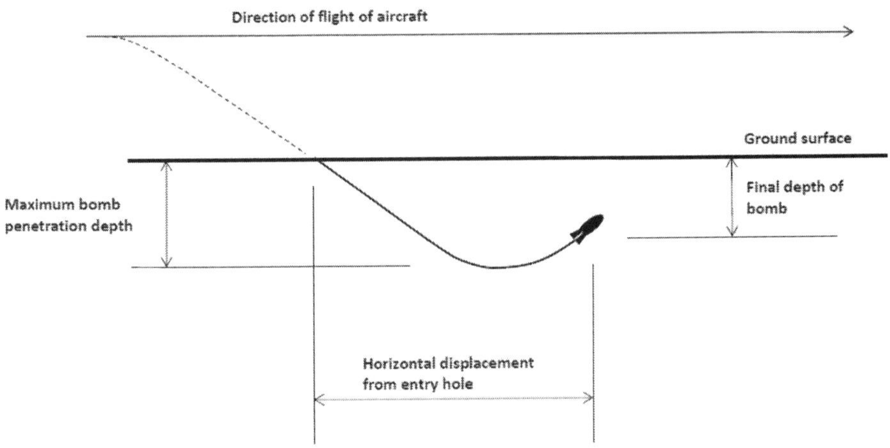

Figure 5. Sub-surface trajectory of a bomb: The 'J-Curve', illustrating the typical path of a bomb passing through soil, and demonstrating both vertical and horizontal displacement from the initial point of impact.

The mapping review included consideration of land use from the point of assessing whether munitions may have been produced, stored, transported or used (e.g. on firing ranges) either on or about the site under review. Consideration was also given at this stage to any known records of services or structures at the site, since they represent non-UXO hazards that may have significant associated risks.

Other associated site factors that are likely to impact on UXO mitigation measures were then considered within stage 3 of the risk assessment: generally, active mitigation of UXO risk on site requires the use of a magnetometer, employed either non-intrusively from the surface (where they can be easily affected by shallow buried ferrous metal contamination), or intrusively into the ground (where they often can deliver a higher degree of resolution). The purpose of mitigation is not to find UXO *per se*, rather it is to show that a location is clear of UXO to permit intrusive ground investigations to proceed without risk of encountering it. However, there is no clear benefit in attempting to use a UXO risk mitigation technique that will be either ineffective, or that results in an overall increase in risk.

Mitigation limitations

During the development of this risk assessment procedure, a trial was undertaken employing non-intrusive magnetometry. The trial site was at a proposed Crossrail borehole location in east London, which was a grass-covered open space (though not a 'green-field' site), at which the near-surface geology consisted of Made Ground and reworked soil. The trial used a FEREX 4.021 system, and indicated that even in this relatively benign (for London) environment, non-intrusive magnetometry was ineffective, since the Made Ground resulted in numerous ferrous contacts being recorded. While the proposed borehole at this site was later cancelled for unrelated reasons, another borehole completed nearby indicated that the Made Ground extended to approximately 6m below ground level, and was composed primarily of sandy clay, but with much demolition waste, including brick fragments. The presence of greater thicknesses of Made Ground (particularly where there was red brick contamination, because "fired-clay" also interferes adversely with magnetometers), sub-surface services and utilities, nearby metallic or reinforced concrete structures (including sub-surface rail tunnels, sheet piling etc) would all cause additional "noise" potentially masking any UXO present.

Moreover, the maximum depth to which surface magnetometry can detect ferrous objects, while variable according to such factors as the mass of the object and the angle of its orientation in the earth, is quite limited, and for most ground conditions is insufficient to "clear" a borehole location to the maximum credible depth of penetration of even a relatively small UXO. The practical limit of detection will rarely exceed 3m below ground surface, and will typically be less.

As a result of this trial, surface based magnetometer surveys, while always reviewed as an option, were considered to be generally unsuitable for all the Crossrail ground investigation locations. However, for shallow intrusive works on green field sites such techniques may be appropriate; expert guidance is often required in order to select the most appropriate option.

Intrusive magnetometer surveys can be undertaken either using so called mag-cone CPT, or as a down-hole magnetometer, where the instrumented probe is lowered within a borehole. Both techniques are able to clear the full depth of credible UXO penetration, but have other limitations: for example, mag-cone CPTs undertaken purely for UXO mitigation increase the number of penetrations into the ground, thus increasing the risk of an (uncharted, especially) service/utility strike. Additionally, mag-cones can be stopped by hard ground and often at a shallower depth than that associated with maximum bomb penetration. Both intrusive magnetometer techniques also provide only a limited radius of detection for ferrous objects. The maximum distance at which a UXO may be detected, whether horizontally or 'ahead' of the instrument at some depth below it, depends on the same factors as for non-intrusive methods, including type of instrument, ground conditions, other nearby ferrous objects, and size of target to be detected. Ranges of up to 6m may be achieved for larger UXO, though in practice, maximum detection radius for UXO is typically 1m to 3m depending on minimum size of target to be detected.

Down-hole magnetometers require non-magnetic casings to support the sides of the preformed borehole. Where down-hole surveys are undertaken, conventional steel casing can be used during boring, but has to be withdrawn when the magnetometer is lowered down-hole, leading to a possibility of collapse in the borehole; where the ground is self supporting this is not such a problem. Alternatively, internal plastic sleeving can sometimes be temporarily emplaced before the steel casing is partially withdrawn, though partial borehole collapse may still occur where the borehole walls are unstable. The use of non-magnetic casings to support the borehole during drilling is feasible, which avoids the risk of borehole collapse, but because it is a non-standard technique, the additional costs involved make this option too expensive for routine works. When employing a down-hole magnetometer, the process generally involves progressing the borehole in short stages. After each stage, the magnetometer is lowered down the borehole to confirm that the soil beneath the base of the borehole is free from ferrous objects and can be advanced without risk from UXO. The instrument is then withdrawn from the hole, and the borehole can then be progressed another step. The maximum depth bored in each step is a function of the detection radius for UXO; 1m steps are typical.

Thus, in reviewing potential UXO mitigation techniques, consideration needs to be given to the expected soil conditions such as whether there will be man-made or natural ferrous content, how stable the soil will be and whether it will stand uncased, etc. Consideration also needs to be given to other non-UXO hazards in the area e.g.: how many underground service/utility runs are know to be present, as well as how likely are uncharted or inaccurately recorded services to be encountered.

Mitigation recommendations

Having assessed all relevant factors as described above, the risk assessment concluded with Stage 4, the recommended mitigation measures. Six broad action levels were available for selection within the assessment process:

(i) **No Action**. Where the UXO risk was assessed to be minimal, no mitigation was justified.

(ii) **Relocate works**. Where the UXO risk was not minimal, the risk might be mitigated by relocating the ground investigation works to an area where there was a lower assessed UXO risk. (Relocation in this instance refers to repositioning works a few tens of metres away). However, the location of ground investigation works was normally driven by the technical requirement for information, limited by the practicalities of obtaining physical and third party consent for access to locations, and in practice the option to relocate works to an area of lower UXO risk never occurred during the project.

(iii) **No Action, non-trivial UXO risk**. Where the UXO risk was assessed as low, but not minimal, the recommended mitigation might still be to take no action other than toolbox briefings for the site staff. This situation would arise where the consideration of on-site conditions indicated that more active mitigation measures would be ineffectual, or would result in an increase in the overall risk (for example due to increased potential for service strikes). The content of the toolbox talk would be the responsibility of the ground investigation contractor undertaking the investigative works, and would be expected to be delivered by a UXO specialist. The talk would usually include a brief description of the UXO type(s) that might be locally encountered, and a warning concerning unexpected obstructions, particularly when boring in what was believed to be natural ground. The toolbox talk should include the depth range over which the UXO risk was considered to occur.

(iv) **Non-Intrusive Mitigation**. Where the assessed risk from UXO was deemed low to moderate, and geological and manmade conditions were such that surface magnetometry was felt to be practical, non-intrusive mitigation would be recommended. For the reasons articulated earlier, this technique was not used at any of the Crossrail ground investigation locations where work has been completed to date.

(v) **Intrusive Magnetometry**. Where the assessed risk from UXO was deemed low to moderate, and non-intrusive techniques where considered impractical, intrusive mitigation was undertaken. The decision to use either mag-cone CPT or down-hole magnetometers was determined on a case-by-case basis.

(vi) **High Risk**: Cancel intrusive investigation. Where the UXO risk assessment identifies an area as high risk, the recommendation was not to progress with proposed investigation works. Such a situation would typically only occur where an abandoned bomb was recorded in very close proximity, or where some other particularly unusual circumstances suggest that there was a significant probability of UXO being on the site (for example, at one site reviewed, 17 number unexploded bombs were dealt with during the war from a single incident; the possibility of other UXB from the same 'stick' of bombs being present, but having not been detected subsequently, was considered high enough to classify this as a high risk area). Where high risk areas were identified, these would be recorded in Crossrail's risk register, and expert UXO advice would be sought before any works were undertaken at that location.

While the assessment procedure was undertaken by the geotechnical engineers responsible for developing the ground investigations, the process required that the completed assessment, including the final recommendations, were reviewed by an external, UXO consultant (in this case 6 Alpha Associates Ltd), whose sign off was required before the assessments could be issued to the ground investigation contractor. This external third party review was to not only to confirm that the process had been correctly followed, but also that the recommendations made were justified in accordance with the evidence presented.

Upon completion of this review process, the completed assessments where then issued to the ground investigation contractor, in accordance with the obligations under the Construction (Design and Management) Regulations 2007 to provide project specific information on significant risks. Only the completed assessments where issued, not the full desk study information utilised in generating the assessments, though this data was available for inspection by the ground investigation contractor if requested.

Notwithstanding the recommendations for UXO risk mitigation measures made, the ground investigation contractor remained responsible for their own safe method of working on site, and could challenge the recommendations of the UXO risk assessment procedure if they believed the recommendations to be inadequate. However, to date, this has not occurred.

Effectiveness of the Crossrail UXO Risk Assessment Procedure.

The Crossrail UXO risk assessment procedure for ground investigations was developed in 2004, and used from November that year on all Crossrail ground investigations undertaken since: in excess of six hundred borehole locations were risk assessed, in locations from Royal Oak and Old Oak Common (to the west of Paddington Station) through the West End and City and on to Pudding Mill Lane in the east, and through the Isle of Dogs and Woolwich to Plumstead in the south east, as well as a number of smaller ancillary sites outside Greater London.

The application of the risk assessment procedure has enabled the vast majority of these works to go ahead without specific UXO mitigation measures, beyond a simple site brief. No confirmed UXO were encountered in any of these locations, though a single piece of unidentified metal was recovered from one borehole near Plumstead. In accordance with their tool box brief, the work crew immediately stopped work and reported the find to the Engineer. While it was initially considered unlikely that this piece of metal was associated with UXO, expert advice was taken which indicated that it could possibly be a fragment of bomb tail fin. Accordingly, the borehole was immediately abandoned and relocated, and the possibility of UXO presence recorded on Crossrail's risk register.

Of the locations that were identified by the assessment procedure as needing specific UXO risk mitigation, the majority were in the docklands area, and all were mitigated using intrusive magnetometer techniques. As at the time of writing, 81 locations have been UXO risk mitigated through such measures, of which 42 were completed using mag-cone, and 39 locations made use of down-hole magnetometers. Of these 81 locations, 35 of the mag-cones and 5 of the downhole probes were undertaken at overwater locations in West India Dock and the River Lea, where the borehole and associated UXO risk mitigation work was undertaken from a work platform or existing structure built over water.

No confirmed UXO were recovered or detected at any location. In two boreholes, ferrous objects were detected at depth, and recorded as possible UXO, resulting in the ground investigation location being relocated *circa* 10m away. In one of these positions, it was suspected that the object detected was a redundant retaining wall tie, but records were not available to confirm this. The nature of the object in the second case remains unknown. In both cases, the unknown ferrous objects were recorded in the Crossrail risk register.

Where the possibility of UXO has been recorded in the Crossrail risk register, this information will be expertly assessed, and appropriate target investigation undertaken, as and when deemed necessary, to permit safe progression of the construction phase of the project. However, such works are well beyond routine ground investigations, and will require dedicated specialist UXO risk mitigation works.

Comparison to CIRIA C681

Figure 1 illustrates the four basic steps for UXO risk assessment and risk mitigation as now defined in CIRIA C681. Figure 4 shows the stages of the Crossrail UXO risk assessment procedure, and indicates how the activities required for each of these stages equates to a step of the CIRIA procedure.

The initial data gathering and analysis stages (stages 1 and 2) of the Crossrail procedure matches closely the CIRIA step 1, as it is in these stages that the potential for UXO to be present on the site is considered and hence the preliminary assessment of the UXO risk is made.

The second step of the CIRIA procedure requires a detailed risk assessment, in which the UXO risk at a particular site is quantified relative to the proposed works at that site. This requires consideration of the probable source of ordnance, ground conditions and penetration depths into the soil, etc, and is comparable to the Crossrail stage 2b, consideration of UXO risk relative to other ground risks and stage 3, UXO survey considerations. The two procedures do not perfectly match; for example, within the CIRIA procedure, ground conditions are not considered until this detailed assessment in step 2, whereas within the Crossrail procedure, these are reviewed earlier in the data analysis stage. Similarly, the Crossrail procedure considers the factors that might influence the feasibility of mitigation methods during the survey considerations stage, but such factors are considered within the CIRIA procedure during step 3, risk mitigation. Nevertheless, there is a good degree of agreement between the two procedures.

Within the CIRIA guide, step 3 is to assess the options for mitigating UXO risk and develop an appropriate mitigation plan. This corresponds to the Crossrail procedure stage 3 and stage 4, where the options for mitigation are considered, and the most appropriate mitigation method for the site is recommended. The Crossrail procedure recommendations fall slightly short of a true mitigation plan, since the recommendation is for the most appropriate general mitigation technique to be applied, but does not detail fully the exact locations and method of mitigation / equipment type to be used.

The CIRIA procedure is completed by step 4, implementation of mitigation measures, including disposal of UXO if encountered and verification. This step lies outside of the Crossrail procedure. However, it is the ground investigation contractor's duty to progress the works in a safe manner, and they were required to address the recommendations of the UXO risk assessment within their site health and safety plan.

Notably absent from the Crossrail procedure was any specific plans for clearance of UXO if encountered. While the site health and safety plans included emergency action plans to enable site works to be shut down and expert UXO advice obtained if suspected UXO were encountered, the purpose of the risk assessment procedure was to allow ground investigation works to progress safely, not to undertake target investigation of suspected UXO followed by disposal. If actual UXO had been encountered, it was accepted that site operations would be seriously disrupted, but the emergency action plan would have ensured the site was safely shut down, and it was felt that the application of the Crossrail UXO risk procedure reduced the probability of actual UXO being encountered during the ground investigation works to a negligibly small level.

It is therefore clear that the Crossrail procedure, developed five years prior to the publication of CIRIA C681, involves the same basic steps of date collection, preliminary assessment, detailed (site-specific) assessment, consideration of appropriate mitigation options and recommendations for mitigation which form the core of a detailed risk mitigation plan. While the matching of the Crossrail procedure stages to the CIRIA procedures steps is imperfect, the similarities between the two processes are clear.

Conclusions

In response to a perceived "unmanaged risk" being posed by UXO to proposed ground investigation works, Crossrail developed a procedure for assessing UXO risk to the ground investigations. This procedure involved a review of wartime records and reported UXO/abandoned bombs, as well as taking into account site geology and local conditions, to enable a holistic assessment of UXO risk to be generated. It enabled areas with low or minimal UXO risk to be identified, so that investigation works could be undertaken without expensive, time consuming and unnecessary on-site mitigation.

It also enabled areas of moderate UXO risk to be identified, and appropriate site mitigation measures to be undertaken. Similarly, it enabled areas of high UXO risk to be identified, where works cannot progress without expert help and advice.

As such, it is consistent in philosophy with the recommended practice set out in CIRIA C681, and demonstrates that a safe and effective UXO risk assessment procedure can be implemented without causing excessive cost or delay to a project.

This paper has illustrated how the Crossrail procedure was developed, and details the stages of the assessment procedure. It shows how an initially unquantified risk was assessed and managed through the application of a logical assessment process that addressed the specific work activity and considered risk to this activity holistically, rather than focus solely on a single risk which was not necessarily the most significant. In this way, the true aim of any risk assessment process – to make the work activity safer – was achieved.

References

CIRIA. *Unexploded Ordnance (UXO): A guide for the construction industry*. CIRIA. London. 2009. Report C681.

The Construction (Design and Management) Regulations, The Stationery Office, 2007. Statutory Instruments 2007 No.320, Health and Safety.

LMA. *The London County Council Bomb Damage Maps, 1939-45.* Publication No.164. The London Topographical Society / The London Metropolitan Archive. 2005.

PRO HO193/53: Bomb Census maps produced by the Research and Experiments Division of the Ministry of Home Security, held at the National Archives, Kew.

Smith, P.D. and Hetherington, J.G.. *Blast and ballistic loading of structures*. Butterworth-Heinemann. Oxford. 1994

Crossrail Project: Infrastructure design and construction
ISBN 978-0-7277-6078-4

doi: 10.1680/cpid.60784.187

Comparison of Drilling Methods used for TaM installations in London Clay

P. Pérez Lupi MSc, Bam Ferrovial Kier JV
E. Lewis MEng, Crossrail Ltd
J. Adetoye BEng, Bam Ferrovial Kier JV
X. Gonzalez TEd, Keller Bam Ritchies
N. Gallego Ramirez MSc, Bam Ferrovial Kier JV
Z. Munn BEng, Keller Getec ITMSoil
D. Campos MSc, Bam Ferrovial Kier JV

Abstract

This paper seeks to compare and contrast observations made of different techniques for drilling horizontal holes within London Clay, and to make recommendations as to how subsequent drilling operations from grout shafts should be undertaken in the future.
The drilling operations discussed in this paper were part of the compensation grouting works for the Crossrail project. The two drilling methods were trialled adjacently, in the first two grout shafts in the area of Bond Street Station.

The first method described is the partially cased, water-flush rotary method. This method of drilling has been previously used on similar projects in central London. The use of this method is also implied by clauses within the works specification, as it has been the standard method of drilling for similar projects in the United Kingdom. The water flush rotary method is known to produce heave while drilling to install the tube à manchettes pipes

The second drilling method discussed is the partially cased, auger rotary method. This method is commonly used in Germany but due to the long horizontal drilling lengths, this has not been used previously in the United Kingdom. This drilling method, in contrast with the water flush method, generates some minor settlement in the drilling area.

This paper also reviews other aspects of the works including; drilling accuracy, asset management, safety, spoil disposal and the control of ground movements. The control of ground movements is based on the monitoring system which has been installed for the project works and has demonstrated reliability and providing a short response time between detecting settlement and user interaction. Monitoring is primarily comprised of Hydrostatic Levelling Cells installed in the basements of the overlying structures.

These findings illustrated by site measurements following the initial trial in these two grout shafts, have been conclusive for the choice of the dry auger method as the preferred method of drilling for the remaining grout shafts.

Acronyms Used:

TaM – Tube- à- Manchettes
SCL - Sprayed Concrete Lining
TBM - Tunnel Boring Machine
I&M – Instrumentation and Monitoring
HLC – Hydrostatic Levelling Cell

1. Introduction

Horizontal Drilling operations to install TaMs (Tube à Manchettes) pipes to facilitate compensation grouting activities for mitigation measures required for Crossrail were undertaken from vertical 4.5 m diameter shafts in Central London. It is known that the drilling operations to install TaMs is a source of settlement and heave and to that end, this paper compares the methods and observations undertaken at the commencement of drilling operations to ensure that the remainder of the work could be completed using a method that was efficient and suitable for the site conditions.

2. Scope

Crossrail contract C300/410, the Western Tunnels section of Crossrail Central, consists of the construction of 6 km of twin bored running tunnels from Royal Oak portal through to Farringdon Station, a crossover tunnel at Fisher Street, and the station and platform tunnels at Bond Street and Tottenham Court Road stations, where connecting shafts, station concourses and adits, are being built by means of excavation and SCL (sprayed concrete lining) in between the Western and Eastern station boxes.

There have been several major tunnelling projects in London in recent years including the Jubilee Line extension, the Channel Tunnel Rail Link and the Heathrow Express. As a result, there is a good understanding of how London Clay reacts to the excavation of tunnels and underground structures. As part of the detailed design, ground movement predictions were calculated from settlement profiles expected from each stage of construction. Short term ground movement predictions are based on 'Green Field' conditions as the stiffness of the buildings and the effect of the stiffness of foundations on the soil are assumed negligible making the assessments conservative.

From the settlement contours calculated, damage assessments are then made to ascertain the potential damage that the effects of tunnelling could have on the aboveground and underground infrastructure, in order to then design mitigation measures as necessary. Building damage is categorised using the tensile strain approach.

As a measure to mitigate the ground movement and the potential damage it could cause, an area in excess of 50'000 m^2 was specified to be controlled by compensation grouting in selected areas around the stations, for the passage of the tunnel boring machines, the SCL tunnelling works, and station box excavations. The extent of compensation grouting for both areas is limited to the 10 mm settlement contour and the duration of the treatment is linked to the works and until the rate of settlement is less than 2 mm per year .

3. Site Constraints

The Compensation Grouting shafts within the C300/410 contract are located in areas of Bond Street, Tottenham Court Road and Holborn. These grout shafts are located in major urban environments which increases the difficulties for the site operations. The working areas are reduced to a minimum in order to reduce the impact in the adjacent third parties. The reduction of the working area has had an impact on the diameter of the grouting shafts which have been reduced to 4.5 meters instead of the usual 6m diameters required. The average depth of the shafts is 15 meters.

Drilling, spoil handling and disposal, limited room available, access restrictions and environmental impacts are some of the key issues to consider when determining the best method to drill TaMs in shafts located in the central London.

4. Compensation Grouting

Compensation Grouting is a method of controlling ground movements in clay. It has been successfully used on projects such as the Waterloo Victory Arch, Jubilee Line Extension and the Heathrow Express. The process of compensation grouting involves the injection of grout into the clay at pressure, in between the level of infrastructure that is being protected and the cause of ground settlement i.e. tunnelling. The grout, under pressure, is injected via small ports at set intervals within the tube à manchettes, this grout will find the natural occurring fissures in the clay and expand them, minimising the settlement and potentially causing heave, depending on the volumes injected. The grouting operations are controlled from a central grout shaft, where radial holes are drilled and TaMs are installed and grouted in position.

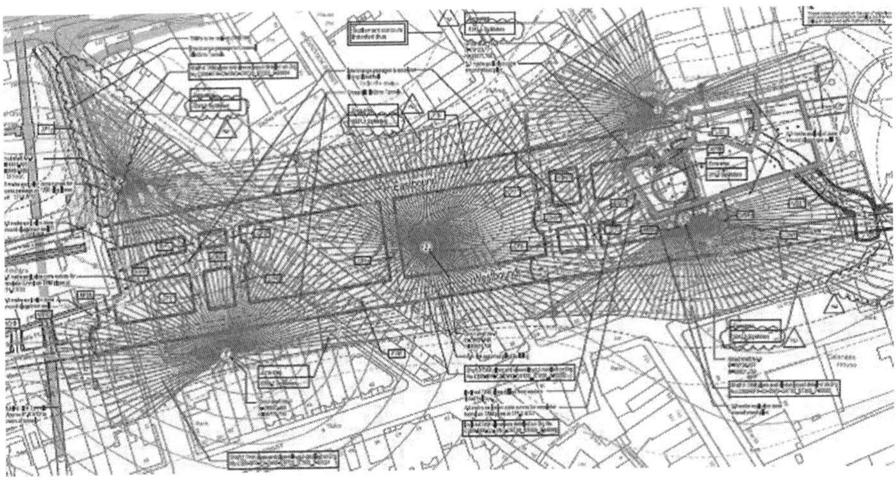

Figure 1: Bond Street Station TaM Layout

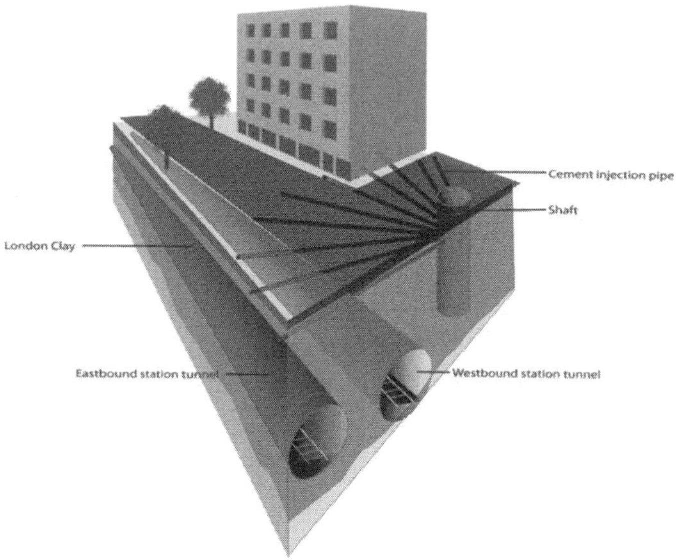

Figure 2: 3D section showing the relation of TaMs, in an urban environment

The TaMs comprise 52 mm diameter steel tubes typically at 2 m lengths, which are screwed together and inserted into the boreholes to the required lengths. TaM lengths on the Crossrail project are typically 50 to 90 m and contain small ports, drilled in the intrados of the steel tube at regular intervals, typically every 500 mm. The ports are covered with a thick rubber membrane that lifts with the pressure of the grout being injected and closes once the grout has been injected, working like a one way valve. During the grouting operations, double packers are inserted into the TaMs and expanded, so that any individual port on a TaM can be grouted from, accurately targeting the settlement real-time, as identified by the monitoring system. Multiple phases of grouting can be carried out from the TaMs, helping to mitigate the effect of tunnelling activities until the rate of settlement has decreased significantly.

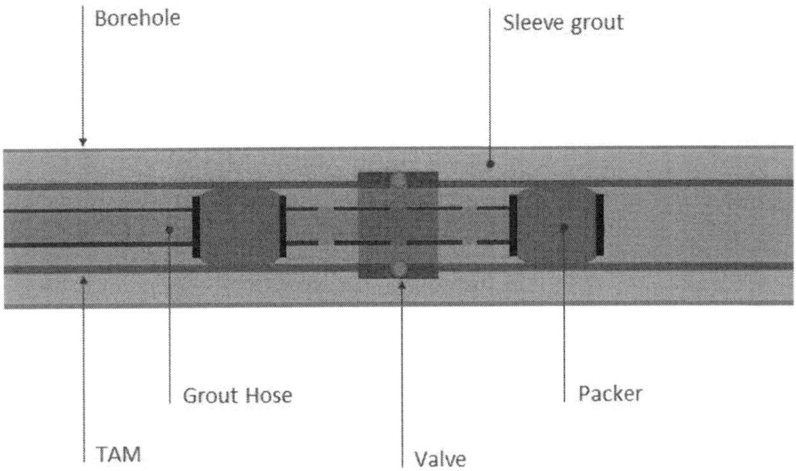

Figure 3: Cross section of TaM with inflation packer

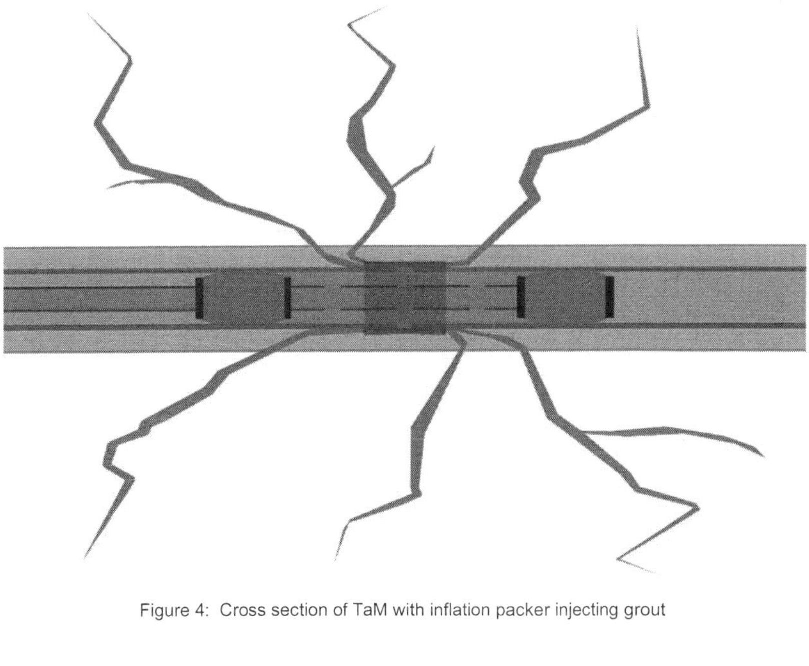

Figure 4: Cross section of TaM with inflation packer injecting grout

Figure 5: Photograph of inflation packer

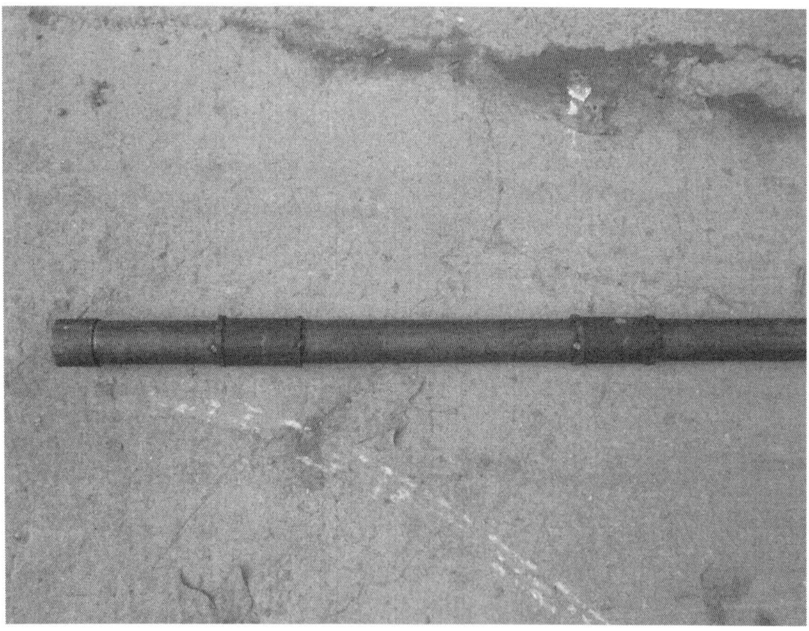

Figure 6: Photograph of tube à manchettes pipe

With a total of 45'000 m of TaMs to drill and install prior to the construction and excavation of the tunnels and stations, a total of 13 grout shafts are being used on contract C300/410.

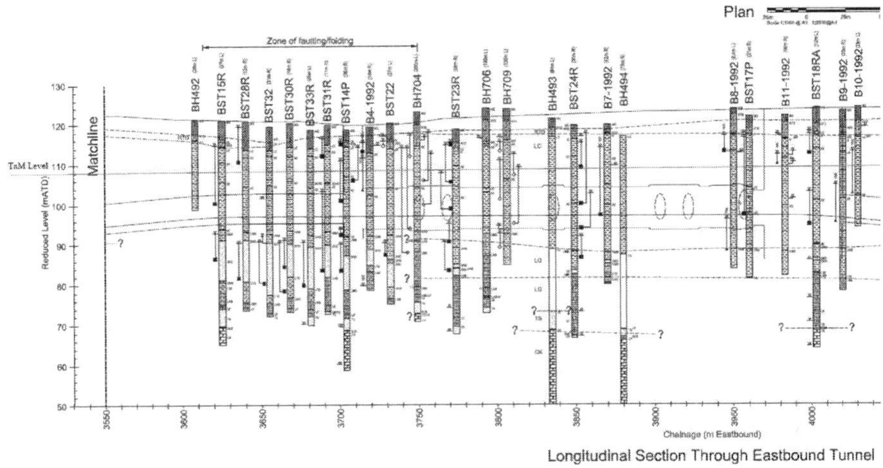

Figure 7: Profile of the clay in relation to the structures and the materials overlying the clay

The drilling of the TAMs required careful control of the placing of the boreholes, with the drilling accuracy and location required to be reported. The accuracy and location was determined by using a Reflex Maxibor instrument that calculates the spatial coordinates along the drill hole path based on optical measurements of direction changes and gravimetric measurements of dip changes.

5. Instrumentation and Monitoring

In order to monitor the ground movements that occur from the TaM drilling, and to subsequently control and manage the compensation grouting operations during the main works a large amount of Instrumentation and Monitoring (I&M) had been specified and installed in the site area.

The I&M used, varies in function from asset protection to ground monitoring. The measurement of heave and settlement during compensation grouting is primarily measured using a hydrostatic levelling cell (HLC) system due to its accuracy and data transfer capabilities, there are also secondary monitoring instruments used to validate and back up the HLCs. Other methods of measurement for the monitoring of the compensation grouting works include precise level points installed in pavements and BRE studs, which are installed on buildings. These points are then read using digital levels. Further monitoring methods include Automated Robotic Total Stations which read prisms installed on the neighbouring building facades. The monitoring management plan for each location provides the primary method of measurement and the associated trigger values.

The hydrostatic levelling system's pressure cells, measure pressure differences versus a reference measuring point. The pressure changes recorded can then be converted into height differences. The reference level is defined by the liquid horizon in a header tank (liquid reservoir). All measuring points are connected together by a water tube which also connects to the header tank and thus the reference level. The sensors utilised are capacitive pressure devices, which are stable and reliable. The analogue signals of the pressure devices are captured and converted into measured values which are displayed in the monitoring system. The cells are installed typically within basements on structural walls or beams. The instruments have a vertical measuring range of 500 mm and a resolution of 0.02 mm. The use of over 800 cells in the Bond Street area requires a complex reference and management system, where cells are read every 15minutes, though they are able to be read every 30 seconds if required.

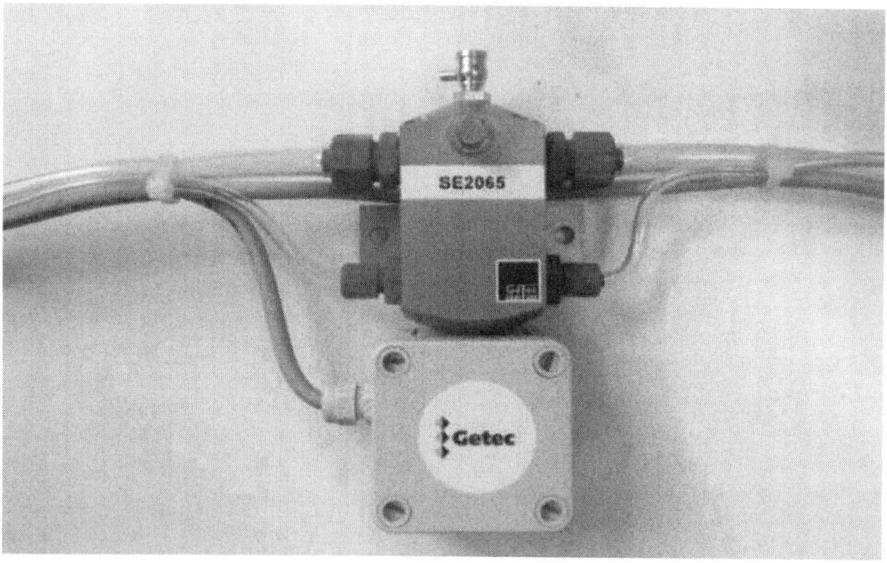

Figure 8: Hydrostatic Cell measurements over time

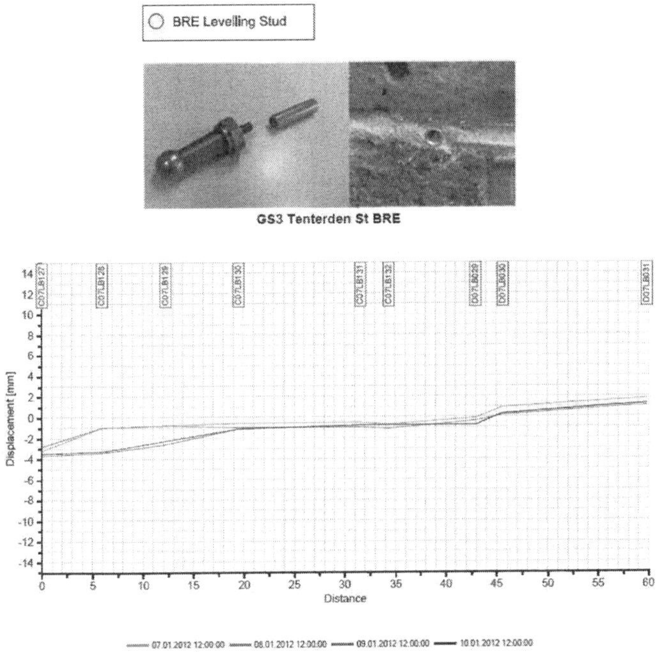

Figure 9: BRE Levelling Stud- Plot BRE

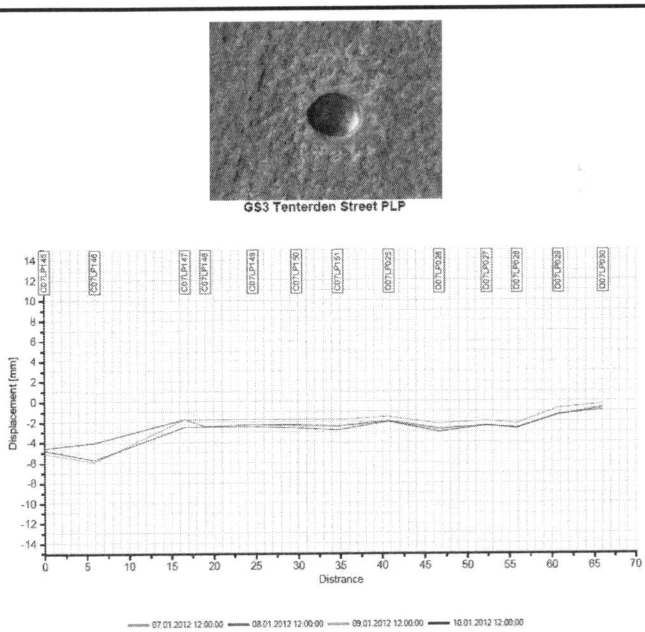

Figure 10: PLP Precise Levelling Point- Plot PLP

6. Comparison of Drilling Methods

Two methods of drilling have been implemented at Bond Street, water flush drilling and auger drilling.

Partially cased, water-flush rotary method – Bond St Shaft 3

Drilling operations commenced in Bond Street from grout shaft 3, using drilling methods used on similar projects from the 1990's, and also as implied by the specification. In variance to these past projects, temporary drill casings were generally installed in accordance with the specification: 50% of the bore for holes of up to 40 m, and all but the last 20 m of longer holes. This varied occasionally with a small number of holes being fully cased (for heave mitigation reasons). Due to the length of casing required, drilling with a polymer flush was initially used, but quickly suspended due to the associated operational difficulties coupled with the lack of available compound space to effectively deal with the flush arising's.

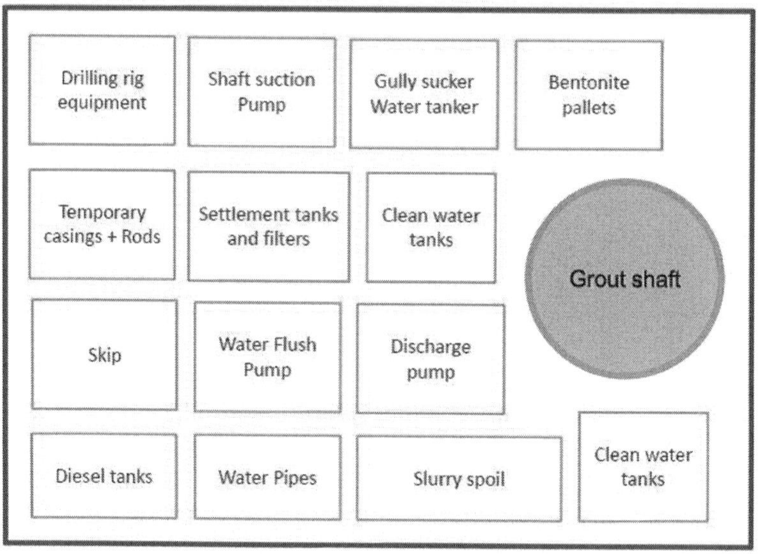

Figure 11: Schematic site layout for the water flush TaM drilling method

Partially cased, rotary auger method – Bond St Shaft 5

Grout shaft 5, was drilled using the partially cased, rotary auger method. The rig was fitted with a double rotary head, and drilling commenced with an air flush to assist with the installation of the drill casing, as often used in central Europe. Following an initial heave issue, drilling continued successfully without the associated air flush. As a result of omitting the air flush, installation of the drill casings was increasingly difficult and was short of the specification requirements, but bore hole deviations were not significantly affected with the reduced casings.

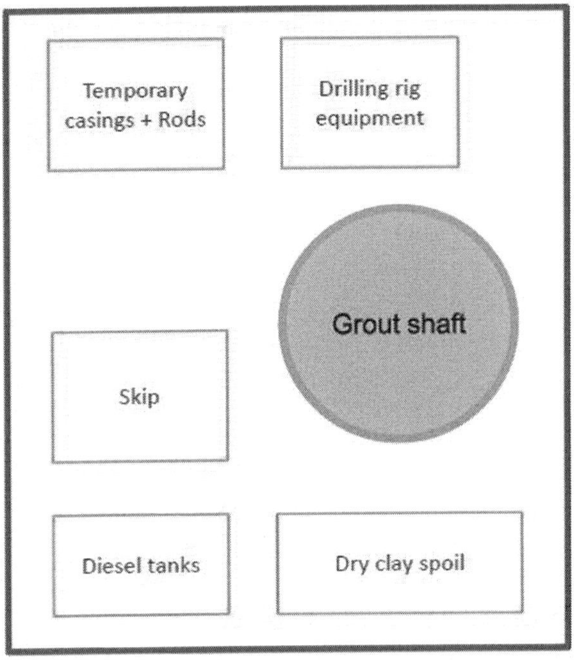

Figure 12: Schematic site layout for rotary auger TaM drilling

Quality / Drilling accuracy

There are significant tolerances placed on the installed location of the TaMs. Exclusion zones have been specified above the crown of the tunnels and below the top level of the clay. There are also maximum spacing tolerances between adjacent TaMs to ensure complete coverage of the identified grouting zones.

The layout of the TaMs means that TaM pipes are being installed to lengths of up to 100 m. Drilling to these lengths means that holes may deviate in location, and in some cases unidentified obstructions are encountered

The accuracy and location of the as built TaM locations are checked using a Reflex Maxibor instrument that calculates the spatial coordinates along the drill hole path based on optical measurements of directional changes and gravimetric measurements of dip changes.

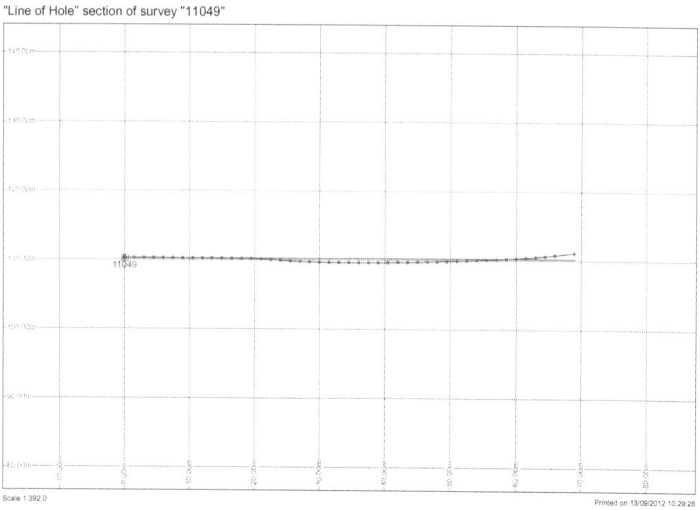

Figure 13: As built TAM showing vertical deviation

The works specification required that the chosen drilling method should have the casing advanced as far as necessary to ensure that the following tolerances on elevation were achieved: ±250 mm at 30 m distance from the shaft wall, ±500 mm at 50 m and ±750 mm at 65 m from the shaft. The control of drilling direction had to also realise an accuracy of 0.5 m in plan of its theoretical location in order to ensure complete coverage of the identified grouting zones. During the drilling operations, 10% of the boreholes had to be re- drilled to comply with the specification requirements on accuracy for both methods.

Ground Movements

During the TaM drilling at Grout Shaft 3, where the water flush drilling system was used, evidence of ground heave generated during the drilling and subsequent sleeve grouting operations were observed. The water flushing pressures during the drilling phases needed to be carefully controlled to minimise the heave generated. The auger method in comparison, generated a gradual, but significant, amount of settlement during the TaM drilling process, and no heave was observed during the sleeve grouting operations. These gradual settlements have been controlled further with the introduction of a secondary phase of grouting. Additional compensation grouting was also used in areas known to have poor ground conditions and where adjacent works were occurring concurrently.

Where settlement has continued, there has been a need to undertake routine corrective grouting operations to recover the gradual settlements during the course of the drilling process. Due to the limited compound space available at all of the grout shaft sites, it has not been possible to undertake grouting operations concurrently with the drilling. Hence, treating this settlement with phases of grouting has had the potential to cause significant delays to the TaM drilling programme.

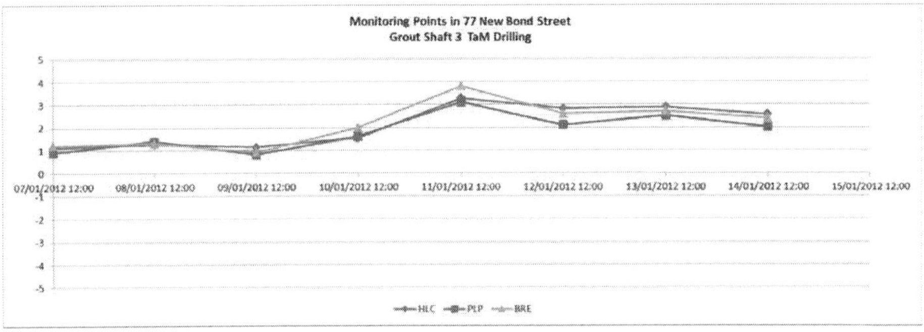

Figure 14: Evidence of heave picked up on the hydrostatic levelling system, BREs and PLPs, caused by the Water Flush drilling at Bond St shaft 3:

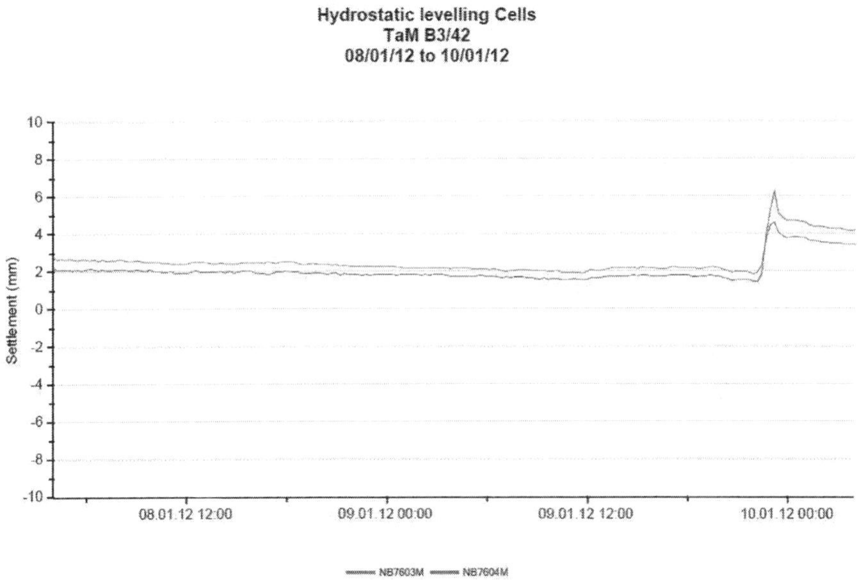

Figure 15: Evidence of heave picked up on the hydrostatic levelling system caused by the Water Flush drilling at Bond St shaft 3:

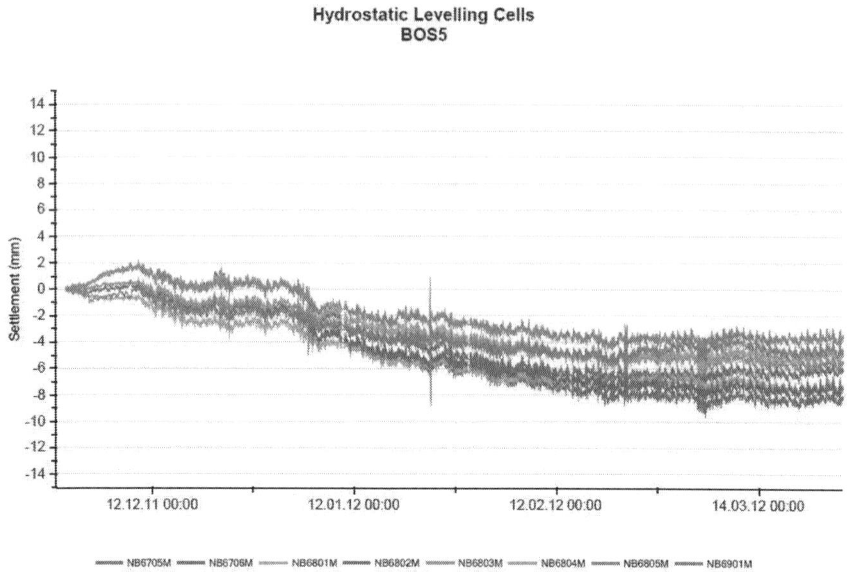

Figure 16: HLC plot showing the gradual settlement at shaft 5, caused by the dry auger drilling method.

Asset Management

Any ground movement, whether settlement or heave, has the potential to cause damage to overlying structures. Though in general, a uniform ground movement poses less risk of damage. The production of differential settlement across a building will generate tensile strain and can lead to cracking and structural damage. As previously described the sudden unpredicted and localised spikes of heave caused by the Water Flush drilling method are unfavourable, the gradual and widespread settlement that has been seen with the Auger drilling method is preferred, as differential ground movement is controlled, and the movement that does arise can easily be treated during phases of grouting, prior to the start of main construction activities.

Safety considerations

The spoil produced by the water flush drilling system consists of a slurry mix of clay and bentonite that requires special equipment for its management and treatment before it can be taken away from site and disposed.

The water flush method uses special flushing pumps with the capabilities of producing high flows at low pressures (necessary to minimise ground heave) and also pumps with the suction capacities to pump the thick slurry mix from the bottom of the shaft to the recycling tanks at ground level. These items of plant are of a considerable size and take up vital room on a site of a very restricted size.

The slurry requires settling and filtering at surface level by means of settlements tanks and a series of filters to eliminate the suspended particles in the mix before it can be returned into the water flush system and reused for drilling. This treatment is necessary because the particles have the potential for damaging and blocking the pipes and pumps of the flushing system causing plant breakdown with the consequent downtime.

With time, the settlement tanks and filters clog up and the recycling system is not capable of filtering the particles from the drilling slurry which thickens up affecting its properties, making the water flush system less efficient. Therefore, maintenance of the settlement tanks and filters is needed on a regular basis, during which activities drilling may need to stop temporarily. The solid materials captured by the filters which settle at the bottom of the settlement tanks are taken away from site by means of gully sucker type lorries on a regular basis, which can add to the pollution and noise generated.

The work force is constantly exposed and is in contact with water when using the water flush method, hence specialised protective water proof clothing is required at all times. This water also has the potential to introduce slippery, wet and frozen surfaces that present a constant hazard to the workforce.

The drilling spoil generated by the auger method is found to be easier to manage due to the dry and solid nature of the clay material withdrawn during the drilling of the boreholes. This dry material is easily handled inside the very restricted space available in the shafts without the necessity of special equipment and is lifted to ground level from where it is easily taken away from the site in lorries.

This solid clay material is free of contaminants and does not require any special chemical treatment which makes it environmentally friendly as the clay can be disposed of and recycled straightaway with no need to store it on site. The traffic requirements for lorries and wagons in order to remove spoil, is minimal with the auger method.

Only standard protective cloths are required to work with this dry spoil and the slips, falls and trip hazards are largely eliminated as the plant and equipment to drill with this method is drastically reduced making the site less congested and easier to maintain.

7. Conclusion

The choice of method for drilling holes for TaM installation in London Clay is dependent on a number of factors; from the depth of drilling and the nature of overlying structures, to the site constraints imposed and the time scale available.

The depth of drilling, the nature of overlying structures, the imposed site constraints and the time scale available are some of the factors to be taken into account when choosing with TaM installation method to use in London Clay.

Following the initial trial, that included drilling with the Water Flush method and the Dry Auger method from the first two grout shafts, the decision was made to progress drilling from the remaining 11 grout shafts using the Dry Auger method.

Limited site constraints, including small diameter shafts and adjacent contracts meant that the utilisation of space and reduced amount of plant required by the Dry Auger method gave it a clear advantage, increasing the free space on site and reducing the number of breakdowns and maintenance required on the plant.

Another clear advantage is the nature of the drill arising's. Dry inert clay is easily removed and disposed of or even recycled, whereas the wet flush method produces a clay slurry which requires significant on site treatment, regular slurry tankers and large volumes of water, increasing health and safety risks on site and increasing the environmental impact.

Given the period of time that grouting operations will need to be active for, and the desire to limit damage to overlying structures, the Dry Auger method also allows for careful control of ground movement. Generating no heave, only gradual settlements which are effectively controlled through phases of grouting. The wet flush method has shown to induce sudden uncontrolled heave, posing possible damage to overlying buildings and utilities.

All of these factors, i.e. easier spoil handling/removal, favoured and controlled ground movements, and less congested sites meant that there were less delays and downtime with the Dry Auger method, overall helping to maintain production rates, ensuring that the TaM installation process would not delay main construction activities.

Though the Dry Auger method has shown to be favoured here. It is highly dependent on the site conditions and should not be taken as a general assumption. On-site trials would be required to determine the preferred method to suit differing locations, geology and constraints.

8. Acknowledgements

The authors would like to thank their colleagues within Crossrail, BFK JV (Bam, Ferrovial. Kier), KBR (Keller, Bam Richies) JV and KGI (Keller, Getech, ITM soil) JV, and especially Paul Thurlow for their guidance and support in the writing of this paper.

9. References

O'Reilly, M.P. and New, B.M. (1982). Settlements above tunnels in the United Kingdom – Their magnitudes and prediction, Tunnelling'82, pp. 179-181

Mair, R.J., Taylor, R.N. and Bracegirdle, A. (1993).Subsurface settlement profiles above tunnels in clays, Geotechnique 43, No. 2, pp. 315-320

Crossrail Civil Engineering Design Standards (Part 7 and 8)

Crossrail Project: Infrastructure design and construction
ISBN 978-0-7277-6078-4

doi: 10.1680/cpid.60784.203

Lessons learnt from the installation of field instrumentation to monitor ground response to tunnelling

Michael S.P. Wan, Geotechnical Engineer (Crossrail), PhD candidate (Imperial College London)
Jamie Standing, Senior Lecturer (Imperial College London)

Abstract

Imperial College in collaboration with Crossrail are performing field monitoring research to investigate the effect of tunnelling in London Clay. This is part of an EPSRC-funded research project investigating how tunnelling affects existing tunnels with segmental cast iron linings. The field research involves a comprehensive instrumentation and monitoring scheme at a site in Hyde Park and the bordering Bayswater Road beneath which the new Crossrail tunnels will be constructed in London Clay below the existing London Underground Central Line tunnels. Surface and borehole instruments have been installed around the Crossrail west-bound and east-bound running tunnel alignments to monitor the ground response to the tunnel construction under "greenfield" conditions. The ground in close proximity to the existing Central Line running tunnels will also be monitored. High resolution and accuracy monitoring systems have been implemented to achieve high quality data to assess with confidence the tunnelling-induced ground response and mechanisms of movement from earth pressure balance machine tunnelling.

The installation of surface and subsurface instrumentation was carried out in the summer of 2011 as part of Crossrail ground investigation Package 16A. This paper describes and discusses the installation of instruments including rod extensometers, in-place inclinometers and multi-level vibrating-wire borehole piezometers. Practical challenges during the installation process and how they were overcome are discussed.

<u>List of Abbreviations</u>

EPBM	earth pressure balance machine
EPSRC	Engineering and Physical Sciences Research Council
ICFEP	Imperial College Finite Element Program
I&M	instrumentation and monitoring
LUL	London Underground Limited
MEMS	micro-electro-mechanical system
TBM	tunnel boring machine
PFA	pulverised fly ash
ppv	peak particle velocity
VW	vibrating-wire

1. Introduction

Research is being undertaken by Imperial College London to investigate the effects of tunnelling on existing tunnels in London Clay. The research opportunity arose with the Crossrail construction of 21km of underground twin-bore railway tunnels through central London where the new tunnels will interface with existing networks of London Underground Limited (LUL) tunnels and other utilities tunnels. The Crossrail western tunnel section involves construction of twin tunnels using 7.2m diameter earth pressure balance tunnel boring machines (TBMs) mostly in London Clay. The tunnel drives below the existing LUL Central Line tunnels near Hyde Park allows a field investigation of the effect of new tunnelling construction on both the "greenfield" ground and the ground in proximity to existing segmental cast iron lined tunnels. The research, involving instrumentation and monitoring (I&M) fieldwork in Hyde Park and Bayswater Road, aims to provide a high quality case study of tunnel construction in London Clay and the mutual interaction of both the new and existing tunnels.

This paper describes the instrumentation scheme and the installation works. Some practical challenges faced during the installation and procurement are also discussed. The intention is that lessons learned can be applied to future similar instrumentation installation works.

2. Instrumentation site

The field research site is located near Victoria Gate, the northwest corner of Hyde Park, where the Crossrail west-bound and east-bound tunnels will pass beneath the LUL Central Line running tunnels near Lancaster Gate. The actual crossing is directly beneath Bayswater Road. The location of the instrumentation site in plan and the Crossrail tunnel longitudinal section are shown in Figure 1 and an aerial photo in Figure 2.

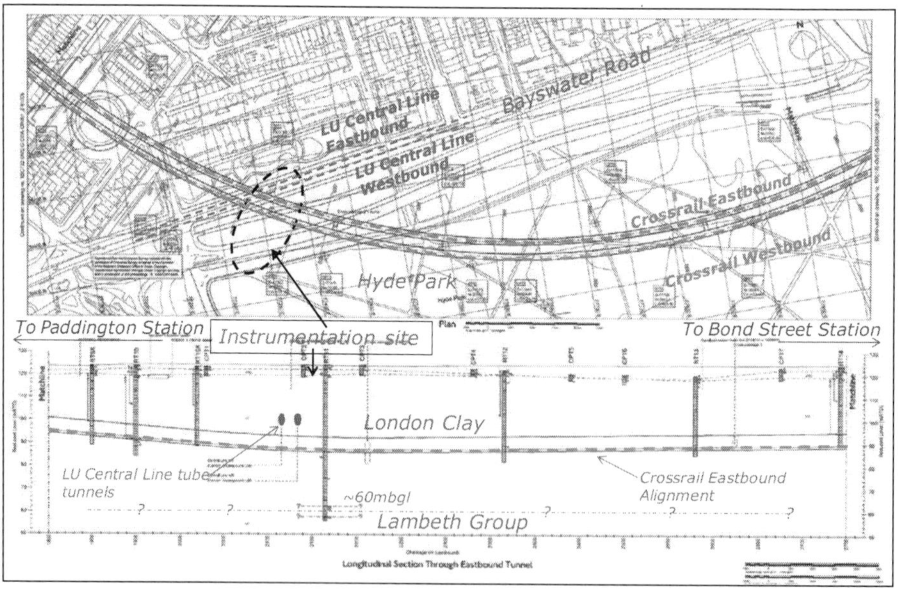

Figure 1 – Location of instrumentation site in plan and Crossrail tunnel longitudinal section.

Figure 2 – Aerial photo of the instrumentation site.

3. Purpose of instrumentation

The Hyde Park instrumentation and monitoring scheme form part of a research activity and so differ from industrial instrumentation projects for construction control or decision making processes involving an observational approach (Peck, 1969). There are no trigger levels or action plans for the I&M scheme described here.

Advanced numerical modelling is a powerful method for estimating and predicting the ground response under specified loading conditions. Both numerical analyses and constitutive models should whenever possible be validated and calibrated using reliable field measurement data. The Imperial College research team are using an in-house finite element programme, ICFEP, to simulate the site conditions and construction details at Hyde Park, both with and without the presence of the existing running tunnels. The field monitoring results will be fed back into the ICFEP model.

Having recognised the purpose of the I&M scheme, it is important to choose the right types of instruments with performance levels such that the anticipated ground movements or stress changes are measured to the required accuracy and precision. Obviously, good workmanship is desirable for high quality instrument installation.

4. Borehole instruments and their installations

4.1 Instrumentation plan

A plan of the instrumentation layout is shown in Figure 3. There are in total 38 boreholes, each accommodating one or more instruments. Extensometer boreholes and inclinometer boreholes are paired so that they can measure both vertical and horizontal ground displacement at the same locations. Piezometers and spade cells have been installed around the Crossrail tunnels to measure pore water pressures and changes in earth pressures.

Most of these instruments are aligned in a main instrument array transverse to the Crossrail tunnels. There is another instrument array on the footpath of Bayswater Road parallel to the LUL Central line tunnels. These boreholes have been positioned as close to the Central Line tunnel as possible in order to see how the existing tunnels affect the ground response, as opposed to "greenfield" condition. An LUL concession application was made so that the boreholes could be drilled at as close as 1.5m away from the existing tunnel edge.
In the main array, the instruments are so positioned that the southwest part of the instruments monitor the "greenfield" ground response whereas the northeast part measure the ground response under the influence of the existing Central Line tunnels. The measured southwest half of the settlement trough would be expected to be different to that of the northeast half.

The positions of the installed rod extensometers and in-place inclinometers along the main array are presented schematically in Figures 4 and 5 respectively. Each cross in Figure 4 represents an anchor at which the subsurface settlements are measured (a maximum of eight anchors can be installed in one borehole). The anchors are largely distributed around the Crossrail tunnels, installing them around the existing Central Line tunnels was precluded by site constraints such as heavy traffic volume and Thames Water pipes on Bayswater Road. Nevertheless, it was possible to install some borehole instruments under the footpaths of Bayswater Road, where the Central Line tunnel will still influence the ground response. As shown in Figure 5, there is a similar arrangement for the inclinometer boreholes in the main array. The inclinometer and extensometer boreholes are paired (about 2.5m apart), to facilitate interpretation of the displacement data: this is particularly useful when determining displacement vectors and ground strains.

In the main instrument array, piezometers and spade cells were installed, as shown in Figure 6, around the Crossrail east-bound tunnel to monitor changes in pore water pressure and total horizontal earth pressure as the TBMs approach and pass. There are two multi-level VW type piezometer boreholes within Hyde Park, with six piezometer sensors at different depths in each. A conventional standpipe piezometer was also installed just 2m away from one VW piezometer borehole, at the same distance from the Crossrail tunnel, to verify the long-term pore water pressures recorded with the VW piezometers.

Along Bayswater Road there is another instrumentation array as shown in Figure 7. The same types of instruments were installed there to measure the response of the ground close to the existing Central Line tunnels. These boreholes are either about 3m from the LUL tunnel extrados or terminate about 1.5m above the tunnel crown. It is envisaged that the piezometers will measure the pore water pressure profile near the existing tunnels and indicate drainage effects on the steady state pore water pressure profile.

The instrument types were selected for the current Hyde Park site, bearing in mind the magnitude of changes predicted by empirical methods. While there may be other instruments and techniques available to perform the tasks, the experience of the Imperial College field monitoring team gained from previous research projects such as the Jubilee Line Extension Project and Channel Tunnel Rail Link Project played an important role in the selection process.

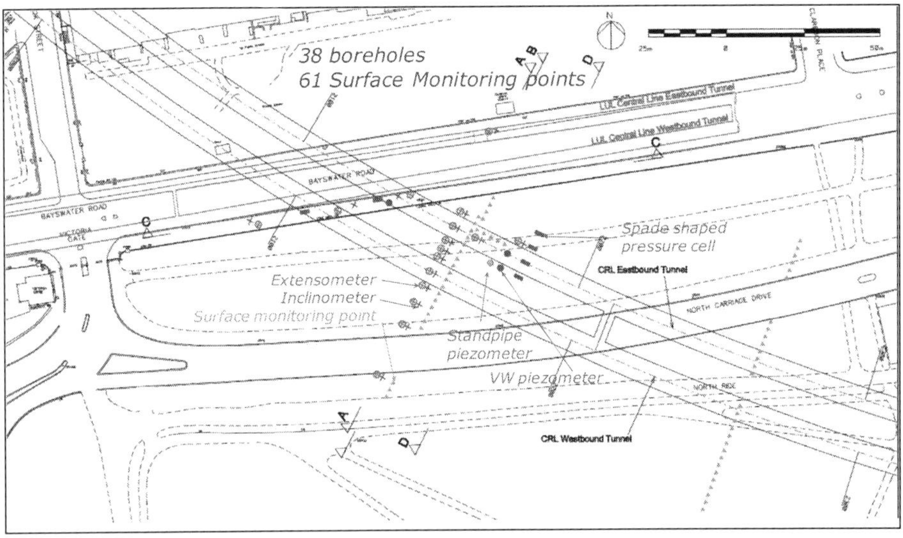

Figure 3 – Plan of instrumentation plan in Hyde Park and Bayswater Road.

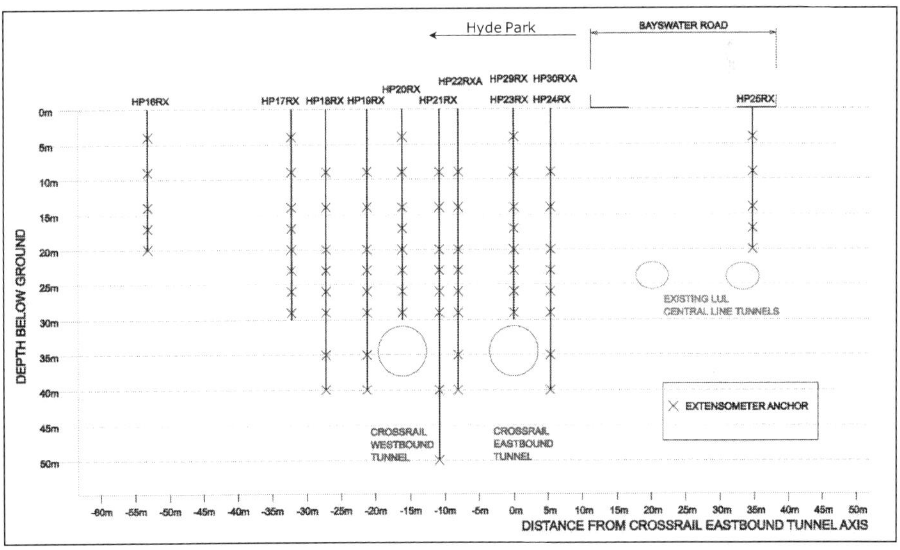

Figure 4 – Cross-section of rod extensometer boreholes in the main instrument array (Section A-A).

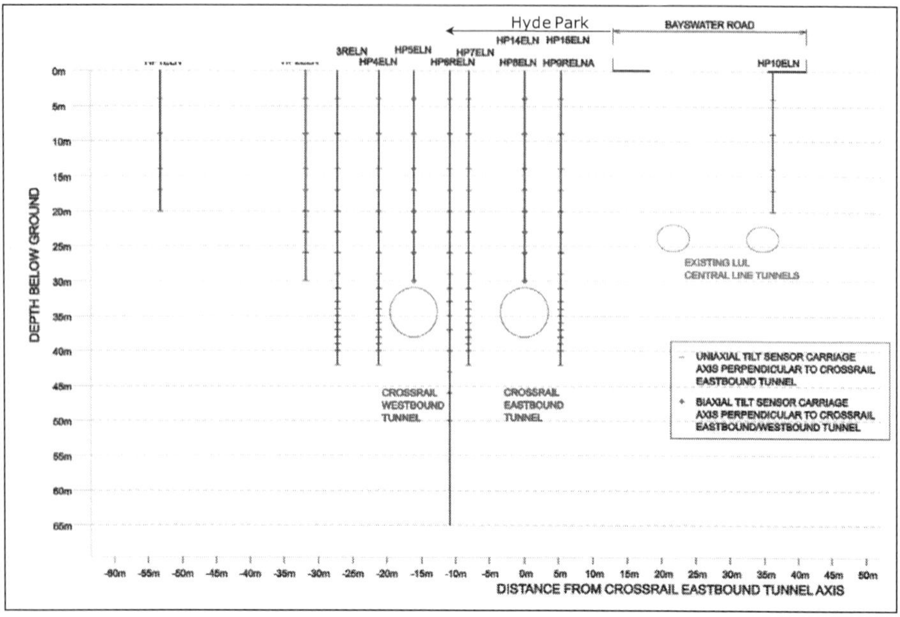

Figure 5 – Cross-section of proposed in-place inclinometer boreholes in the main instrument array (Section B-B).

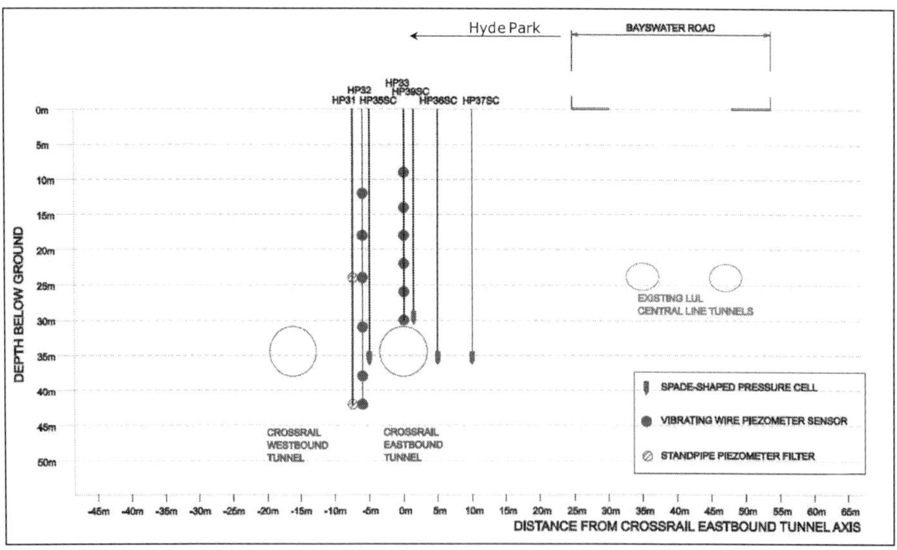

Figure 6 – Cross-section of piezometer and spade cell boreholes in the main instrument array (Section D-D).

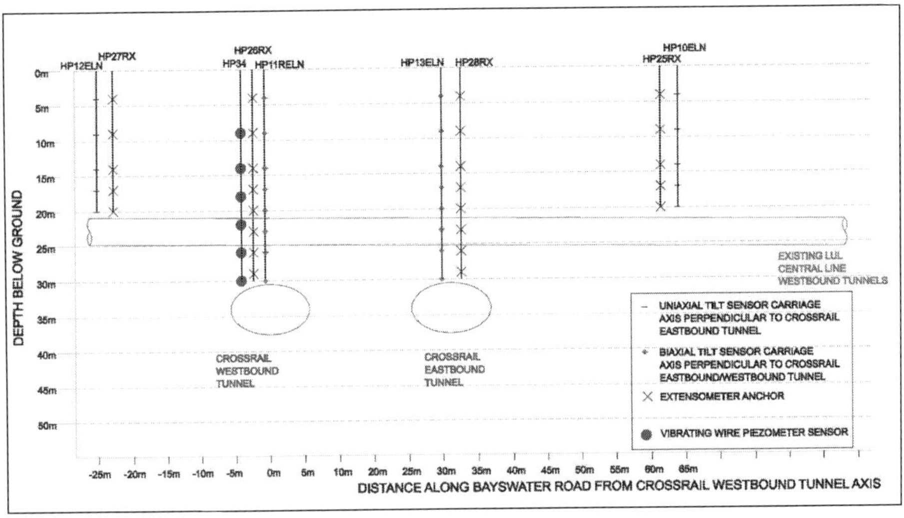

Figure 7 – Cross-section of borehole instruments in the array parallel to Bayswater Road (LUL Central Line tunnels) (Section C-C).

4.2 Rod extensometers

4.2.1 Measurement principle and performance

Subsurface vertical displacements are to be measured using multi-level rod extensometers. A maximum of eight anchors, positioned at different depths, attached to stainless steel rods were installed in a single borehole. The steel rods were adjusted so that the rod heads were just below a reference head plate at the ground surface (Figure 8). The depths of the rod tips below the reference head are measured by using dial gauge, while the reduced level of the reference head plate is measured by precise levelling. Combining the dial gauge and precise levelling measurements, the reduced level of each anchor at each measurement time can be determined.

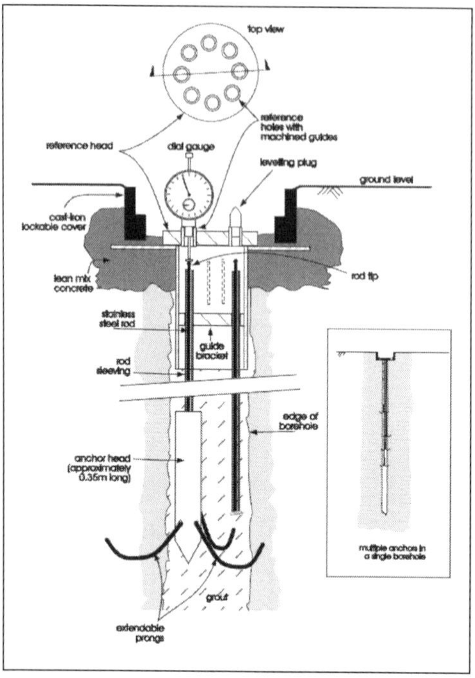

Figure 8 – Details of multi-level rod extensometer (Nyren, 1998).

Rod extensometers generally have a much higher accuracy than magnetic extensometers. As the dial gauge to be used has an accuracy of +/-0.05% full scale range (100mm), the accuracy is +/-0.05mm, compared with the +/-1mm accuracy normally achieved with a typical magnetic extensometer. The overall measurement accuracy is governed by that of the precise levelling which is about +/-0.3mm at best. Sub-millimetre accuracy is desirable at the current research site as the maximum anticipated settlement is only about 10mm.

4.2.2 Installation

The rod extensometers were installed in boreholes formed by cable percussion drilling. During the borehole drilling, the stainless steel rods, PVC sleeves and hydraulic tubing for the anchors at different depths were laid out and prepared on the ground surface by the instrumentation specialist contractor. The rods, sleeves and anchors were colour coded with duct tape to identify the anchors at different depths. Immediately after the borehole was drilled, a tremie grout pipe was put down the borehole. Then the whole assembly of the bottom-most anchor was first inserted into the borehole. Once at the required depth, the prongs of the anchor were jacked out hydraulically via the tubing to secure the anchor into the wall of the borehole. This procedure was repeated for each of the remaining anchors.

After all the anchors were fixed in the required depths, the borehole was ready for backfilling with a cement bentonite grout. Grout mix trials were conducted in the contractor's workshop before commencement of the site work, where the grout properties and pumpability of different design mixes were tested. The grout mixes used for different types of instruments are discussed in Section 5.1. Once the backfill grout had hardened, the reference head was installed and the headworks constructed. Finally, the top of the stainless steel rods were replaced with adjustable rod heads which were set to about 50mm below the reference head. The dial gauge has a range of +/-50mm which allows measurements of maximum 50mm settlement or heave of the rods and hence anchors relative to the reference head.

4.3 In-place inclinometer

4.3.1 Measurement principle and performance

Subsurface horizontal ground displacements are measured by in-place inclinometers. Each inclinometer consists of a series of micro-electro-mechanical system (MEMS) accelerometer sensors mounted on short carriages positioned at selected intervals down a typical inclinometer casing. Each accelerometer measures the capacitance change in response to tilt and transmits the voltage change to a data-logger. The MEMS tilt sensors are left in place during and after the tunnel construction so that automated data-logging is possible. A maximum of 16 sensors were installed in one single inclinometer casing. The horizontal displacement profile along the inclinometer casing can be deduced if appropriate gauge lengths for the sensors are assumed, as demonstrated in Figure 9.

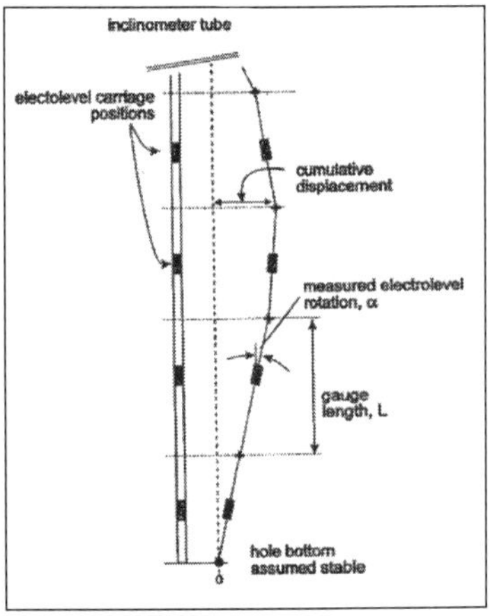

Figure 9 – Schematic presentation of in-place inclinometer (Nyren, 1998).

The MEMS tilt sensors have a measurement resolution of 2 arc seconds. Baseline monitoring after installation shows that repeated readings on a number of tilt sensors over a few days fall within a range of about 4 arc seconds, which is equivalent to about 0.08 mm for a sensor spacing of 4m (i.e. 0.02mm/m). Since that is a random error and is independent for each individual sensor, the overall repeatability for a maximum of 16 sensors would equate to individual repeatability x $\sqrt{16}$, which is 0.16mm over 32m (2m spacing), or 0.32mm over 64m (4m spacing).

4.3.2 Installation

Typical inclinometer casings were installed in the boreholes formed by cable percussion drilling. The inclinometer casings are ABS plastic of 59mm internal diameter and 3m length with snap couplings at the ends. Each joint is sealed with two O-rings and leak-proof to prevent ingress of the backfilling grout. Prior to placing the inclinometer casings into the borehole, a water test was performed to demonstrate that the casing joints do not leak. As the inclinometer casings were lowered down into the borehole, the casing joints were covered by two layers of duct tape as an additional seal. After aligning the casing grooves to the desired orientation (perpendicular/parallel to the Crossrail tunnel alignment), the borehole was backfilled with a cement-bentonite grout (see Section 5.1). Again prior grout mix trials were performed in the contractor's workshop. In order to counter the buoyancy of the casing in the wet grout, they were filled with water. For deeper boreholes, the grouting was completed in two stages so that the grout at the bottom section had semi-hardened before further wet grout was pumped to complete backfilling. Finally a headworks was constructed to accommodate a data-logger. After the installation, the casing was mapped by a torpedo probe for an initial measurement of the casing inclination.

In May 2012 (almost a year later), in-place tilt sensors were installed in the inclinometer casings. Each MEMS tilt sensor is fixed within a protective aluminium tube, with O-ring seals at its ends, which is mounted on a short carriage with wheels on one side and a leaf spring on the other. The carriage could then be pushed along the keyways of the inclinometer casing and stays in position after pushing. Each aluminium tube contains one MEMS sensor measuring tilt in one plane only. For most inclinometers, only tilt measurements transverse to the TBM movement are required, and therefore a series of uniaxial tilt sensors were installed. For some inclinometers, measurements on both the longitudinal and transverse directions are required, and so two carriages were fixed orthogonally to each other.

Consecutive tilt sensors within a single borehole are connected by a signal cable so that all the sensors are connected in series with just one signal cable coming out from the inclinometer casing. The carriages were pushed down the inclinometer casing to required depths using PVC rods of different lengths. Finally the sensors were connected to a data-logger which was placed within the headworks. The frequency of the data logging was set and the data can be downloaded to a laptop from the data logger.

4.4 Vibrating wire (VW) piezometer in fully grouted borehole

4.4.1 Measurement principle and performance

The technology of vibrating wire (VW) type instruments has advanced within recent years and they are considered to have better longevity and durability than before. Historically single VW piezometers were installed within a borehole, surrounded by a sand filter and sealed for the remainder of the borehole with bentonite (Figure 10b). Recently the technique of installing them using the "fully-grouted" method, in which the VW piezometer units are embedded in a suitable cement bentonite grout, has become popular (Dunnicliff, 2008). The conventional installation method was originally for Casagrande type standpipe peizometers which require large groundwater inflow to or outflow from the standpipe (Figure 10a).

Vaughan (1969) suggests that for diaphragm type piezometers which require only very small amounts of water flow to operate, granular filters can be omitted and the piezometer sensors can be fully backfilled by low permeability grout. Mikkelsen (2002) discusses the practical side of the fully-grouted method and recommends good practice and appropriately proportioned cement bentonite grout mixes. Dunnicliff (2008) and Contreras et al (2008) report successful installation and monitoring of piezometers installed by the fully-grouted method at various sites around the world. A schematic of borehole piezometers installed by the fully-grouted method is shown in Figure 10c.

The fully-grouted method has a major advantage in that the straightforward backfilling procedure allows multi-level piezometer sensors to be installed in one single borehole. This makes it possible to obtain a pore water pressure profile with depth at a single borehole location. This significantly reduces the number of boreholes required.

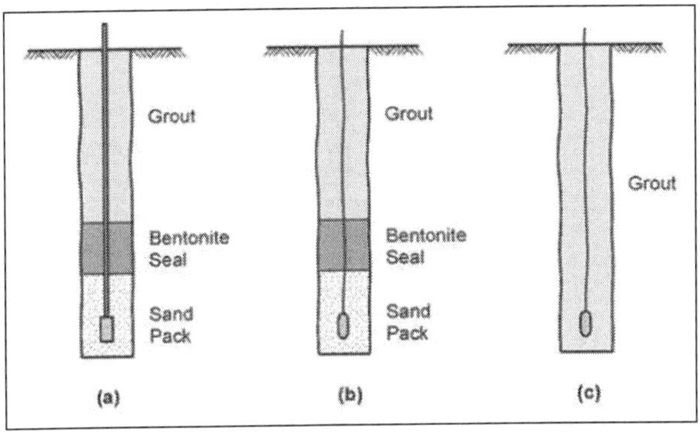

Figure 10 – Schematic presentation of (a) traditional standpipe piezometer, (b) diaphragm type piezometer with traditional installation, (c) diaphragm type piezometer in fully-grouted borehole (Contreras et al, 2008).

4.4.2 Installation

Prior to installation, the VW piezometer sensors were placed in de-aired water at ground level and pre-installation 'zero' readings taken. This allows future readings from these sensors to be referenced to atmospheric pressure. The porous filters of the piezometers were de-aired by boiling for at least 2 hours before the installation. On site, six sensors were installed in a single borehole. PVC grouting pipes of 75mm diameter were used for backfilling and also for attaching the sensors at different depths. With the grouting pipe lying on the ground, the depths of the sensors were marked and small slots were cut in the pipes for the signal cables to pass inside the grouting pipes. The sensors were then fixed onto the grouting pipes by cable ties, the signal cables fed through inside the pipes and the cut slots sealed by at least two layers of duct tape. As the grouting pipes were lowered down into the borehole, the joints of the grouting pipes were connected and sealed by duct tape. Low permeability cement bentonite grout (see Section 5.1) was then pumped via the grouting pipe to the base of the borehole and the piezometer assembly was fully grouted. During backfilling, the sensors were monitored by a hand-held VW readout unit to check that they were not over- stressed by the wet grout. The signal cables were labelled with duct tapes of different colours to identify the depths of the sensors. Finally, the cables were connected to data-loggers housed within a headworks.

5. Technical challenges

Achieving research-quality instrumentation installations at times led to additional technical and sometimes contractual challenges to those for more standard ground investigation and instrumentation projects. Some of the more salient challenges and their solutions are now discussed.

5.1 Grout mixes for borehole backfilling

Depending on the instrument types, the backfilling grouts should have different mechanical properties relative to the surrounding ground. Mikkelsen (2002) argued that for extensometers, the grout should have stiffness equal to or smaller than the surrounding ground, while for inclinometers, a stiffer grout is desirable. For piezometers fully grouted in a borehole, Vaughan (1969) proved analytically that the error associated with pore water pressure measurement is insignificant if the grout permeability is not larger than 20 times the ground permeability. More recently, Contreras et al (2008) demonstrated using finite element analyses that the grout permeability could actually be 1000 times higher than that of the ground before significant errors in pore water pressure measurements occur. Therefore, three types of grout mixes were chosen for backfilling the extensometer, inclinometer and piezometer boreholes. In addition to the stiffness and permeability requirements, the grouts should also be readily mixable and liquid enough to be pumped through the hose and tremie pipe to the borehole.

Grout properties such as stiffness and permeability are affected by many factors such as the material composition (cement type, fineness of bentonite powder, water acidity etc), mixing equipment, mixing sequence and even mixing temperature. The initial grout mixes determined from the pre-installation trial mixes in the contractor's workshop were found to be too thick to be pumped through the tremie pipe on site. Checks were made to ensure that the materials were the same for the trial mixes at workshop and the actual mixes on site; the same sequence of the mixing was followed; the same model of the mixer was used too. It was believed the difference in the grout consistency could have been due to small temperature differences (winter to summer) and perhaps more significantly the trial grout mixes in the workshop were not pumped through the same length of grout pipe as on site to check the pumpability. This latter effect was further compounded on site as the grout pump had additionally to overcome the hydraulic head of the wet grout in the borehole.

Further grout mix trials were performed on site, using the same materials and mixer but under the site environment and using a stronger piston pump. Successful revised grout mixes were determined but this took time and led to delays in the overall programme of work.

5.1.1 Properties of trial grout mixes

Ordinary Portland cement, sodium-bentonite powder and pulverised fly ash (PFA) were used for the grout. For both the initial workshop trial and the final site trial, a mixing sequence as recommended by Mikkelsen (2002) was followed: mixing bentonite at the end after cement and fly ash. The mixing procedure started by adding a measured quantity of clean water into a mixing pan. The proportioned cement was then gradually added and mixed thoroughly, followed by adding and mixing the fly ash. Finally the bentonite powder was slowly added, stirring the mixture constantly, to break down any lumps of bentonite that formed.

The initial workshop grout trial involved twelve mixes, of which six were not pumpable using a progression cavity pump that the contractor proposed to use on site. Samples of 100mm diameter were prepared for the other six pumpable grouts and tested in laboratory for unconsolidated undrained shear strength and permeability using a triaxial apparatus. The details of the grout mixes and test results are summarised in Table 1 and Figures 11 and 12. Although it is compatibility of stiffness of the grout and the ground that is sought, usually comparisons are made in terms of undrained strength for simplicity (because of the varying nature of stiffness with strain).

	Mix No.	1	2	4	6	9	12
	Mix proportion by mass (Water:Cement: Fly ash: Bentonite)	4.0 : 1.0 : 0.0: 0.72	4.0 : 1.0: 0.0: 1.0	2.5 : 1.0 : 0.0 : 0.74	2.0 : 1.0 : 0.0 : 0.5	4.0 : 1.0 : 0.5 : 1.0	2.5 : 0.8 : 0.9 : 0.55
*UU triaxial compression test	Curing time (days)	59	58	38	38	32	32
	Average sample void ratio	5.14	4.93	3.50	3.03	4.16	2.76
	(Confining pressure)/ UU triaxial shear strength (kPa)	(100) 91	(100) 152	(200) 117	(200) 702	(100) 145	(200)398
		(200) 56	(200) 156	(400) 421	(400) 691	(200) 146	(400) 410
		(300) 114	(300) 166	(600) 420	(600) 489	(300) 166	(600) 375
**Permeability test in triaxial apparatus	Sample void ratio	N/A	N/A	3.57	2.84	3.72	2.34
	(Consolidation pressure, kPa)/ Coefficient of permeability (m/s)	N/A	N/A	(200) $6.3x10^{-9}$	(200) $1.5x10^{-9}$	(100) $9.4x10^{-10}$	(200) $1.3x10^{-10}$
		N/A	N/A	(400) $4.4x10^{-9}$	(400) $1.2x10^{-9}$	(200) $6.0x10^{-10}$	(400) $7.0x10^{-11}$
		N/A	N/A	(600) $3.7x10^{-9}$	(600) $8.9x10^{-10}$	(300) $4.2x10^{-10}$	(600) $6.0x10^{-11}$

Table 1 – Details of trial grout mixes and properties.

*UU = Unconsolidated undrained triaxial compression test on 38mm specimens

**100mm diameter specimens were consolidated under different cell pressures before permeability tested under a constant pressure gradient. The tests were performed in accordance to BS1377 Part 6: 1990 Clauses 5.3, 5.4, 5.5.2.2 to 5.5.2.7 and 6.

As indicated in Table 1 and Figure 11, generally the shear strengths of the grouts are insensitive to the confining cell pressures. In the cases of mix numbers 1, 4 and 6 there was one sample in each set of tests that had a significantly lower strength than the others. As there is no trend in the lower strength with cell pressure and as only a single sample was prepared for each cell pressure, it is assumed that these lower strengths are not representative. In general, the grout shear strength decreases with increasing water-cement ratio (n.b. mixes with higher water-cement ratios (and PFA) produce grouts with higher void ratios). Similarly, grouts with a lower water-cement ratio and hence lower void ratio generally have a lower permeability, although the bentonite proportion also influences permeability as can be seen from Figure 12.

Based on the results from the initial grout trials and ground investigation reports, Mix 2 and Mix 4, respectively having smaller and larger shear strengths than the surrounding soils, were considered suitable for backfilling extensometer boreholes and inclinometer boreholes respectively. Mix 6, having a coefficient of permeability of $1x10^{-9}$ m/s, was considered less than 100 times more permeable than the surrounding London Clay (for which the lowest permeability was considered to be about $1x10^{-11}$ m/s) and therefore considered suitable for backfilling VW piezometer boreholes.

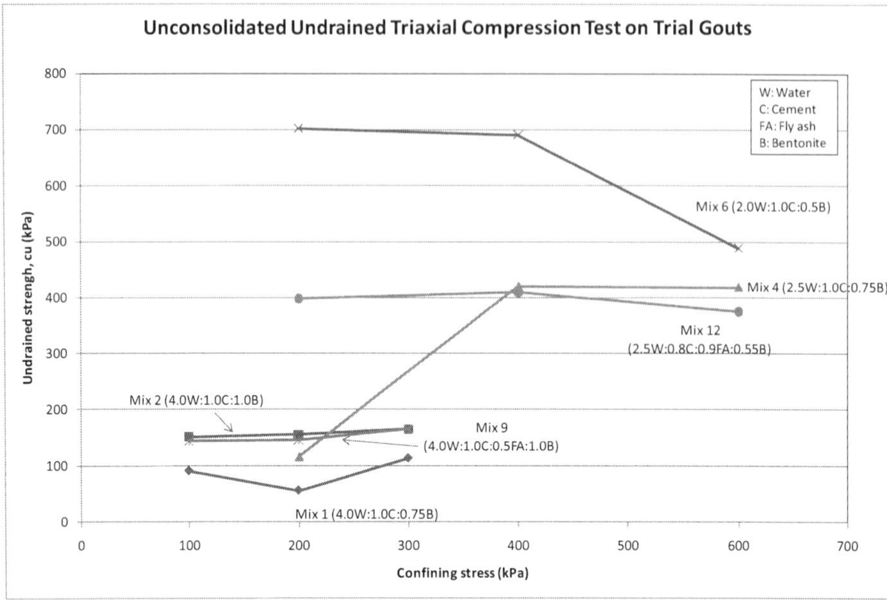

Figure 11 – Undrained shear strengths of 38mm specimens of trial grout mixes.

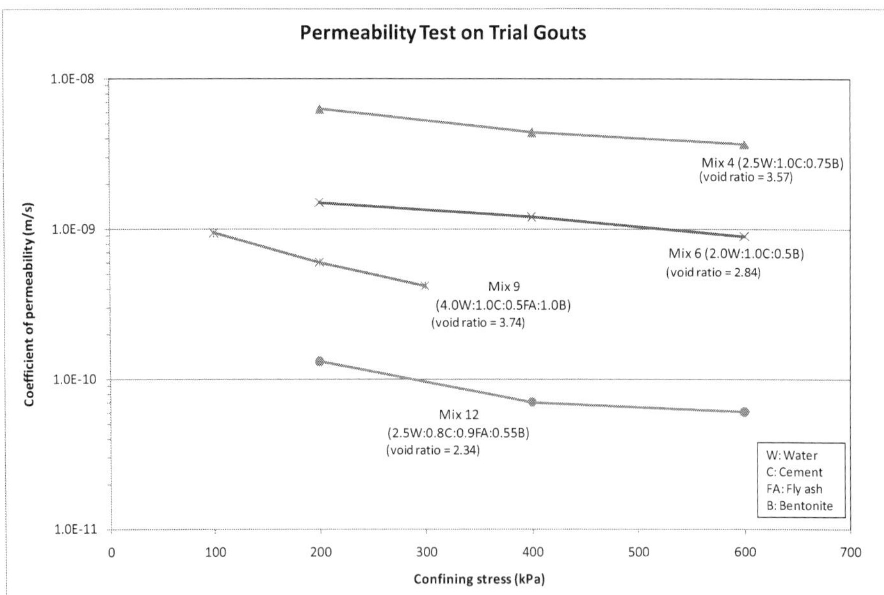

Figure 12 – Coefficient of permeability of trial grout mixes.

The lack of pumpability of the initial trial mixes on site using the original progression cavity pump led to modifications being made to the grout mixes for extensometer and inclinometer boreholes primarily involving reducing the bentonite quantity and adding PFA. The pumpability of the revised mixes was tested by pumping the grouts into an abandoned 30m borehole using a stronger piston pump. The final grout mix proportions used are shown in Table 2.

Borehole backfill	Water	Cement	Fly ash	Bentonite
Extensometer	4.0	1.0	0.5	0.9
Inclinometer	2.5	0.8	0.9	0.5
VW piezometer	2.0	1.0	0.0	0.5

Table 2 – Final grout mixes proportion by mass.

Ideally grout mix trials should be performed using the same materials and equipment on site prior to the grout backfilling. This would involve careful planning of site activities as it could take weeks to obtain representative laboratory testing results. Allowance in grout pumpability should also be made to take account of the need to overcome the wet grout hydraulic head.

5.2 Encountering obstructions of Thames Water pipe

During the borehole drilling, shallow ground obstructions were encountered below the Bayswater Road pavement. Prior to drilling on the southern side of Bayswater Road, several inspection trenches were excavated to expose an existing water pipe as required by Thames Water to determine the exact size and location of the pipe. It was found to be a cast iron pipe of 600mm diameter, with the depth of the pipe crown ranging from 0.2m to 0.6m below ground level in the vicinity of the boreholes. The pipe is a major water main and so Thames Water required the boreholes to be drilled in a controlled manner outside a set distance from the pipe. As the water main runs exactly above the LUL Central Line west-bound tunnel along Bayswater Road, it was not possible to install the boreholes and instruments in close proximity to the existing tunnel.

In order to assess potential effects on the water main of borehole drilling, a trial was performed within Hyde Park, measuring ground vibration. Two 0.5m deep pits were excavated at 1.5m and 3.0m from the borehole. Ground vibrations in terms of peak particle velocity (ppv) were measured using geophones at the bottom of the pits during various steps of borehole drilling. Potential mitigation measures to reduce vibration were investigated during the measurements, including digging a deeper inspection pit before starting cable percussion drilling, advancing the casing by surging instead of hammering, reducing the rate of hammering and using more water to facilitate penetration of the casing close to the pipe. It was found that the maximum vibration occurred when the casing was hammered into the top of London Clay at about 5m depth. A ppv of 4.4m/s was measured at 1.5m from the borehole when the casing was rapidly struck. However, if a 2-3 second delay was allowed before each casing strike, reduced ppv's of 3.9m/s and 3.5m/s were measured at 1.5m and 3.0m respectively from the borehole. The results indicated that risk of damage to the water main should be sufficiently low at 3m from the borehole with the above-mentioned precautionary measures. No damage to the water main was reported during the work. Clearly, when utilities pipes are expected, early engagement and close liaison with utilities companies is vitally important.

5.3 *Time of completion of instrument installation*

Ideally, drilling, installation of instruments and backfilling of the borehole should be completed in as short a period of time as possible, preferably within one day, to minimise disturbance to the ground around the borehole due to stress relief and desiccation. This helps ensure that a high instrument conformance is achieved.

Finishing a 30m or deeper borehole with instrument installation and backfilling within a day was not usually practically feasible, as there was a strict control on the working hours (8am to 6pm) within Hyde Park. A compromise was to complete the installation over a two day period. A borehole would be started and left open overnight down to 30m below ground, with the borehole being cased over the depth of Terrace Gravels, taking account that the London Clay sub-unit B2 to this depth is relatively more stable than the underlying A3ii London Clay sub-unit. The borehole drilling was completed the next morning and installation of the instruments and grout backfilling of the borehole was finished by the end of the afternoon.

Clearly, early discussions with the contractor about specific project requirements and any site restraints help achieve workable solutions.

5.4 *Borehole diameters*

Smaller diameter boreholes usually are more advantageous than larger ones as faster, more effective drilling can be achieved with less disturbance to the ground. On site, 200mm diameter boreholes were drilled initially, although the instruments were able to fit into 150mm boreholes. The main reason was to cover the risk of abandoning the borehole in case of borehole wall collapse: with a 200mm diameter borehole it is possible to continue drilling with 150mm casings. If the borehole were started with 150mm casing, there would be a risk of abandoning the whole borehole if a borehole collapse occurred which would be beyond the contractor's control.

Drilling rates are much slower for 200mm compared with 150mm boreholes, especially when stiffer strata or claystones are encountered. In order to minimise the ground disturbance and improve the construction progress, it was agreed that an Engineer's instruction be given to the contractor to drill boreholes of 150mm diameter. Any abandoned hole and redrilling due to borehole collapse induced by factors uncontrollable by the contractor would be payable works. This was a re-allocation of the risk from the contractor to the client. Based on earlier records and experience of drilling in the Hyde Park area, the chance of encountering water strike was deemed to be low. In the end no collapse of boreholes occurred at the site.

6. Summary and conclusions

As part of the work of Crossrail ground investigation Package 16A, high-quality borehole instrumentation was successfully installed in Hyde Park and Bayswater Road. Reasons for instrument selection are given and discussed in the paper. An important factor for achieving good conformance of the instrumentation to the surrounding ground conditions is the choice of grout for backfilling the boreholes. Ideally for extensometers and inclinometers the grout should have a compatible stiffness to the ground, although usually it is the grout strength that is compared with ground strength. Equally grout and ground permeabilities should be compatible for piezometer installations. A number of grout mixes were trialled and the final choices were made according to strength and permeability criteria as well as practical mixing and pumping considerations. Other issues that had to be taken into account are also described (sensitivity of a cast iron water main, time constraints and optimum borehole diameter). At the time of writing this paper, more than 12 months of baseline monitoring work has been completed, confirming the accuracy of the various instruments and measurement systems and allowing thermal and seasonal effects to be assessed.

Acknowledgements

The authors would like to acknowledge Crossrail and the Engineering and Physical Sciences Research Council (EPSRC) who are major sponsors of the research project at Imperial College. Many thanks are due to the Imperial College research team, in particular Professors John Burland and David Potts, for their contributions in discussing the choice and layout of the instrumentation. A thoughtful review and comments on the instrumentation plan from Mr. John Dunnicliff is also greatly appreciated. The support provided by the Royal Parks and Westminster Council during the installation work is gratefully acknowledged. The successful Hyde Park instrumentation would never be achieved without the diligent work by the Package 16A Contractor, Fugro Engineering Services, and specialist sub-contractors Construction Monitoring Control Systems (CMCS) Limited and ITMsoil. The Engineer, Arup Atkins JV, in particular Ms. Vicky Hope and Mr. Paul Braddish, made enormous efforts in both the procurement and supervision of the work. Special thanks are reserved to the experienced crew of drillers, especially Mr. Ian Shortland, whose vast experience and great technique of drilling in the London Basin proved to be very essential to achieving the high quality installation.

References

Contreras, I. A., Grosser, A. T. & Ver Strate, R. H. (2008) The use of the fully-grouted method for piezometer installation, Part 1 and Part 2. *Geotechnical News.* (June), 30-40.

Dunnicliff, J. (2008) Discussion of "The use of the fully-grouted method for piezometer installation" by Contreras et al (2008). *Geotechnical News.* (June), 38-40.

Mikkelsen, P. E. (2002) Cement-bentonite grout backfill for borehole instruments. *Geotechnical News.* (December), 38-42.

Nyren, R. (1998) *Field measurements above twin tunnels in London Clay.* PhD thesis. Imperial College London.

Peck, R.B (1969) Advantages and limitations of the observational method in applied soil mechanics, *Geotechnique*, 19, (1), 171–187.

Vaughan, P. R. (1969) A note on sealing piezometers in boreholes. *Geotechnique.* 19 (3), 405-413.

Crossrail Project: Infrastructure design and construction
ISBN 978-0-7277-6078-4

doi: 10.1680/cpid.60784.221

The Use of Shape Accel Array for Monitoring Utilities during Urban Tunnel Drives

Brendan Bradley MEng (Hons), Monitoring Engineer, BamFerrovialKier JV
Roberto Prado Garcia MEng, Monitoring Engineer, KGi JV

Abstract

The protection of third party assets with reliable monitoring instruments is essential for the running of a successful urban tunnelling project. With a twin bore tunnel being excavated below the Ranelagh Sewer, in Paddington for the Western Tunnel drive of Crossrail, a reliable method was required to measure displacements (convergence) in near real time. The design of the system required careful planning and an installation method that would be safe and comply with the stringent requirements of the asset owner.

The introduction of Shape Accel Arrays (SAA) for the measurement of convergence within large sewers is a new concept in the UK construction industry. The instrument is successfully being used for the monitoring of slopes, embankments and diaphragm walls. SAA are very quick to install and base read, with the data output configurable for various types of monitoring software.

This paper documents the use of Shape Accel Arrays within the Ranelagh Sewer. It will describe the method of installation, the results obtained and lessons learnt during the first large tunnel drive directly below the shape accel arrays.

INTRODUCTION

Third party asset protection is an area of critical importance to the success of Crossrail as a project. For this reason, a vast amount of monitoring systems are being employed across London to ensure that potential ground movements caused by Crossrail works are understood and monitored so that appropriate actions can be taken should unexpected movements occur. This paper will focus on the monitoring of the Ranelagh Sewer, a Victorian brick sewer located approximately 230 m east of the Royal Oak Portal – the launch site for the C300 Western Running Tunnels.

Figure 1 Location plan of the Ranelagh Sewer with respect to the Royal Oak Portal

The Drive X Western Running Tunnels have been launched in sequence from the Royal Oak Portal (west of Paddington Station) and are boring eastward toward Farringdon Station, passing by Paddington, Bond Street and Tottenham Court Road Stations on the way. The tunnels are being excavated by a Tunnel Boring Machine (TBM) using Earth Pressure Balance Mode (EPB) as a control mechanism to reduce potential ground surface movements. Each tunnel has an excavated diameter of 7.1 m, although the concrete lining of the tunnel will only have an external diameter of 6.8 m. Although EPB method of tunnelling is designed to reduce ground movements as much as possible, movement is still expected and hence provisions must be made for third party assets that may be affected by this movement.

One of the first major obstacles identified by Crossrail's designers for the Drive X tunnel construction was Ranelagh Sewer. Ranelagh sewer is situated over the centre line of the proposed tunnel drive with a clearance of 4.6m from the invert of the sewer and the top of the TBM. The condition of the sewer was also a larger risk, which led to initial proposals to install a protective liner within the sewer to prevent collapse. However, as the expected damage was considered minimal by the designers, it was agreed that near real-time monitoring instrumentation should be installed to allow precise monitoring of the sewer during TBM passage.

Detailed discussions with Thames Water highlighted restrictions and long term safety concerns that would have led to the standard method of monitoring convergence and settlement using 3D geodetic methods being an unacceptable risk, Thames water also required any system that would be installed semi-permanently would have to cause minimal ragging and provide no restriction to flow.

The Shape Accel Array (SAA) was eventually proposed as it would satisfy all needs and restrictions but providing near real-time precise convergence and settlement monitoring of the sewer as well as minimise sewer man entry and obstruction to flow. This was the first time such an instrument was to be installed in a Sewer in the UK.

ASSET DESCRIPTION

The Ranelagh sewer is a Victorian brick tunnel. The sewer runs from the north-west to south-east and approaches the proposed Crossrail works from the North West. Above the east bound tunnel, it turns approximately 30^0 before exiting the interface area in a southerly direction. It is an egg shaped sewer with an approximate height of 2450mm and width of 2300mm. Local sewers serving the area are predominately large egg shaped sewers of varying sizes, with some smaller local sewers connecting into the main system. Although the existing ground levels slightly vary, the invert of the sewer is approximately 5m below the existing ground surface.

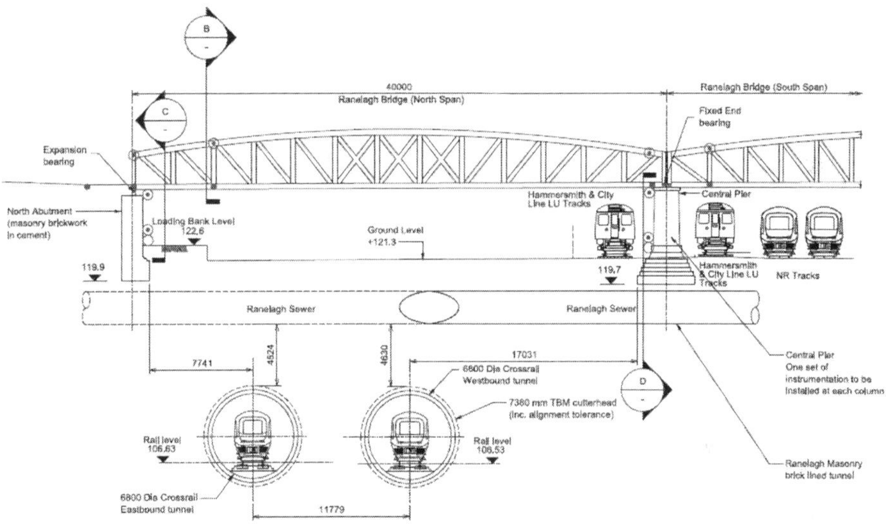

Figure 2 Section of Crossrail Running Tunnels showing the Ranalagh Sewer and Bridge in close proximity to the alignment

SHAPE ACCEL ARRAY

Shape Accel Arrays (SAAs) are manufactured in Canada by Measurand Inc. It is a rope-like array of sensors and microprocessors used to measure deformation or vibration. It is made up of a number of rigid segments either 500mm or 200mm in length (depending on the location to be monitored) and has a diameter of 25mm. Each segment has three MEMS (Micro Electro Mechanical System) accelerometers and the segments are connected by flexible joints that can move in any direction but cannot twist. There is one processor and temperature sensor for every eight segments. The in-array microprocessors collect and send signal data through a single wire output. Typically, the instruments arrive on a reel and can be up to 100m in length.

Figure 3 Shape Accel Array on a reel delivered to site

The SAAs are installed in 27mm diameter PVC casing and can be installed either vertically or horizontally, or in the case of the Ranelagh Sewer, both. Any deformation that moves the casing is accurately measured as a change in shape of the SAA. They are also re-usable and are able to withstand huge deformations.

The specifications of the instrument are:

- Resolution: 2 arc seconds
- Accuracy: ±0.5mm/32m when the casing is grouted in place
  ±1.5mm/32m when the casing is not grouted
- Pressure: tested to 100m depth water
- Vibration: can measure ±2G

Data

Data can be collected from the SAAs either manually by use of a laptop or tablet computer and a SAA-PC Interface Box or in real time wirelessly via a data logger. The data logger can be powered wither by a 12v DC battery or by 110v AC.

Calibration

After production, every sensor in every SAA is calibrated through a range of multiple temperatures and multiple orientations. Calibration constants of offsets, gain g/count, tcoeff cc/degc and temp gain factor are calculated and are programmed into the data logger specific to the SAA. The calibration constants are downloaded with the SAARecorder software for manual readings and programmed into the individual data logger boxes for each SAA. The only calibration procedure required after installation is to zero the SAA's when installed to produce a zero value for the SAA which will be used to compare all future readings to. As the SAAs are factory calibrated and sealed, they require no field assembly or calibration.

REQUIREMENT FOR MONITORING

In order to determine the expect ground movements on the asset due to tunnelling works, an assessment of ground movement effects on the Ranelagh Sewer was carried out by the Crossrail C122 contract. The purpose of this assessment was to define the predicted ground movements associated with the tunnelling and calculate the expected induced bending and axial strains on the sewer concluding with defining the required mitigation.

The ground movements were assessed based on empirical methods using Mair et al. (1993). The modelling parameters used in the assessment are considered to be moderately conservative and are shown in the table below:

Modelling Parameter	Value
Volume Loss	1.0%
Trough width parameter, k	0.5

The assessment considers 'greenfield' conditions, which do not take into account the stiffness of the sewer or adjacent assets, which includes the Ranelagh Bridge and its foundations.

Based on this assessment, the maximum calculated settlement was found to be 26mm along the Sewer and the lateral displacement was predicted to be 4 mm maximum. The maximum longitudinal displacement (along the axis of the sewer) was calculated to be 5 mm. Figure 4 below shows the results of this assessment.

The assessment recommended that a monitoring system should be installed in order to observe the response of the sewer due to ground movements and also to repair resulting damage if it occurred. The report also suggested that a Volume Loss Control Zone be defined in the Contractor's specifications.

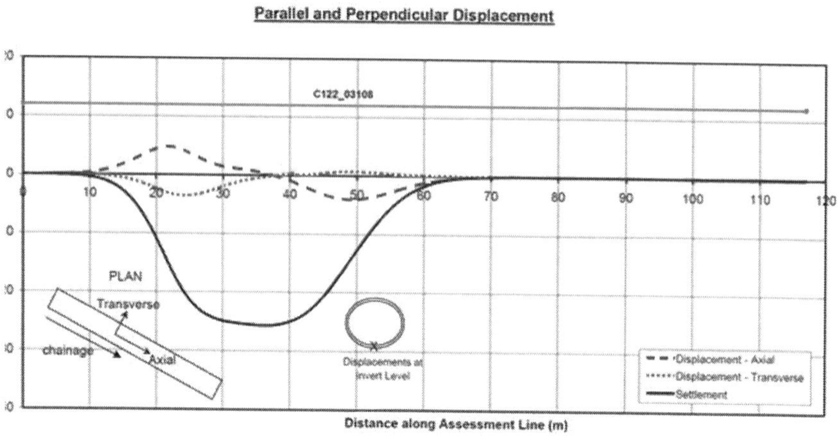

Figure 4 Results of the C122 (Arup/Atkins) ground movement assessment on the Ranelagh Sewer

As a result of this assessment, the Specification for the Control of Ground Movements for the C300 tunnel drives sets a Volume Loss control Zone of 0.5% in the area of the Ranelagh Sewer. This control zone requires that the contractor increase TBM control in order to further reduce surface ground movements (by increasing face pressures and tail-grout pressures, for example).

BFK ASSESSMENT FOR 0.5% VOLUME LOSS

A further assessment of ground movement was undertaken by the contractor Bam Ferrovial Kier JV in order to determine the expected movements of the Sewer due to the passage of the Westbound TBM only beneath the asset assuming the contractual requirements – a maximum of 0.5% volume loss. The results of this analysis would help understand the contractual limits to which the contractor was to control the works.

With the exception Volume Loss and the trough width parameter (k), the model used assumed the same greenfield parameters as those for the above mentioned assessment by the Client's designers. An xDisp model was run based on the above parameters and the following plots were produced:

Figure 5 below shows the calculated settlement for both the crown and invert of the Sewer, based on VL = 0.5%. Also given is the calculated longitudinal elongation of the asset. Figures 6 and 7 below show the calculated vectoral movement of the Sewer itself, at a transect directly over the Crossrail Westbound TBM alignment. Figure 6 shows the absolute movement whereas Figure 7 gives the differential movements.

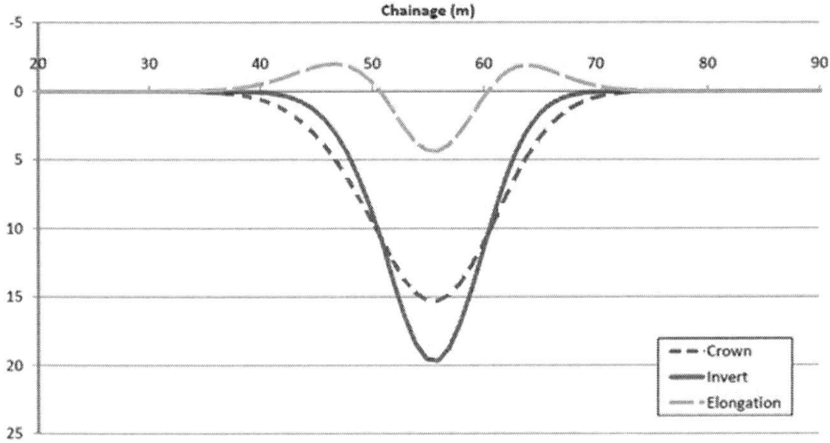

Figure 5 xDisp results from BFK assessment of ground movements with VL = 0.5%

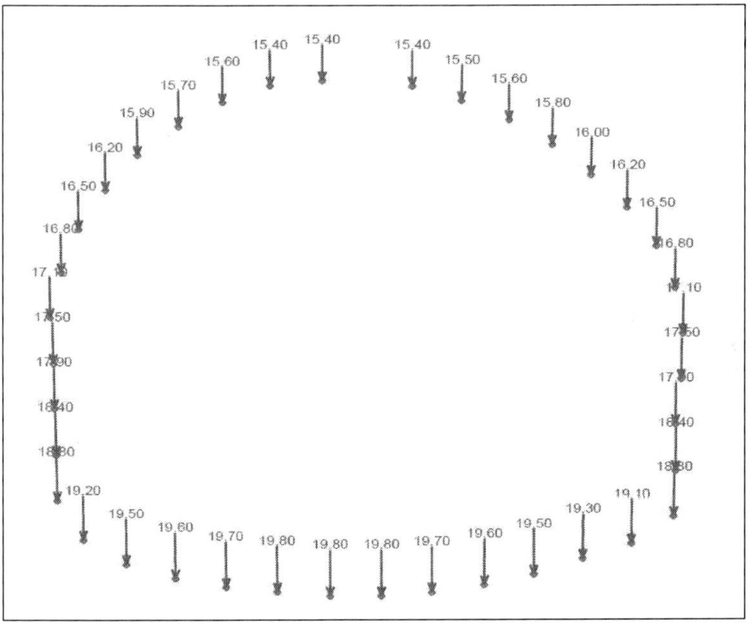

Figure 6 Absolute movement of Ranelagh Sewer at section over WB TBM centreline, VL = 0.5%

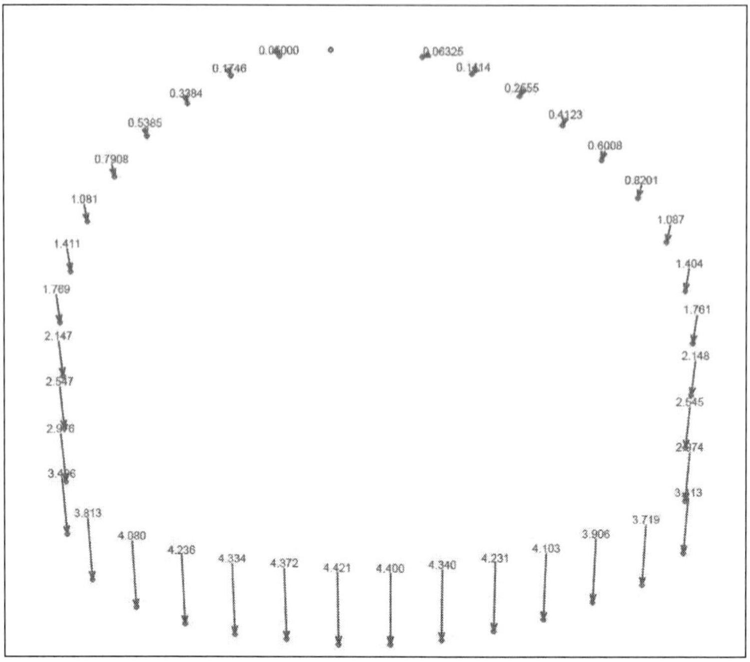

Figure 7 Differential movement of Ranelagh Sewer at section over WB TBM centreline, VL = 0.5%

The results show maximum settlements of approx. 20 mm and differential settlements between sewer crown and invert of less than 4mm.

Although the assessment shows a reduction in calculated movement, and minimal differential movement, the requirement for monitoring was based on an assessment of 1% volume loss. A real-time automated system was therefore still required to help determine whether or not damage was being caused to the sewer. Furthermore, the accuracy of the SAA, as well as its ability to be installed circumferentially within the sewer would allow a comparison of actual movement and calculated movements (as per Figure 7) to be carried out.

TRIGGER VALUES

No specific trigger values were defined for this asset by either the Designer or Thames Water. However, some Alert values were defined for construction control purposes. A values of 25 mm displacement was defined by the Designers as being an Alert, as was a slope of 1:2000 to be measured in the crown of the Sewer. Furthermore, a slope of 1:5000 was given as an Alert value for the surface movement directly above the Sewer, which was being measured by Precise Levelling Points (PLP).

INSTALLATION

A total of three SAA were installed in the Ranelagh Sewer; 2No 7.4 m long installed in the intrados of the sewer and 1No 43 m length installed horizontally in the crown of the sewer, directly above the two Crossrail Running Tunnel centrelines. Each SAA consists of rigid segments, of lengths that depend on the location in which they are installed.

Due to the internal profile of the sewer, a SAA with shorter segments was used, (200 mm segments). This was to enable the SAA to form the internal profile of the sewer as closely as possible, to obtain the most accurate data in relation to any possible movement within the structure of the sewer. The 200mm segment lengths were manufactured especially for this application.

A third SAA of 500mm segments length was installed along the crown of the sewer, again by setting out an equal distance on each side of point C. This is also illustrated in Figure 8, whereby each green square represents a sensor location in the 43 m long horizontal. A data logger was also installed, to allow the data from the SAAs to transmit to an on line monitoring system.

The SAAs in locations B and D (see Figure 8) were installed directly over the centrelines of the Westbound and Eastbound Crossrail tunnels respectively. The installation team was advised on the required locations prior to the installations and these were subsequently set out measuring from point C which is the centre point between the two tunnel drives.

Figure 8 SAA Locations

Figure 9 SAA fixed to crown of Sewer. Image Credit: Gary Evans

The installation of the SAA on the crown of the sewer was carried out in the following manner.

Once the location of the SAA was set out and marked, holes were drilled in the sewer lining. This was achieved by using a PAT tested battery powered drill. The next stage was to install the 27mm (internal diameter), (1" trade size), electrical conduit, to the crown of the sewer, which will be referred to as casing, in this document. This casing was supplied in 3m lengths and was joined in the sewer using the flared couplers at each end, which were glued together. The casing was fixed along the crown of the sewer at intervals between 0.5m and 1m, using marine grade stainless steel P – clips.

The SAA was then pulled through the casing, while ensuring that the X axis on the instrument was always facing in the direction in which the expected movement was likely to occur.

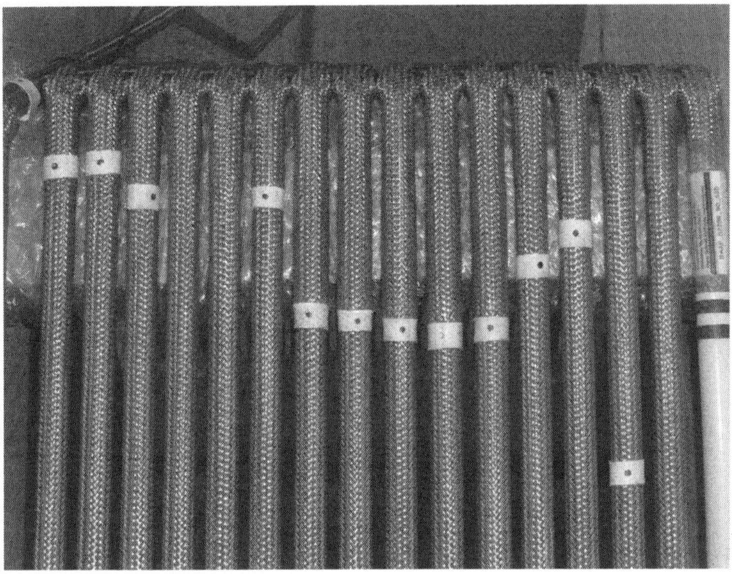

Figure 10 Black dots indicating x-axis of the instrument

The installation of the SAAs around the inner lining of the sewer was carried out as follows:

Upon establishing the locations where the SAAs were to be installed within the sewer – these being locations B and D in Figure 8 – holes were drilled in the same manner as in the previous SAA installation.

The principal difference for the installation of the circumferential SAA was that the instruments were supplied in a special reinforced clear plastic tubing, from the manufacturer. This was to ensure that the SAA was protected once installed, as it was not possible to use the rigid casing in this situation due to the circular shape of the sewer. The SAA was put in place and fixed directly to the sewer lining as previous.

Upon completing the fixing of the instruments around the inner lining of the sewer, a further layer of 3mm thick butyl rubber sheeting was placed over the flexible casing of the instrument as shown in Figure 11.

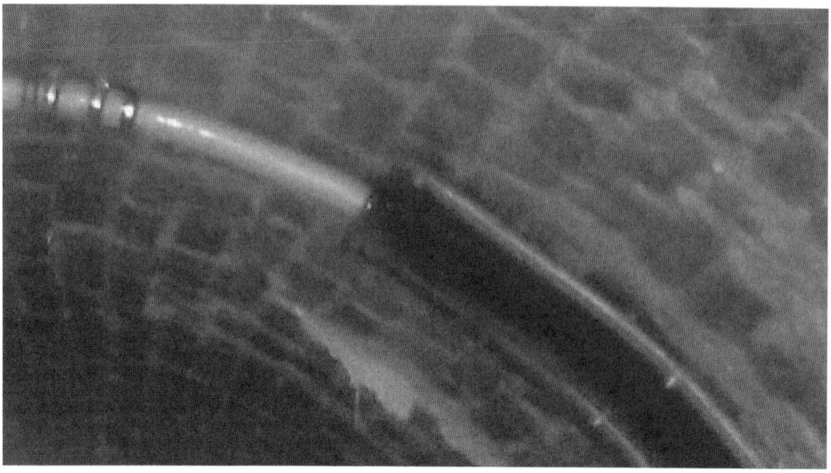

Figure 11 3 mm thick butyl rubber sheeting fixed over SAA Image Credit: Gary Evans

The purpose of the rubber sheeting was primarily to mitigate against the possibility of build-up of debris behind the instrument time, as well as ensure Thames Water requirements to avoid any restriction to flow would be met (see Figure 12).

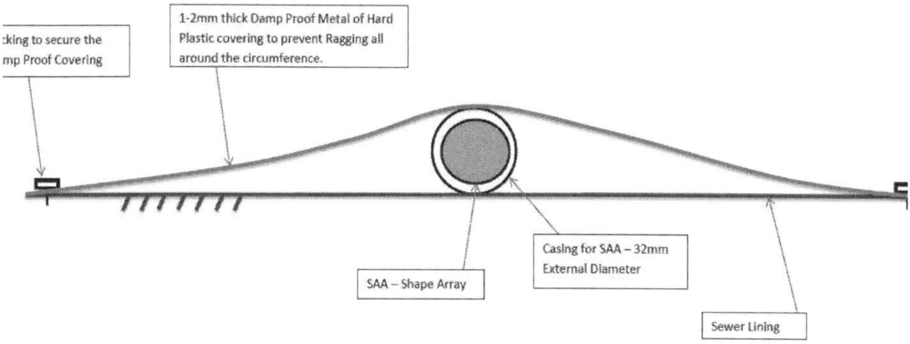

Figure 12 Cross section view of butyl rubber sheeting fixed over SAA

Upon completion of the installation of the three SAAs in the sewer, the data cable was extended from each SAA and fixed along the inside of the sewer as far as the entrance to the sewer. The data cables were brought up vertically inside the manhole chamber and exited the chamber beside the manhole cover onto a secure area. The baseline values for the SAA where taken a few weeks after installation, to allow for the instrument to adjust and to observe any background interference that may have occurred due to the location.

MONITORING RESULTS

TBM Control

In order to ensure the specified volume loss of 0.5% was achieved in the vicinity of the Ranelagh Sewer, the contractor took the decision to gradually increase its control parameters to help reduce surface settlements. The face pressure (EPB) and the volume of tail grouting injected at ring build were increased slightly and maintained at higher rates while the TBM passed beneath the Sewer.

The zone of influence of the sewer was defined from ring number 144 and the pressure was maintained until ring 162 (a distance of approx. 30 m) when a gradual decrease was implemented once the sewer was no longer within the TBM Zone of Influence. This increase and decrease of the EPB can be seen in Figure 13.

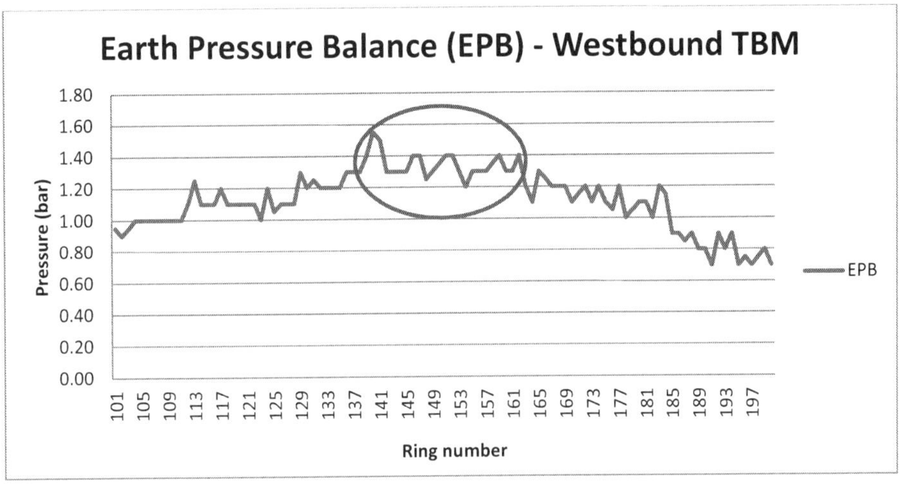

Figure 13 Earth Pressure Balance (EPB) – Westbound TBM

SAA Monitoring

Monitoring of the Sewer was carried out continuously during passage of the TBM, with two-hourly updates being sent to the TBM control team as well as the contractor's engineering team. In addition, the monitoring results would be reviewed twice daily (once a shift) by the Shift Review Group – a committee comprising of representatives from the contractor, designer and client.

SAA in the Crown

The data from the SAA located in the crown of the Sewer provided particularly useful data with respect to the rate of settlement during the TBM drive. Figure 14 below shows the vertical movement of selected sensor points over a period of two weeks including the time of

TBM passage. The accuracy and consistency of the readings is such that the movements can be related back to precise tunnelling activities, giving a clearer understanding of how TBM operations cause ground movement. The data shown in Figure 14 shows three main 'movement's': a first initial settlement (30/06/2012) followed by a more sustained and steeper settlement (01/07/2012) before a final slight heave (02/07/2012). These movements correspond to settlement caused by the cutter head, followed by settlement over the shield and at ring build location and finally the heave caused by pressurised grouting behind the finished ring.

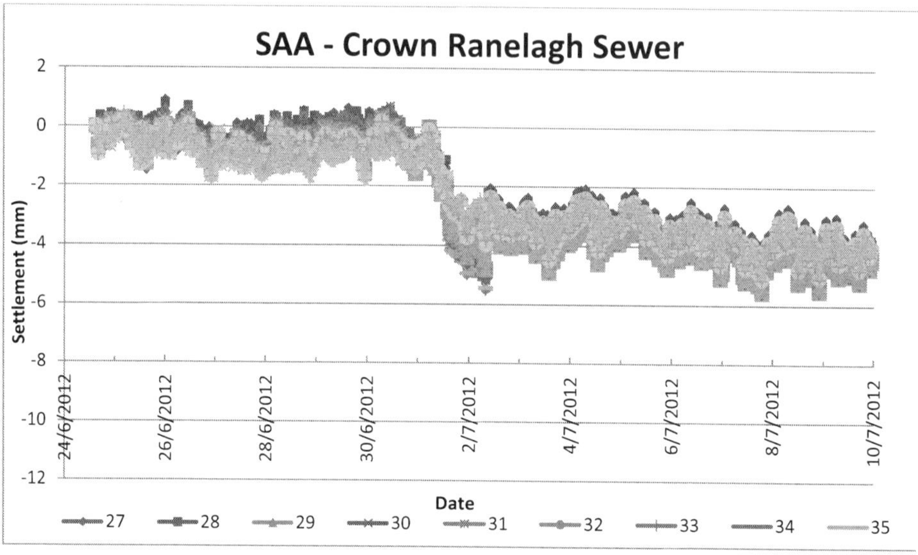

Figure 14 Time plot of settlement recorded on the SAA installed on the sewer's crown.

Figure 15 shows a profile of the whole SAA giving the settlement at different times, allowing the progression of settlement of the length of the sewer to be studied

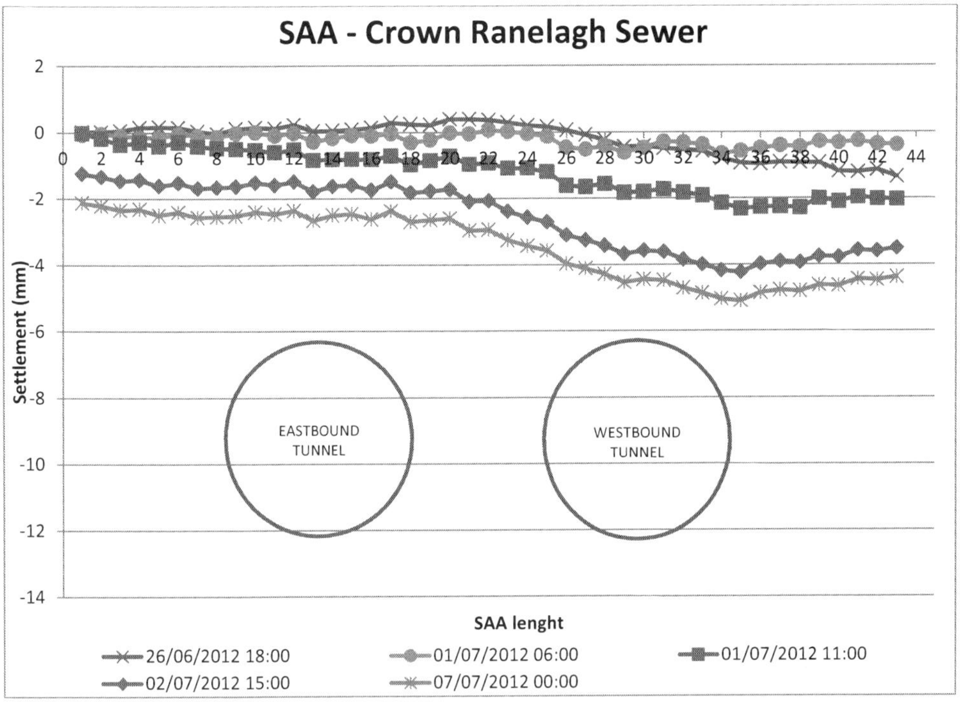

Figure 15 Profile of settlement recorded on the SAA installed on the sewer's crown.

SAA in the Sewer Intrados

The expected relative (or differential) movements of the intrados of the Sewer were calculated to be minimal. The data presented below in Figure 16 shows clearly that little or no relative movement was observed in the circumferential SAA. The scale of the graph, which was chosen to help show graphically not only the whole Sewer shape, but also to 25mm Alert value (represented as the green line). This method of reporting was used during construction as a means of determining any potential trends of movement towards the Alert value.

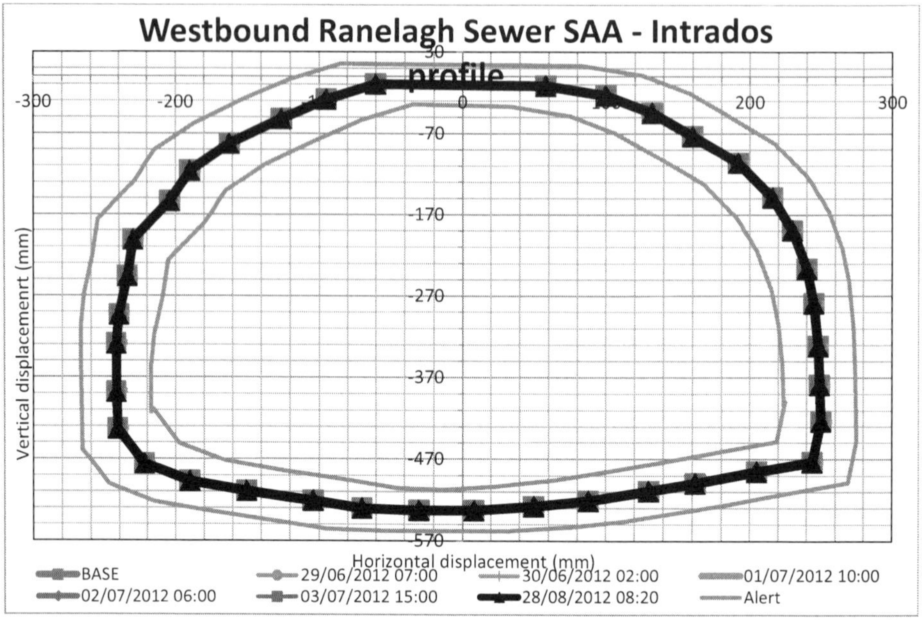

Figure 16 Intrados profile – Westbound SAA at Ranelagh sewer

One important aspect of the monitoring of the intrados is that only relative movement could be observed. As the reference cell of the SAA shown in Figure 16 was in fact in the crown and not located externally, it too was subject to movement. Absolute movement can therefore be obtained by adding the observed movement in the crown of the Sewer (given by the crown SAA) to the relative movement of the intrados SAA.

CONCLUSION

The use of SAA as a method of providing near real-time internal monitoring of large diameter sewers has proved extremely successful. This form of automated monitoring provided a suitable solution to overcome the limitations of monitoring inside a hazardous confined space.

The information received from the SAA not only provided and insights into utility movement, but also the effect of TBM parameter adjustments in lower cover situations. The use of the SAA data and the ease and success of the installation has led to further assessment of services along the route of the TBM drive. It is hoped that the data will be back analysed by asset protection designers and the utility companies to allow more informed decisions on future installations and alarm limit parameters. Moreover, detailed and precise monitoring of utility movement will enhance considerably the understanding of how similar assets react to tunnel induced movement. Such monitoring data can therefore help design engineers produce more efficient designs based on realistic empirical data,

Acknowledgements

The authors would like to acknowledge the support of the following companies and persons while conducting the fieldwork and preparing the manuscript: Crossrail, Thames Water, Bam Ferrovial Keir JV, Keller Getec ITMsoil JV and Geo Observations. Particular acknowledgment should be made to Gary Evans, who prepared documentation and methods to ensure compliance, and David Harris, John Roseler and Paul Thurlow for their technical help and guidance.

REFERENCES

1. Mair R J, Taylor R N and Bracegirdle A (1993) *Subsurface settlement profiles above tunnels in clays,* Geotechnique, Vol. 43 No. 2, pp 315-320
2. Ridley A., Kavanagh J., Lardi M., Saffari N., *'Results from a Trail of a New Displacement Monitoring Tool (Shape Accel Array)'*
3. C122 *Assessment of ground movement effects on the Ranelagh Sewer,* C122-OVE-C2-RAN-CRG01-00061 Rev 1.0
4. C300 *Installation Report for the Installation of Shape Accel Array (SAA) in Ranelagh Sewer* C300-BFK-C2-RGN-CRT00_ST005-50001 Rev 1.0

Crossrail

Crossrail Project: Infrastructure design and construction
ISBN 978-0-7277-6078-4

doi: 10.1680/cpid.60784.239

ice Institution of Civil Engineers publishing

The Use of Shape Accel Arrays (SAA) for Measuring Retaining Wall Deflection

R. Lipscombe, Keller Getec ITMSoil JV
C. Carter BEng, Keller Getec ITMSoil JV
O. Perkins MEng, Bam Ferrovial Kier JV
P. Thurlow, Keller Getec ITMSoil JV
S. Guerrero MICE CEng MBA, Bam Ferrovial Kier JV

Abstract

The measurement of deflection of designed diaphragm walls is critical for near real time information to assess wall and soil behaviour as the internal area is excavated in a controlled manner. Data enables the site team and designers to check performance of the design, which is linked to safeguarding the site and the external stakeholders.
This paper documents the use of Shape Accel Arrays at the Tottenham Court Road Western Ticket Hall during excavation and propping within the diaphragm wall for the Crossrail project and how it is used within the site settlement management plan to assist in the control of the works and the information to be used for Value Engineering purposes that benefitted the project.

Introduction

It is usual practice to instrument diaphragm walls with inclinometers and these are read either manually or automatically as required by the specification or excavation progress. The deflection is measured as a cumulative plot between each reading. Manual inclinometer readings are usually taken once a day at pre-set times and the data interpreted and displayed on specific software. The readings are taken at half metre intervals, working from the bottom upwards. In place inclinometers are usually installed or fixed at specifically designed centres within the reservation tube installed while the diaphragm wall is cast. The readings are taken automatically via a data logger and sent for processing.

These systems of measurement have been sufficient in the past but are limited by several factors which include, repeat ability of measurements using manual techniques, and the spacing of sensors using in place inclinometers.

The introduction of Shape Accel Arrays for the measurement of deflection within a diaphragm wall is relatively new in the UK construction industry. The instrument is successfully used for real time monitoring of slopes and embankments and has been used for reviewing positions of jet grouting drilling positions and sewer monitoring in the UK previously. The SAA is very quick to install and base read with the data output configurable for various monitoring software programmes.

Sensor Design and Specifications

Shape Accel Arrays (SAAs) are manufactured in Canada by Measurand Inc. An SAA is a rope-like array of sensors and microprocessors used to measure deformation or vibration. It is made up of a number of rigid segments either 500mm or 305mm in length and has a diameter of 25mm. Each segment has three temperature-calibrated MEMS (Micro Electro Mechanical System) accelerometers and the segments are connected by flexible joints that can move in any direction but cannot twist. There is one processor and temperature sensor for every eight segments. The in-array microprocessors collect and send signal data through a single wire output.

The 3D shape of SAA in near-vertical casing is determined from static accelerations of X and Y accelerometers. In near-horizontal mode, the Z accelerometers are used to determine 2D shape in a vertical plane.

The SAAs are installed in 27mm diameter PVC casing and can be installed either vertically or horizontally. Any deformation that moves the casing is accurately measured as a change in shape of the SAA. SAAs are re-usable and are able to withstand huge deformations. Data can be collected from the SAAs either manually by use of a laptop or tablet computer and a SAA-PC Interface Box or in real time wirelessly via a data logger. The SAA features the following :

- Resolution: 2 arc seconds
- Accuracy: ±0.5mm/32m when the casing is grouted in place, ±1.5mm/32m when the casing is not grouted
- Pressure: tested to 100m depth water
- Vibration: can measure ±2G

After production, every sensor in every SAA is calibrated through a range of multiple temperatures and multiple orientations. Calibration constants are calculated and are programmed into the data logger specific to the SAA. The only calibration procedure required after installation is to zero the SAAs when installed to produce a zero value for the SAA which will be used to compare all future readings to.

Figure 1 – An SAA on the reel, prior to installation and view of logger box and inclinometer.

The Site

Tottenham Court Road Station is located in London's West End, where Oxford Street meets Tottenham Court Road. Crossrail is working alongside Transport for London on a £1bn transformation of the station, which includes the upgrade of the existing London

Underground station, as well as the construction of a completely new station specifically for Crossrail. The construction works involve the building of two brand new ticket halls; the Eastern and Western Ticket Hall (WTH) (see Figure 2).

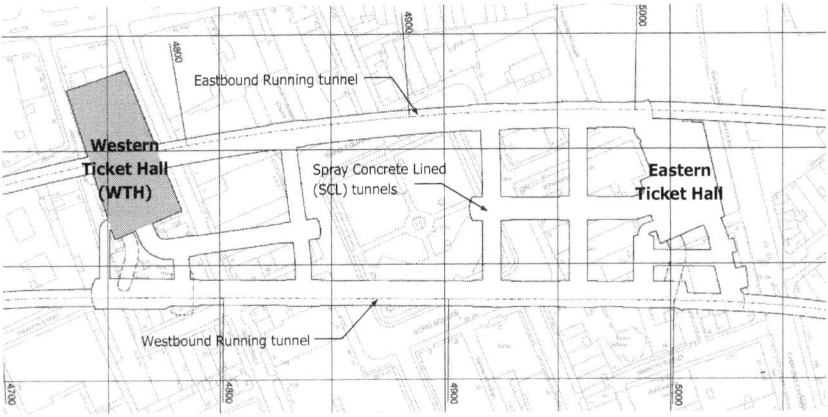

Figure 2 – Location Map showing the new Tottenham Court Road Station

The excavation of the Western Ticket Hall is divided into three zones; a 29m deep southern section, a 13m deep northern section, and a shallow excavation at the northern end of the site associated with the over-site development. The diaphragm walls and piles for both the southern and northern sections were completed in early 2012 by BBMV (a joint venture between Balfour Beatty, Morgan Sindall and VINCI Construction). The excavation is being undertaken in two phases: the southern (deep) box and the northern box, with the southern box being constructed first to allow Spray Concrete Lined (SCL) tunnels to be constructed prior to the arrival of the Tunnel Boring Machines (TBMs). The southern box is approximately 41m by 31m and the northern is approximately 27.5m by 22m.

This paper will document the use of SAAs within the diaphragm wall for monitoring deflection in the southern section of the WTH during excavation and temporary propping works. This work falls under C300/410 contract; the principle contractor for these works is Bam Ferrovial Kier – Joint Venture (BFK) and the sub-contractor for Instrumentation and Monitoring (I&M) is KGi JV (a joint venture between Keller Getec and ITMSoil). Excavation of the southern box commenced in April 2012.

Plate 1 – View of props within the excavation

In the original scope of works it was proposed that there would be five levels of temporary props to be installed by C300/410 during the excavation. BFK submitted a Value Engineering (VE) proposal to use the observational method at the TCR WTH south box excavation to omit the P5 propping level. This was accepted by Crossrail. A cross-section through the TCR WTH south box is shown in Figure 3.

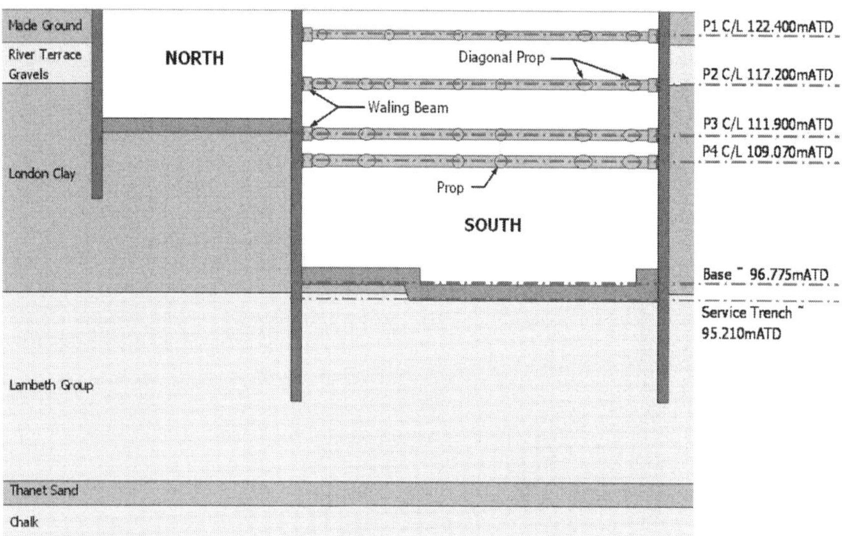

Figure 3 – Cross-Section of the TCR WTH

Installation of SAA into the Diaphragm Wall alongside Manual Inclinometers

At the Tottenham Court Road Western Ticket Hall site eight no. SAAs were installed at locations around the diaphragm wall. The SAAs were installed in vertical 27mm diameter PVC tubes alongside standard inclinometer casing into 140mm diameter reservation tubes. The casing was grouted into place within the reservation tubes, see Figure 4. The reservation tubes had been pre-installed into the diaphragm walls to facilitate installation. The SAAs installed were of varying lengths, dependant on the panel length of the diaphragm wall. The SAA's were installed to the bottom of the panel and not drilled through the toe of the diaphragm wall as it was considered that the requirement of a 10m anchor would not be sufficient in the ground conditions. The SAA's were made up of rigid segments of length 500mm. In all instances the SAA were installed with the x-axis facing into the excavation. Eight no. data loggers were also installed, one for each SAA, to allow the data to be sent via GPRS to a central FTP server and downloaded to the site monitoring database.

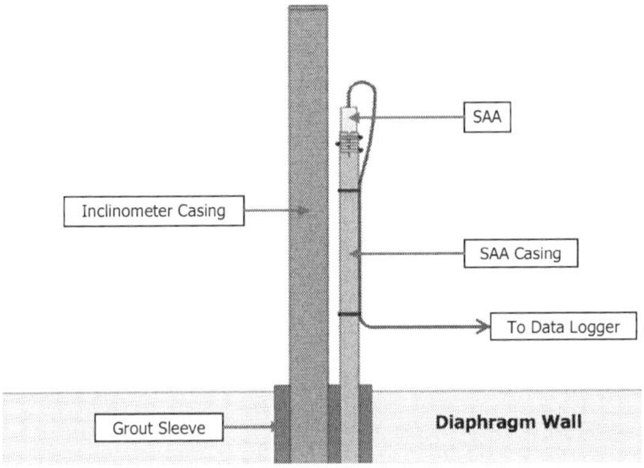

Figure 4 – Diagram of an SAA installed alongside Inclinometer Casing

The excavation and installation of temporary props was divided up into nine stages. After the VE proposal to omit the temporary P5 props was accepted these stages were redefined as shown in Table 1.

Stage:	**Description of Works:**	**Completion Date:**
2	Excavate to 121.400 mATD.	30/04/2012
3	Prop P1 and excavate to 116.200 mATD.	P1 propping complete 22/05/2012 Excavation to 116.200 mATD complete 15/06/2012
4	Prop P2 and excavate to 110.900 mATD.	P2 propping complete 06/07/2012 Excavation to 110.900 mATD complete 15/07/2012
5	Prop P3 and excavate to 108.070 mATD.	P3 propping complete 27/07/2012 Excavation to 108.070 mATD complete 05/08/2012
6	Prop P4 and excavate to 100.970 mATD.	P4 propping complete

		14/08/2012 Excavation to 100.970 complete 04/09/2012
7	Excavate to 97.285 mATD (500mm above hold level).	04/09/2012 – 19/09/2012
8	Excavate to 96.775 with structural binding strut.	19/09/2012 – 27/09/2012
9	Excavate service trench to 95.210 mATD.	23/09/2012 – 26/09/2012

Table 1 – Stage summary and completion dates

As part of the Management Plan for the Control of Ground Movement (MPCoGM) for the WTH excavation, Amber and Red triggers of displacement of the diaphragm wall were provided by Crossrail. The Red trigger was calculated by the designer, Arup Atkins (C134), based upon the design expected deflection at a percentage of the load, and the Amber trigger was calculated as 75% of the Red trigger. Observed displacement is compared directly with the trigger values. Any trigger breach results in immediate verbal notification of the BFK Monitoring Engineer and the BFK Sub-Agent by the KGi JV I&M Engineer. If the trigger breach is confirmed all BFK members of the Contract Technical Committee (CTC) are notified with the BFK Sub-Agent instigating any actions required to ensure the safety of the works, the public and third-party assets. E-mail notification detailing the specific trigger breach and rate of movement, as well as any specific actions taken or planned, was sent to all CTC / Shift Review Group (SRG) / Engineering Review Panel (ERP) members. An emergency CTC or Emergency Engineering Response Team (EERT) meeting is held dependant on the risk associated with the trigger breach. Target response times of 4 hours and 1 hour for Amber and Red triggers respectively were detailed in the Specification for the Control of Ground Movement (SCoGM).

The following actions and communications at trigger levels are detailed in Table 2.

Trigger:	**Communication:**	**Action:**
Green	Review at SRG meeting. No further action unless the rate of change of movement is sufficient to warrant a special CTC meeting.	Formal review of instrumentation data compared to excavation progress. Pre-emptive action to be taken if trends indicate that the Amber trigger is likely to be exceeded. Potential impact on buildings and utilities to be reviewed.
Amber	Verbally notify all Monitoring and members of the Contract Technical Committee. all members of CTC/SRG/ERP to meet within four hours..	Inform Project Manager immediately providing records of construction activities prior to the trigger breach. Implement contingency plans. Review data for consistency and identify trends with respect to trigger levels at each stage of excavation. Investigate the possible impact of the environment and any other works in the vicinity. Review the wall deflections and decide if changes to the construction method or sequence are necessary to ensure the Red trigger is not breached. Review potential impact on buildings and utilities.
Red	Verbally notify Monitoring Engineer and members of the CTC. All members of CTC/SRG/ERP to meet within one hour	As per Amber trigger breach, but stop all construction activities that may lead to further increase in movement around trigger breach. Backfill area around trigger breach with spoil. Seek instruction from the Project Manager on how to proceed.

		Review potential impact on buildings and utilities.
Black	Call emergency services as appropriate. Phone select on-call representatives and contractors reaction line.	Immediately put in place any actions required to ensure the safety of the works, public, and third-party assets. As necessary; evacuate excavation, call emergency services, backfill against walls from stockpile, place structural binding strut, mobilise priority props to site.

Table 2 – Communication and action to be taken in case of Trigger breach

The Black trigger has no numerical value, but is reserved for when the stability of the excavation is deemed at risk due to a dangerous condition being observed.
The VE proposal to omit the P5 props was based upon the use of the observational method. The designer undertook a back-analysis of the observed behavior of the diaphragm wall during excavation to P4 level. From this the calculations were refined and new triggers defined for the excavation to base slab level. A back analysis with the 'most probable' soil parameters is currently being treated as a Green trigger, the Amber trigger is mid-way between the 'most probable' and 'moderately conservative' soil parameters, and the red trigger is based upon 'moderately conservative' soil parameters. The most recent stage triggers and the calculated back analysis for stages 1 to 9 are shown in Figure 5. The superseded triggers are also shown on the graph, along with the levels of strata, appropriate prop levels, the top of the diaphragm wall and capping beam, and the stage excavation level for reference.

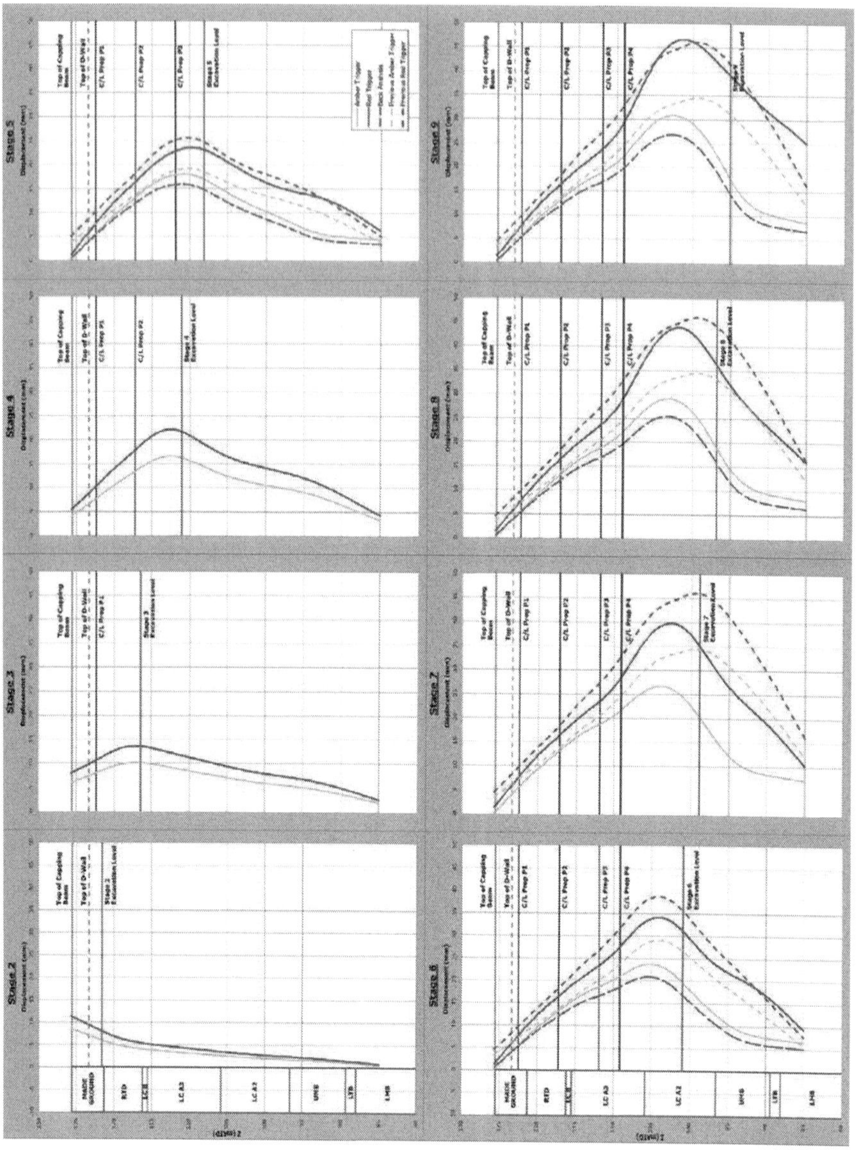

Figure 5 – Graphs showing the Stage Triggers for Stages 2 to 9

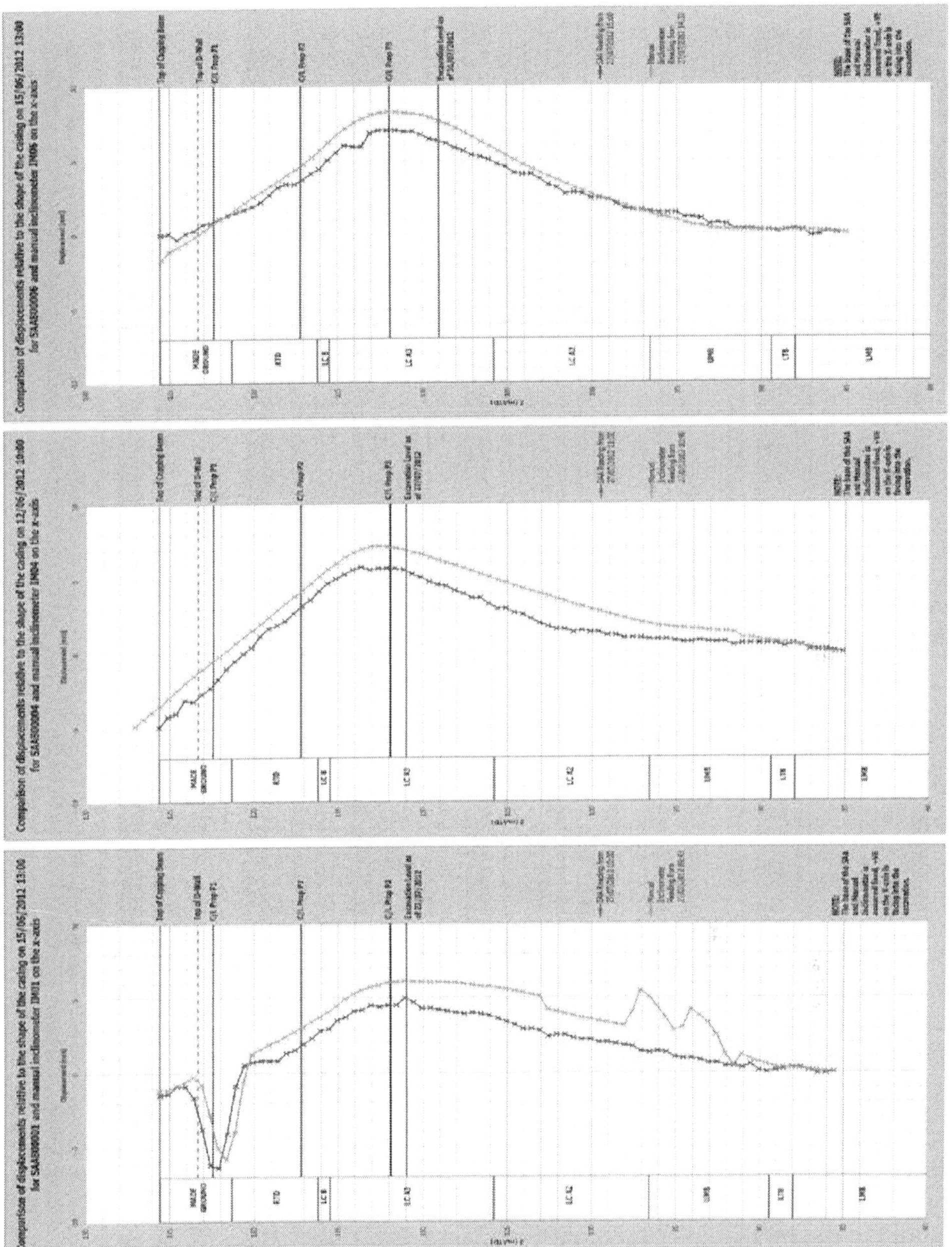

Figure 6 – Comparison of displacements of the SAA and Manual Inclinometer
in three locations around the Diaphragm Wall

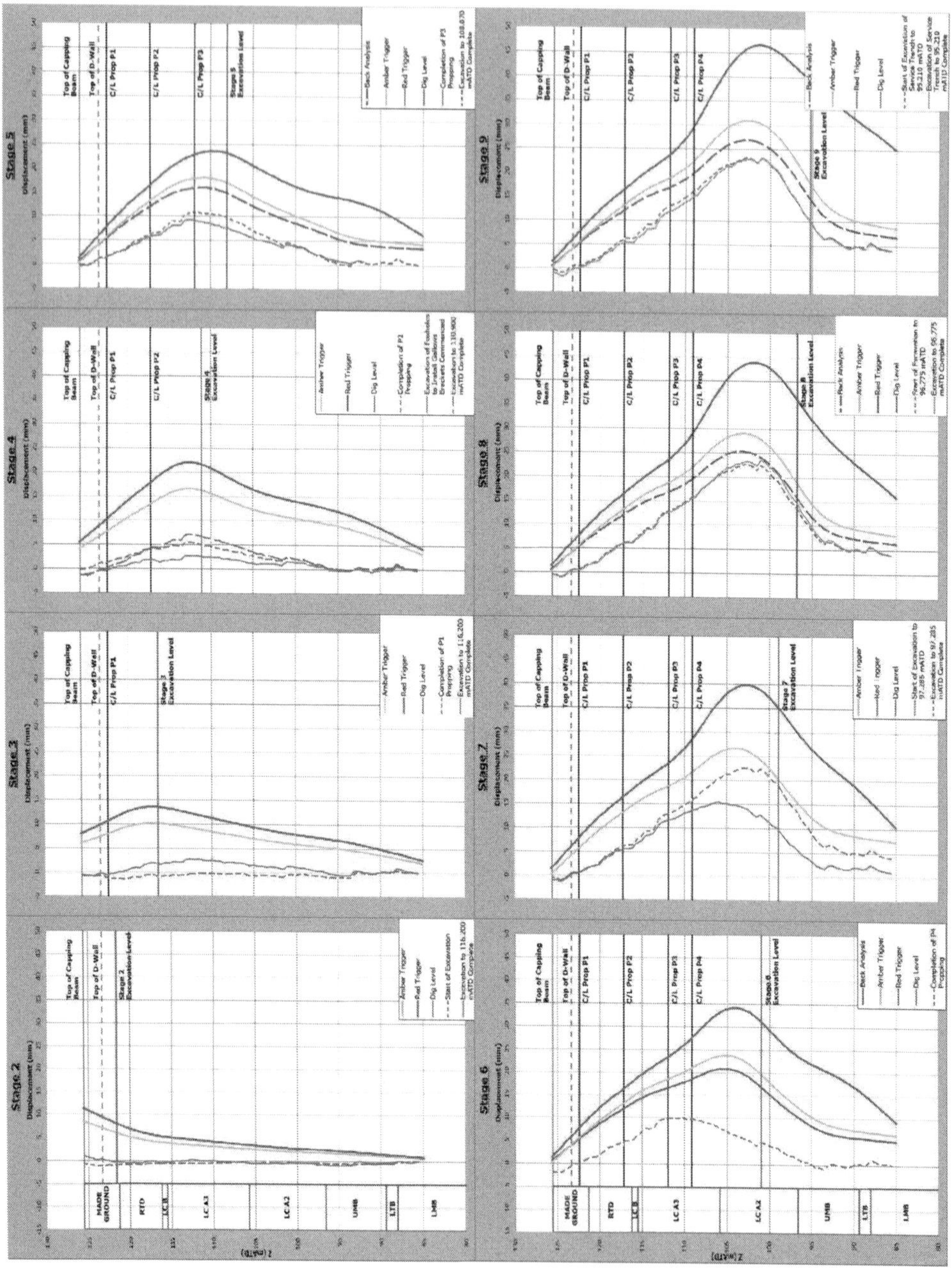

Figure 7 – Graphs showing the movement of the Diaphragm Wall throughout Stages 2 to 9

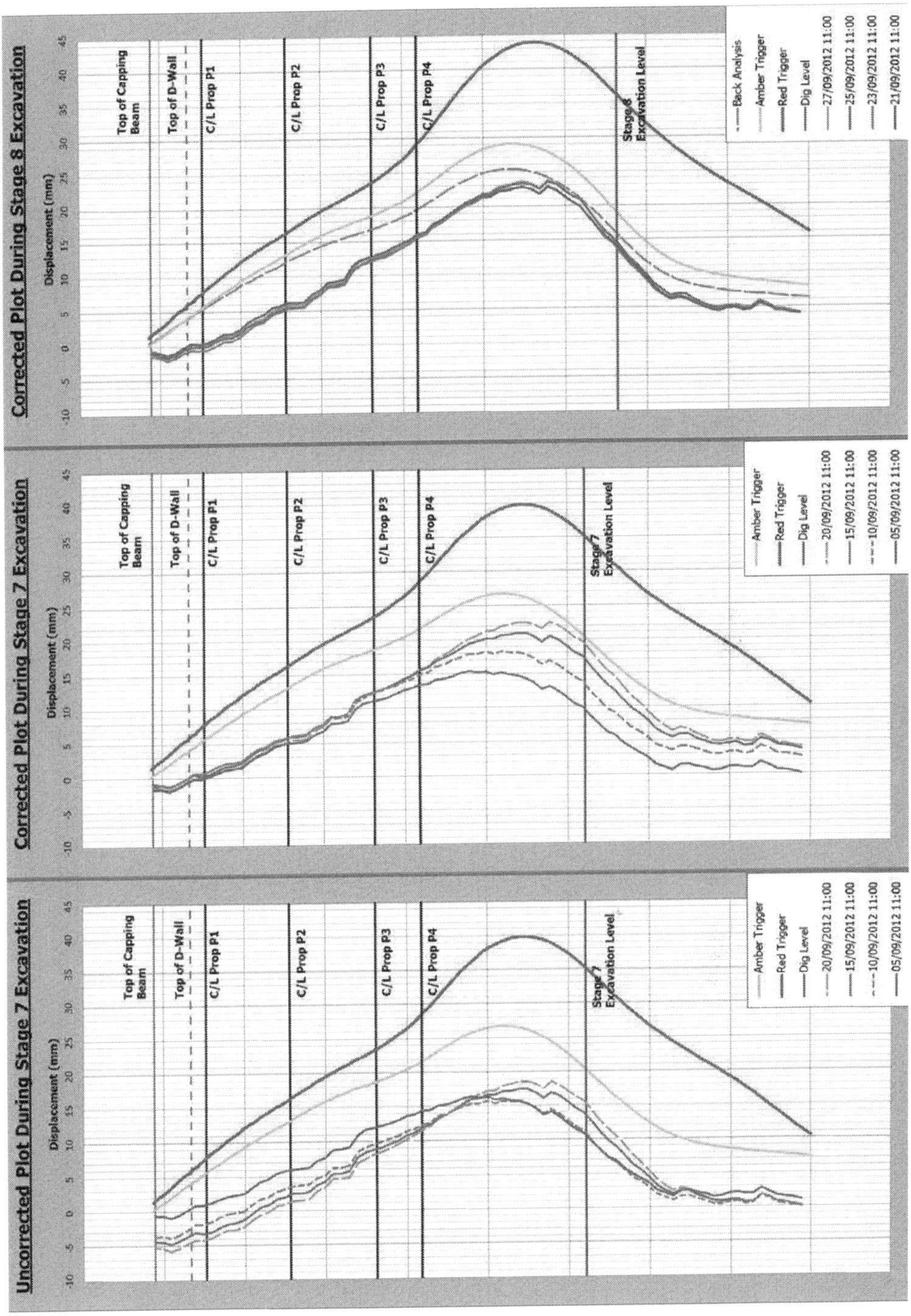

Figure 8 – Graphs showing the movement of the Toe of the Diaphragm Wall

Results and Discussion

Figure 6 shows a comparison of the displacements of the diaphragm wall, as read by both a manual inclinometer probe and the in-place SAA, on the same date and time, taking the same reference time for both the manual inclinometer readings and the SAA readings. The centre line of the props installed and the excavation level at the time of the readings has been indicated on the graph as well as the top of the diaphragm wall, the top of the capping beam, and the different strata.

A good comparison is shown on Figure 6 between the manual inclinometer readings and the SAA readings, showing the accuracy and reliability of the relatively new Shape Accel Array compared to the manual inclinometer probe. The SAA seems to show a more accurate profile of the casing, with the graphs showing even the slightest deformation of the casing due to the high resolution of the instrument and the snug fit of the instrument within the casing. As can be seen from the comparison for SAA80001 and IM01, the manual inclinometer readings between the Z values of 91mATD and 98mATD show a significant difference of up to 5mm from the SAA readings. This is most likely due to an error in the manual inclinometer readings, possibly due to site conditions and the inclinometer probe not being given a chance to stabilise before the reading was taken.

Figure 7 shows the movement of the diaphragm wall as shown by SAA800006 at the start, completion of propping (if relevant), and the completion of the excavation for each stage as detailed in Table 1. As with Figure 5, the Amber trigger, the Red trigger, and the back analysis (Green trigger) are displayed on the graphs. The graphs show movement of the diaphragm wall into the excavation with the dig, roughly following the shape of the trigger lines as would be expected.

As the base of the SAA are assumed fixed, all plots of the SAAs are expected to show movement from the excavation at the top. In reality the SAA is moving into the excavation at its base and the top of the SAA is not moving. A manual survey of 3D Geodetic Prisms fixed at the top of the diaphragm wall, in line with the SAAs, takes place concurrently with the SAA readings to monitor the position of the top of the diaphragm wall. Interpretation of these readings show that little or no movement has occurred at the top of the wall, as was predicted. From the movement of the 3D Geodetic Prims and the movement of the SAAs at the top of the diaphragm wall, the correction factor that must be applied to the SAA readings can be calculated. Due to the manual survey of the 3D Geodetic Prisms being lass accurate and at a lesser frequency than the SAA readings, intervals of 0.5mm were used as corrections to the SAA readings.

Figure 8 shows the movement of the diaphragm wall during Stage 7 with no corrections applied to the SAA readings, with corrections applied to the SAA readings, and the corrected SAA readings through Stage 8, which overlaps with Stage 9. As can be seen from Figure 8, after corrections were applied to the Stage 7 readings, a movement of around 4mm can be seen at the base of the diaphragm wall and the top has moved less than previously shown. The displacement of the SAA at the Stage 7 excavation level is also a lot closer to the Amber trigger. Movement over the two week excavation period is shown to be 10mm as opposed to 5mm. Based on this observation, concrete for the construction of the base slab was poured around the foot of the diaphragm wall first in order to provide extra support and slow the movement at the toe of the wall. The concrete pour commenced on 23/09/2012 and after completion the movement of the diaphragm wall at excavation level and at the base was observed to slow and no further increase towards the Amber trigger was observed. The remaining excavation was re-scheduled to start from the centre of the box, to allow the movement of the diaphragm wall additional time to stabilise. The graph showing the corrected movement through Stage 8 shows that there were no trigger breaches and any movement was negligible due to the remedial actions that were taken.

Conclusions

This paper has documented the role the Shape Accel Arrays have played in the monitoring of the diaphragm wall of the southern section of the Tottenham Court Road Western Ticket Hall during excavation and propping, and how they have been used within the Management Plan for the Control of Ground Movement to assist in the control of the works. It has shown the huge advantage they have over the previously used in-place inclinometers and manual inclinometer probes, in terms of cost, time, and the quality of the data and the frequency of the readings.

The use of the SAAs at the WTH enabled the Value Engineering proposal for the omission of the level P5 props to be drawn up, and the subsequent acceptance of this proposal due to the quality and frequency of the SAA readings taken from the start of the excavation to P4 level. The data obtained from the SAAs allowed the back-analysis of the readings and the refinement of the triggers. The omission of the P5 level props has provided the contract with significant cost saving and reducing the programme by 26 days. The use of the observational method, and the continual evaluation of the SAA data throughout the dig to the base slab, have provided confidence that wall movements were not breaching the trigger levels and thus not exceeding the structural capacity of the diaphragm wall.

As the SAA were installed to the bottom of the panel of the diaphragm wall, and not drilled through the toe of the wall with a 10m anchor, the 3D Geodetic Prisms at the top of the wall were manually monitored and the data obtained from this used to provide correction values to the SAA data. This showed an increase in movement into the excavation of the base of the wall, as opposed to a movement in the top of the wall as was previously shown when the base of the SAA was assumed fixed. By noticing this movement the preventative pouring of concrete binding around the diaphragm wall, and the commencement of the next stage of excavation from the centre of the box, allowed the movement in the wall to stabilise ensuring that no triggers were breached.

Unlike traditional In-Place-Inclinometers (IPIs), SAAs fit snugly within their casing with the joints or 'knuckles' between segments expanding outward when downward pressure at the top of the SAA is applied and then clamped. Traditional IPIs have probes at set lengths (usually 500mm) with sprung loaded jockey wheels at the top and bottom of each probe. The jockey wheels can be prone to failure over time, reducing the accuracy of the instrument. These can be time consuming to replace and involve the removal of the instrument

.
Manual inclinometer readings were taken alongside the SAAs at the TCR WTH, and because of this the clear advantages of the SAA were highlighted. A trained technician is required to install and take readings with manual inclinometer probes, and process the data obtained. This is costly and considerably more expensive than using SAAs, which only require a trained technician to install and set up the instrument. There was also the issue of access to the inclinometer casing to take readings with the inclinometer probes, as some of the inclinometer casing was within the gantry crane beam and hence required the gantry crane to be stationary for upwards of half an hour per manual inclinometer to allow time for reading. This resulted in loss of time for the contractor, BFK, and was not desirable.

Acknowledgements

The authors wish to thank the following companies and persons for their input into this paper; Bam Ferrovial Kier JV, Crossrail, Nicola Bicocchi – Monitoring Engineer Bam Ferrovial Kier JV, Dominic Kisz – Monitoring Engineer ITMSoil for site information and analysis and Dr Boulent Imam, of the University of Surrey for proof reading of the manuscript.

References

- Harris D.I., 2012, *'Management Plan for the Control of Ground Movements: Addendum 5: Tottenham Court Road Station – Western Ticket Hall Excavation'* Revision 4.
- Harris D.I., 2012, *'Tottenham Court Road – Western Ticket Hall, Instrumentation and Monitoring Proposals'* Revision 2.
- Ridley A., Kavanagh J., Lardi M., Saffari N., *'Results from a Trail of a New Displacement Monitoring Tool (Shape Accel Array)'*
- SAAF Specifications, [Online], Measurand Geotechnical, Available: http://www.measurandgeotechnical.com/docs/saaf_specification_5sep2012.pdf
- *'Is this the shape of things to come?'*, [Online], Railway Strategies, Available: http://www.railwaystrategies.co.uk/article-page.php?contentid=11389&issueid=346
- Akram N. Alshawabkeh, Krishna R. Reddy, Milind V. Khire, 2008, *'Characterization, Monitoring, and Modeling of Geosystems: Proceedings of Selected Sessions of Geocongress 2008'* pp.534-540.
- Abdoun T., Bennett V., Dobry R., Thevanayagam S., Danisch L., 2008, *'Full-Scale Laboratory Tests Using a Shape-Acceleration Array System'*

Crossrail Project: Infrastructure design and construction
ISBN 978-0-7277-6078-4

doi: 10.1680/cpid.60784.253

Lord Hill's Bridge Mitigation Works

Brendan Bradley MEng (Hons), Monitoring Engineer, BFK
Daniel Callaghan BSc (Hons), Field Engineer, Crossrail
Thomas Moore HND, Sub Agent, BFK
Aodhán Reaney BSc, M.E Section Engineer, BFK
Jaime Sevilla BEng, Section Engineer, BFK

ABSTRACT

Lord Hill's Bridge (LHB) was the first piece of infrastructure that the Western tunnels TBM's passed beneath, at a depth no greater than 4 m below current ground level. The shallow tunnelling depths are a result of the 100 year old bridge being located 50 m east of the Royal Oak Portal head wall. The bridge itself is a simply supported two span structure passing over the existing Network Rail (NR) main lines, London Underground (LU) Hammersmith & City tube lines and the proposed Crossrail Western tunnels. Royal Oak tube station entrance is positioned at the central pier of the bridge and 3 utilities span the bridge deck in the form of: 1 x 600 mm gas main, 2 x 450 mm water main and 2 x 66KV power cables. Assessments undertaken in recent years highlighted that the bridge was in a poor condition, the latest report in 2005 recommending that the bridge be re-built in the next 4 years due to the deterioration of the structure. A combination of the condition of the bridge and the shallow tunnelling depths led to the implementation of programme critical mitigation works. A design for ground replacement below the bridge was developed; this involved the installation of vertical piles beneath the bridge and raking piles beneath the north abutment of the bridge. In total 1225 piles up to 15m in length were to be placed below the bridge with a working headroom of 5.0 m. Upon the bridge deck diversions of the water main and gas main were undertaken while the 66Kv cable chamber was exposed to allow for monitoring installations. All works were carried out whilst the bridge deck was live to traffic and pedestrians, as well as the adjacent Network Rail and London Underground transport services beneath the bridge.

INTRODUCTION

Critical mitigation works were required at LHB prior to the passage of the TBM's to minimise settlement of the bridge structure. Not only were the planned works complex due to the location and structure of the bridge but they also had to be completed to a challenging programme, to avoid delay of the TBM's launch.

LHB is an NR asset which is over 100 years old and situated 50 m east of Royal Oak Portal (ROP). Beneath the bridge passes the NR main line tracks and LU Hammersmith and City line tracks, whilst the entrance to the LU Royal Oak station is situated on the central pier of the bridge, see Figure 1. The proposed Crossrail tunnels pass under LHB at shallow depth due to the constraints on positioning ROP. The portal is bound by NR & LU assets on the south, Harrow road wall to the north, LHB / Ranelagh Bridge and Westbourne Bridge to the east whilst sufficient room was required to the west of the portal to allow space for the logistics of a tunnelling operation and the final operational railway.

Crossing the bridge is a 600 mm steel gas main, twin 450 mm water mains and twin 66kV power cables. The bridge construction consists of a two span simply supported structure with a fixed central bearing and two outer guided bearings. The outer guided bearings are situated on top of masonry brickwork abutments which in turn are situated on a concrete footing spread foundation, as is the central pier foundation. Both spans of the bridge consist of two steel truss main girders with the road and footpath supported on under slung steel cross girders. The north span of the bridge is 27 m in length and the south side is 44.5 m in length.

Figure 1 – Lord Hill's Bridge Location.

The foundations of the bridge are founded upon London clay which is situated a couple of meters below the ground level. Above the London clay is a layer of Alluvium and made ground. Adjacent to the north abutment was an area of existing Brunel foundations.

Due to the deteriorating condition of the bridge restrictions were implemented in

2003 to limit the number of vehicles crossing the bridge, this was achieved by reducing the bridge from three lanes to two with the introduction of a central reservation. Further assessments in 2005 by Westminster County Council (WCC) highlighted that the bridge showed signs of further deterioration and it was recommended that the bridge be rebuilt within 4 years.

DESIGN

Predicted Movement

The C300 western running tunnels running from ROP to the east side of Farringdon station are being constructed using two Earth Pressure Balance (EPB) TBM's. The TBM's passed beneath LHB barely 4.0 m below the ground surface and with less than 3.0 m clearance to the North Abutment during construction of the E/B drive.

Due to the shallow nature of the tunnels below LHB an assessment of ground movement effects as a result of tunnel construction was undertaken. The ground movement assessment was based on greenfield ground movements and modelled on a volume loss of 1.0 % from the bored tunnelling works. The assessments gave max vertical displacements of 32.6 mm at the north abutment footings with corresponding horizontal movements of 10.8 mm with a maximum differential settlement of 28 mm.

LHB foundations were modelled based on the aforementioned greenfield settlement values and the following assumptions:

- The bridge abutment was modelled as a linear elastic rectangular beam.
- It was assumed that the calculated horizontal strains were transferred directly to the structure being assessed.
- Induced ground strains are based on a greenfield approach, i.e. no stiffness was attributed to the foundation, and bridge loads were not included.

Due to the predicted differential settlement at the footings it was envisaged that the North Abutment would tilt towards the central pier and thus increase horizontal movements on the North Abutment bearings. The calculated movements at bearing level were; 58 mm horizontally and 26 mm vertically.

The poor condition of the bridge, meant it had limited capability to accommodate further horizontal movement at the North Abutment. Visual inspections of the bearings at the North Abutment indicated that the bridge could likely accommodate 34 mm horizontal movement (if a movement joint was installed on the bridge deck).

Structural assessments also indicated that the bridge could accommodate settlement of up to 50 mm and differential settlements of up to 30 mm.

As the bridge was considered to be able to accommodate 34 mm horizontal movement and movements of 58 mm were expected it was proposed that a mitigation scheme was put in place to minimise the impact of tunnelling on the bridge structure.

Mitigation Proposals

The first proposal for mitigation works was termed Option A, (Appendix 1 drg no: C122-OVE-S-DDB-CR001_Z-42003). Option A involved constructing 1800 mm dia mass concrete piles with 600 mm dia concrete infill piles across the area of both the eastbound and westbound tunnels, including up to 10 m either side of the east and west flanks of LHB. The 1800 mm dia piles were intended to be installed to a depth of 1 m below tunnel invert level with further inclined piles installed below the north abutment. It was intended that the piles would be used as a form of ground replacement.

After consultation between the designer (C122), contractor (BFK) and Client (Crossrail) the mitigation works were refined to Option B, (Appendix 2 drg no: C122-OVE-S-DDB-CR001_Z-42031). Option B works consisted of mass concrete piles placed above the eastbound drive to tunnel invert depth, and raking piles installed below the north abutment, see Figure 2. A further two rows of reinforced contiguous piles were planned for either side of the westbound tunnel with a reinforced concrete capping beam to act as a prop across the two rows of piles. The raking piles were designed to fully encapsulate the eastbound running tunnel, whilst the slab above the westbound tunnel was implemented to prevent uplift should the TBM be operating at a pressure greater than overburden.

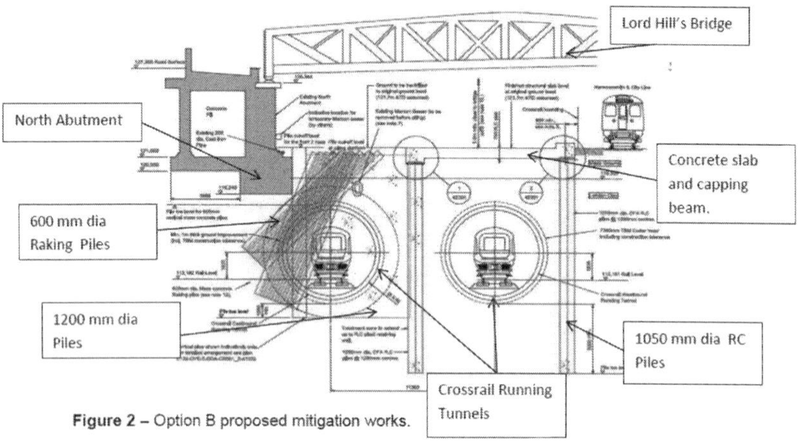

Figure 2 – Option B proposed mitigation works.

It was anticipated that during construction works ground movement would be caused as a result of installation of the piles and reduction of the ground level. As part of the Option B mitigation works there was a requirement to reduce the ground level below the bridge, this created the required 5 m of headroom for the piling rigs to construct the mass filled concrete piles.

A number of considerations to the works on site were suggested by the designer to ensure that ground movement was kept to a minimum, they included:

- No adjacent piles were to be built until the previous piles had achieved the stiffness and strength of the ground. It was therefore stipulated that adjacent piles would not be installed within a 72 hour time frame. Sequencing was of paramount importance – especially for the raking piles under the North Abutment.
- Temporary casing was to be installed to support the upper section of the pile in the made ground during construction.
- Piles were to be backfilled with concrete as soon as excavation and spoil clearance was complete.
- During piling platform construction, bay widths were not to exceed 4 m in width.
- Piling on the outer flanks was to be carried out with a CFA (Continual Flight Auger) piling rig to reduce ground settlement.

Ground Settlement Predictions Post Mitigation Works

Once a design for mitigation works had been agreed, ground settlement predictions were run for a second time to establish predicted movements due to tunnelling.

The resulting movement predictions at bearing level of the North Abutment gave values of < 1 mm horizontally and 2 mm movement vertically, and movements at the central pier of < 1 mm horizontally and 8 mm vertically.

The combination of settlement due to tunnelling and settlement due to undertaking the mitigation works was combined and gave predicted movements for the bridge post passage of both TBM's – Table 1.

Location	North Abutment		Central Pier	
Stage Of Construction	**Max. Horizontal Movement (mm)**	**Max. Vertical Movement (mm)**	**Max. Horizontal Movement (mm)**	**Max. Vertical Movement (mm)**
Mitigation Works	<3	<3	<3	<3
Tunnelling	<1	2	<1	8
Total	<4	<5	<4	<11

Table 1 – Predicted ground movements post TBM passage

A design was produced for a movement joint on the north side of the bridge should excessive horizontal movement be recorded.

Utility Diversions

Utility diversion works were required as they essentially removed the risk of the existing mains being damaged, due to settlement induced by piling and subsequent tunnelling activities. This was achieved by replacing the existing utilities which were in a very poor condition and replacing them with new diversions which acted independently of the bridge structure.

Assessments based on a 0.5% volume loss indicated that mitigation works were required on the existing 600 mm steel gas main which crossed the bridge on the west side. A number of proposals were considered with the final solution consisting of a 450 mm HDPE replacement main running across the bridge deck and being connected to the existing gas main behind the north and south abutments.

On the east side of the bridge is located two 450 mm water mains encapsulated in a steel box conduit. Mitigation works were required as assessments indicated that there was the possibility of inducing excess strain in the pipe, this was most prevalent at the intersection between the North Abutment and bridge deck. For these reasons it was decided that a 300 mm steel rider main was to be constructed and placed on top of the bridge deck on the east side. Connections were to be made on the north and south side of the bridge into the existing main, see Figure 3.

Figure 3 – North Abutment rider main connection – looking south.

The final utility crossing the bridge was a 66 Kv cable situated under the footpath on the east side of the bridge. Assessments indicated that the cable would be unaffected by ground movements and that induced strain in the cable would be negligible. For these reasons it was decided to install monitoring in a cable joint bay on the north side of the bridge to closely monitor for un-expected movements.

All of the aforementioned diversionary works were subject to an intense public interface, as the entrance to the LU Royal Oak station is situated at the centre of LHB. In addition to the location of the station, road and footpath closures were required to facilitate the diversion works.

CONSTRUCTION

Pre Piling Works

Before any of the main works could commence at LHB there were a number of work activities that had to be completed; these consisted of removal of old Brunel foundations from underneath LHB, construction of a piling platform and the removal of the Marcon sewer that ran through the area which was to be piled. The Brunel foundations were removed under a strict sequence due to the depth of the excavations underneath LHB. The piling platform was constructed using imported recycled type 1, depth of which reached 800 mm at times. The Marcon sewer, an old Victorian sewer which ran the length of the site was removed using a proprietary drag box system, as the brick sewer was in close proximity to the bridges North Abutment.

The main constriction with the piling works was the low headroom under the bridge, for this reason the piling platform was constructed at a lower level than original ground level. The platform was constructed to give the piling rigs 5.1m clearance, which was the minimum required by Keller/Martello (the piling subcontractor) who modified their rigs to suit the restricted clearance. In addition to the low headroom issue, a great deal of attention was paid to the London Underground Hammersmith and City Line, which was only 3 m from the piling area. This required close liaison with LU on a weekly basis to ensure the works did not have an adverse effect on the LU assets, due primarily to settlement being caused by the piling activities.

Piling Beneath LHB

The first piles consisted of 1200 mm vertical mass filled concrete (MFC) installed by the restricted headroom Martello rigs, see Figure 4. These were installed using a 3.5 m long casing; the reason for the casing was to give a seal into the London clay to prevent ground water entering the pile during excavation. A strict sequence was incorporated for these piles, which entailed installing 1, missing 3 and not piling adjacent piles for 72 hours. The sequencing was carefully planned by the site team with forecast drawings being produced to ensure that the restrictions with installation were met. On a daily basis these sequence drawings were as built to confirm that the correct sequence was completed.

For tunnelling purposes the strength of the concrete in the piles needed to be less than 10 N/mm^2 but more than 4 N/mm^2. The reason for the upper and lower bound concrete strengths was to ensure that the mitigation works provided an effective mitigation to ground movement during tunnelling but didn't prove an obstacle to tunnelling beneath the bridge. Due to the required characteristics of the concrete strict quality control by the concrete supplier and the site team was required to ensure compliance to the specification. The concrete had no coarse aggregate and contained either 80% or 90% PFA to meet the required specification, as well as being designed to be easily pumpable from the on site re-mixer to the pile location. Due to the strict parameters & unusual mix requirements of the concrete strength there were issues with conformity and a number of piles achieved a higher strength than required. Due to a strict quality control regime enforced and managed by the site team the high strength concrete piles could be easily identified, removed and replaced to ensure all works were compliant with the specification.

Figure 4 – Martello 5000 Series low headroom piling rig working beneath LHB.

Issues with ground water also became apparent upon boring 1200 mm dia piles near the bridge abutment, which was caused by an existing drainage system from the Harrow Road retaining wall. Due to the presence of the water the site team were unable to obtain a seal into the clay, which is required to prevent water ingress into the bored excavation. The site team proposed and implemented a solution which entailed installing a 1050 mm dia pile in the area affected by water. This enabled a longer casing to be installed, thus providing a sufficient seal into the clay as required.

Simultaneous to the 1200 mm dia piles the 900 mm and 1050 mm dia piles were installed around the area of the Marcon sewer. Due to the presence of backfill from the removal of the existing Marcon sewer an outlet had been formed for water from the surrounding made ground to drain into, this meant that the 3.5 m casings used for 1200 mm dia piles were not long enough to provide a seal into the clay in the area of the existing sewer. Smaller diameter piles were used to enable segmental pile casings to be installed which would ensure a suitable seal into the clay, casings ranged from 3.5 m to 5.5 m in depth. Due to the low headroom under the bridge there was great difficulty in manoeuvring the casings, a mini crane was used for lifting and manoeuvring the casings. The CFA rig installed the 600 mm piles on the flanks of the bridge and the reason for these was to infill the areas that were not covered by the larger piles. The sequence for the 600 mm piles allowed 48 hours between each pile, again the concrete used had to provide strength of less than 10 N/mm^2.

The reinforced piles and the slab was the next phase of works which essentially created an underground portal for the westbound TBM to pass through. The piles flanking the slab were 1050 mm reinforced concrete (RC) piles. Due to the minimal clearance underneath the bridge the rebar cages needed to be installed in 5.0 m sections (Figure 5) and spliced together in situ, once in place C32/40 concrete was then placed to form a 1050 mm RC pile. The reinforced concrete slab and capping beam were poured in 2 sections, the first section was against the Asset Protection Barrier (APB), which was in place to form a physical barrier between the C300 site and LU assets. Excavating close to the APB was difficult as the dig was approx 1.5 m in depth and in close proximity to the LU track; therefore requiring a temporary ACROW propping system to support the APB during excavation and construction works, see Figure 6. A strict programme for these works were in place with only 2 weeks to complete the construction capping beam, this was achieved through working extended hours and utilising 2 shifts to complete the works on time. Upon completion of the capping beam along the APB the temporary props were removed and the main RC slab was constructed using 32 mm dia and 30 mm dia reinforcing bar. Once all the rebar and formwork shutters were fixed for the slab, $400 m^3$ of C32/40 concrete was pumped into the area to complete the RC slab.

Figure 5 – Mini crane lowering 5.0 m reinforced cage in 1200 mm dia pile under LHB.

Figure 6 – ACROW props supporting APB as capping beam reinforcement is placed.

Figure 7 – Klemm 709 piling rig installing raking piles below LHB.

The raking piles were the last piles to be constructed and 3 specialist KLEMM piling rigs were utilised, see Figure 7. A total of 597 raking piles were installed with a very stringent sequence - 72 hours between piling an adjacent pile. The reason the sequence was so stringent was that the piles were directly beneath the North Abutment foundations and there was a high risk of settlement. Leaving the pile open for a period of time before concreting was considered unacceptable, as there was a potential for settlement to occur, for this reason piles were concreted immediately upon completion of boring. The sequence was changed a number of times to ensure maximum efficiency for the installation of the piles along with minimising the risk of settlement. The piling rigs were modified to allow for working within the restricted headroom of the bridge and the site team also fabricated special gates around the auger area of each rig to keep operatives away from the moving parts.

Tolerances constructing the piles were as follows - horizontal ±75mm and vertically 1:75. All tolerances were met through a stringent and proactive quality control procedure established on site, with every pile checked at setting out stage and when the casing was installed.

The highest daily output recorded from pile installation was 25 in one day, this was in part achieved thanks to a well planned site layout. Coordination between the different construction teams allowed the ever changing interface of the piling activities to run smoothly, as at times there were up to 5 piling rigs working in close proximity to one another.

All key dates/milestones were met, with works completed 1 week ahead of programme – Figure 8. All of the aforementioned was achieved due to the good working relationships on site and with no compromise to safety, quality or the environment.

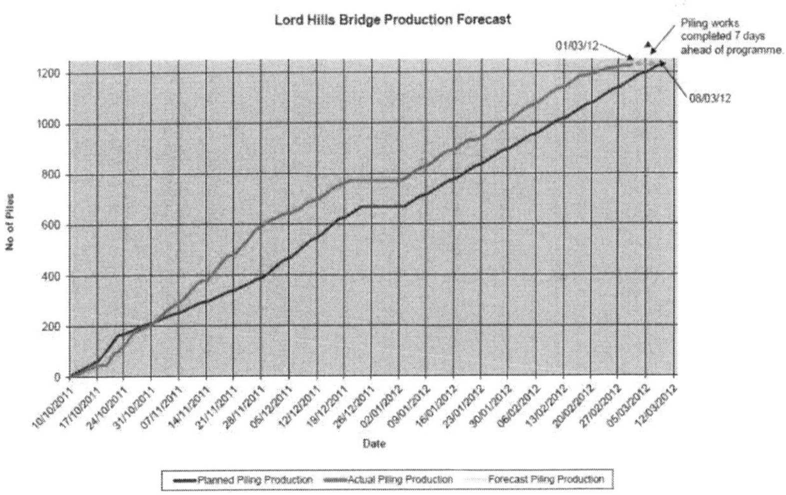

Figure 8 – Graph of planned vs actual piling production.

LORD HILL'S BRIDGE MONITORING

In accordance with the Assessment of Ground Movement Effects for LHB, an observational approach was implemented to monitor the piling construction works and subsequent tunnelling works.

Installed Monitoring

Many different types of monitoring instruments were used on LHB and the associated assets - Table 2.

Instrument	Quantity	Location	Purpose
3D Geodetic Prism	10	North Abutment (2 Rows of 5No.)	To provide horizontal displacements and Rotation or Tilt
3D Geodetic Prism	12	Girders (both Main and Tertiary)	To provide horizontal displacements for bridge deck compression
3D Prism	8	Central Pier (North side)	To provide horizontal displacements and Rotation or Tilt
BRE type Levelling Studs	5	North Abutment	Precise manual levelling
Invar Levelling strips	4	Central Pier	Precise manual levelling
RVDT	3	North Abutment/Tertiary Girders	Rotational movements of the north abutment from the deck
HLC (Hydrostatic Levelling Cells)	4	North Abutment	Real time precise levelling
Crackmeter	1	UKPN Cable Joint	Differential movements between
Strain Gauges	3	NGG Gas main	Accurate monitoring of the old gas main
PLPs	Many	Bridge deck and off North Abutment	Precise levelling of the bridge deck and ground behind the bridge.

Table 2 – Summary of instrumentation installed on LHB.

In addition to the minimum I&M requirements, hydrostatic levelling cells (HLCs) were installed on the North Abutment to provide real-time and extremely precise settlement readings. This allowed the monitoring engineers to gain a more complete understanding of how the piling works affected the structure.

Monitoring of the works was reviewed during the Shift Review Groups (held for each shift), during which the monitoring data would be reviewed and discussed along with the construction works. By bringing this information together during each shift, the SRG (made up of a construction representative, an engineering representative, the designer and the client) could determine the most appropriate way forward with regards to sequencing of works so as minimise movements caused by the works.

Expected Ground Movements

The Assessment for the Control of Ground Movements for Lord Hill's Bridge concluded that no more than 3mm (both Horizontal and Vertical) of ground movements should occur as a result of the installation of the mitigation works. Based on expected ground movements a number of trigger values were set.

Trigger Values

Trigger values on Lord Hill's Bridge were set based on the expected movement from the Assessment. Table 3 shows the Trigger Value originally set prior to the commencement of the works.

		Trigger Values (mm)[1,2]		
Location	**Parameter**	**Green**	**Amber[3]**	**Red**
North Abutment	Settlement	4[4]	6	20
Central Pier	Settlement	10	13	20
North Abutment and Central Pier	Differential vertical displacement between supports	10	13	20
Tops of North Abutment and Central Pier	Combined (differential) horizontal displacement parallel to span (inward)[5]	3	5	10

[1] Triggers based on planned mitigation works being carried out (Dwg Nos. C122-OVE-S-DDA-CR001_Z-41031, C122-OVE-S-DDB-CR001_Z-42031 & C122-OVE-S-DDB-CR001_Z-42032).

[2] Recorded displacements shall be adjusted to exclude thermal and transient load effects (established from baseline readings) prior to comparison with trigger values.

[3] An additional Amber trigger shall be any new sign of damage from visual inspection (i.e. cracking or deformation).

[4] Green trigger value set at minimum 2mm below Amber value.

[5] Triggers to limit compression in bridge deck (applicable if bridge bearings locked).

Table 3 – Trigger values prior to works commencement

The most notable values in the table are those for settlement of the North Abutment due to the considerable range between the Amber Trigger and the Red Trigger. The large variation between the two triggers was because the Amber Trigger was set based on anticipated movements, whereas the Red Trigger was based on tolerable movement limits for the bridge.

Further to these trigger values, construction Alert values were set, separated into the mitigation works and the tunnelling works - Table 4. The purpose of these Alert values was to provide control for construction – to provide targets by which each activity should aim to adhere.

Location	Parameter	Displacement Allocation per Activity[1] (mm)			
		Mitigation Works - Alert		Tunnelling Works - Alert	
		Green (mm)	Amber (mm)	Green (mm)	Amber (mm)
North Abutment	Settlement[2]	3	5	3	5
Central Pier	Settlement[2]	3	5	8	10

[1] Recorded displacements shall be adjusted to exclude thermal and transient load effects (established from baseline readings) prior to comparison with alert values.

[2] Where calculated values are very small and therefore measured values expected to be sensitive to measurement accuracy, Green and Amber alert values are set at a minimum of 3mm and 5mm, respectively.

Table 4 – Construction alert values.

Results – Movements Due To Implementation of Mitigation Works

Within two weeks of commencing the piling works, 3mm settlement and 3mm horizontal displacement of the North Abutment had been recorded, thus constituting Green Alerts. It was clear from the early stages that the Amber Trigger for Settlement (6mm) would likely be reached. The initial sequencing, which concentrated on completing the piles closest to the abutment first, was designed to provide an early restraint to ground movements from the construction of subsequent piles. Although the rate of movement did decrease from the initial rate observed, the North Abutment continued to settle gradually whilst piling works continued.

By completion of the vertical piles, the maximum settlement of the North Abutment had reached 5mm, which constituted an Amber Alert for the activity. Considering the high number of piles now installed in front of the structure, it was generally assumed that little or no further horizontal movement would be observed and that this would remain stable at 5mm. This value was of particular importance as the compression of the deck was considered critical for the structural integrity of the bridge. On the other hand, it was expected that the implementation of the 597 raking piles would indeed cause continued vertical displacement of the Abutment.

Figure 9 shows the progression of settlement of the North Abutment during the construction of the raking piles. Point HLC_04M and HLC_05M were located closest to the centre of the abutment and show the maximum observed settlement. The graph clearly shows the reaction of the abutment to the piling works and how the rate of piling corresponds to the rate of movement. Indeed, during the Christmas break when no piling works were being carried out, no significant settlement was observed.

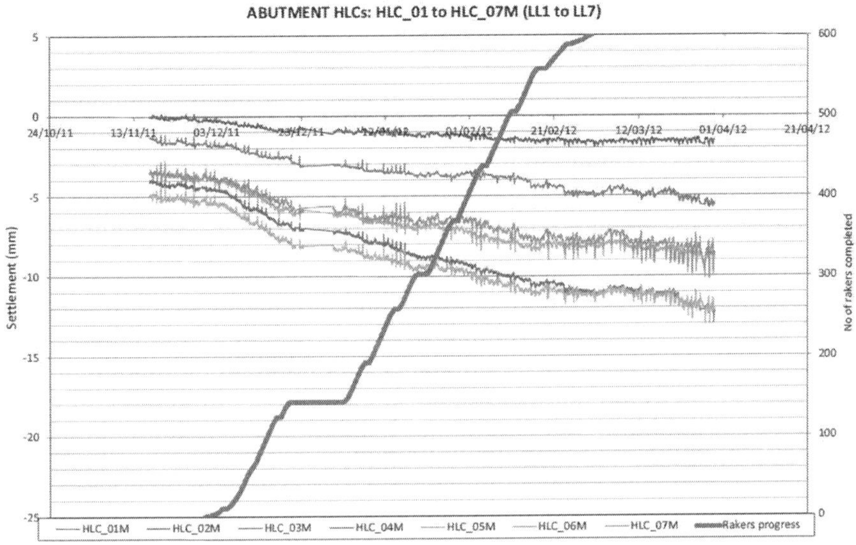

Figure 9 – Chart showing settlement progression during raking pile construction works.

The total movement once all mitigation works were complete is summarised in Table 5. The Amber Triggers reached had required at various times for an emergency Contract Technical Committee (CTC) to be called, during which the Triggers were discussed and actions determined. Typically, the Amber Triggers would then be revised, as was the case for all those below. As a result, prior to the passage of the TBMs beneath the bridge, the Amber Triggers mentioned below were considered to be Green and the value of the Amber Triggers was raised in accordance with the CTC's decision.

Asset	Movement	Value (mm)	Trigger
North Abutment	Settlement	12	Amber
	Horizontal	5	Amber
Central Pier	Settlement	2	
	Horizontal	0	
Bridge Deck	Vertical	9	
	Compression	5	Amber
UKPN Joint Bay	Expansion	0	

Table 5 – Total movement due to mitigation works.

CONCLUSION

The conclusion to the mitigation works carried out beneath Lord Hill's Bridge can be best described through the monitoring results following the passage of the two TBMs beneath the bridge. Initial assessments of ground movements caused by tunnel construction estimated that the bridge would settle by approximately 33 mm, causing differential settlements of nearly 28 mm which were considered detrimental to the structure. A further assessment was then carried out assuming the mitigation works described in this report would function as per their design. This assessment concluded that the North Abutment of the bridge should move no more than 6mm in total, including an allowance of 3mm for the construction of the mitigations works. Figure 10 shows the observed vertical movements of the North Abutment from the beginning of Raking pile construction until the end of September 2012, three months after the westbound TBM's passage through the piled mitigations works and only a couple of weeks since the eastbound TBM's passage beneath the bridge.

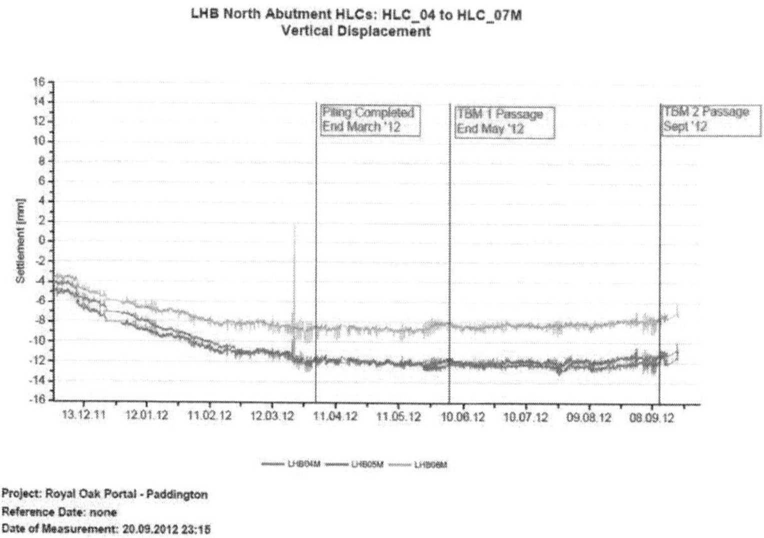

Figure 10 – Graph showing vertical movement, readings start from raking pile construction to TBM transit beneath LHB.

It is clear from the graph that settlement of the abutment ceased shortly after the completion of the mitigation works at the end of March 2012. The Hydrostatic Levelling Cells show that movements continued to be stable during the passage of TBM 1 (end of May 2012) and, more importantly, during the passage of TBM 2 almost directly beneath the North Abutment and through the mass concrete mitigation work.

It was thought initially that the fluctuations observed at the beginning of September (around the 08/09/2012) may be due to vibrations cause by the TBM, however this was more likely due to temperature effects, as this period happened to also be one of the hottest of the year. Furthermore, the apparent heave of the abutment shown on the graph following the passage of the TBM may potentially be due to the TBM tail grouting. However, this is more likely to be settlement of the reference cell, which is located on a retaining wall beyond the Bridge and mitigation works.

In addition to there being no adverse effect to the bridge during construction it is also worth noting that no delays or damage were recorded during the advancement of the TBM through the piled zone – testament to the desired concrete strength values being achieved.

The complete stability of the structure throughout the tunnelling process is testament to the design of the works, as well as to the quality and control of the build. In addition works were completed to programme, causing no delay to the TBM's.

Crossrail Project: Infrastructure design and construction
ISBN 978-0-7277-6078-4

doi: 10.1680/cpid.60784.271

In Situ Stress Strain Measurements at Tottenham Court Road Platform Tunnel

Jessica Yu BE MSc DIC MICE, PhD student, Imperial College & Geotechnical Engineer, Crossrail Ltd
David Potts, PhD DSc FREng CEng FICE FCGI, Professor of Geotechnical Engineering, Imperial College London
Jamie Standing BSc MSc PhD DIC MICE, Senior lecturer, Imperial College London.
Robert Vollum MSc PhD DIC CEng MIStructE, Reader in Concrete Structures, Imperial College London

Abstract

The redevelopment of Tottenham Court Road Underground Station started in 2011 as part of the Tube Upgrade Plan to improve and increase the capacity of the existing facility. The plan is to upgrade the station by 2016 to meet an estimated demand of more than 200,000 journeys per day once Crossrail is built.

During April to November 2011, major structural work was carried out on the Northern Line platform tunnels as part of the station upgrade. This included removing grey cast iron tunnel lining segments on the platform side to allow for construction of new cross passages to improve access to the platforms.

The upgrade works provided an opportunity to trial in-tunnel instrumentation prior to implementation in other London Underground (LUL) tunnels which interface with the Crossrail project. Mechanical and electrical resistance strain gauges were installed on tunnel segments to make discrete measurements of changes in strain due to unloading as the segments were removed from the tunnel rings. Linear variable differential transformer type displacement transducers were installed to make continuous measurements of the opening and closing of circumferential and longitudinal joints on trackside segments which are left in situ and affected by adjacent excavations.

This paper describes the installation process and highlights the lessons learnt for future applications. The in situ strain measurements are presented and compared to the expected response based on laboratory tests conducted on grey cast iron tunnel segments in the 1970s. The changes in strain measured by both types of strain gauges agreed well with the estimated changes assuming full overburden unloading.

Notations

$\mu\varepsilon$	microstrain
Ω	Ohms

Introduction

The redevelopment of Tottenham Court Road Underground Station started in 2011 as part of the Tube Upgrade Plan to improve and increase the capacity of the existing facility. The plan is to upgrade the station by 2016 to meet an estimated demand of more than 200,000 journeys per day once Crossrail is built.

During April to November 2011, major structural work was carried on the Northern Line platform tunnels as part of the station upgrade. This included removing tunnel lining segments on the platform side to allow construction of new cross passages to improve access to the platforms.

The works presented an opportunity for the Imperial College research team to trial their field instrumentation.

The main aims of the exercise were:

- To make discrete measurements of the strains in selected platform side segments before and after the segments are removed from the ring.
- To make continuous measurements of the opening and closing of circumferential and longitudinal joints on trackside segments which are left in situ and affected by adjacent excavations and tunnel segment removal.

Terminology

Figure 1 shows the terms relating to tunnel lining that are adopted in this report. Circumferential strain and longitudinal strains are defined as shown in Figure 2 . Positive bending refers to the mode of bending having a straightening effect on the segment, and negative bending increases the curvature of the segment (Figure 3).

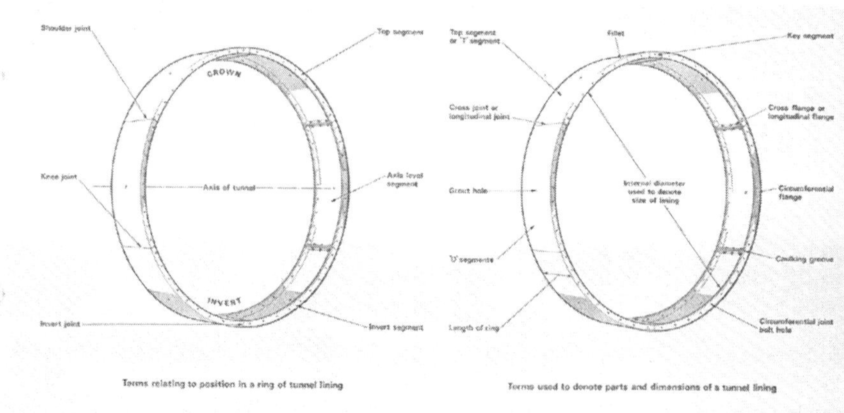

Figure 1 Terms relating to tunnel lining (after Thomas, 1977)

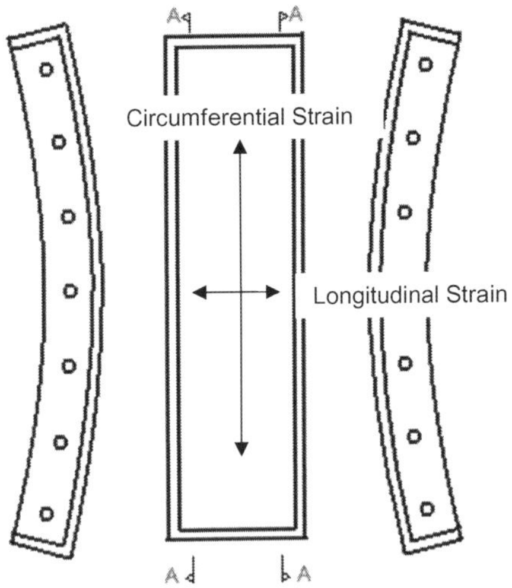

Figure 2 Directions of circumferential and longitudinal strains

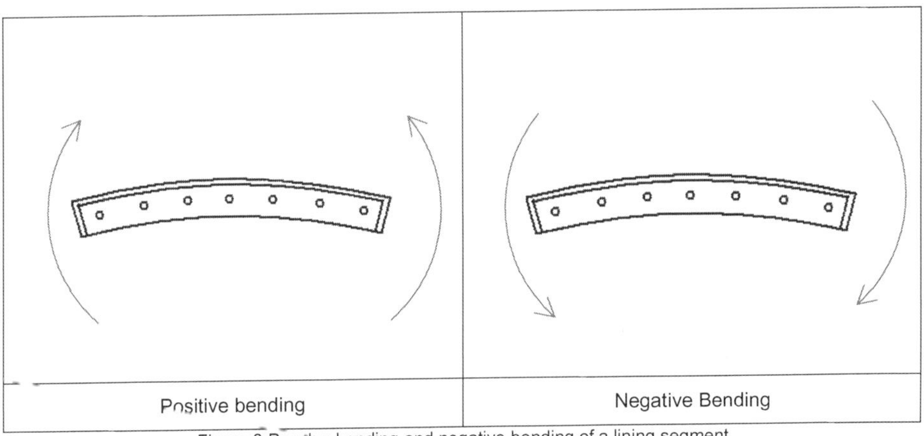

Figure 3 Positive bending and negative bending of a lining segment

The Instrumentation

DEMEC gauge

The DEMEC gauge is a portable device (Figure 4) which gives measurements of relative movement over a short gauge length. The studs were affixed to the surface of the structure using composite mortar. Once the studs were fixed and the mortar was set, the two points of the DEMEC gauge were carefully placed into the studs to take a zero reading. Further readings were taken subsequently to give the relative movement.

The selected gauge length was 150 mm. The gauge used was a digital gauge with a resolution of 0.001 mm. The accuracy under field conditions was given to be ±25 to 50 με, compared to an accuracy of ±5 με under ideal laboratory conditions (Dunnicliff, 1988).

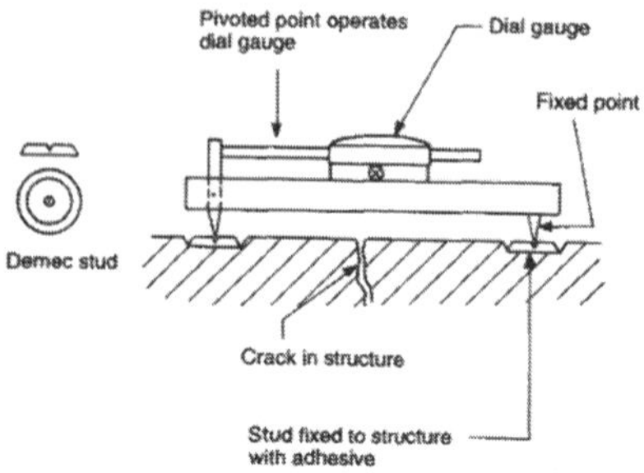

Figure 4 DEMEC gauge

Electrical resistance strain gauges

350 Ω rosette type electrical resistance strain gauges with leadwires attached (Figure 5) were bonded directly onto the prepared cast iron surface using cyanoacrylate. Each strain gauge rosette was 16.5 mm by 20.3 mm in plan and less than 1 mm thick.

After the strain gauges were installed, an initial reading was taken with a portable, battery-powered datalogger, the P3 strain indicator (Figure 6).

A quarter-bridge network with three wires was used in strain monitoring. This, along with the higher resistance 350 Ω resistance gauges, was selected to reduce lead wire effects. The strains caused by temperature change at the gauge were corrected during data processing. An AC excitation voltage was used for the portable strain indicator.

Temperature on the surface of the cast iron segments was measured with a laser thermometer probe to facilitate adjustment of strain gauge measurements.

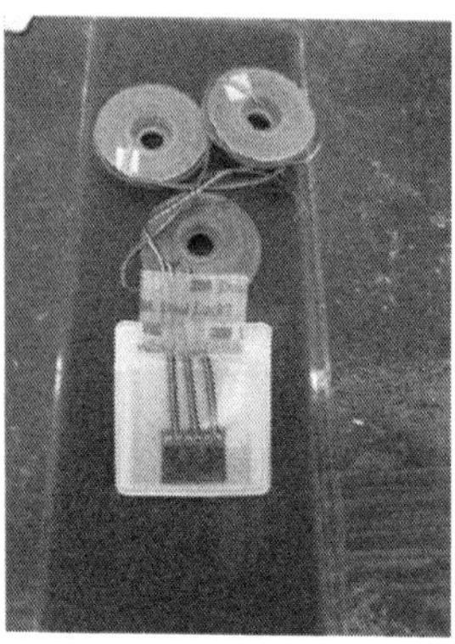

Figure 5 Rosette type electrical resistance strain gauge

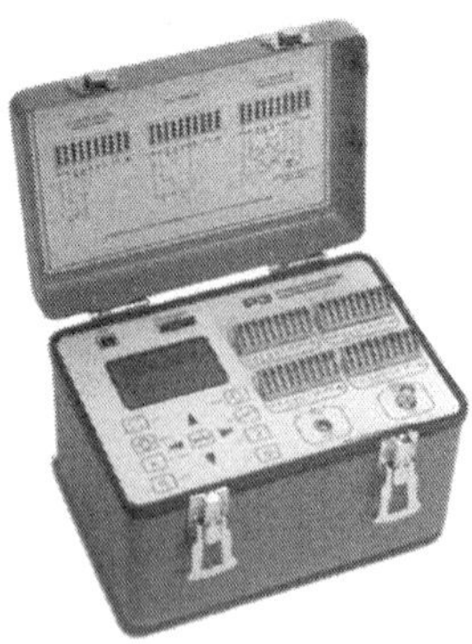

Figure 6 P3 Strain Indicator

Displacement transducers

LVDT (Linear Variable Differential Transformer) type displacement transducers were used to measure movement across the circumferential and longitudinal joints of trackside segments.

The displacement transducers were fixed onto the segments using a brass holder (Figure 7) which was initially positioned using magnets and then fixed in place using composite mortar.

The measurements were taken using mini-dataloggers. The mini-dataloggers were housed in a 120 mm x 120 mm x 90 mm ABS plastic box (Figure 8) which were fixed onto the skin of the trackside segments using composite mortar. There were two boxes in total, one box with two transducers set to log at 3 minutes intervals and the other box with two transducers set to log at 10 minutes intervals.

Figure 7 LVDT displacement transducer in metal holder

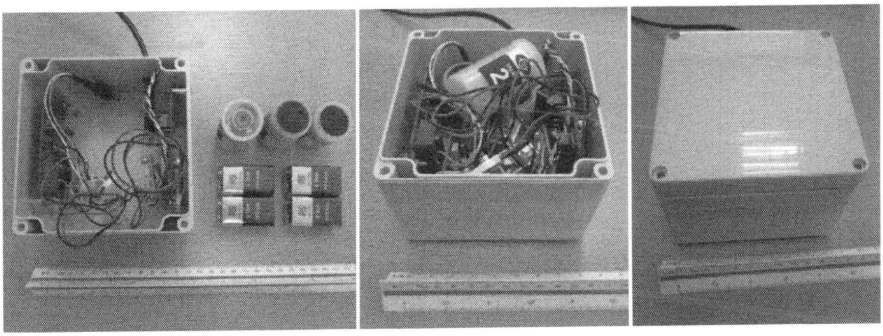

Figure 8 Dataloggers for LVDT and temperature gauge housed in plastic box

Temperature gauge

There was one temperature gauge with its own datalogger located in each plastic box (Figure 8).

The installation

Platform side installations

Mechanical and electrical resistance strain gauges were installed on the top and axis segments of one ring on the northbound Northern Line platform tunnel. The positions of the two segments and the approximate locations of the strain gauges are shown on Figure 9 and Figure 10.

Figure 11 and Figure 12 show the as-built installations. The protective coating seen in the figures consisted of a base layer of Teflon to insulate all bare connections, a layer of butyl rubber sealant, a layer of neoprene rubber and a top layer of aluminium foil tape.

On the inside of the circumferential flange, a pair of DEMEC studs was placed close to the intrados where the maximum strains are expected. Another pair of DEMEC studs was placed close to the skin to allow bending strains to be calculated.

Two pairs of DEMEC studs were located on the inside of the skin, to measure circumferential and longitudinal strains and allow comparison with the electrical strain gauge rosette.

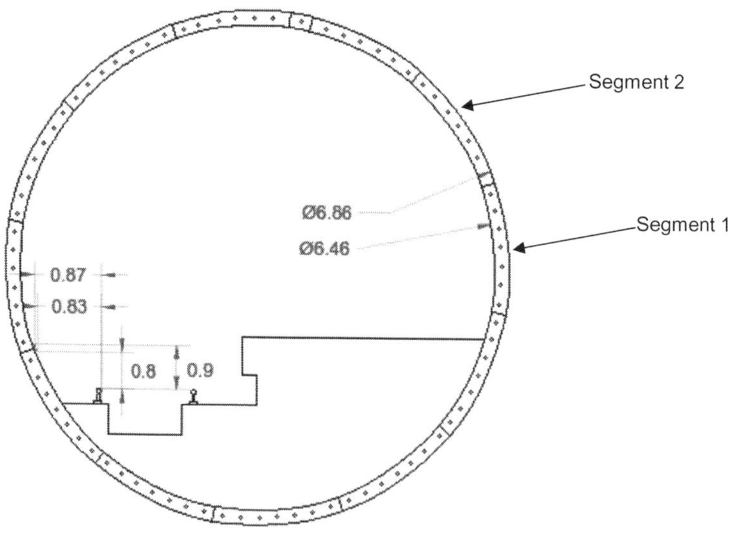

Figure 9. Northern Line northbound platform tunnel looking north

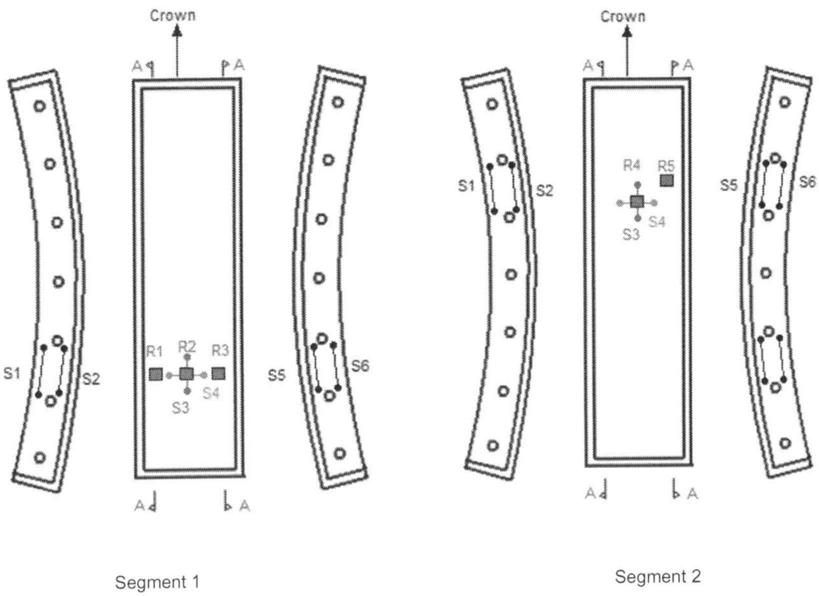

Figure 10. Approximate locations of strain gauges. (S=DEMEC span. R=strain gauge rosette)

Figure 11. Strain gauges on segment 1. Without protective coating (left) and with protective coating (right)

Figure 12. Strain gauges on segment 2.

Trackside installations

Figure 13 shows the location of the two data logging boxes on the trackside segment. Each box housed three data loggers, one for a temperature gauge and two for the LVDTs. One LVDT from each box was placed across the longitudinal joint and the other LVDT was placed across the circumferential joint.

The loggers in Box 1 were set to log every three minutes and the loggers in Box 2 were set to log every 10 minutes (refer to Figure 13).

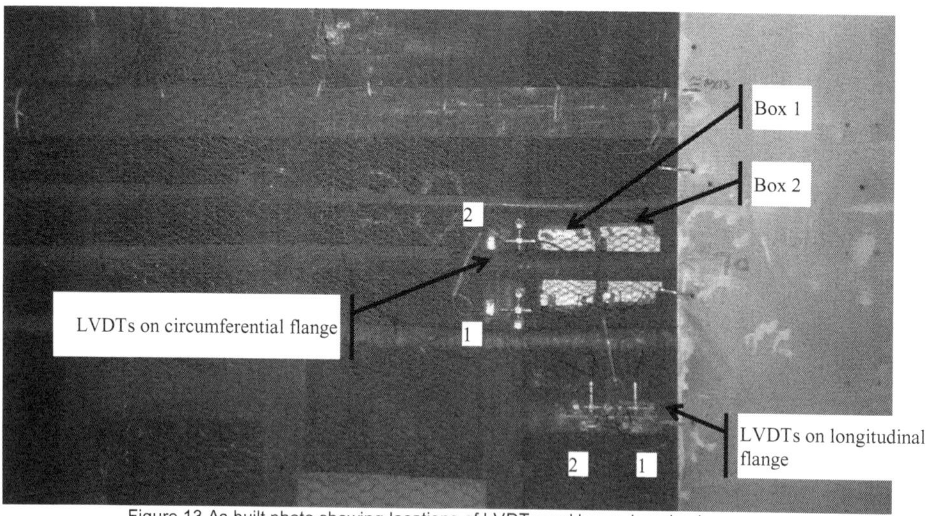

Figure 13 As built photo showing locations of LVDTs and boxes housing loggers.

Estimation of unloading strain

It was assumed that when the platform tunnel segments were removed from the tunnel lining they were completely unloaded. The circumferential unloading strain assuming the full ground overburden was estimated to provide a comparison with the strains measured using the mechanical and electrical strain gauges.

The ground level was estimated to be at 25.0 m OD obtained from spot heights shown on drawing 1D0101-G0G00-G00-P-01016 Rev B from Geotechnical Sectional Interpretative Report 1&2: Royal Oak to Liverpool Street, Volume 3: Drawings, Report No. 1D0101-G0G00-00551 (GCG, 2009).

The tunnel axis for the LUL Northern Line platform tunnel was estimated to be at -5.0 m OD from drawing HAG-N105-8742-TUN-D-SEC-5-03426 by Halcrow for the Tottenham Court Road Station Upgrade project. The axis level coincided with the London Clay and Lambeth Group interface (Halcrow, 2008).

Information on the geometry of the lining segments was obtained from LUL Archive Drawing No. 146 for a 21'2½" (6.46 m) internal diameter tunnel. The section properties were calculated and used to estimate the tensile strain in the segment from the release of the hoop stress. The hoop load was estimated assuming 30.0 m of overburden. The resulting strain was calculated to be about 400 $\mu\varepsilon$ when the Young's modulus was taken to be 100000 MPa for the grey cast iron.

Previous laboratory tests

Thomas (1977) investigated the structural behaviour of bolted ordinary grey cast iron (GCI) and spheroidal graphite iron (SGI) tunnel segments in the laboratory. The segments were dimensioned such that six GCI segments plus a key piece would join to form a ring of 3.83 m ID while 12 SGI segments plus a key piece would join to form a ring of 3.86 m ID. Since GCI behaves hysteretically upon loading while SGI exhibits little hysteresis, Thomas' experiments were mostly performed on the SGI segments.

Figure 14 shows an SGI segment loaded at its quarter points. The loading was applied to the circumferential flanges and not onto the skin. The experimental investigation suggested that positive bending distorts the segment as shown in Figure 15. As the circumferential flanges straighten, the relatively flexible skin dishes radially outwards, which results in a distorted cross-section with the inside skin experiencing longitudinal compressive strain and the outside skin experiencing longitudinal tensile strain.

Thomas (1977) also investigated the effect of ground pressure on the bending characteristics of an SGI segment. A uniform pressure was applied to the back of a segment as shown in Figure 16. This pressure caused the skin to dish inwards and this dishing inwards of the skin induced longitudinal strains in the skin in the opposite manner to positive bending.

Figure 14. Bending test on SGI tunnel segment (Thomas, 1977)

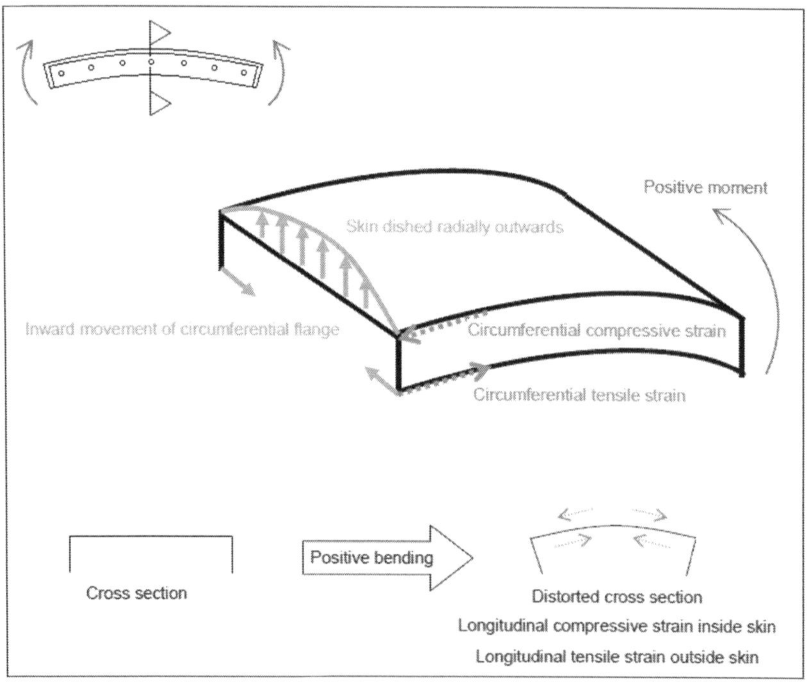

Figure 15. Schematic diagram showing distortion associated with positive bending (distortion is exaggerated)

Figure 16. Rubber bag strapped to the back of a SGI segment for the application of a uniformed pressure (Thomas, 1977)

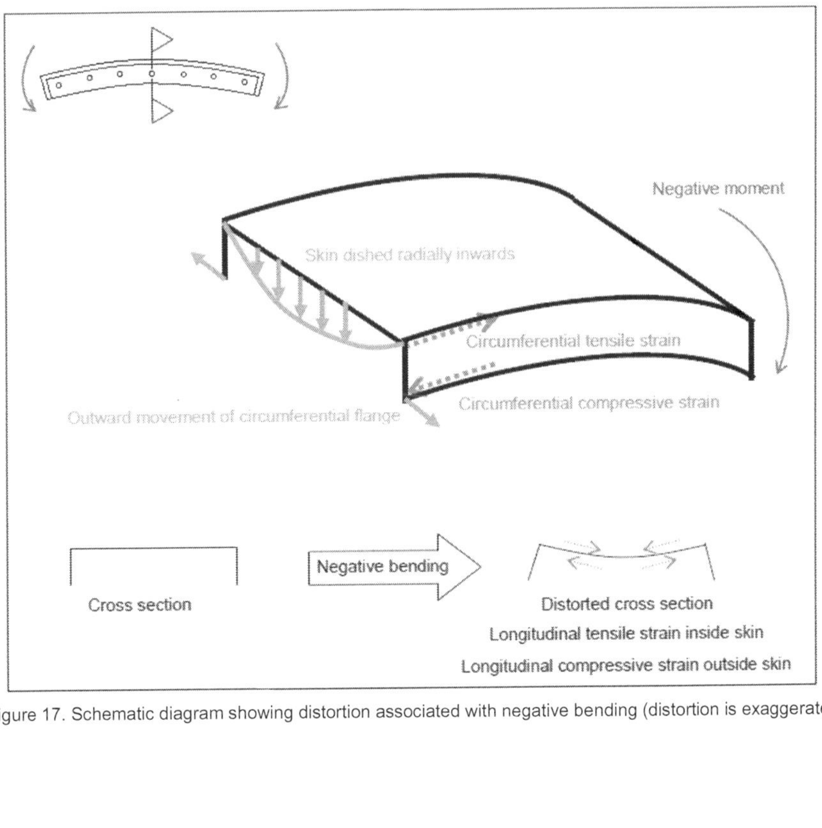

Figure 17. Schematic diagram showing distortion associated with negative bending (distortion is exaggerated)

Results from DEMEC measurements

Unloading strains due to removal of segments from ground

Using the baseline and post-segment-removal DEMEC measurements on spans 1, 2 and spans 5, 6 on the circumferential flanges (Figure 10), the axial and bending stresses due to unloading were estimated. In the calculations, the strain was assumed to vary linearly between spans 1 and 2 and spans 5 and 6. A constant modulus of 100 GPa was adopted. The results are presented in Table 1, along with the stress blocks, presented as variation in stress with distance away from the extrados, for each circumferential flange.

The measurements from all four circumferential flanges indicated circumferential tension and a positive bending moment upon unloading, which was expected since the segments would have experienced circumferential compression and negative bending moment (see Figure 17) while underground. It was not clear why the circumferential tension was higher on span 1-2 for both segments. However, the magnitude of circumferential tension from span 1-2 on both segments compared well with that calculated assuming full overburden unloading, which was estimated to be 940 kN.

On the inside of the skin, DEMEC span 3 measured tensile circumferential strains from unloading, and DEMEC span 4 measured compressive longitudinal strains, for both segments. Referring to Figure 15, these distortions could be associated with the dishing outwards of the skin related to positive bending. The ratio of the longitudinal strain to circumferential strain was 1.2 for segment 1 and 1.0 for segment 2. Given that grey cast iron has a Poisson's ratio of approximate 0.26, this distortion of the skin was not due to Poisson's effect.

Span 1-2 Segment 1	Span 5-6 Segment 1	Span 1-2 Segment 2	Span 5-6 Segment 2
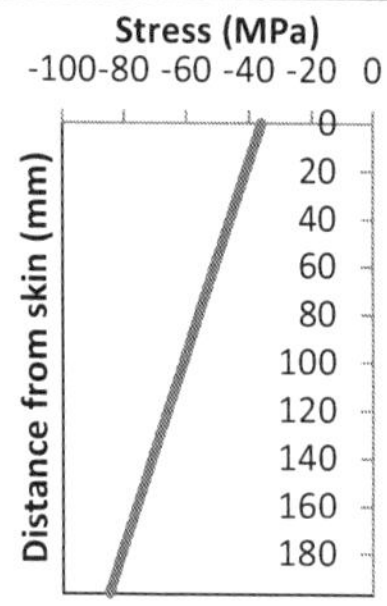	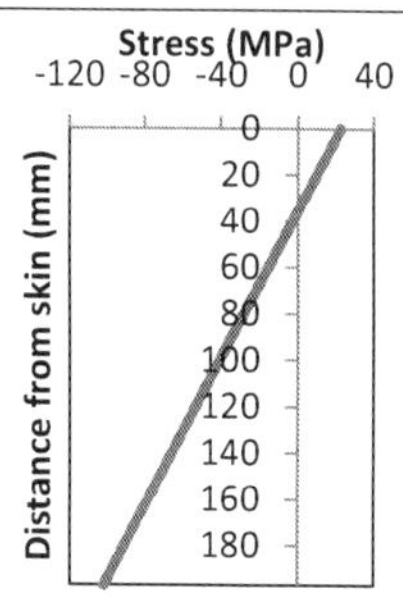	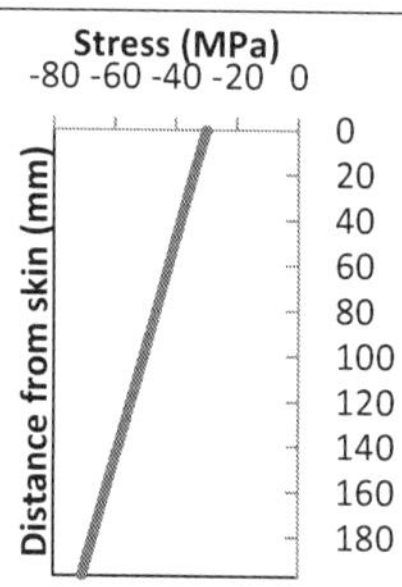	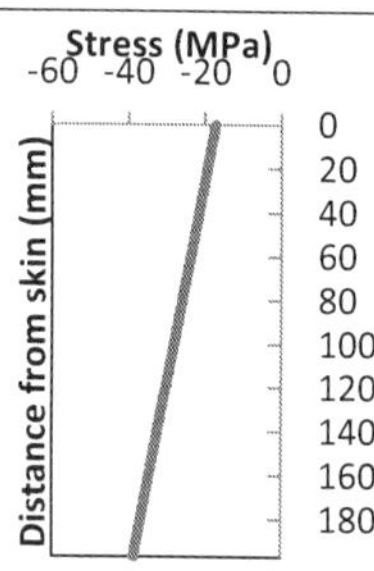
Internal actions Axial force: -1213 kN BM: 21 kNm	Internal actions Axial force: -347 kN BM: 53 kNm	Internal actions Axial force: -1009 kN BM: 17 kNm	Internal actions Axial force: -565 kN BM: 9 kNm
Axial force=>Compression positive, Bending moment=>compression in skin positive			

Table 1. DEMEC results for circumferential flanges

Intermediate DEMEC measurements

A wire bracket was attached onto the skin to protect strain gauge rosettes from mechanical damage (Figure 18). Due to the presence of the wire bracket, only DEMEC spans 1 and 6 (see Figure 10) were available for intermediate monitoring between day of installation and the day of segment removal. The results are given in Figure 19.

For segment 1, span 1 and 6 measured increasing tensile strain, which is in line with the segment experiencing positive moment as the excavation works around the segment releases the ground loading. However, for segment 2, span 1 and 6 measured increasing compressive strains during the excavations works around the segment, while the segment was still part of the lining ring.

Figure 18. Wire bracket to protect strain gauge rosettes

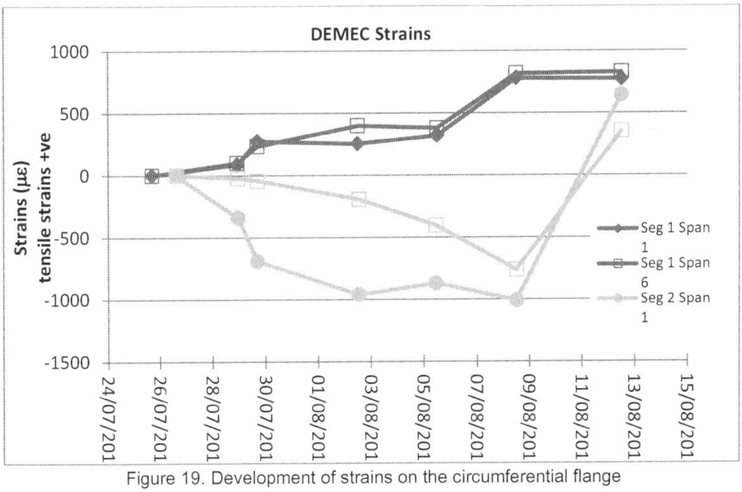

Figure 19. Development of strains on the circumferential flange

Results from strain gauge rosettes

Unloading strains due to removal of segments from ground

Using the baseline and post-removal strain measurements from each strain gauge in a rosette, the principal major and minor direct strains and the orientation of the principal direct strain directions on the surface of the cast iron were calculated for each rosette location. The calculations included corrections for strain gauge thermal output and gauge factor variation with temperature.

Refer to Figure 10 for the locations of the rosettes. The results for segment 1 are presented in Table 2. The results for segment 2 are presented in Table 3. Each plot shows the direction and magnitude of the major and minor principal direct strains on the surface of the cast iron. The y-axis represents the circumferential direction and the x-axis represents the longitudinal direction as illustrated in Figure 2.

The strain gauges measured tensile strains in the sub-circumferential direction and compressive strains in the sub-longitudinal direction. This is consistent with measurements from DEMEC spans 3 and 4. The circumferential strain and longitudinal strain at the locations of rosette 2 and rosette 4 have been resolved from the major and minor principal direct strains to allow comparison with the DEMEC measurements. This is given in Table 4.

From Table 4, it can be seen that there was less than 5% difference in the circumferential unloading tensile strains measured using the electrical and mechanical strain gauges. However, the difference between the longitudinal strains measured by the two instruments differed greatly, well over 20%. This larger difference corroborates with the positive bending mode of distortion the segment was expected to experience upon unloading, with the skin dishing radially outwards, compressive strains was expected on the inside of the skin in the longitudinal direction. Since the DEMEC instrument had a longer gauge length compared with the rosette, the compressive strain measured by the DEMEC was also larger. In the circumferential direction, axial tensile strain dominated over bending strains, and the difference in gauge lengths had lesser effects on the measured strains.

Rosette 1	Rosette 2	Rosette 3
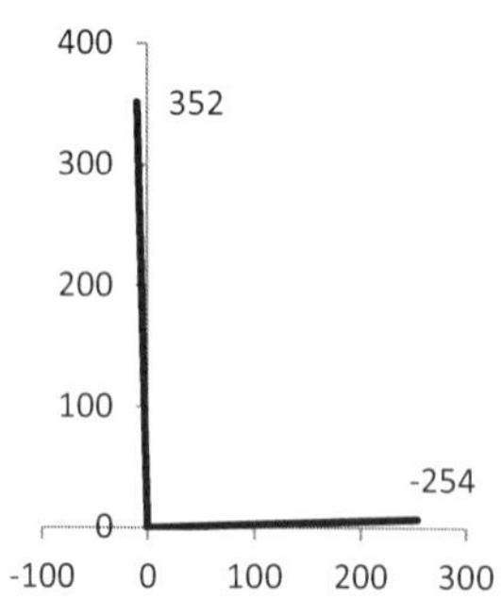 	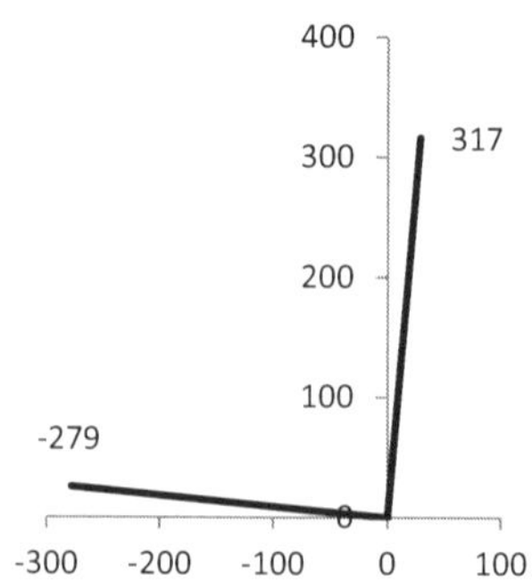 	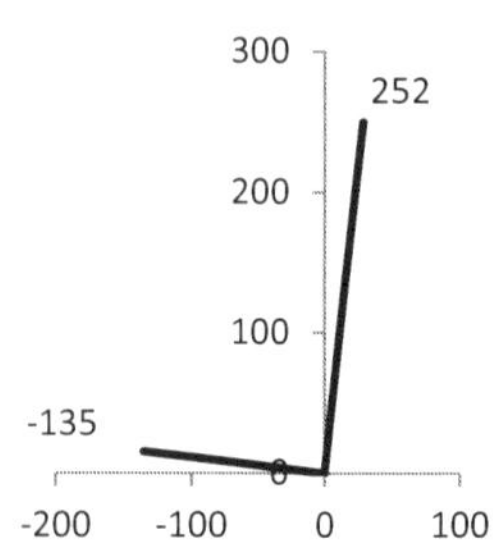
Vertical axis = circumferential direction; horizontal axis = longitudinal direction; tensile strains positive; unit is microstrain		

Table 2 Strain gauge results for segment 1

	Rosette 4	Rosette 5
	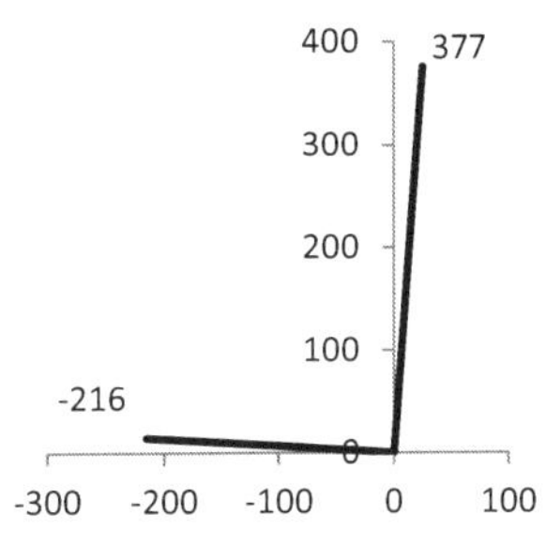	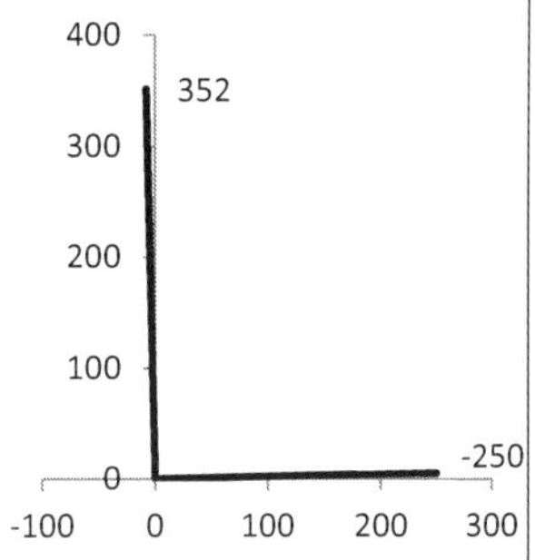
Vertical axis = circumferential direction; horizontal axis = longitudinal direction; tensile strains positive; unit is microstrain		

Table 3 Strain gauge results for segment 2

Segment		Circumferential strain (με)	Longitudinal strain (με)
1	Rosette 2	312	-274
	DEMEC	300	-353
2	Rosette 4	374	-213
	DEMEC	365	-370
Tensile strains positive			

Table 4 Comparison of DEMEC and strain gauge rosette results

Intermediate strain gauge rosette measurements

The baseline readings and the last set of readings taken before segment removal were used to calculate the changes in principal major and minor strains in the skin. The intermediate DEMEC strains near the intrados of the circumferential flanges suggest that excavation works around the tunnel lining caused positive bending in segment 1 and negative bending in segment 2, before the segments were removed from the tunnel. The results for segment 1 are presented in Table 5. The results for segment 2 are presented in Table 6. The plots show the magnitude and direction of the principal major and minor direct strains on the surface of the cast iron. The strains in the circumferential and longitudinal directions have been calculated and are given in the tables.

For segment 1, the measured strains are predominantly slightly compressive in the longitudinal direction, and slightly tensile in the circumferential direction. The compressive strains in the longitudinal direction are consistent with the assumed mode of positive bending as suggested by the DEMEC readings on the circumferential flange.

For segment 2, the measured strains are tensile in the longitudinal direction, consistent with the negative mode of bending suggested by the DEMEC readings and the associated dishing inwards of the skin.

Rosette 1	Rosette 2	Rosette 3
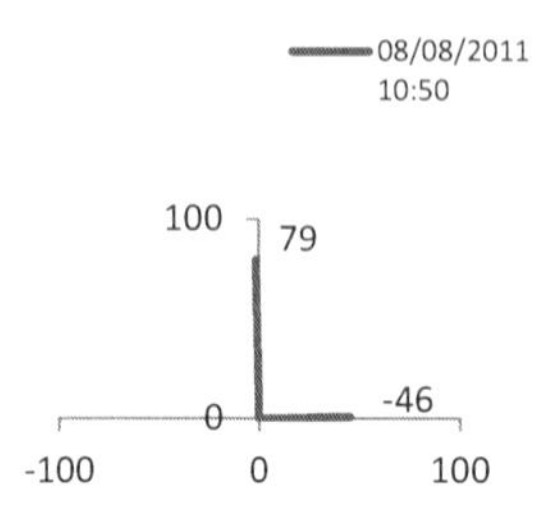	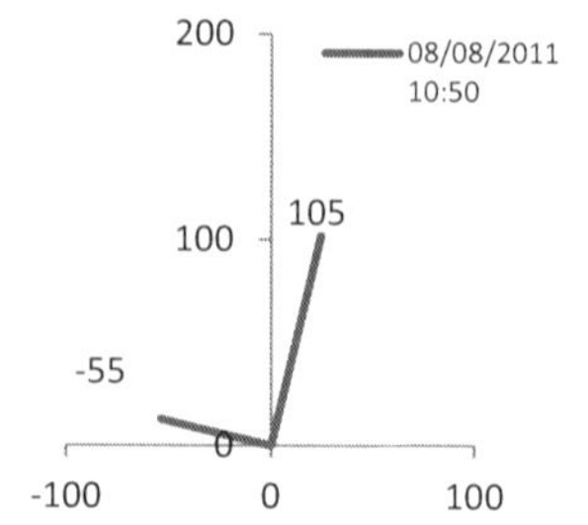	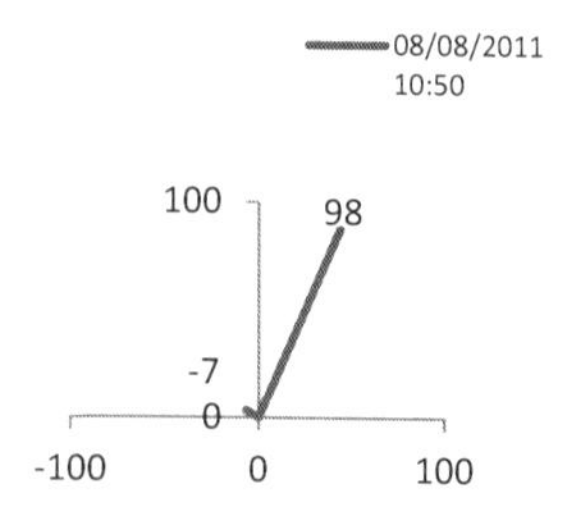
Circumferential strain: 79 με Longitudinal strain: -46 με	Circumferential strain: 96 με Longitudinal strain: -46 με	Circumferential strain: 76 με Longitudinal strain: 13 με
Vertical axis = circumferential direction; horizontal axis = longitudinal direction; tensile strains positive		

Table 5 Intermediate strain gauge results for segment 1

	Rosette 4	Rosette 5
	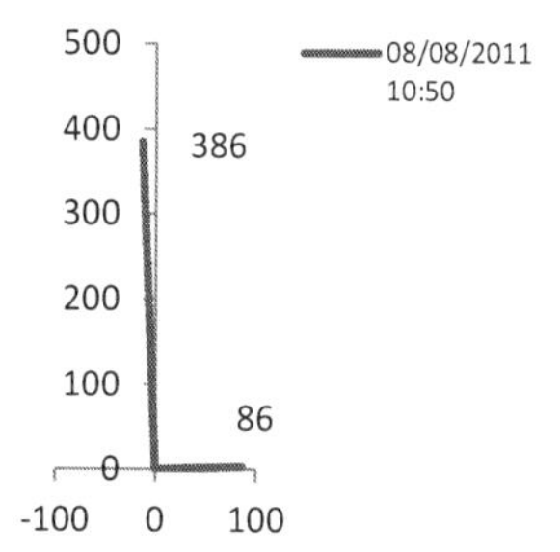	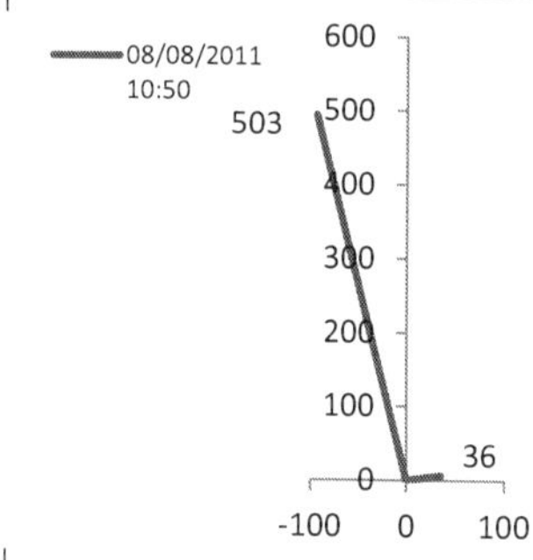
	Circumferential strain: 385 με Longitudinal strain: 87 με	Circumferential strain: 487 με Longitudinal strain: 52 με
Vertical axis = circumferential direction; horizontal axis = longitudinal direction; tensile strains positive		

Table 6 Intermediate strain gauge results for segment 2

Results from trackside displacement transducers

Continuous measurements of joint movement were taken by the displacement transducers over a period of six weeks. The LVDTs were affected by the diurnal temperature changes in the tunnel. Figure 20 shows the changes in voltage of the four LVDTs with the diurnal fluctuations smoothed out by using 24 hour running averages. While the voltage output was still clearly influenced by the overall temperature trends, there was also a downward shift in voltage output for three of the four LVDTs. This downward trend was steeper before the platform side segments were removed (11th August 2011) and flattened out post removal.

The LVDTs were calibrated so that a reduction in voltage related to increase movement, i.e. the joint opening up.

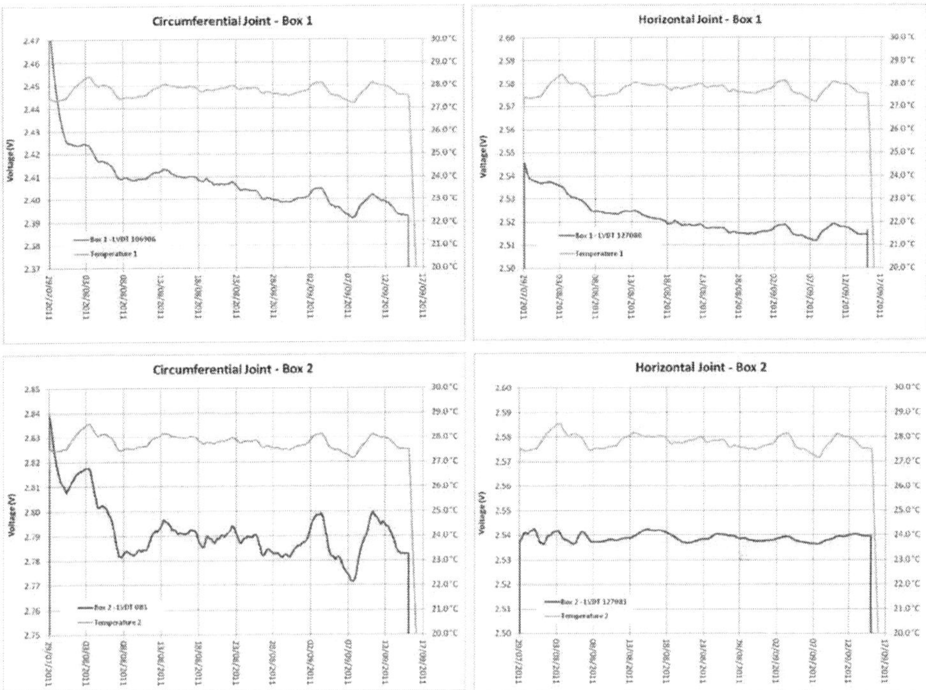

Figure 20 Comparison of LVDT trends

Conclusions

The strain release when cast iron segments were removed from a tunnel ring was measured by electrical and mechanical strain gauges. The magnitude of change in strain was similar to that predicted assuming full overburden unloading and a constant modulus of 100 GPa. Installation of strain gauges on the circumferential flanges of the segment allowed the computation of change in axial stress and bending moment of the segment due to unloading. The use of electrical strain gauge rosettes allowed the principal major and minor direct strains on the surface of the cast iron segment to be estimated, and provided insight into the mode of deformation of the segment.

Measurements made by the LVDT displacement transducers were affected by the diurnal temperature fluctuations in the tunnel.

Acknowledgements

The authors wish to thank Chris Barnes, Dr. Keith Bowers, Neil Moss, Jeremy Pallant and Duncan Parker for all their efforts in making this instrumentation trial possible.

References

GCG (2009) *Geotechnical Sectional Interpretative Report 1&2: Royal Oak to Liverpool Street. Volume 3: Drawings*. Report No. 1D0101-G0G00-00551

Halcrow (2008) *Northern Line cross passages/staircases long sections through staircase No.10.* Drawing no. HAG-N105-8742-TUN-D-SEC-5-03426 Rev 00. Dated 15/12/2008.

Thomas, H. S. H. (1977) Measuring the structural performance of cast iron tunnel linings in the laboratory. *Ground Engineering*, 10 (5), 29-36.

Crossrail

Crossrail Project: Infrastructure design and construction
ISBN 978-0-7277-6078-4

doi: 10.1680/cpid.60784.289

ice
Institution of Civil Engineers
publishing

Royal Woolwich Arsenal – Crossrail Box Sewer Diversion

Peter Buckley BEng Civil Engineering, Deputy Project Manager for Berkeley Homes

Abstract

In a unique arrangement Berkeley are collaborating with Crossrail Ltd (CRL) to deliver the Royal Arsenal Woolwich Station Box. In order to construct the Box initial ground investigation was required as well as site clearance and multiple utility service diversions. The most complex of the utility diversions was the diversion of 325m of diameter live private sewer serving the Woolwich Arsenal estate. In February 2011 Berkeley instructed Waterman to undertake a feasibility assessment and propose a permanent design solution.

Waterman's design anticipated trenchless methodologies would need to be utilised to minimise settlement of the pipeline when the Crossrail tunnels are bored beneath the realigned sewer in 2013 as disturbances in the ground may increase the ground settlements and consequently the deflections in the pipe. Detailed analysis of ground settlements as a result of the Crossrail Box were undertaken along with the resultant pipe deflections and the ability of the pipework used to accommodate the anticipated movements. Due to the stage of tunnel design, the ground settlement from the tunnelling operation was unknown and as such in the future further analysis will be required by Crossrail of the pipe's ability to accommodate settlements from the tunnelling operation. The revised alignment included 9 new manholes and micro-tunnels varying in diameter from 225 mm to 1200 mm. In preparation for tunnelling a number of neighbouring buildings required strengthening to prevent any damage if subsidence was to occur. Remote 3D monitoring with robotic total stations by Soldata was established to monitor the buildings and existing services prior to tunnelling and during the construction of the Station Box. This provided cycles of readings every 15 minutes providing assurance that the buildings were not suffering any structural damage.

Following successful tender, Joseph Gallagher Ltd was awarded the contract to undertake the sewer diversion. Their experience in tunnel boring in testing ground conditions proved essential to the successful completion of the works in line with the constrained project programme. The tunnels were bored through the shallow aquifer at depths of between 4.4 m and 8.4 m below ground level. Gallagher commissioned a bespoke tunnel boring machine (TBM) & pipe jacking rig from Iseki. The TBM was equipped with ground water pressure sensors. The TBM utilised a slurry pumping system to remove the arisings from the tunnel face and to maintain a positive pressure against the ground water, negating the need for any dewatering.

The completed foul and combined sewers needed to meet stringent adoptable standards requiring design approvals and regular inspections by Thames Water during construction. Temporary over-pumping of significant flow rates was required during the diversion works and will also be required during the Crossrail tunnelling operation in case any significant settlement occurs. Crossrail's tunnelling team, Hochtief Murphy JV, is examining temporary support options to prevent any significant settlement of the diverted pipeline.

Introduction

Berkeley Homes (East Thames) Ltd. (BH) are providing The Secretary of State (The Department for Transport) with a Crossrail Station Box within the Royal Arsenal site at Woolwich. On completion, this will be handed to Crossrail Limited (CRL), the nominated undertaker under the Crossrail Act. Waterman Transport & Development Ltd (WTDL) are acting as lead consultant to Berkeley Homes (East Thames Ltd) for the civil and structural concept design.

The Station Box comprises a West End with minimum internal dimensions of 45 m by 26.6 m width, a Central Box of 173 m by 20.4 m width and an East End of 38m by 26.6 m width. It is planned to construct a station entrance at the West End with a single level structure above ground level and a 2-3 storey above-ground Oversite Development at the East End, but details have not yet been finalised. Over the Central Box and to the south of the West End and Central Box, Berkeley will be constructing a separate multi-storey Oversite Development. Arsenal Way will be reinstated in its current position between the Berkeley oversite development and the East End.

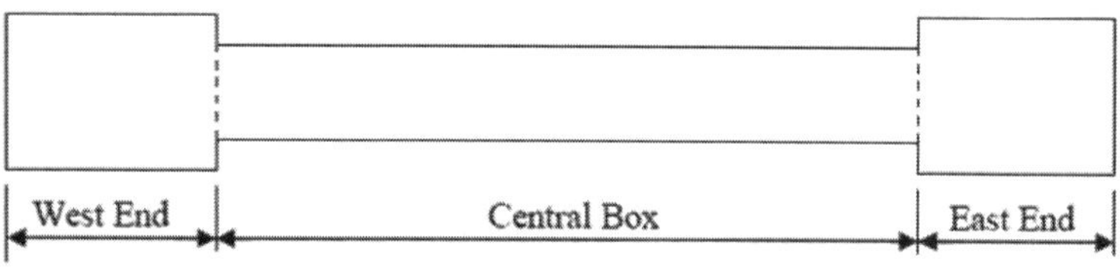

Figure 1. Station Box Components

As part of the Enabling Works, the existing sewers in Arsenal Way, Cornwallis Road and generally within the area of the proposed Station Box have been diverted through a new 1.2m diameter sewer on the north, east and south sides of the Station Box. The diverted sewer was installed in advance of the construction of the Station Box and Crossrail tunnels. The new diverted 1.2 m diameter sewer was assessed for potential ground movements from the following activities:

- Installation of the perimeter diaphragm wall
- De-watering of the station box to permit the safe excavation within the box
- Excavation within the station box and installation of quasi-permanent props
- Construction of base and roof slabs
- Release of de-watering
- Construction of the bored tunnels at either end of the station box

Figure 12. Launching the TBM

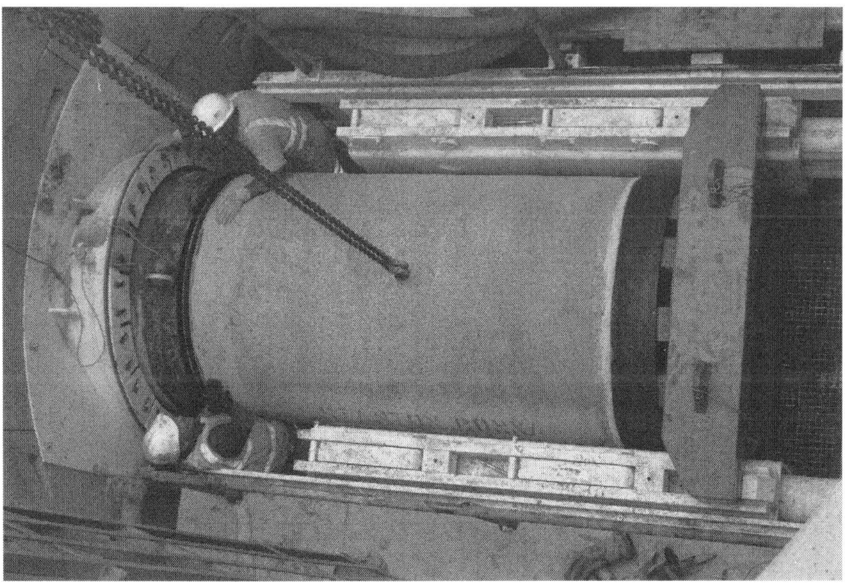

Figure 13. Pipe jacking behind the TBM

Figure 14. TBMs recovered

Construction Challenges & Conclusions

Generally there were very few unforeseen events during construction. The employment of a specialist contractor such as Joseph Gallagher Ltd ensured that risks were identified and mitigating measures put in place in advance of the works to safeguard the delivery of the works and the programme.

A small quantity of white asbestos was uncovered which required specialist contractor removal. The area was surveyed to ensure no asbestos remained so that works could continue.

Dewatering systems were in place to cope with any significant influxes of water into the caissons however only minimal water ingresses were experienced.

Overpumping of the flows through the existing pipework was installed to facilitate the new pipework installation. This was in place throughout the works and Berkeley have facilitated a further four week overpumping operation for the period of the TBM passing beneath the diverted route. This is to provide resilience to the Royal Arsenal estate sewage outfall in case any significant deflections occur whilst tunnelling.

The emergency services attended site to familiarise themselves with the site and what scenarios they may be faced with. The works were completed without incident, accident or injury.
Berkeley continues to work closely with Thames Water, Crossrail and their tunnelling contractor HMJV to mitigate any impacts that the tunnelling operation may have on this key element of infrastructure.

Acknowledgements

Grateful acknowledgement is made to the following companies for their contribution to the project and to this report:

Crossrail Geotechnical Team

Berkeley Project Management Team

Waterman Group Design Team

Joseph Gallagher & Sons Construction Team & their specialist business unit Johnston Trenchless Solutions

Iseki Tunnel Boring Machines

Thames Water

References

1. CIRIA C580, 2003, 'Embedded retaining walls – guidance for economic design'.
2. CRL Document No. CR-STD-303-7, Version 5.0, August 2009, Civil Engineering Design Standard, Ground Movement Prediction.
3. London Underground Manual of Good Practice G-058.
4. Crossrail Information Paper D-12 – Ground Settlement, Version 5, February 2008.
5. WTDL document no. C-11331-GR-0001-C01, Royal Woolwich Arsenal Civil and Structural Concept Design Statement, April 2011.

Crossrail Project: Infrastructure design and construction
ISBN 978-0-7277-6078-4

doi: 10.1680/cpid.60784.303

The Role of Operations in a Mega Project

Charles Devereux, Crossrail Ltd
Jim Claxton, Systra UK

Abstract

Only when it opens will it be possible to judge the success of Operational involvement within Crossrail. There is an inherent conflict between the engineer's desire to deliver a project at lowest first cost, and that of the operator to enjoy a robust product when complete. Crossrail has had operator involvement from the outset. This paper examines both the benefits and identifies things which would be done differently, and where time has been wasted in project development.

The paper discusses the need for robust governance principles from the outset; who will own and manage the new infrastructure, and the involvement of a robust and informed, operational client who has an interest in the completed railway.

The key operator input is to develop and agree requirements that define **what** the project is expected to deliver but not **how** it is delivered. These increase in detail over time.

Stressed is the importance of equipping Operators with tools that allow them to deliver the best result to the project - the importance of defining requirements, and the crucial distinction between scope and requirements. It is not appropriate to parachute an operator into a mega project without providing a grounding in how they will inform the design process, and in turn understanding from engineers how operations can contribute in a timely manner to design.

The corollary is the need for the engineering function to understand where an operator needs to be involved, often in seemingly trivial matters. The paper highlights the breadth of operations- encompassing train planning, passenger modelling, reliability modelling, rules and procedures and system control.

The paper underlines need for the engineering function to respect and trust the operator as a credible, competent body who can add to a detailed debate. There is nevertheless a paradox; the process of operators becoming familiar with the engineering constraints, and acquiring a robust trustful working relationship will take a matter of years; during this time the operator will become less familiar with the operational world beyond the project and risks losing competence.

There is a need to distinguish between the civil engineering elements and the systems elements of a project and the requirement for each to have realistic 'requirements freeze'.

The value, timing and paramount importance of Operations Concepts, standards selection and close working with systems engineers, but equally the recognition by the operator of when to step away from the design process either because the designers need no more operational input or indeed (particularly at the early stage of a project) not all operator input is relevant or timely.

Introduction

It is a truism that all railway projects require "operational input" throughout their lifecycle. This paper defines the role of the operator, and identifies the inputs which an operator could (or should) make during project development.

For the purposes of this paper 'Operator' is the entity which plans, monitors, resources and executes the train plan. The Operator implements recovery processes to return the railway to plan after a failure. The Operator is the custodian of the rules and procedures for the completed railway. For the completed railway the role of the operator may be accommodated within several different organisations (for example Network Rail and Train Operating Companies). Therefore the operational 'guiding mind' may vary during the course of the project. For example the 'guiding mind' for the Crossrail train operations rests with Rail for London, but will in due course rest with a concessionaire train operator albeit working to a framework determined by RfL. The disciplines within operations vary –strategic system view, passenger flow modelling, train service modelling, application of rules and standards, resourcing, incident management, and the emphasis will change over the course of a project.

Much in this paper also applies to the Maintenance Engineer. Operator and Maintainer will 'own' the completed system, and have an incentive to ensure that the output of the project can be delivered, and that the operational cost is understood. They will be there long after the engineers and designers have left for other projects.

'Operations Input' to a major project can vary from a high level view of the operability and feasibility of a project to agreeing and implementing the detailed rules and procedures for a completed system. The paper describes these varying roles and the differing operational skill sets required during project lifecycle.

Some major projects have suffered in the past through insufficient operator involvement at an early enough stage; for example although the engineering design for Thameslink 2000 was well advanced in the early 1990s, no detailed modelling is believed to have been done during Network South East (NSE) management to validate the core assumption of the scheme, that 24 trains per hour (tph) was feasible within the design assumptions then being made. The original designs for the Portbury Dock freight development in Bristol would have been inoperable if executed in their original form.

The Inception of a Project

Mega projects such as Thameslink, East London Line, HS2, tram schemes and Crossrail all depend upon political support. Political support is engendered, especially in the Treasury through presentation of a business case. The business case is primarily made by looking at the scheme, and by using established (and proven) techniques estimating the likely usage of the new infrastructure based upon assumptions of varying train service levels. This is influenced by the journey time (and counting both the benefits to existing users of the time saving as well as new journeys likely to be thus generated). The costs for the scheme will have been generated by engineers, and may well underestimate the requirements for operating the planned level of service, possibly underestimating enhancement to the existing network.

In the case of Thameslink 2000, and Crossrail, the enthusiasm of the Transport planners ran well ahead of the operators-see case study 1.

Case Study #1:

Crossrail's initial consultation assumed one branch to Kingston and another to Heathrow Airport. Transport planners had identified significant suppressed demand in South West London, and the benefit:cost case for Crossrail extension to Kingston via Richmond was good. There was also an eastern leg proposed to Ebbsfleet, not discussed here.

The branch to Kingston would have required interworking with SW trains, the North London Line and the District line, and would have been close to inoperable. The alternative (ceasing operation of either NLL or District Line to Richmond) was politically unacceptable.

At the time, Crossrail was being led by the Strategic Rail Authority in general and transport planners in particular.

Although there was a small Crossrail operations function in place it's role was reactive and not strategic- in other words it could react to a proposed scheme, but was insufficiently influential to say that, however desirable, the consulted scheme would simply not deliver a reliable railway. At the same time as the demise of the SRA in 2004, the Montague Report was commissioned with a specific remit to (inter alia) examine the operability of Crossrail.

Unsurprisingly, the report concluded that Crossrail as configured would not perform or meet the intended Central area throughput without the trains operating on more dedicated infrastructure.

This conclusion would have been apparent to any competent operator.

In case study 1, the Montague Report identified that the Richmond/Kingston extension did not work, a conclusion which could have been achieved far earlier if an outcome with success criteria (e.g " a railway that can deliver 12 tph per hour to Richmond with 24tph in the central area and continued operation of 8 tph on the District line with 95% punctuality") had been laid out and an "operator" invited to validate the feasibility of the proposal against these requirements.

For those within an organisation such as the embryonic Crossrail, identifying that a scheme is unworkable (and therefore condemn the scheme to oblivion) is not necessarily in their personal or organisational best interests, especially if stakeholder pressure has built up for a particular option. 'Operability' at this stage of the project is to some extent judgemental and proving inoperability can be difficult, since absolute proof requires a modelling, which to be accurate would require a host of detailed assumptions about track layout, signalling, rolling stock and other services which is simply not feasible to assemble at an early stage of a project and committing about 4 months to it. Stakeholders will often demand 'proof' that their chosen scheme will **not** work. Proving a negative is very hard, especially as there is often a subliminal belief that with a bit of (unspecified) tweaking the chosen scheme will 'work'

Case Study #2:

Network Rail is seeking planning permission for a station at Low Moor (near Bradford). The incumbent operator, Northern Rail, is objecting, on the grounds that without (unfunded) injection of rolling stock and crew resource, the scheme will not work, despite Local authority support.

It is inevitable that a 'mega project' will at its' inception examine a variety of options. However if 'operator involvement' is to be meaningful (and potentially save a significant amount of money) any options for further development should be assessed against a formal set of requirements. Using the example above, it is possible (albeit unlikely) that if Crossrail had been conceived as a railway which only needed to convey 15 trains per hour through Central London, the outcome would have been different.

Lesson Learnt #1:

Trust the Operator. Don't waste time on fanciful options

Ensure that a competent and senior operator has independently validated any proposed scheme and confirmed that the scheme is operable against success criteria, trusting that judgement without requiring excessive modelling

The inception of a scheme requires three operating disciplines;

1) Planning expertise; to understand the likely flows and to dimension the stations and prepare an outline of the train service

2) Strategic-identifying the likely impact on the network, and suggesting how any proposed scheme can most robustly be operated, and identifying scope that will be required to support the requirements.

3) Detailed operating knowledge (such as the likely signalling system capability) to inform high level design decisions.

Project Development and Requirements

What follows on Requirements could be taken out of a Project Managers' manual. But in hindsight, although Crossrail has been better than most projects, some of the hard learned lessons were not addressed from the outset. It is hard to understate the value of robust Requirements Management from an early stage, and in particular an understanding that any changes, however trivial, will impact on the project and the cost. Requirements define the technical output that operations (and maintenance) functions would expect at the end of the project in order to realise the client's anticipated benefits .Requirements must be change controlled, and complemented by an assumptions register, also change controlled.

Crossrail's requirements were articulated explicitly only after the Crossrail Bill had been deposited. Previously, there was an apparent consensus about what Crossrail was seeking to achieve, but nowhere was it written down in anything other than motherhood statements ("a significant increase in London's transport capacity**").** When the time came to write down the requirements in detail, it was clear that there was a wide range of understandings within the project; for example some people understood that the benefit of two way signalling was that a 24 hour service could be operated. If that had indeed been the intention then it would have changed numerous systems, for example ventilation fans would have to be able to support dirty works in one tunnel and a passenger service in the other.

Case Study #3: The Impact of Seemingly Trivial Change: End Doors

Crossrail trains could be evacuated from a gangway dropped down from the front of the train (like East London Line class 378s) or side evacuated onto a walkway. This has a significant impact on both train operation in the case of a fire emergency & evacuation of a failed train, on rolling stock design, and on the profile of the tunnel and therefore the route alignment. However, until this had been evaluated by all disciplines, and a decision reached, even the most basic ***tunnel*** *design is impossible. Conversely a unilateral decision by the tunnelling engineer could have significant long term implications for other disciplines*

Case Study #4: The Impact of Seemingly Trivial Change: 2 doors or 3 doors per coach

A Crossrail requirement is platform screen doors at sub surface stations. Once the configuration of these doors is established, they cannot be changed to a different door position on the coach without the entire fleet and all the doors being taken out of use at once for a 'big bang' transition. Door density predetermines door configuration forever. A low door density will have a longer dwell time, and so the ability to deliver 24tph is compromised and delivery of 30tph impossible. However if procurement of the rolling stock envisages that it may be leased to another party, residual value is important and a 2 door per coach train more saleable. Therefore this decision underpins other assumptions about the ultimate performance of the system; there is no point providing ventilation for a 30tph train service if the door configuration prevents a 24tph service and no credible upgrade path.

It is imperative that everyone in the project is aware of, and works to, the Requirements which the operator endorses. This may appear to be self-evident; however a mega project such as Crossrail will be predominantly composed of civil engineering disciplines which may not be aware that some of their decisions will have an operational impact. In a mega project, it is a fact of life that operators will be many times outnumbered by engineering disciplines all anxious to 'get on with the job'.

It is equally important that key strategic decisions (such as door configuration and escape strategy) are decided early because upon this a whole raft of other decisions sit. There is often huge pressure to 'firm up' requirements for the letting of long lead and civil contracts. In reality, these contracts are required for space proofing the finished railway. For a mega project the bulk of the detailed issues that an operator has relates to systems, which need to be finalised far later in the project life. Requirements which are concerned with system fit out are able to follow significantly later, when the project has built up sufficient systems expertise to allow the operators an appropriate professional dialogue.

Lesson Learnt #2:

Distinguish Core Requirements from Systems Requirements

Recognise that Requirements for space-proofing and civil works evolve at a different rate to those for systems, so there will be initial Requirements and final Requirements. Understand what is important and what can wait.

Case Study #5:

The Impact of Seemingly Trivial Change: Capability

The capability of the Crossrail central section to handle up to 30 trains per hour of 240m in length has a significant and apparent bearing on systems (for example signalling, train performance and station systems to minimise dwell time). However, less obvious is the impact on the ventilation system; because there can only be two trains in a ventilation section, the spacing of the shafts must be such that the planned train service can be operated. If the planned train service had been lower, then the ventilation configuration would have been different; a decision at the last minute to increase train frequency may become unaffordable because a different spacing of shafts is required.

Requirements have to be tangible from the outset, so that designers have a clear remit. Use of phrases such as "the ability to increase the train service from 24tph to 30 tph *shall be considered*" or "the ability to separately operate the station *shall be considered*", become entirely inappropriate early in the project lifecycle. Much better to make a requirement (" the system shall without closure be capable of upgrade to accommodate a service of 30tph within 10 years of opening") which can if need be challenged during value management, so that the cost of the requirement can be assessed.

The most difficult of all requirements are those which are aspirational; Crossrail Ltd is required to deliver a "world class railway". This can mean that facilities not strictly required to make the railway work can nevertheless be justified as being 'world class'.

Case Study #6:

'World-Class': A Matter of Opinion.

To an architect this means a line of route with inspiring stations (for example the Jubilee Line extension)

To an operator this means a railway which operates reliably with minimum waste of resources, but no particular 'ambience' to the line (PATCO in Philadelphia/Camden)

To a maintainer this may mean a railway which is straightforward to maintain with excellent data on the 'as built' railway (MTR-Hong Kong)

To a passenger it may mean a railway that operates so reliably that they do not notice the journey, even though the facilities are basic (Toronto)

To railway commentators who systems, it is easily overlooked that LU is very much 'world class'

Case Study #7:

What does "World Class" mean in practice?

When the bill for Crossrail was deposited the output requirements were not specified, so the bill was deposited with the maximum possible range of facilities likely to be required- this is normal, and indeed prudent, practice. At Custom House, the bill scheme showed escalators. The final (value managed) scheme does not, since the standards and the passenger flows do not demand them; unsurprisingly the local authority are challenging this, and the notion of 'world class' is a factor in this challenge.

Requirements should not presume a solution; for example the length of a Crossrail train at opening should be expressed as 200m (nominal) and not as (for example) an 8 coach train, since the latter predetermines a rolling stock solution.

Keeping Options open costs money. The DfT as sponsors of Crossrail wanted to have the option of managing stations by a separate entity to LU (this was at the time that the PPP was still in place). Whilst this was only an option the subsequent integration of the Crossrail demise with that of LU at Farringdon, Whitechapel and Moorgate stations could have been avoided if a firm output (the stations shall be operated by LU) had been expressed form the outset.

With hindsight, it is clear that a project wide understanding of Requirements-their ownership, their control, and the way in which they were expressed was a slow learning curve for all. For many of those operators likely to be involved in a project such as Crossrail, the importance of this is probably not well understood, and may not be a discipline with which they have been familiar. In particular the difference between requirements and scope is ill understood. Requirements define the outcome. Scope defines how the engineer plans to deliver the Requirement.

There is also a subtle change in the skill set needed. Initial project consideration requires strategic and modelling skills, described in section 2.7. As design progresses, there is also a need for operators with recent 'front line' experience to contribute to the requirement setting process to ensure that the project is grounded and workable. Equally these individuals will not have been exposed to a major project and so the means of engaging with engineering colleagues in this structured manner needs to be formally trained.

Lesson Learnt #3:

Train Operators in Requirements Management

Have a clear, controlled set of requirements whose content is agreed by your professional operator. Make sure that your operator understands how to express those requirements, and make sure that the project understands the importance of them. You should not let an operator anywhere near a project until they have been formally trained to do this.

Recognise that the requirements will become more detailed over time, but also recognise that any changes, no matter how trivial may have a latent cost.

Seeking authority

Once the project has developed sufficiently, it will need to seek powers- an Act or a Transport and Works order. A high profile project such as Crossrail, attracts petitions which impact on operations as the Bill progresses through Parliament.

These petitions can either be upon interested parties wishing to impose a solution (for example petitioners who wished to extend Crossrail west of Abbey Wood) or upon those who have an existing interest, which would be affected by the scheme (freight operators who felt Crossrail would impact upon their ability to grow the business).

The consent process is key because the process can impose a solution upon the project that is unsatisfactory; for example the creation of an International Station at Stratford not served by International trains. Crossrail was required to provide a straight and level alignment at Kensal Green for a future station, even though providing such a station would arguably not be in the best interests of the overwhelming majority using the route. Sometimes the operator and engineer are forced to respond to a petition overnight on a matter which at other times would be agonised over for several weeks.

In hindsight, there was an underlying assumption that because Crossrail was so important for London that other railway interests would be broadly supportive. Instead the scheme was used as a means for these interests to adopt a very combative position, partly to protect their existing rights, partly to secure enhancement that would benefit them, and partly to maximise their gains if Crossrail acquired their rights. This is an understandable consequence of the railway industry as it is now structured. However for the promoter of a project it requires significant operator input to think through the effect on existing users, and to some extent challenge the impact that the project will have on other users.

Lesson Learnt #4:

Operators Engagement with the Planning Process

For the consent process anticipate heavy petitioning, and be prepared to defend the project position robustly even though it may appear self-evident that the project is of wide social benefit. Expect your operators to be involved in this.

Value Management

A project will need to demonstrate value for money, and value management will challenge every element. Experience suggests that the elements of the project most likely to be proposed for value management will be those which are most difficult to build. In some cases VM will challenge the requirements (the Fisher St example below), but unless the requirements exist in a robust form, the operator may be on weak ground to defend removal of scheme elements. However, even where a requirement does exist a robust operational response is still required, often explaining operational matters from first principles, in order to retain key project elements.

Case Study #8:

Value management 'ad-absurdam'

One suggestion was to make the Crossrail tunnel single through the Thames tunnel. To an operator, this immediately imposes a constraint on the timetable (painfully demonstrated at Heathrow where the single line branch to terminal 4 predetermines much of the Crossrail timetable). It significantly reduces capacity and materially worsens reliability. Despite this, some considerable work had to go in to 'prove' this, without the view of the professional operator being taken on trust- see case study 1

Case Study #9:

Requirements vs Scope

Crossrail as conceived, specified a pair of crossovers at Fisher Street, as a firm requirement. This would have been almost impossible to build, and in fact the underlying requirement subsequently adopted (by having a single Crossover at Bond Street and Whitechapel respectively) gave much more flexibility both to the operator and to the designers;

System Requirements

A railway is a uniquely difficult system to manage; unless a completely isolated system is built, any new project must integrate with the existing railway, either by sharing a (possibly) enhanced route, sharing expanded stations and interfacing at the boundary with legacy systems. How the railway is operated is influenced by who the ultimate duty holder will be; the development of Crossrail has highlighted numerous differences between National Rail and London Underground methods and approaches to operation. It is imperative that the planned duty holders are identified early on, and that they do not change, and that the embedded operators within the project correctly reflect the view of the final operator. The failure to nominate RfL as the body responsible for letting the concession for Crossrail (as opposed to the DfT letting a franchise because the DfT wished to keep options open) means that the project developed the specifications for the rebuilt stations on the existing network based on a 'Best guess' rather than by embedding the views of the final operator in the project.

Case Study #10:

Define the End-Game early.

Crossrail did not know the final governance of the project (the relative involvement of RfL/LUL) and therefore appointed an experienced team to provide proxy operational input. Once governance was established, it was apparent that there were differences in approach between this proxy input, and the approach that the (finally) nominated duty holders would take.

Whilst the Requirements describe the output that is required, this does not provide the information on how it is expected that the railway will operate. Crossrail has described this in a set of so called "Operational Concepts" which describe in plain English how the systems are expected to integrate with one another. They do not supersede the requirements, although their authorship has in some cases influenced the Requirements.

Case Study #11:

Requirements versus Concepts.

There is a requirement for trains to be able to operate in 'auto reverse' with a driver on the train, but not supervising movement as the train moves to a stabling berth before reversing and returning on the opposite line.

The operations concept describes how it is expected that the driver will change mode, how the train will be dispatched, how the driver will intervene if required, and how the driver then takes over the train in the cab at the other end of the train. It is possible that a manufacturer wishes to challenge this modus operandi . If upheld, the operations concept will alter (through a controlled process) but the requirement does not change

The concepts are intended to have a profound influence through the project;

1) They evolved from step by step analysis of routine procedures- a so called "day in the life" that went through the procedures in minute detail- describing for example how a train is despatched- when does the driver know it is time to go, how is time recovered, what do staff on the platform see etc

2) They inform the procedures to be adopted in the event of certain system failures (for example how it is expected to evacuate a failed train). This is crucial to the systems integration function which needs to understand how it is intended that the systems fit together.

3) The concepts have proved to be an excellent means of checking the requirements, and validating as a 'proof of concept'.

4) The concepts have been shared with all potential operators (including LU and NR) so there is confidence that if the railway is integrated in accordance with these concepts it will be readily accepted into use. Crucial to this process is understanding who the 'end operator' will be (see above)

5) The concepts have exposed some of the cultural and operational differences that exist between LU and main line railways- in some cases the same term means something different. They have allowed all those involved to have a deep insight into both metro and main line types of operation. It has allowed a cross check so that if there are several ways of doing a task (Driver only operation, see below) the optimal way can be chosen and specified. Equally if there is no compelling reason to change existing practice this is made clear.

6) They also reduce the system integration risk; because by their nature they describe how the systems will integrate in plain English. A governance process is then put in place to identify any gaps between the emerging designs (and their interfaces to other systems) and the operations concepts and associated

requirements. The process involved in creating the concepts has, in itself required a forensic approach to every step of the operation

Crossrail is currently at the stage where the Requirements and Operational Concepts have been completed, and will be provided to systems contractors. For those contracts already in the course of discussion, the requests for information appear to indicate engagement, but their real worth during systems fit out has yet to be tested.

Paradoxically, it is whilst system design and execution are underway that the role of operations within the project should becomes less prominent; all the modelling to validate performance at the reference stage has been done and the requirements articulated. If the Operations Concepts and requirements are to have any value they should be stable, and only change if there is a compelling reason to do so. Endless tinkering only causes confusion. Whilst there is clearly a need for a 'wise head' to respond to operational issues, operations during system design should be responding not leading.

As the designs conclude, hazards will be identified which can only be mitigated by operational measures; some of these (such as fire evacuation) are already documented. However others (for example how tolerant the system can be to a pump failure) can only be determined as the designs emerge and the operator agrees the mitigations and procedures to be put in place for the completed system. The process intensifies as the project nears handover, with the operator being almost exclusively concerned with rules and procedures. In all probability the people who will be undertaking this work will be those who will be operating the finished system. Their experience will be in day to day operation; clearly there will need to be the continuity within the project to narrate why.

The process of developing operations concepts frequently identifies gaps in scope in the intended design, which have to be funded through change control. There is a danger that operations is seen as 'gilding the lily' or 'changing their mind' when in reality the gaps in scope frequently occur because explicit operations demands have not been met, or because the project was insufficiently mature to have defined the requirements.

Lesson Learnt #5:

Operational Agreement

When specifying systems, have a set of operations concept agreed by the ultimate duty holders, whose content aligns with the Requirements

The completion of Operations Concepts marks the conclusion of the design and specification phase of the project. For a mega project such as Crossrail, this was the culmination of 10 years work. The individuals involved have become very close to their engineering and maintenance counterparts, especially those involved in systems. It is inevitable though that those same individuals, well versed in the role of Requirements and working with engineers have, by definition become more distant from operators who have been more recently involved in the day to day operational railway. As the completed designs emerge, it is inevitable that there is a gradual migration to a different profile of operator- less theoretical, and more 'hands on' as the proxy to the final operator. This progressive migration requires a sensitive handover. It is important to avoid the 'new blood' wishing to revisit decisions embedded in the systems design, so there is a need for continuity; a project as complex as Crossrail has only a tiny number of people versed in end to end operation, and the reason why a given path has been adopted.

Summary

The paper has tried to make explicit the role of the Operator in the project. 'Operations' covers a variety of skills which are identified in Figure 1 below:

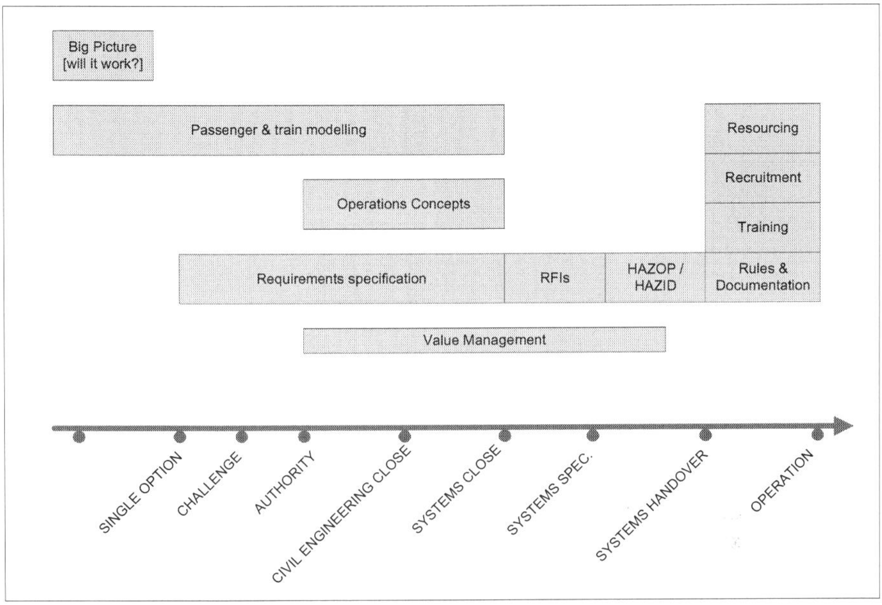

Figure 1. Operational Input vs Project Phases

If the paper has only one message it is that disciplined Requirements Management in which the operator is deeply involved is imperative from the outset. A strong operations function will support engineering and avoid rework because of unforeseen operational consequences. The profile of operations support over time will vary, but the need for professional operations support aligned with those who will ultimately operate the railway is imperative.

Crossrail Project: Infrastructure design and construction
ISBN 978-0-7277-6078-4

doi: 10.1680/cpid.60784.317

Crossrail Art Programme

Robert Wood BSc(Hons)(UCL) DipArch(Cantab) RIBA L'Ordre des Architectes Francais D&AD, Interface Manager C100 Architectural Components

ABSTRACT

At the heart of the Crossrail project is a commitment to support London's position as a world-class city by delivering "a world class railway that is fit for the 21st century."*

The construction of Crossrail is an immense feat of engineering, which will transform the city's congested streets. But it is also a project of great cultural significance, with the potential to transform the world's view of London.

London is a city of culture. It understands that 'culture' is a complex mix of art, business and the wider social realm that forms a tangible asset, which can attract inward capital investment, tourism and the world's most talented people and successful companies. The design and construction of a new railway in the heart of London presents a major opportunity for a new cultural programme that can dramatically increase the regenerative force of Crossrail.

The golden era of London Transport, under the stewardship of Frank Pick, showed how public art can help to enhance the passenger journey experience, improve the public realm, forge engagement with staff and local communities and encourage ownership of a network.

This paper discusses the creation, development and structure of Crossrail's art programme examining the overall vision, the creation of a project delivery team, the curatorial process and policy, funding and sponsorship proposals and the evolution of the programme into a comprehensive project that sought to achieve the following:

- The integration of permanent artwork into the fabric of the network
- The creation of temporary artworks, artist's residencies and cultural events
- The creation of an artistic record of the project
- A programme that will attract the foremost artists, galleries and curators
- A programme that will help to build the Crossrail brand

* Crossrail Architectural Design Guidelines: Vision for the System Pt.2.0, 2009 edition.

INTRODUCTION

Crossrail had an obligation to deliver an art programme set by the Sponsors' requirements for Central Section stations to display sponsored public art.

> *A programme for the sponsorship of artworks at new stations shall be established by CRL, with display to be considered during the detailed design of these stations in order not to compromise operational, safety or security requirements.*
>
> Sponsors' requirements v.4.1

This obligation represented a great opportunity. It was very clear that a project of such enormous social, economic and historical significance for London and the South East should aspire to deliver a cultural component as part of the overall project programme that would deliver the widest possible benefits for the project, local communities and the surrounding regions.

It was with this broad and ambitious vision that C100 and the Crossrail team embarked on the project for the development of an art programme.

BUILDING THE ARTS PROGRAMME

Prior to commencing the project the team consulted widely to gather opinions and advice from experts in the field, explore the many precedents and case studies available of existing and established art programmes and begin to broker potential partnerships.[1]

This initial research was used to inform and structure a development strategy for the Crossrail arts programme. The strategy evolved from asking five key questions, which acted as cornerstones from which to begin building:

- Art Programme Model — What do we want to achieve?
- Art Programme Structure — How will we do it?
- Curatorial process & policy — How do we engage artists?
- Art Programme Funding — How will we pay for it?
- The Crossrail Project context — What are the benefits?

There was no funding for art within the original Crossrail project budget. The overriding challenge for this project was to establish an art programme that could become a properly grounded, and fully funded, cultural programme that could be delivered to the project timeline. This had to be achieved during the worst economic crisis since the nineteen-thirties, in a market that was seeing public funding of the arts cut dramatically.

ART PROGRAMME MODEL

What do we want to achieve?

In order to establish a structure for the Crossrail Art Programme the team first had to define *a vision* for the programme, the *key principles* that would underpin it and the *key objectives* that it would set out to achieve.

Vision

The vision, key principles and key objectives for the art programme were predicated upon the ambitions for the Crossrail project as a whole, which were defined in the vision for the new rail network.

Crossrail had a unique opportunity to create a public art programme of global significance. The design and construction of new stations, infrastructure and environments presented major opportunities for the integration of art into the project's development and realisation. The strategy that was evolved by the project team sought to achieve the following:

- Involving artists in the conceptual development of the new stations
- The integration of permanent artwork into the fabric of the network
- Temporary artworks, artist's residencies and cultural events
- Creating an artistic record of the project

Crossrail's vision to support the development of London as a world-class city, and its remit to deliver a world-class railway fit for the twenty first century, could be amplified by an ambitious public art programme whose vision was to enhance the passenger journey experience, improve the quality of the public realm, forge engagement with staff and local communities and encourage ownership of the network that will stand the test of time.

Key principles

The Vision for Crossrail [2] details the rail network's ambitions in terms of the following components:

- Perception
- 21st Century Design
- Responsiveness
- Integration
- Sustainability and the Environment
- Engineering Design
- Lasting Simplicity and Coherence
- Responsibility
- Value for Money

A programme for art was capable of supporting all of these ambitions but one particular aspect of this vision - integration - offered the team an insight into how they could embed the new network into the cultural identity of its location, and simultaneously open up a range of artistic, financial and organisational opportunities that would support the creation of a public art programme.

Integration

The project will integrate with and be respectful of the fabric of the city, ensuring that the Crossrail facilities and their urban context form one coherent environment. The project will encourage the development of public realm initiatives by other stakeholders to maximise public benefit.

Figure 1 – Integration: Crossrail Architectural Design Guidelines: Vision for the System Pt.2.0

Crossrail has a specific role to play in the development of London as a city, and the South East region of the United Kingdom. It is the specificity of this location that can be celebrated and articulated at the heart of a vision for the art programme.

Founding the art programme upon the idea of 'Place' opened up a rich vein for exploration in the development of a structure for the programme that linked it with many of the principles and objectives that Crossrail promoted. It also aligned the project with the common agendas of many of Crossrail's partner organisations and stakeholders.

Figure 2 – A cross-section of public realm initiatives in 2011

Putting place-making at the centre of the Crossrail art programme gave it a powerful narrative as well as a firm commercial foundation for development. It was a position that others could decide to become involved with and support politically, financially and creatively.

The stakeholders and potential partners considered at an early stage included public bodies, government departments and NGOs, organisations and institutions and private companies.

The establishment of the line-wide of place-making provided a structure for the art programme whilst still allowing for individual proposals to be developed for each location that could be based on the particular nature of that place. This built flexibility into any specific project proposals whilst ensuring that a compelling story could be constructed for the project as a whole.

Place-making, Art and Crossrail

The idea of place-making, using the history of the station's location as a tool for generating visual difference, can prove as rich for architectural design as it can for art. When considering using history and location as catalysts for architectural design and artistic expression that is seamlessly integrated within the station design, the team were very conscious that poor design could lead rapidly to pastiche.

In Paris, on the walls of Concorde station, each tile carries an individual letter, which when read together form the words to the Declaration of the Rights of Man from the French Revolution of 1789. This poetic use of historical association delights and intrigues as its true meaning is slowly discovered.

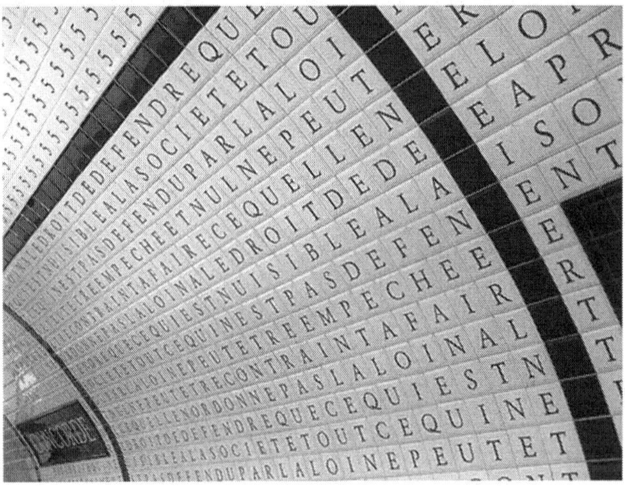

Figure 3 – Tiled walls at Concorde Station, Paris Metro

This type of artistic intervention was one of many used to guide the early stages of development of the nascent art programme.

Sites for art

The design and construction of the nine new stations along Crossrail's route, together with five portal structures, were the focus of the development of the art programme. Each of these offered an opportunity for integrating artwork permanently into the fabric of the buildings, the creation of temporary pieces and events that might be associated with the sites' construction or life in use and the recording of the construction programme by artists.

Of these nine stations the greatest potential for the programme lay within the seven Crossrail stations that form the Central Section area of the line. Each of these seven stations has an historically rich environment from which to draw inspiration for an artistic brief, and within which to forge local partnerships.

ART PROGRAMME STRUCTURE

How will we do it?

An evolving concept

The primary requirement, of the art programme was to broker partnerships between artists and station designers that would enable the creation of public art of an appropriate vision, type and scale for the network.

As the team began discussing the potential structure of the art programme, the concept of a tiered structure for the art programme began to emerge. This structure envisaged a programme that would allow for, and enable, a system whereby a diverse range of artists could be commissioned to provide art for Crossrail across a wide range of opportunities within the project.

The commissioning policy would seek to engage internationally recognised artists, new and emerging talent as well as recent graduates. Commissioning globally recognised artists would give the project a high profile, but would bring with it a higher fee base. Commissioning emerging talent would involve smaller fees, but with good curatorial advice could lead to the project being the host of work that achieved global significance. And commissioning recent graduates, and new talent, would complete the project's profile as a patron of significance, one that is supportive of the development of the public realm and the global community of artists.

This tiered structure would allow the project team to have far greater 'volume control' in how it developed and managed the public art programme, creating a varied portfolio of projects of all scales and types, which would in turn create a sliding scale of sponsorship requirement. This structure had the dual benefit of allowing the project to engage a wide range of artists in the art programme and allow a wide range of businesses, and individuals, to become patrons of the programme via different levels of funding.

Creating a commissioning hierarchy

The concept of a tiered structure for the programme led the team to the idea of developing three types of commission, which would embed the principles of diversity and inclusion into the programme. These three types of commission were as follows:

- Permanent works: works that would be part of the architecture and built fabric of the network
- Temporary works: works that would include the use of site hoardings, events, performances and installations during the project construction
- Project record: works that would record the gestation and construction of the project forming an historical record of the project

These three types of commission would have different roles to play within the project and programme, but combined together they would define Crossrail's reputation as a patron of the arts and a sophisticated infrastructure client.

In-house, outsourced or a hybrid structure

Once the team had agreed upon the basic idea of a tiered structure for the art programme, for all the benefits it would bring in delivering the project values of inclusion, diversity and engagement, the next challenge was to consider how best to structure the long-term team that could develop and deliver such an ambitious programme.

The team's research had identified two well established models for structuring a public arts programme, and then a number of variants on these models. The team analysed these precedents in terms of their 'pros' and 'cons' in order to decide on the best model for the art programme that Crossrail wanted to build.

In-house art programmes

Many modern regions, cities, and developers create programmes for public art that are run by dedicated teams embedded within the organisations. In New York the Design Commission retain an overview of any permanent works of art proposed for city-owned property, and act as caretaker and curator of the city's public art collection.

Pros
Managing a public art programme within the organisation gives direct control over the whole process involved with creating public art, which has benefits in cost control and the delivery of a consistent and coherent vision.

Cons
Recruiting and maintaining a team that has the right abilities and experience takes time, and it brings with it a fixed overhead in terms of staff salaries and associated costs.

Outsourced art programmes

Some large organisations treat the creation of a public art programme as they would any other package of work that can be tendered for, and outsourced to an external organisation. The DLR in London has done this and UK Property developers Hammerson followed this route for the Cabot Circus development in Bristol.

Pros
Arts consultancies offer a source of specialist knowledge, experience and project management skills that can help to create and run large-scale public art programmes.

Cons
External consultancies can command high consultancy fees, and can reduce the control that a client can exert over a project. Many consultancies have rosters of artists that they regularly use, and this can restrict the ability of the client to commission more widely.

Hybrid art programmes

A number of public sector organisations have used a hybrid version of these models, looking to maximise the benefits of each. An example of this is the way that the Olympic Delivery Authority (ODA) for London 2012 created their art programme.

The ODA formed a small in-house team that developed and led the high-level arts strategy, setting an agenda for all of the works commissioned. This team decided on a project-by-project basis how best to manage and deliver the individual commissions. The delivery can include external arts consultancies, members of the local community or local authority and funding bodies.

Pros
Control of the overall project is maintained within the organisation by a core team of specialists directly answerable to the Project Executive ensuring that a coherent and consistent vision is developed and delivered.

Cons
The in-house team need to establish clear management structures for controlling what can be a large number of sub-contracted individual projects.

Crossrail art programme team structure

The team developing the art programme decided that a hybrid structure best suited the needs of the Crossrail project. The team reached this conclusion for the following reasons:

- The complexity of the project's design programme, architecture, engineering and construction meant that only a fully committed, small, focused in-house team would be able to understand, drive, deliver and manage public art as a component of the scheme.
- There were immense financial pressures on the project, and these were likely to increase during the project timeline. The best value for money would be delivered by creating a small specialist in-house team who could draw on the expertise of external specialist consultants.
- Management of individual art projects can be tendered to external consultancies and organisations, so that competitive fees can be secured from the market for shorter fixed term commissions.

The team suggested that Crossrail create a dedicated post for a Head of Arts and Culture. Managing an art programme of this scale and potential is a full time role. It was envisaged that this role could mirror that of Head of Stations, and work in parallel with the whole client design team to ensure an integrated approach to the work.

It was key that leadership should be given to the art programme from the very top of the organisation, and thus it was recommended that a Crossrail board member should be identified to champion the Crossrail art programme. The culmination of this process of structuring the art programme resulted in the following recommendation for an Arts Programme Team [3]:

Arts Programme Team (APT)

- A Head of Arts & Culture
- A small in-house project management team
- An external specialist arts advisor
- Representatives from the employer
  - CRL Board member
  - CRL Design Management (Head of Stations)
- Representatives from the project sponsors
  - Transport for London
  - London Underground Limited
- C100 Consultant (representing pan-station design integration)

This structure was proposed as an ideal scenario. However, it was not possible at the time to create a position for a Head of Arts & Culture, and so the C100 team acted as the in-house resource, with specialist knowledge and support provided by external consultants.

The APT were supported by the station designers and were responsible for:

- Selecting artists appropriate to each location and station designs
- Preparing artist's briefs as appropriate
- Manage designer/artist cooperation
- Monitoring development of and integration with station design
- Identifying opportunities for sponsorship and funding

The APT would be responsible for developing and delivering the art programme over the course of the project. It was also recommended that in addition to the Arts Programme Team the project would benefit from the creation of a carefully selected Board of Trustees, drawing its constituents from the worlds of culture, the arts and finance. Such a board would add tremendous value to the project's profile and depth of resource.

CURATORIAL PROCESS AND POLICY

Establishing the creative direction

Creation of the Master Brief and securing curatorial expertise

Crossrail needed to build its art programme incrementally, establishing a cumulative process that would over time increase funding and raise the profile of the project.

The first step was taken with the launch of the project to the Crossrail design teams and the foundation of the Crossrail Arts Programme Team. The APT quickly moved to a process of competitive interview to secure expert curatorial experience. Futurecity, an innovative UK based arts consultancy won the contract to help the APT build the programme.

The Crossrail design teams had all presented initial concept proposals for their ideas on integrating art within their buildings, naming artists that they would be interested in working with. Crossrail's APT gave guidance on the general theme of 'place' as an idea that would coordinate the teams' individual proposals for art so that they could build to create a collection of works under the umbrella of the Crossrail brand.

To move the project forward we needed to create a detailed Master Brief and assemble a roster of artists and galleries that could become involved in the programme. The Master Brief provided a clear and detailed description of the artistic and cultural programme, a schedule of the opportunities that existed on the project and a process for their realisation. It was intended to guide the artists without constraining their creativity, providing an intellectual framework for their proposals.

The Master Brief

The Master Brief identified three principal areas for artistic intervention within Crossrail. [4] These were described in great detail within the Master Brief but can be summarised as follows:

Crossrail Sculpture project
The main material for the construction and finish of the stations, platforms, walls and tunnels is to be glass reinforced concrete. This material is malleable and can be coloured, patterned, extruded, moulded and textured allowing images and text to be set inside the architecture of the stations.

River of Light
This proposal was for a digital arts programme, which would see large digital screens installed prominently at each station. There would be designated areas where film and art projections could be shown, and digital wallpaper on the walls for art and advertising with LED lighting that could be programmed to change colour and movement.

Urban Gallery
This proposal explored the public art opportunities in the areas in and around the stations and portals. A menu of projects was established as a guide for commissioning artwork for each place, with an emphasis on local character and community engagement. This included sculpture, photography, seating, wayfinding, lighting and events.

Developing initial proposals

With the apointment of Futurecity the station design teams teams were briefed to undertake a *charrette*, a short period of intense creative activity, in which they elaborated initial proposals for art within the station schemes.

The results of this charrete were used to identify artists who could be engaged to partner with the designers, and to explore holding a competition organised with major art schools. The student prize would be a chance to work with the integrated teams, working alongside the established lead artists.

Project selection

Proposals tabled by the teams were assessed by the APT. The team decided which proposals were suitable for development and had the potential to attract corporate or individual sponsorship.

Sponsorship prospectuses and presentations could be designed and made for each station, and for the project as a whole, to provide material with which to begin to atttact seed funding from potential programme sponsors.

Securing sponsorship

The final step in establishing the project on a firm foundation came at the end of 2010 with a presentation to the City of London, organised in partnership with the London First group. This presentation launched the programme and sought to secure initial sponsorship pledges for all of the proposals that the teams had elaborated. This initial funding would allow for the development of the proposals, prior to a second round of funding for implementation.

ART PROGRAMME FUNDING

How will we pay for it?

In 2010 central government funding for arts and cultural activity in the UK was cut by £60m. In it's place the new government promised to create a landscape where personal and corporate philanthropy would fill the void. In 2009 £655m was made available to arts and cultural organisations in the UK through the funding efforts of personal and corporate philanthropy. However, the financial crisis was creating an intensely competitive landscape in which the challenge of securing a slice of that investment would only be achieved by projects that could excite the interest of sponsors by the level of the project's ambition.

The principle of a tiered structure that had been used to underpin the overall vision for the arts programme was applied to the ideas for creating a funding structure. The team decided that the best prospect for securing funding of the programme was to target a variety of potential sources. These were classed as primary, secondary and tertiary sources of project funding.

Primary funding

A firm baseline funding needed to be established to make the project viable. Primary funding would act as a catalyst to attract other sources of funding from the public and private sectors; it would give the project credibility.

Secondary funding

Secondary funding can come from a variety of sources.

Grants and awards
Grants and awards are available for public art, such as those from the Gulbenkian and Clore Foundations, or those offered by Arts Council England. These sources of funding often like to use their comparatively small grants to support and engage with a larger project.

Statutory funding
Local authorities use Section 106 planning agreements as a source of funding for public art. Many of the counties and London Boroughs that Crossrail passes through have established arts policies and Art Advisory Panels. Partnership with local authorities was identified as a potentially useful tool in developing the art programme for Crossrail.

Philanthropy
Private benefactors or institutions could be encouraged to become involved in the Crossrail Art Programme. There is a long tradition of philanthropy in the United Kingdom but it faced extraordinary demands as the financial crisis deepened. The idea of creating the Board of Trustees for the art programme was designed to engage directly with this audience of philanthropists.

Sponsorship
Alongside philanthropy, or gifting contributions to the project, there was scope to consider using sponsorship as a funding mechanism for the programme. Controlled sponsorship could be a useful mechanism to leverage funds for investment in the art programme.

Revenue offset
The Art Programme Team were very clear that the programme needed to be considered as a living entity, not as a project that stopped when Crossrail's construction phase was complete. Funding was an on-going issue.

A way to leverage continued investment in the art programme would be to consider offsetting revenues from income such as advertising. Partnering with a media company to manage the advertising network was identified as a way to generate income from advertising to fund artistic content.

Tertiary funding

Crossrail's Central Section stations run through the densely populated heart of London and will become prime generators for London's continued growth. The stations have the potential to directly engage with local boroughs, businesses and stakeholders and support their plans for the development of their area. And in turn these stakeholders can support Crossrail and its art programme.

Cultural Mapping
Arts organisations and local boroughs have become increasingly inventive to realise the funding required for projects that they want to undertake.

Creating 'Cultural Maps' of the areas surrounding the new Central Section stations helped to identify key attractions – museums and galleries, colleges and universities, theatres and clubs, restaurants and bars and tourist attractions – and 'mapped' the potential beneficiaries of the new stations. These beneficiaries are potential partners for the development of art projects.

Institutions and galleries provided a specific target for partnership in the development of the Crossrail arts programme Many institutions, such as the Barbican and the Whitechapel Gallery, have programmes whereby they look to take their artistic programmes beyond the walls of their buildings and into the public realm. Their initiatives embrace traditional conceptions of public art, such as sculpture, but include the full gamut of artistic expression including theatre, event, music, digital and experiential art.

Funding summary

The process summarised above gave the Crossrail Arts Programme Team a clear direction in which to develop the Crossrail art programme. It took the form of simple, incremental steps. Each step facilitated the next step. Each step built the project's profile, and engaged the attention of potential sponsors.

Seed funding from the project set the ball rolling. It got the programme to the point where proposals for art could seek funding. With that funding in place it will be possible to develop and realise the proposals.

The initial development phase of the programme secured early support from Arts & Business, the Arts Council, The Photographer's Gallery, the GLA, a number of independent art galleries and a host of other organisations.

THE CROSSRAIL PROJECT CONTEXT

What are the benefits?

The art programme delivers significant benefits to Crossrail. These include:

- Helping support London's status as a world-class city
- Creating a memorable cultural identity integral to the rail network
- Marketing Crossrail as a cultural network and destination
- Connecting Crossrail to its staff, passengers and communities
- Supporting the project's aims for sustainability and integration
- Improving the public realm of the city of London
- Capitalising on the growth of cultural tourism
- Partnerships with cultural organisations and business sponsors
- Forming a unique part of the public record of building Crossrail

Public art programmes make a significant contribution to the cultural identity, brand perception and marketing of a city, region or service. They add tangible value. Day-to-day they enhance the passenger journey experience, improve the public realm and encourage network ownership. This ownership has been shown to lead to a decrease in vandalism, and a 'policing' of the system by its staff and users.

Crossrail has had a long gestation period and is being designed, and built, in a period of immense economic turmoil. This places the project under great financial and political scrutiny. The benefits of the project to London, the South East and the nation as a whole are well recognised, but the project needs to embrace and promote every element within its scope that can engage the imagination of the body politic and the public at large.

CONCLUSION

There is a powerful and engaging story to be built, and communicated, about the Crossrail public art programme. There is a chance to create a network that has a unique quality within the world of modern infrastructure development. The challenge that remains for Crossrail is to develop and deliver the full potential of the project that the Crossrail Arts Programme team began in 2009.

Acknowledgements

The parties consulted were:

- Eric Parry RA
- Sarah Weir, Head of Arts & Cultural Strategy for the ODA
- Amanda Smethurst, Director, Arts Council England.
- Louise Trodden, Head of Art for Open House
- Piers Masterson, Arts Development Officer, LB of Camden.
- Sarah Collicott and Sam Wilkinson, Directors of InSite Arts
- Vivien Lovell, CEO of Modus Operandi Art Consultants
- Tamsin Dillon, Head of Art for London Underground.
- Munira Mirza, Director of Policy for Arts, Culture, Creative Industries
- Justine Simons, Head of Cultural Policy for the Mayor's Office.

Crossrail Project: Infrastructure design and construction
ISBN 978-0-7277-6078-4

doi: 10.1680/cpid.60784.331

Delivering a Safe Integrated SCL Design: The Challenges and Successes of Crossrail Contract C300/C410

Adrian St.John, Eur Ing MICE CEng BEng (Hons), BAM Ferrovial Kier (BFK) Joint Venture, London, UK
Andrew Pickett, MICE CEng Tunnels Division, Mott MacDonald, London, UK
Andrew Kendall, BEng (Hons) CEng MICE MCIHT Crossrail, London, UK
Vicky Potts, MA MSc PhD Donaldson Associates Ltd, London, UK
Damian McGirr, MICE CEng BEng (Hons) Donaldson Associates Ltd, London, UK

Abstract
The design delivery for the sprayed concrete linings (SCL) for Crossrail contract C300/C410 Western Running Tunnels and Station Caverns has involved a unique set of circumstances.

The SCL primary lining was designed to withstand the various temporary conditions and also formed part of the composite permanent works design. Retaining design responsibility for the composite SCL primary and secondary lining design with one party (Mott MacDonald (MM)), whilst ensuring that the Contractor retained responsibility for excavation and temporary support presented both benefits and challenges. These are discussed within the contractual framework between the Employer's SCL designer (MM), the Contractor (BAM Ferrovial Kier (BFK) Joint Venture) and the Contractor's SCL designer (Donaldson Associates).

The paper investigates how an integrated and collaborative approach drove efficiencies in the design through all stages of design development. This commenced with the "Optimised Contractor Involvement" (OCI) process through to a series of workshops to ensure that buildability, sequencing and the Contractor's preferred methods of working were embodied within the final SCL design. As the project has evolved, the team has provided innovative solutions to meet the changing construction programme and interface constraints, in particular the AP3A Wraparound at Bond Street providing early access for the platform tunnel construction.

The process for delivering a safe, consistent and integrated SCL design for both temporary and permanent works is described in detail. This encompassed a "Due Diligence" exercise, the design of a set of Contractor's "Temporary Measures" compatible with the primary lining design, and culminated with the issue of a comprehensive and assured SCL design "for construction".

The paper concludes that, whilst not necessarily suitable for all projects, the C300/C410 project team overcame the challenges and successfully delivered an assured, efficient and integrated SCL design. This has relied upon a clear division of responsibilities and considerable co-operation and collaboration between all parties.

List of Notations

BFK Bam Ferrovial Kier JV (Contractor)

DAL	Donaldson Associates Ltd, in conjunction with specialist sub-consultants IGT Geotechnik und Tunnelbau, and Bernd Strobl Ingenieurleistungen (Contractor's Temporary Works Designer)
ICE	Institution of Civil Engineers
HSE	Health and Safety Executive
MM	Mott MacDonald (Permanent Works Designer)
SCL	Sprayed Concrete Lining
TBM	Tunnel Boring Machine
URS	URS (Category III checker for Permanent Works Designer)

Introduction

Crossrail is Europe's largest construction project, and will transform rail transport in London, increasing capacity by 10%, supporting regeneration and cutting journey times across the city. The Crossrail route will run over 100km from Maidenhead and Heathrow in the west, through new tunnels under central London to Shenfield and Abbey Wood in the east. There will be 38 Crossrail stations including 9 new stations at Paddington, Bond Street, Tottenham Court Road, Farringdon, Liverpool Street, Whitechapel, Canary Wharf, Custom House and Woolwich.

Crossrail contract C300/C410 "western running tunnels and station caverns" was awarded to a joint venture of BAM Ferrovial Kier (BFK) in January 2011 with a value of approximately £500m. The project comprised the construction two 6.2km tunnel drives between Royal Oak and Farringdon and the construction of circa 75 sprayed concrete lining (SCL) tunnels for Bond Street and Tottenham Court Road Stations. The design for the SCL works (Primary and Secondary linings) is by Mott MacDonald (MM) as part of the C121 Framework design contract.

This paper focuses on the design of the SCL works on C300/C410 and, in particular, the way in which the project team managed a unique set of circumstances and overcame considerable technical and organisational challenges.

The SCL primary lining was designed to withstand the various temporary conditions and also formed part of the composite permanent works design. It was considered beneficial to retain design responsibility for the composite SCL primary and secondary lining design with one party (the Employer's SCL designer, MM), whilst ensuring that the Contractor (BFK) retained responsibility for excavation and temporary support. However, such an approach also presented potential difficulties.

The approach to SCL design and construction was heavily influenced by the Health and Safety Executive (HSE) report into the collapse of NATM tunnels at Heathrow Airport (HSE, 2006). The findings of this report, combined with other HSE and ICE publications (e.g. HSE (1996) and ICE (1996)) and Crossrail Information Paper D23 (Crossrail, 2007), had a profound impact on the contractual arrangements and responsibilities for the design and construction of these tunnels.

The challenge of delivering a safe, assured and integrated SCL solution incorporating both permanent and temporary loading conditions has been overcome through close collaboration and co-operation, despite no formal contractual relationship existing between the various parties.

Contractual Arrangements and Division of Responsibility

The C300/C410 contract was based on the NEC Engineering and Construction Contract (NEC3), target contract with activity schedule (Option C). This form of contract encouraged collaboration and communication between the parties, and incentivised cost and time savings through a painshare / gainshare arrangement.

The Works Information stated that the Contractor (BFK) was not responsible for designing any of the permanent works, and was responsible for designing all temporary works. However, in relation to the SCL design, this specifically included the Employer retaining responsibility for primary SCL design at all stages of construction, including portions of lining prior to ring closure.

As part of the tunnelling industry's drive towards delivering an efficient one-pass lining, Crossrail's SCL design philosophy used composite action between the primary and secondary linings, and used the primary lining as temporary ground support for the part-constructed tunnels. This new approach was in contrast to previous projects which considered the primary lining as purely temporary works and, accordingly, its beneficial effects were disregarded in the permanent works design. Whilst delivering an innovative and efficient SCL design solution, this new strategy introduced a potential conflict since BFK were not only responsible for the design of all temporary works but also responsible for the safety and security of excavations at all times.

The above arrangement posed an interesting challenge for the teams; BFK were unquestionably responsible for stability and safety of the excavation, yet the temporary works consisted of a partially-constructed SCL primary lining which had been designed and checked by the Employer. This unusual arrangement caused a great deal of debate between BFK, the Project Manager, MM and the Contractor's SCL designer (Donaldson Associates Ltd (DAL)). Clarifying and resolving this interface was one of the early challenges for the project team.

The contract included an "Optimised Contractor Involvement" (OCI) period which took place immediately after award. OCI was a Crossrail-wide initiative to gain early alignment between BFK, the Crossrail Project Manager, Employer's designers and key members of the Contractor's supply chain.

The key objectives of the OCI process were as follows:

- to review the buildability of the Employer's design.
- to identify opportunities for refinement of the Employer's design.
- to ensure alignment between temporary and permanent works design.
- to improve understanding of the contract scope and risk.

A key element of the OCI therefore involved extensive workshops and discussions between BFK's SCL engineering and SCL construction teams, MM and DAL. These initial workshops had two distinctive threads; firstly, debating the precise wording and intent of the contract in order that all parties fully understood the division of responsibilities; secondly, to seek improvements in the Employer's SCL design in respect of buildability, sequence and methods.

A key driver was to make most efficient use of the SCL designers available within this specialist field, and keep the number of design parties involved to an absolute minimum, thereby reducing interfaces. The Employer had already appointed MM as his SCL Designer, and a significant amount of design had already been completed. For the SCL design it was necessary to consider all of the temporary and permanent loading conditions acting on both the primary and secondary lining. To provide an efficient design with load sharing between the primary and secondary linings, finite element modelling was used to ensure a more complete understanding of the lining behaviour. This entailed a number of assumptions about BFK's preferred sequence and timing of the SCL works; these sequences were reviewed, optimised and validated during the OCI period in order to gain alignment between the parties.

In order to produce an assured design, the Employer (via MM) had engaged an independent Category III checker (URS) for the SCL design. The Category III check was predicated on an assumed sequence and, again, it was essential that this sequence was understood and agreed between all parties.

The key reasons for the approach adopted on C300/C410, illustrated in Figure 1, were thus:

- The Employer had already engaged the services of a competent SCL designer and Category III checker with sufficient resources to deliver a complete SCL design "for construction". Employing additional designers and checkers would increase demand on scarce SCL design resource in an already over-heated market.
- In order to deliver a safe, assured design, the Employer's design already needed to consider the primary lining at every stage of construction, both temporary and permanent.
- If the Contractor were to be held responsible for the design of all temporary works (including the SCL primary lining in its temporary state), what value would be added by employing an additional designer and a Category III checker? The main impact would be wasteful duplication of effort and resources.
- Employing additional designers and checkers would introduce confusion and potential mis-alignment of models, geometry and design information during design development, in what was already a complex and challenging interface.

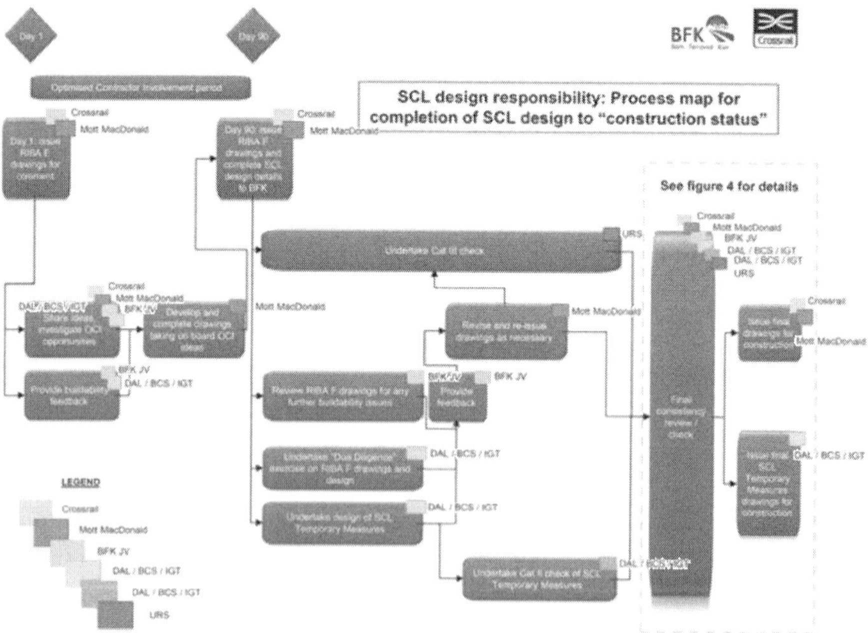

Figure 1: From OCI to IFC – delivering a safe, integrated and assured SCL design

In order to clarify the design delivery philosophy, a Division of Responsibilities document was jointly prepared and signed by the relevant parties. The key elements of this were:

- the Employer retained responsibility for the design of the primary SCL lining at every stage of construction, both temporary and permanent.
- BFK had to "understand and accept" the Employer's SCL design. This entailed a "due diligence review" (DDR) process carried out by DAL.
- DAL (on behalf of BFK) had to design and Category II check of a set of "toolbox items" which could be deployed in response to adverse conditions encountered in the field.
- BFK had to declare that the Employer's design was buildable, that the sequences shown were consistent with the BFK's methods, sequences and plant, and that the "toolbox items" could be deployed without detriment to the primary SCL lining.
- DAL would provide highly-qualified "Senior SCL Engineers" and "SCL Shift Superintendents" to be embedded into the BFK Engineering team to oversee excavation stability at all times through the SRG/RESS process described later in this paper.

The division of responsibilities in the context of advancing a typical SCL tunnel is shown indicatively in Figure 2, and a summary of the integrated design process is included as Figure 3; this was successfully adopted as the design delivery strategy for the C300/C410 SCL works.

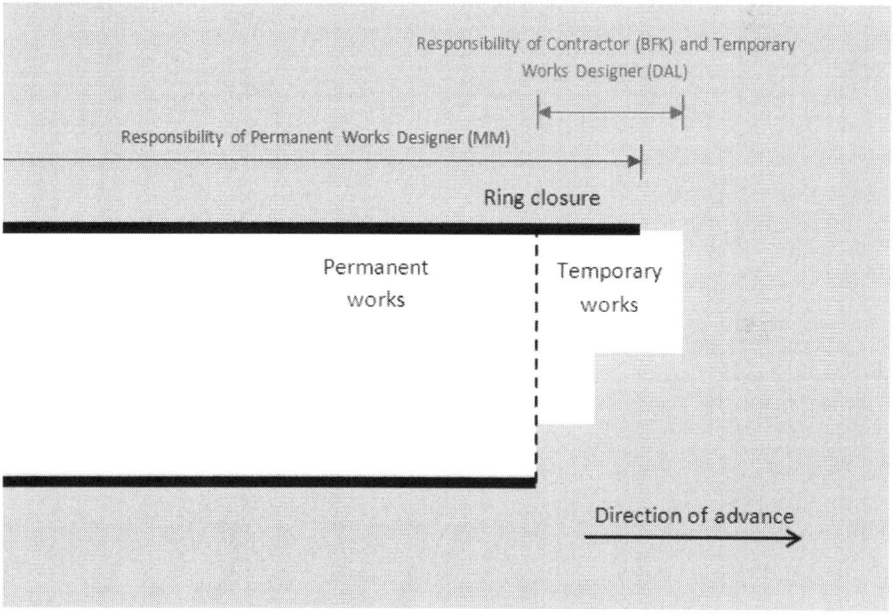

Figure 2: Typical SCL excavation sequence illustrating division of responsibilities

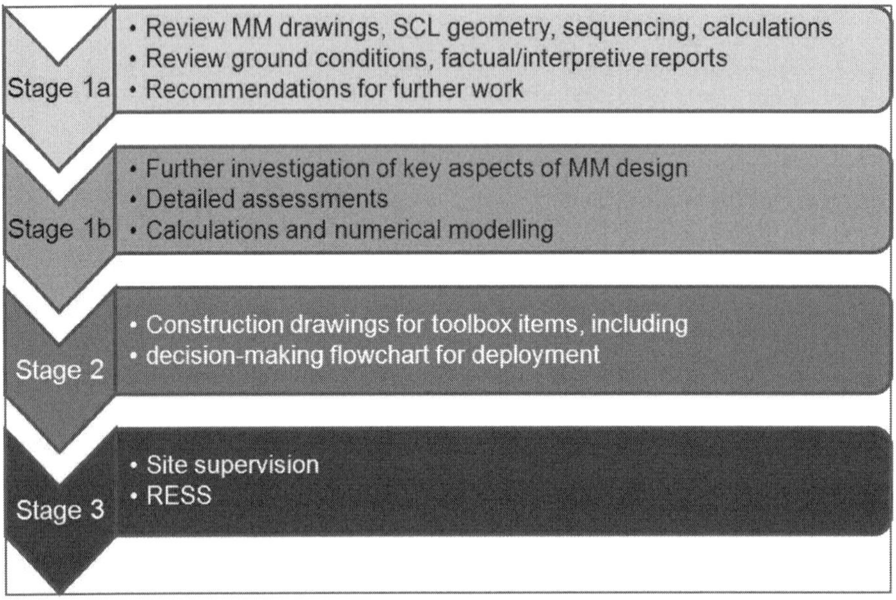

Figure 3: Integrating the permanent and temporary works via a "Due Diligence Review" (DDR) process, backed up with experienced site supervision

Design Development

At contract award the design of the SCL structures (Bond Street and Tottenham Court Road station tunnels) were at RIBA F stage (detailed design), which had been developed from RIBA D.

At this stage the design underwent a significant change to incorporate BFK's vision for how the stations would be built incorporating the C300 running tunnel contract, which was originally a separate contract.

The original strategy was to build the station platform tunnels prior to the running tunnels. BFK however devised a solution to increase efficiency by constructing the running tunnels through Bond Street and Tottenham Court Road stations prior to the SCL platform construction, thereby using the segmental running tunnels through the station as platform pilot tunnels. This delivered considerable cost and programme savings, reduced settlement by 30%, took 90,000 lorry movements off the road and was a key factor in the BFK winning bid combining C300 with C410 into a single integrated contract.

The programme however did allow for some of the station SCL works to be carried out prior to the TBM running tunnel arrival. The OCI period following RIBA F design therefore worked towards ensuring that the original design met the aspirations of BFK and also provided new solutions to carrying out the SCL works under the new overall construction sequence and for the pre-TBM SCL advance works.

As a consequence of the above, the design was developed in two strands:

1. Contractor's preferred excavation sequences and advance rates
2. Design options for SCL advance works (discussed in more detail in section 7 of this paper)

Due to the complex nature of SCL design in clay, arising from the time dependent natures both of loading from the ground and the strength gain of shotcrete and the interaction of these effects, MM had to form a basis of design in relation to both the general excavation sequence and advance rates as follows:

Following industry recommendations (ICE, 1996), excavation faces were limited to a maximum of $30m^2$. The tunnel excavation sequences were categorised as follows:

- <5m diameter:- full face
- 5m-7m diameter:- Top heading / invert
- 7m-9m diameter:- Top heading / bench / invert
- 9m-12m diameter:- Pilot tunnel / top heading / bench / invert

During the OCI process, a number of refinements were proposed by the contractor to improve efficiency and to suit his preferred plant and method of working.

An example of this was optimisation of the enlargement of the platform tunnels. The "TBM first" scheme provided a much larger pilot tunnel than was originally envisaged. This reduced the excavation required for enlargement sufficiently that the platform tunnel excavation sequence could be optimised to a pilot / top heading / invert sequence, i.e. eliminating the bench stage. This refinement was proposed by BFK and then verified as acceptable by MM to the SCL design.

BFK's proposed advance rates were generally faster than those originally assumed by MM. MM therefore carried out 3D numerical modelling to check how the load being experienced on the tunnel varied with the early age of the SCL. In most cases MM were able to adopt BFK's proposed advance rates.

Changes incorporated into the design had to be re-issued at RIBA F and were subject to Crossrail's design assurance process, which included two main areas:

- Coordination with interfacing Framework Design Consultants (e.g. the designer of the Bond Street Station box and Station architecture (including the spaceproofing of tunnels))
- Revised Category III check

Co-ordination issues mainly related to ensuring that the space provided inside the tunnels remained the same (to satisfy functional requirements) and adapting the SCL – station box/ticket hall connections to suit the programme changes e.g. tunnelling out of a station box as opposed to tunnelling up to a station box.

Due Diligence Review

The purpose of the Due Diligence Review undertaken by Donaldson Associates Ltd (DAL) was to develop a detailed understanding of:

- the SCL design basis/philosophy
- the design assumptions and any constraints or limits these might impose on the design
- the anticipated behaviour of the excavation and lining (assessment of convergence of the tunnel lining, lining loads and ground movements)
- the sensitivity of the design, for example to changes in advance rate or ground conditions
- design risks associated with the SCL tunnelling for each tunnel

This would then trigger a buildability review undertaken by BFK, inform the subsequent design of "toolbox items", describe in Section 6 of this paper, and assist personnel (Senior SCL Engineers and SCL Shift Superintendents) on site during construction in interpreting observed behaviour. A complete Category III check on the design had already been undertaken and it was not intended to repeat this exercise. For this reason the DDR was limited to considering the behaviour of the ground and lining up to the point of ring closure for the lining, as indicated in Figure 2 (i.e. the primary lining has been completed around the full perimeter of the excavation in accordance with MM's design, drawings and specification).

The DDR process is broadly outlined in Figure 3. The review was based on the following information, supplied by MM:

- Drawings
- Design basis reports
- Numerical modelling reports
- Geotechnical factual and interpretative reports
- Defined advance rates
- Defined performance and strength criteria

Initially DAL undertook a high level review of this information to identify areas of the design that were complex or higher risk. For each tunnel the geometry, ground conditions, excavation sequence and key risks were collated during this phase of work, together with technical queries on the MM design. A series of workshops were then held between DAL, BFK, MM and Crossrail to discuss the issues identified during the initial review, again on a tunnel by tunnel basis. This allowed DAL to evaluate the conditions across the various sites and identify key issues, as well as critical tunnels (based on geometry, excavation sequence and ground conditions), for further consideration. These were selected to ensure that a full range of design scenarios were assessed to develop a detailed understanding of the lining behaviour and design constraints, without analysing each individual tunnel. For example, the geometry and ground conditions for all parent-child junctions were considered to identify a worst case for further assessment. Similarly critical tunnel sections for each of the excavation sequences listed in section 3 of this paper were selected on the basis of the ground conditions and the design primary lining thicknesses and geometry.

To complete this review DAL undertook their own numerical modelling of those tunnels considered critical, as well as looking at key issues across the full contract (e.g. junction design, reinforcement at joints). The numerical models created by DAL were used for parametric studies to develop an understanding of the lining response to various conditions (for example changes in advance rate, strength criteria, poor ground conditions) and assess the sensitivity of the design to these different factors. Where necessary, additional meetings and workshops between all parties were held in relation to specific aspects of the design. These meetings generally related to the most complex, and consequently highest risk, tunnels, and the workshops allowed all parties to collaborate to find safe methods of addressing the risks that were considered acceptable from both a design and a construction perspective.
At the end of this process the concerns identified during the initial screening were reviewed on a tunnel by tunnel basis to ensure that all had been addressed. DAL produced a DDR certificate, declaring understanding and acceptance of the Employer's design for each tunnel.

Design of "Toolbox Items"

Having completed the DDR for a given tunnel, DAL were then in a position to design temporary measures or "toolbox items" to improve excavation stability, without compromising the permanent lining, as indicated in Figure 1.

"Toolbox item"	Notes
Spiles	To prevent blocks of clay falling into the excavation; to be used in conditions where potential block sizes are considered likely to exceed the capacity of the initial lining indicated on the MM drawings. Spiles must not intrude into permanent lining.
Face dowels	Needed to ensure face stability in poor ground conditions (soft clay, or other materials in the face).
Pocket excavation	
Depressurisation	To prevent water ingress and/or heave of the invert (if water-bearing soils present in sand channels below tunnel) or face (if water-bearing soils present ahead of face),

Table 1: "Toolbox items" used in C300/C410

At Bond Street and Tottenham Court Road stations, these "toolbox items" would only ever be utilised if ground conditions proved worse than anticipated. In order to deliver the philosophy of an assured engineering design prior to construction, these "toolbox items" were fully designed and Category II checked for each tunnel. This provided the site team with the flexibility to deploy these items if worse than expected conditions were encountered (see section 6 on Construction and Supervision). This meant that an assured and integrated temporary works design was available at all times, even in unexpectedly adverse conditions, and without causing a delay to works. The "toolbox items" are briefly summarised in Table 1.

The most adverse conditions considered in the temporary works design were defined by the Geotechnical Baseline Reports for Bond Street and Tottenham Court Road Stations.

Inevitably the design process could not eliminate all construction risks. The "toolbox items" described above mitigated some of the risk inherent in the SCL design for the permanent works; however residual risks remained that could not be addressed through either permanent or temporary works design.

One further advantage of the DDR process was that risks were identified at an early stage and discussed extensively throughout the process at meetings and in workshops. Residual risks were collated into CDM risk registers, and highlighted on construction drawings. In order to make this risk and knowledge transfer more effective, BFK instigated a series of workforce briefings which were delivered jointly by DAL and MM. These covered residual design risk and SCL design intent, and ensured a high degree of understanding of these key issues by the construction and supervisory teams from Crossrail, BFK, DAL and MM.

Construction and Supervision

Upon completion of the DDR process, BFK prepared a "buildability declaration" confirming that the DDR was complete, that BFK were satisfied that the Employer's design was buildable and compatible with the "toolbox items" and that, from a design perspective, a complete, integrated, assured "for construction" SCL design could be released to the site team. This was ratified by the Crossrail Project Manager. The full process of completing this design assurance is included in Figure 4.

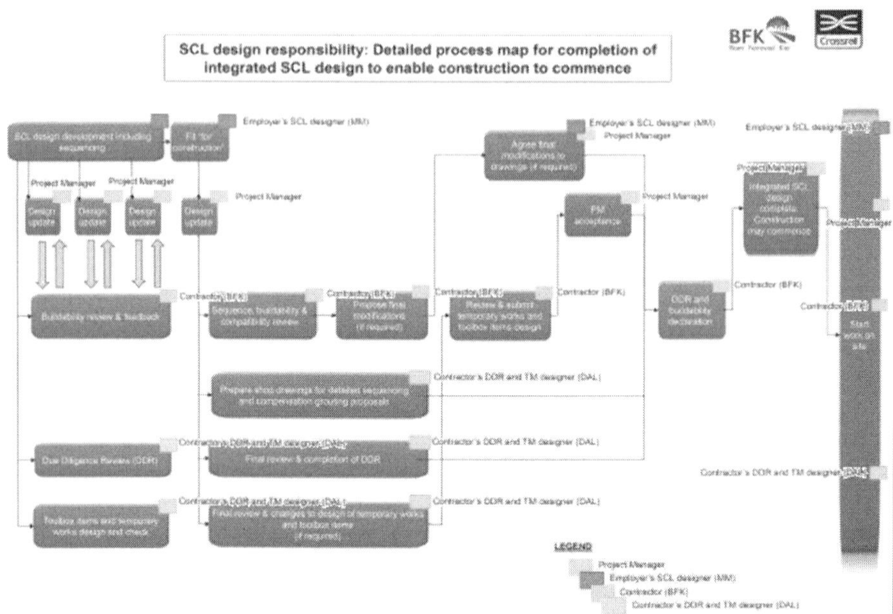

Figure 4: Detailed process map to deliver an integrated and assured SCL design incorporating both temporary and permanent works

The "Shift Review Group" (SRG) and "Required Excavation Support Sheet" (RESS) constituted a critical aspect of the safe construction of the SCL works. The members of the SRG are illustrated in Figure 5.

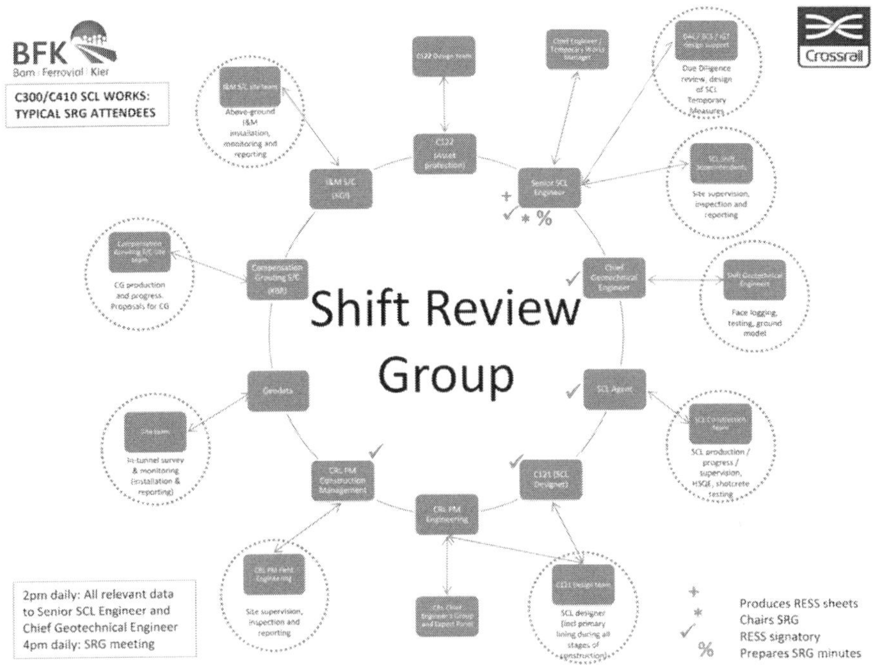

Figure 5: Typical attendees and flow of information for the SCL "Shift Review Group"

The purpose of the SRG was to review the relevant data from the previous day's activities and to agree a plan to safely advance the tunnels for the coming 24 hours. Excavation stability and asset protection were paramount, and excavation works could not proceed without a valid RESS.

This entailed detailed co-ordination between the surface and sub-surface monitoring team, the compensation grouting team, the SCL primary lining designer (MM), the geotechnical team, surveying team and construction team. In order to achieve an effective and efficient sharing of information, each party uploaded its relevant data onto a "collaboration server" two hours prior to the SRG; this information was reviewed by all parties and, on that basis, a preliminary RESS was prepared by the Senior SCL Engineer. The Senior SCL Engineer then chaired the SRG meeting, and made the required changes to the RESS to obtain the necessary acceptance and signatures to validate and issue the RESS to the site.

Key elements of this process included:

- The Crossrail SCL philosophy was that the assured design should not be changed in the field; any changes should be limited to the deployment of additional "toolbox items" to improve excavation stability.
- Certain changes to the SCL design were permitted, but were documented carefully with due regard to the designer and Category III checker. Such changes, by their nature, were comparatively slow to implement and thus were thus avoided if possible.
- The performance of the primary SCL lining was monitored on site against in-tunnel convergence trigger levels that had mitigation actions to be implemented if they were exceeded. Both Crossrail and MM employed field engineers and inspectors to monitor the SCL works.

- Careful communication between the compensation grouting team and SCL construction team was essential to ensure that the grouting exclusion zones were respected.
- The Senior SCL Engineer reported through the BFK Engineering team to maintain a degree of independence from the BFK SCL construction team. The Senior SCL Engineers covered dayshifts only (12/7) but were supported on a 24/7 basis by the SCL Shift Superintendents.
- "Toolbox items" could be deployed at the direction of the Senior SCL Engineer and/or the SCL Shift Superintendent.

Despite the complexity of the arrangements, the site teams rapidly established themselves into effective working relationships, and this model has been successfully used throughout the SCL works on C300/C410.

Example Case Study - AP3A and AP1 Wraparounds

As mentioned in section 3 of this paper, as a result of the change in overall sequence (adoption of the "TBM first" scheme), the SCL station tunnel design had to be adapted to provide additional design solutions to carry out the pre-TBM SCL advance works.

In each case MM met with Crossrail, BFK and DAL to review BFK's aspirations and to review best these might be implemented whilst minimising the impact on the following interfaces:

- Station and spaceproofing interface
- Third party assessment (through the impact of ground movement induced by tunnel construction)

This case study discusses one of these pre-TBM initiatives; the AP3A wraparound (Bond Street) was successfully developed from concept to detailed design. The wraparound design met BFK's requirements, and was acceptable to interfacing parties, the CAT III checker and subsequently also the client's Expert panel review.

AP3A Wraparound Design

Figure 6 shows the layout at the Eastern Ticket Hall prior to the revisions due to pre-TBM advance works.

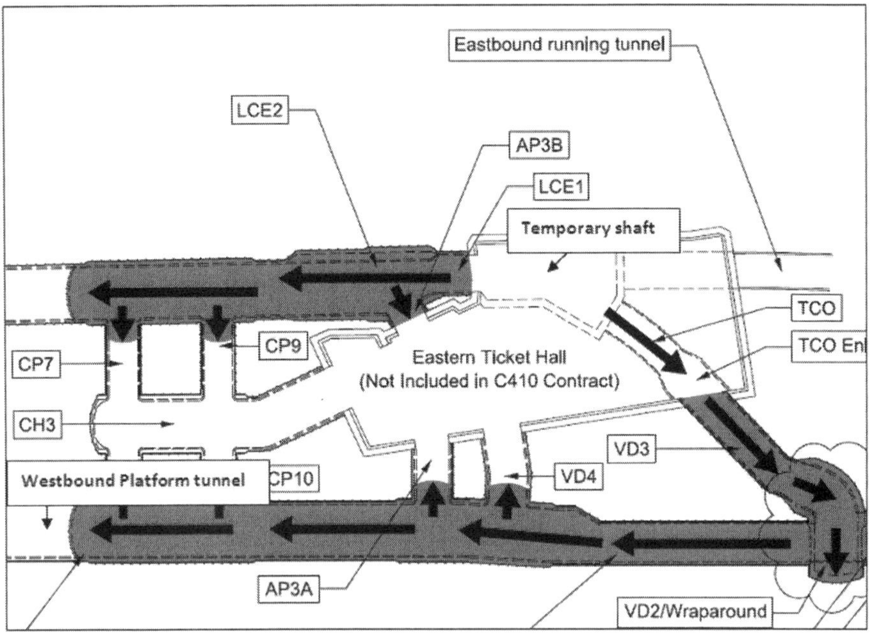

Figure 6: Eastern end of Bond Street Station – original scheme

The layout shows the westbound platform tunnel being constructed via TCO and VD3, with AP3A driven back towards the Eastern Ticket Hall.

In collaboration with the Bond Street Eastern Ticket Hall contract, early access was made available to C300/C410 from the "Master Plan Shaft", at the western end of the Eastern Ticket Hall shown in Figure 7. This meant there was alternative access to construct the platform enlargements, which led to the conception of the AP3A wraparound solution. Figure 7 shows the plan of the wraparound; this gave BFK flexibility to construct the westbound platform from an alternative access.

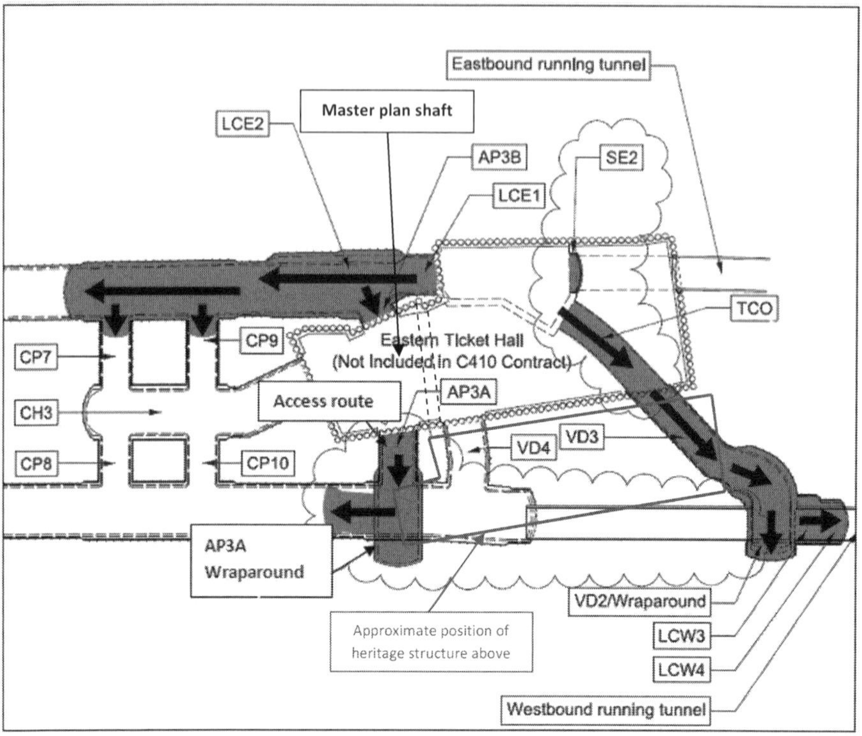

Figure 7: Eastern end of Bond Street Station – showing the AP3A wraparound

The wraparound design was subject to the following constraints and challenges:

1. Short design period
2. Design assurance – Category III checking and Crosssrail "gate" within the short design period
3. Stringent settlement criteria
4. Westbound TBM programme

The short design period meant that the design had to consider how quickly it could be designed and Category III checked. The preferred solution may have been to design a wraparound large enough to build the platform tunnel from a typical parent-child relationship. With the typical maximum ratio between the parent and child tunnel height being 0.7 this would equate to the wraparound cross section being a minimum of 14-15m in height (the platform tunnel being 11m high). MM had experience of a wraparound this large on the C510 Crossrail contract but this required extremely time-consuming and sophisticated numerical modelling. This was not a viable option given the programme constraints. This coupled with the increased settlement due to the larger excavation, and the close proximity of a critical and sensitive Grade II* listed structure (20 Hanover Square) above meant an alternative solution was required.

The preferred solution was to provide a smaller wraparound which was suitable for a child tunnel large enough to accommodate the TBM, which could then be enlarged to full platform size at a distance from the wraparound. This would provide a chamber for commencing the future platform enlargements once the TBM had reached Farringdon. The principal benefit of the AP3A wraparound solution was that it provided BFK with a safe and convenient means of commencing platform tunnel enlargement. It also allowed some flexibility to suit the TBM advance rate.

The main challenges to overcome with this design were as follows:

1. Provide access for platform enlargements
2. Safe break-in of the TBM
3. Once platform enlargements had occurred to the west of the advance chamber, how to construct the platform tunnel and permanent works junction to AP3A back through the temporary wraparound structure

This was solved through the following excavation steps, summarised in Figures 8, 9 and 10:

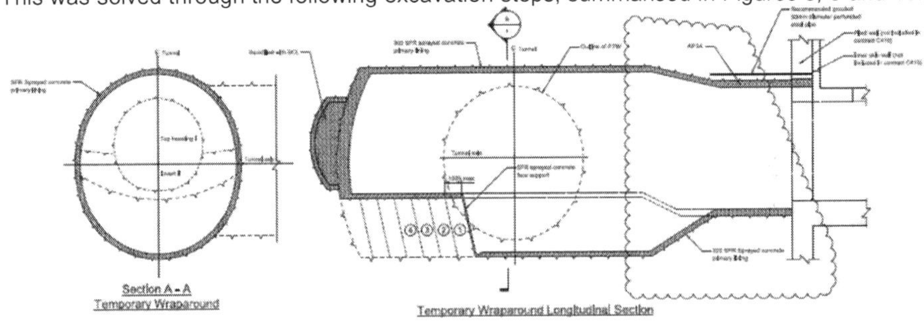

Figure 8: AP3A wraparound construction

Figure 8 shows the general excavation sequence for the wraparound from the Eastern Ticket Hall, comprising pilot tunnel, top heading, bench/temporary invert and finally permanent invert. The temporary wraparound structure provided an ideal starting point for the advance chamber SCL works.

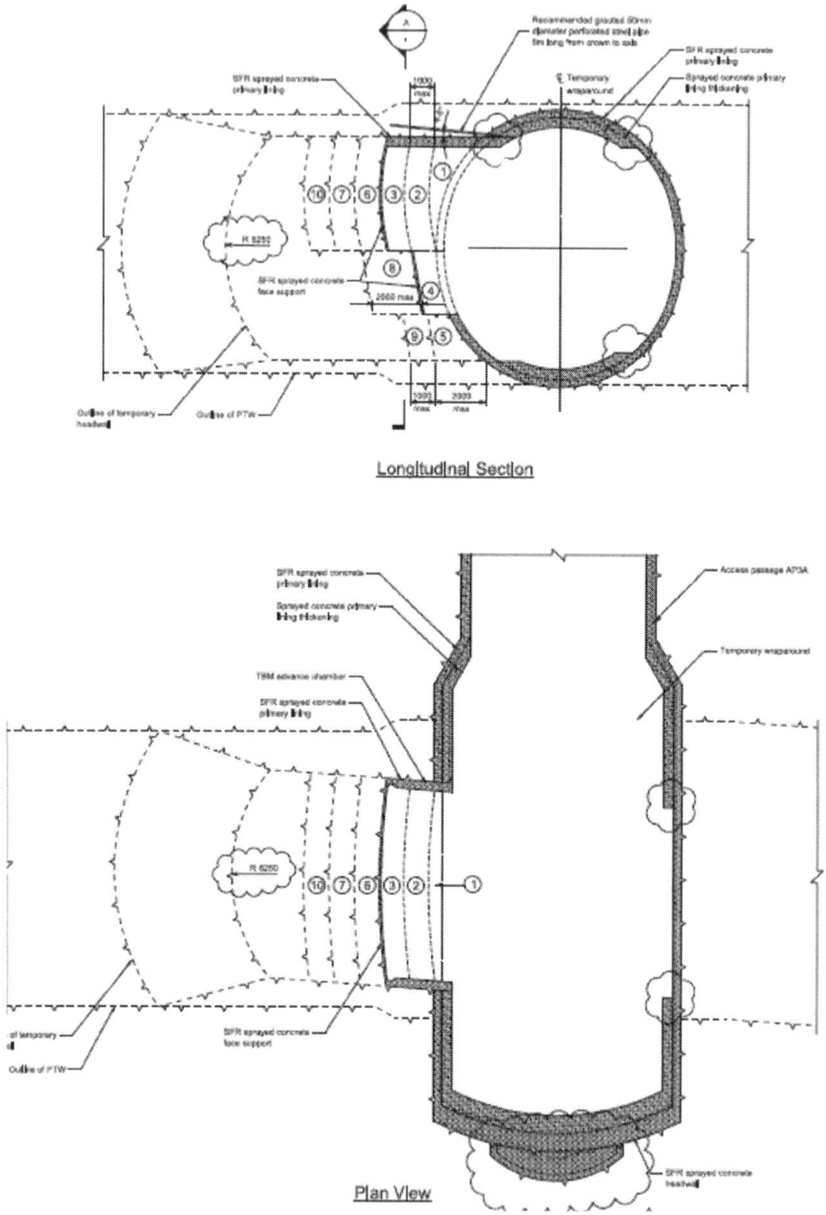

Figure 9: AP3A wraparound advance chamber construction

Figure 9 shows the excavation steps for the advance chamber out of the wraparound structure which flared to the full platform section. Once completed, the advance chamber was backfilled with foam concrete to allow the TBM to pass through rapidly and without interruption, creating a 'pilot' tunnel (see Figure 10).

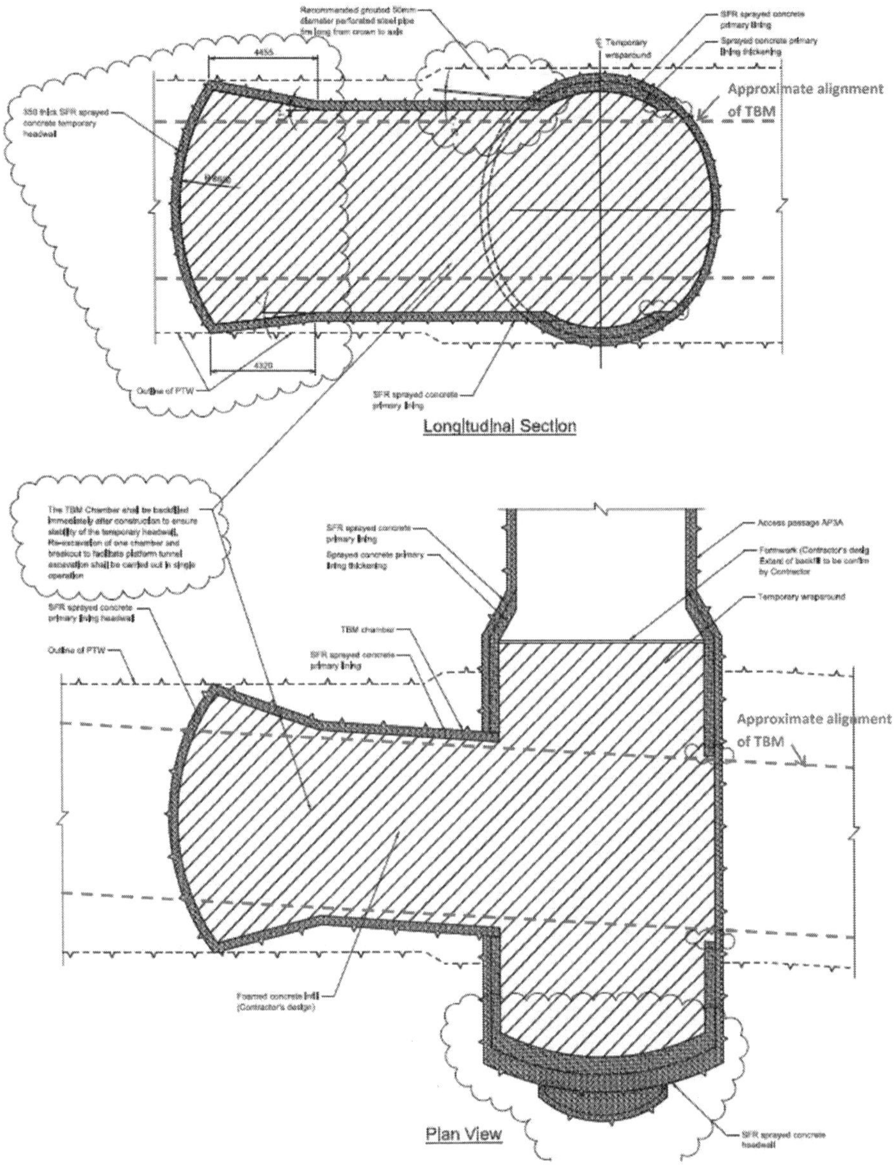

Figure 10: TBM passage through AP3A wraparound and advance chamber

Discussion - AP3A Temporary Wraparound Design

The temporary wraparound design was an excellent example of collaboration between the Client, two adjoining Crossrail Contractors, the permanent works Designers and the

Contractor's temporary works designer. This produced innovative SCL design, meeting the needs of all parties and delivered significant cost and time savings.

Subsequently this design provided enough flexibility to be employed elsewhere on C300/C410, e,g, Tottenham Court Road AP1, providing further benefits and flexibility to the Crossrail construction programme.

Conclusions

A novel approach to design management has been developed for the SCL works under Crossrail contract C300/C410 "western running tunnels and station caverns". This arose because the use of the primary lining in the permanent scheme made it desirable for the designer to retain responsibility for the lining in its temporary state, however the contractor and his designer clearly retained responsibility for excavation stability.

To deliver an assured and integrated design whilst satisfying the requirements of all parties in view of their responsibility, and avoiding duplication of effort given the assured nature of the permanent works design, a bespoke integrated design management procedure was developed. This involved a due diligence review of the final SCL design by the contractor's designer to develop a detailed understanding of the SCL design, development of assured "toolbox item" design to complement the permanent works, construction supervision and a clear definition of responsibilities during construction.

This unusual division of responsibility necessitated a high level of interaction between the designer (MM), the contractor (BFK) and the contractor's designer (DAL). In spite of no formal contractual relationship existing between some of these parties, and the complexity of the process developed to satisfy the division of responsibilities, the process has been successfully implemented for C300/C410. This is illustrated by the example of the AP3A temporary wraparound at Bond Street.

It is recognised that this solution arose from an unusual set of circumstances and that adopting the same approach may not be appropriate for other projects. Nonetheless continued advances in tunnel lining design mean that similar scenarios in terms of using the primary lining as permanent works will arise again, and the integrated design management process described in this paper is one successful approach that should be considered and developed further.

References

Association of British Insurers and the British Tunnelling Society (2003). The Joint Code of Practice for Risk Management of Tunnel Works in the UK.

Crossrail (2007). Crossrail Information Paper D23 – Sprayed Concrete Linings.

ICE (1996). Sprayed Concrete Linings (NATM) for Tunnels in Soft Ground, an ICE Design and Practice Guide. ThomasTelford.

HSE (1996). Safety of New Austrian Tunnelling Method (NATM) Tunnels: a review of sprayed concrete lined tunnels with particular reference to London clay.

HSE (2000). The collapse of NATM tunnels at Heathrow Airport: A report on the investigation by the Health and Safety Executive into the collapse of New Austrian Tunnelling Method (NATM) tunnels at the Central Terminal Area of Heathrow Airport on 20/21 October 1994.

Crossrail Project: Infrastructure design and construction
ISBN 978-0-7277-6078-4

doi: 10.1680/cpid.60784.349

Geotechnical Risk Management for Sprayed Concrete Lining Tunnels in Farringdon Crossrail Station

Angelos Gakis, Dr. Dipl.-Ing, MSc DIC, Chief Geotechnical Engineer, Dr. Sauer & Partners Ltd
Eur Ing Petr Salak, MSc, CEng MICE, Design Manager, Dr. Sauer & Partners Ltd
Eur Ing Adrian St.John, BEng (Hons) CEng MICE, Chief Engineer – SCL, BFK Joint Venture

Abstract

The proposed Crossrail station will comprise two ticket halls, two escalator/concourse tunnels, two platform tunnels, numerous cross passages and ventilation tunnels. All station tunnels are constructed using sprayed concrete lining (SCL) techniques.

The geological conditions were the principal challenge for the design and construction of the Farringdon station tunnels. The tunnel horizon was mainly within the Lambeth Group, a heterogeneous formation comprising stiff to very stiff, overconsolidated clays with interbedded sand lenses of unknown orientation, size and water pressure regime. In addition, four major geological faults crossed the area of the station, increasing the level of complexity and risk.

This paper illustrates the approach to geotechnical risk management adopted from design through to the construction phase as additional geotechnical data became available. The risks were managed through additional surface investigation, a comprehensive in-tunnel probing strategy and a "live" 3D ground model, which provided a robust framework for effective on-site decision making.

Initial studies assessed the probability of encountering water bearing units during tunnelling. This enabled a strategy for the implementation of toolbox items to be developed and refined for the construction phase.

The report demonstrates how dynamic modelling of the geological features encountered has reduced project risk by continuously increasing the level of knowledge and understanding of the complex ground conditions at this station.

Glossary

BFK BAM Ferrovial Kier Joint Venture, main contractor
CRL Crossrail, the Client and Project Manager
DSP Dr. Sauer & Partners, specialist SCL designer for BFK
MMD Mott MacDonald, the Client's permanent works SCL designer
RESS Required Excavation and Support Sheet
SHE Safety, Health and Environmental information
SCL Sprayed Concrete Lining
SRG Shift Review Group
TBM Tunnel Boring Machine

Project Overview

Crossrail is a new railway system which will provide, for the first time, a connection between the South-East and the South-West railway network through the centre of London. The Crossrail route will run from Maidenhead and Heathrow in the west through 21km of twin-bore tunnels under central London – connecting key London stations including Paddington, Bond Street, Tottenham Court Road, Farringdon, Liverpool Street, Whitechapel and Canary Wharf – and out to Shenfield and Abbey Wood in the east.

Crossrail will be integrated with the Tube and national rail networks and will include interchanges with nine London Underground lines. From 2018, Crossrail trains, each capable of carrying up to 1,500 people, will run at metro-style frequencies every two and a half minutes, in each direction, during peak periods. Farringdon station is situated at the heart of London's rail network and will be one of London's major rail interchange stations. Millions of people travelling to and through London will benefit from Farringdon station, which links Crossrail, Thameslink services and London Underground trains. It will also provide links to four of London's airports – Heathrow, Gatwick, City and Luton.

The station layout comprises two ticket halls, two platform tunnels, connecting cross passages, concourse tunnels, two escalator inclines, escape and ventilation adits. The chosen tunnelling system for Farringdon Station is sprayed concrete lining (SCL), which is a sequential open face tunnelling method. The primary sprayed concrete linings and cast in-situ/sprayed concrete secondary linings will be constructed using steel fibre reinforced shotcrete in combination with a sheet waterproofing membrane. In total, approximately 1,000 linear metres of SCL tunnels will be built with excavation cross-sections varying from 25m^2 to 110m^2 and will be constructed at depths of up to 35m below ground level. The use of compensation grouting is a key aspect of managing settlement caused by tunnelling works. The running tunnels between the stations are constructed by other Crossrail contractors using tunnel boring machines (TBMs).

Crossrail awarded the contract for the construction of the Farringdon Crossrail station to a joint venture of Bam Nuttall, Ferrovial Agroman and Kier (BFK) in 2011. BFK appointed Dr. Sauer & Partners (DSP) as specialist consultant responsible for all sprayed concrete lining (SCL) tunnelling works prior to ring closure. The Employer's SCL designer (Mott MacDonald) was responsible for the design of the permanent works including the composite SCL tunnel linings post-ring closure.

Tunnelling Works

In general the SCL components at Farringdon Station comprised the eastbound and westbound platform tunnels, escalator inclines, lower concourse tunnels, cross passages, ventilation adits, stub tunnels and temporary connection tunnels (Figure 1). The platform tunnels were constructed using SCL methods to enlarge a 6.2m diameter precast concrete TBM pilot tunnel to the full platform tunnel size (circa 11.5m wide by 10.5m high).

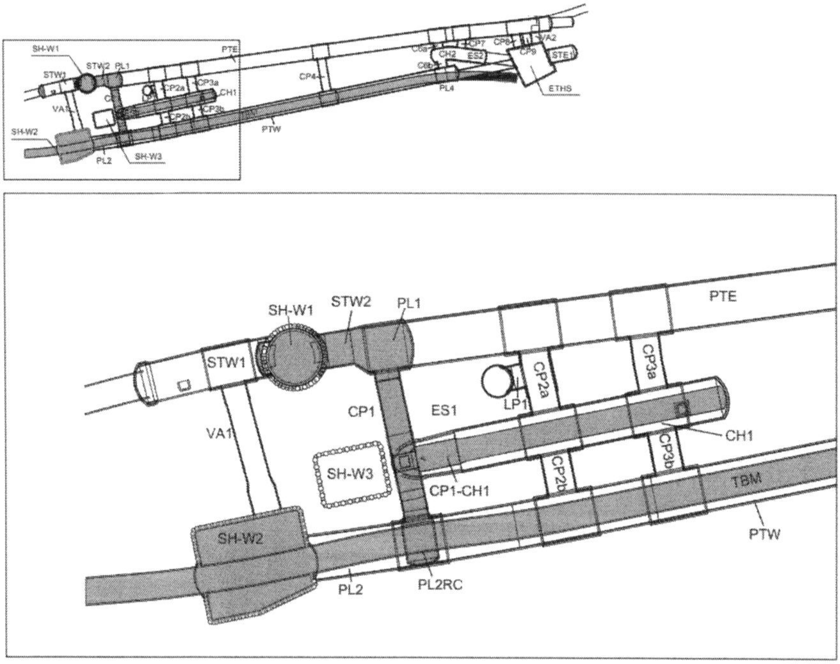

Figure 1. Plan view of the tunnel structures in Farringdon Station, with a close up of the completed SCL works. Principal access for the SCL works was via shafts SH-W1 and SH-W2

The SCL structures at Farringdon Station completed by October 2013 were:

- Stub tunnel STW2 (enlargement from SCL pilot tunnel),
- Platform extension tunnel PL1 (enlargement from SCL pilot tunnel),
- Cross passage CP1,
- SCL soft eyes for TBM reception and transit,
- Lower concourse CH1 pilot tunnel,
- Temporary tunnels – PL2RC Wraparound and CP1-CH1 Connection. Westbound TBM segmental pilot tunnel completed.

Construction commenced with the breakout of the piles in shaft SH-W1 to create a soft eye for TBM reception, followed by breakout and construction on the opposite side for STW2 and PL1. PL1 was approximately 12m wide and 11m high, with a total excavation area of 110m^2 . The next step was the breakout and construction of CP1 tunnel. At the end of CP1 temporary tunnel, a wraparound structure (PL2RC) was constructed to provide safe and rapid future access to the TBM segmental tunnel. In order to provide the construction team with additional programme flexibility and benefits, a temporary connection tunnel was designed and constructed between CP1 and CH1. This allowed the excavation of CH1 ahead of the construction of ES1, thus de-linking the SCL production from the works at SH-W3. In the SH-W2 shaft, two soft eyes were constructed to allow for the passage of the TBM. All of the SCL caverns constructed pre-TBM, were backfilled temporarily with foam concrete to facilitate rapid and safe transit of the TBMs without the need for temporary works or interrupting the

TBM mining cycle. The tunnels constructed by October 2013 are highlighted in green colour in Figure 1.

Geotechnical Description

The typical London basin geological formations were encountered in the area of Farringdon Station. The upper strata comprised Made Ground, River Terrace Deposits (mainly in the Eastern part) overlying London Clay, the Lambeth Group, Thanet Sands and Chalk. The Harwich Formation beneath the London Clay was encountered in very few boreholes. An important feature was that the thickness of the London Clay layer was almost constant to the east of the Smithfield Fault (circa 20m to 22m thick), but was significantly thinner in the western part of the station (circa 4m to 15m thick).

The tunnels in Farringdon station were expected to encounter mainly the Lambeth Group formations (approximately 83%) and to a smaller extent London Clay (approximately 15%) and Thanet Sand (approximately 2%).

The main characteristics of the geology in Farringdon station were the following:

- the presence of multiple vertical / sub-vertical faults which affected the thickness, the elevation and the continuity of the soil layers.
- the thickness of the London Clay which varied between approximately 4m and 22m.
- the very low pore water pressures in the deep aquifer (Upnor Formation, Thanet Sand and Chalk) due to historical water abstraction. This had an under-draining effect (downward seepage and pressure relief) on the Lambeth Group pore water pressures (and partially on the London Clay), which were lower than hydrostatic from the water level at the top of the shallow aquifer.
- the spatial distribution of water bearing sand lenses within the top of the Lambeth Group (mainly Upper Mottled Beds), which varied in thickness and horizontal extent.
- the presence of the buried valley of the former course of the historical Fleet River in the area of the West Ticket Hall (along Farringdon Road). This resulted in a local deepening of the surface of the London Clay, an absence of River Terrace deposits and the localised presence of alluvium deposits (erosion effects).

The station can be separated geologically into two different sub-areas, west and east of the Smithfield fault (Figure 2). The main difference between these two areas was the thickness of the London Clay layer; this was affected by the multiple faults through this area and had a significant effect on the pore water pressure profile. Thus, in the eastern part of the station, where London Clay was significantly thicker, the pore pressures were increased in both this formation and in the Lambeth Group, hence, the under-draining effect from the deep aquifer was smaller than in the western part.

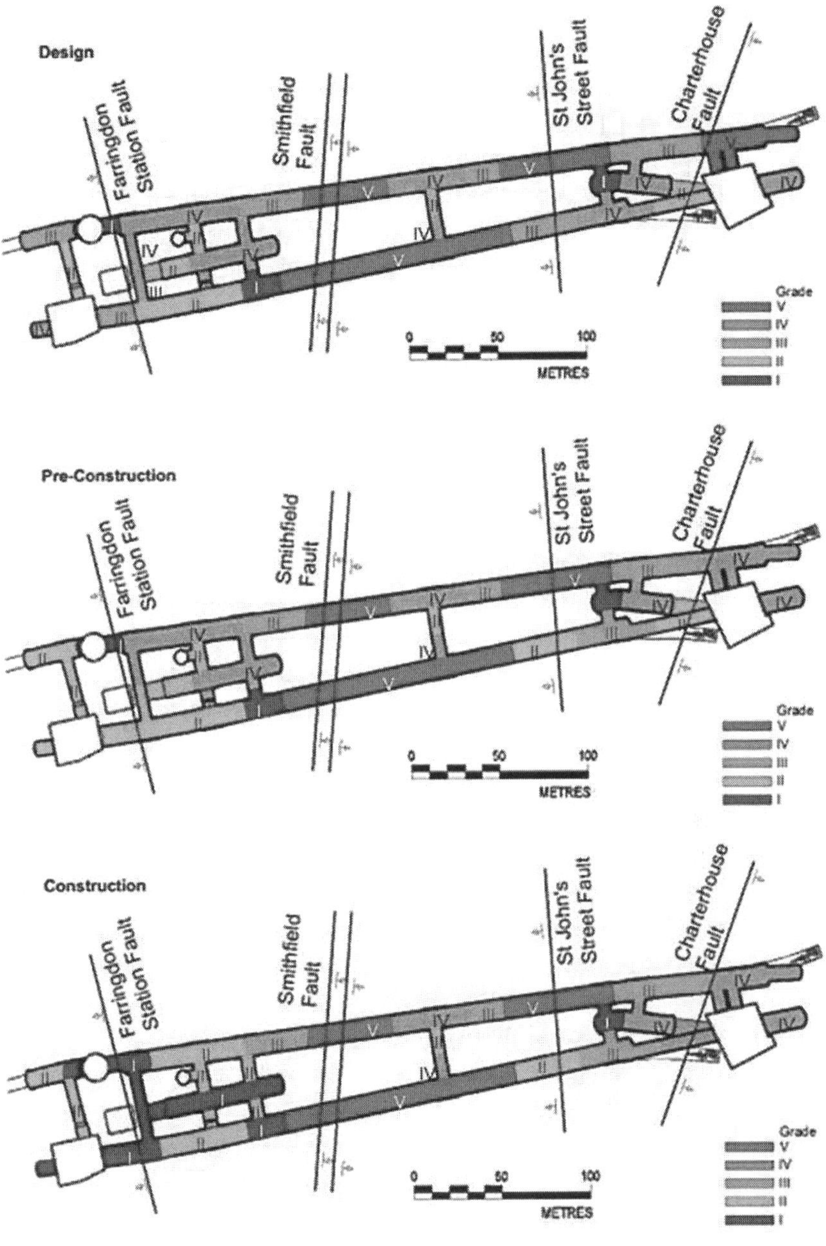

Figure 2. Mapping of Geotechnical related Hazards in 3 phases of the project: Design, Pre-Construction and Construction

Two main aquifers could be distinguished; the upper aquifer located in the Made Ground and River Terrace deposits; and the lower aquifer comprising the lowest part of the Lower Mottled Beds, Upnor Formation, Thanet Sands and Chalk deposits. These two aquifers were separated by the London Clay aquitard. The pore water pressures in the deep aquifer were very low (almost zero) due to historical pumping and thus had an under-draining effect on the Lambeth Group deposits and partially on the London Clay. However, localised presence of sand bodies within the upper Lambeth Group was also expected and this was the feature that had the greatest potential impact on the stability and safety of the open face SCL tunnelling. After a thorough examination of the available data, DSP concluded that the most probable scenario was that these formations would be encountered as water bearing sand lenses rather than as continuous sand channels (i.e. without significant horizontal continuity).

Tunnelling and geotechnical risk assessment and mitigation in that challenging and variable geotechnical environment required a sophisticated and robust strategy to be in place.

Risk Management

Geotechnical risk management was a key aspect of successful delivery of this complex station in a complex urban environment surrounded by critical infrastructure and Third Party assets. Excavation stability, rapid ring closure and minimising settlement were the primary drivers for the site team, yet this had to be performed in some of the most challenging and complex geological conditions on the whole of Crossrail.

The risk management strategy adopted has been heavily influenced by published reports (see BTS & ABI (2003), HSE (1996) and ICE (1996)). The actual management and control of risks was delivered through the use of risk assessments and risk registers which originated from the Client's permanent works designers (MMD). Their registers rigorously required to identify and clarify the ownership of risks and detailed clearly how these risks were to be allocated, controlled, mitigated and managed. The Risk Registers were 'live' documents which were reviewed on a regular basis and revised as appropriate. Where possible, these risks were eliminated at source through the design. The residual risks were transferred to the BFK, and were addressed in a variety of ways including; BFK methods and sequence of working; BFK construction management arrangements; temporary works design.

The CDM risk register provided by the client at Contract Award was updated by DSP to ensure that the design mitigation actions related to the actual design measures proposed by DSP, including implementation of depressurisation and toolbox items. The risk matrix scoring was reviewed to reflect the Contractor's actual control measures and detail was added to the register to address the ground conditions detailed in the original design intent document provided by the permanent works designer (MMD).

At all times, the project team's focus was on hazard identification and reducing risks as low as reasonably practicable (ALARP), and assigning those risk to the party most able to control them.

The CDM Risk Register demonstrated that the anticipated risks and hazards were identified and defined in a formalised risk management process and that the appropriate mitigation measures were in place and that the residual risks were ALARP. Communication of these residual risks was via the "SHE boxes" on the DSP temporary measures construction drawings; this highlighted to the BFK production team any unusual or unexpected risks associated with a location or activity which the team may not have reasonably anticipated. The residual risks were addressed through the risk assessments included in the BFK method statements and/or via the BFK "SCL Action Plan" and Emergency Preparedness Plans. Therefore, the design risks were ultimately addressed and mitigations briefed through the

Task Briefing and Method Statements to the construction and supervisory staff and to the workforce.

The net effect of this process was a demonstrable chain of ownership of each of the identified risks, which showed how these risks had been eliminated or reduced ALARP by either the permanent works designer, the temporary works designer or the contractor.

Hazard Mapping

The primary hazard related to the SCL tunnelling works in Farringdon was the presence of non-cohesive, water bearing units in proximity to the excavated face. This hazard was mapped and presented in Figure 2 for three distinct stages:

- The design phase: dated April 2012 when the Optimised Contractor Involvement (OCI) period started with preliminary geotechnical investigation data being made available by the Client and MMD.
- The pre-construction phase: dated May 2013, when the excavation of the first tunnel (STW2) commenced, with additional ground investigation data becoming available. This additional information was obtained by BFK as a by-product of installing instrumentation and monitoring devices within 25 new boreholes around the station. This innovative idea provided the project team with a wealth of additional information to augment the existing ground model.
- The construction phase: dated October 2013, when part of the Early Western Tunnels was already constructed (the area in green on Figure 1)), with additional data acquired during the excavation of the tunnels.

The following assumptions were made in order to derive the presented mapping:

- Grade V [red] applied to either "low confidence" in the knowledge of the geological environment and/or evidence of water bearing sandy units in the tunnel horizon.
- Grade I [blue] applied to parts of the station that had already been excavated or thoroughly investigated, thus providing a high level of confidence.
- Larger tunnels posed a higher risk.
- The area east of Smithfield Fault was assigned an increased risk by default, as the pore water pressure regime was higher and tunnels were excavated within more risk-prone geological formations (the Upper Mottled Beds).
- Equally effective in tunnel investigation (probing) would be carried out for all the structures, reducing the risk by the same amount.

As illustrated in Figure 2, the geotechnical risk mapping has significantly changed through these three phases, with the area east of the Smithfield Fault exhibiting a higher residual risk compared to the west.

Geotechnical Risk Management

DSP developed an optimised geotechnical risk management framework which assessed all of the available data in order to minimise the residual risks at the tunnel face. In addition to the investigation already carried out from the surface, a pattern of in-tunnel probing was prescribed throughout the SCL works. In addition to the data acquired through probing, face mapping data was integrated into an existing 3D Geological model that was originally developed in 2009 by the British Geological Survey (BGS). This 3D model provided a useful day-to-day visual geological database, and allowed the site team to make accurate predictions for forthcoming SCL works. These tools and the way they were implemented are detailed below.

In-tunnel Probing

The in-tunnel investigation scheme, prescribed an optimized probing pattern ahead of the tunnel face. The implementation of this strategy was performed as follows:

- Investigation and depressurisation for “small” tunnels, such as pilot tunnels for STW1, STW2 and PL1, and CP1. This was performed ahead of the tunnel face from within the advancing face, by drilling inclined and horizontal probe holes and,
- Investigation of “enlarged” tunnels, such as STW1, STW2 and PL1 for the advancing enlargements by drilling radial probe holes within the previously installed pilot tunnels.

Typical results from in-tunnel radial probing for PL1 tunnel are illustrated in Figure 3.

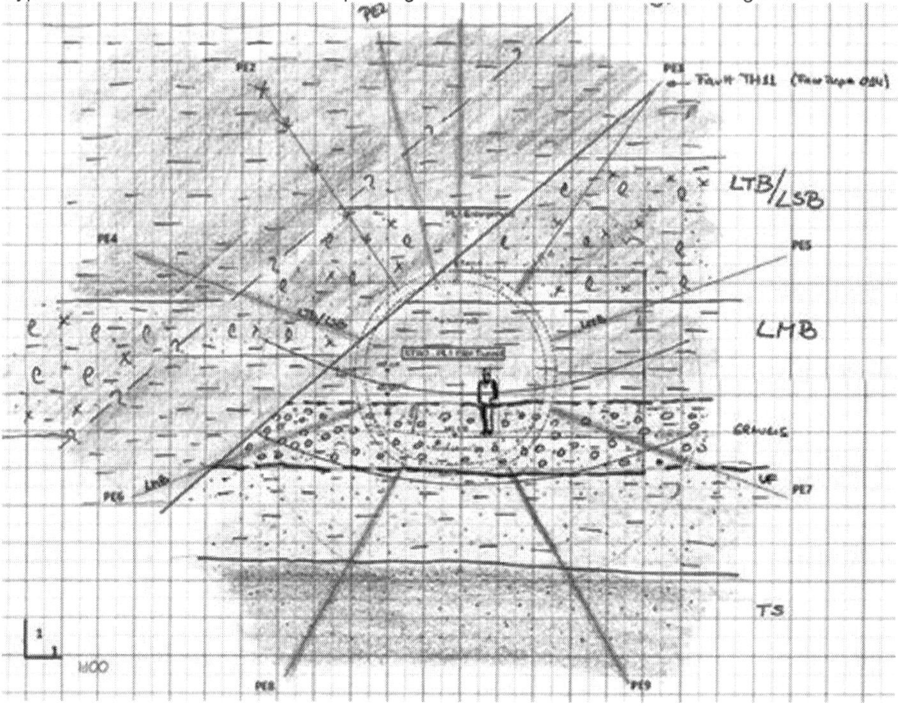

Figure 3. In-tunnel radial probing results for PL1 tunnel, performed through the primary lining of the pilot tunnel, predicting the presence of the Farringdon Fault

DSP developed a flow chart to help to rapidly determine the future actions based upon the findings of the advance probing (Figure 4). This included criteria for water flow and pore-pressure control to incorporate individual or combined measures (in-tunnel water management, passive drains, or vacuum wells).

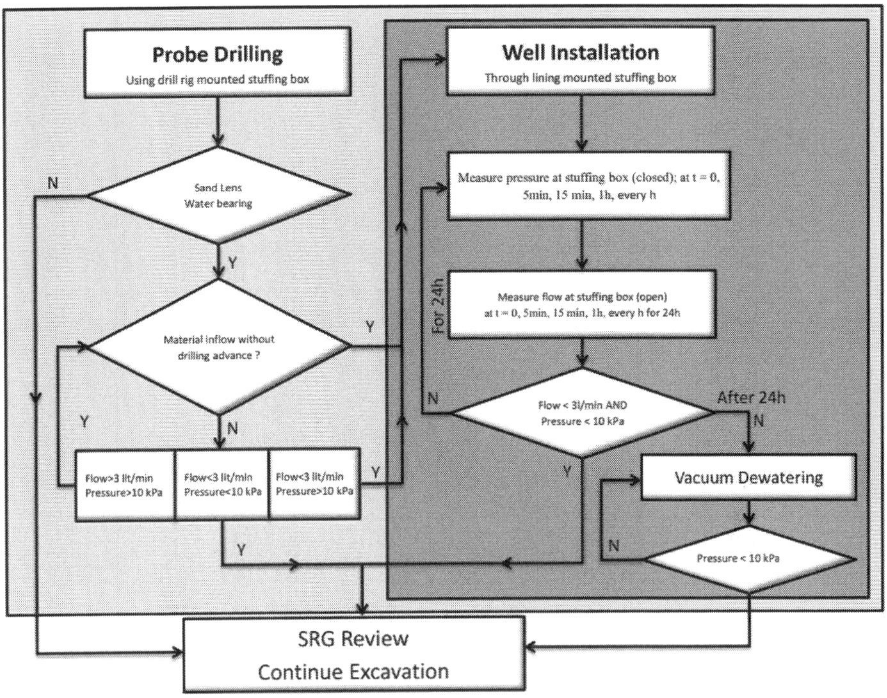

Figure 4. Performance criteria for groundwater flow and pore-pressure control

For the early western tunnels, (comprising the soft-eyes for the TBM reception at shafts SH-W1 and SH-W2 and tunnels STW2, PL1 and CP1), this scheme started from the circular shaft SH-W1 initially targeting the Lambeth Group prior to the excavation of the pilot tunnels for STW1 and STW2 and PL1, as well as CP1 (the "small" tunnels). The in-tunnel investigation was completed successfully, without requiring any depressurisation measures to be implemented. In addition, the roof support measures that according to the base design had to be installed in several locations as the soil was expected to be of inferior strength due to the presence of faults, were de-scoped as the geological information determined that they were not required.

Tunnel Face Mapping

The open face excavation using SCL, allowed for accurate geological observations to be made and documented through the face mapping process. Face mapping was carried out for every single excavation step, and provided valuable, continuous data that confirmed the predictions made through the probing steps. This data was also presented in longitudinal geological sections. A typical representation of the geology encountered in the open face of PL1 tunnel, is presented in Figure 5.

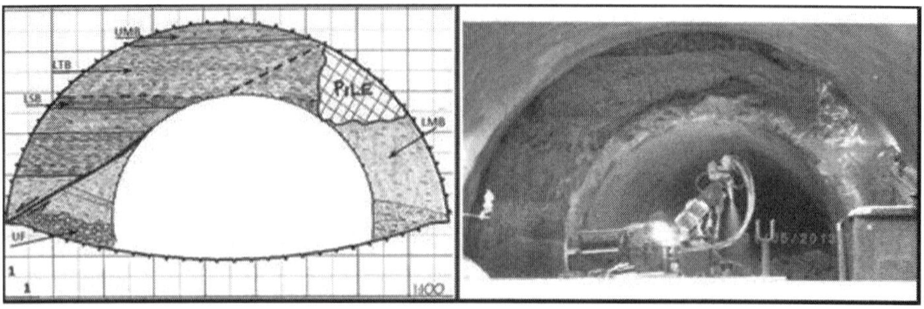

Figure 5. Face mapping of the excavated top heading of PL1 Enlargement, encountering a historical pile and Farringdon Fault

3D Geological Model

The 3D geological model for the area of Farringdon station was initially developed by the British Geological Society (BGS) in 2009, based on historical data and Crossrail specific site investigation. In 2013, the model was handed over to BFK/DSP and was integrated in the site supervision workflow. Initially, the additional data from borehole drilling and shaft excavation, received between 2009 and 2013 was used to update the model. Data from the face mapping has been progressively integrated into the model on a daily basis, thus providing a geological database for predictions of constantly increasing accuracy for future tunnels.

The cycle of risk reduction presented in Figure 6, started with the prediction based on the 3D geological model. After the excavation of each SCL tunnel, the data from the face mapping was integrated in the 3D model. Thus, the risk was reduced as the level of knowledge and confidence was constantly increasing. At the same time, in-tunnel probing provided additional data, validating at the same time the prediction from the 3D geological model.

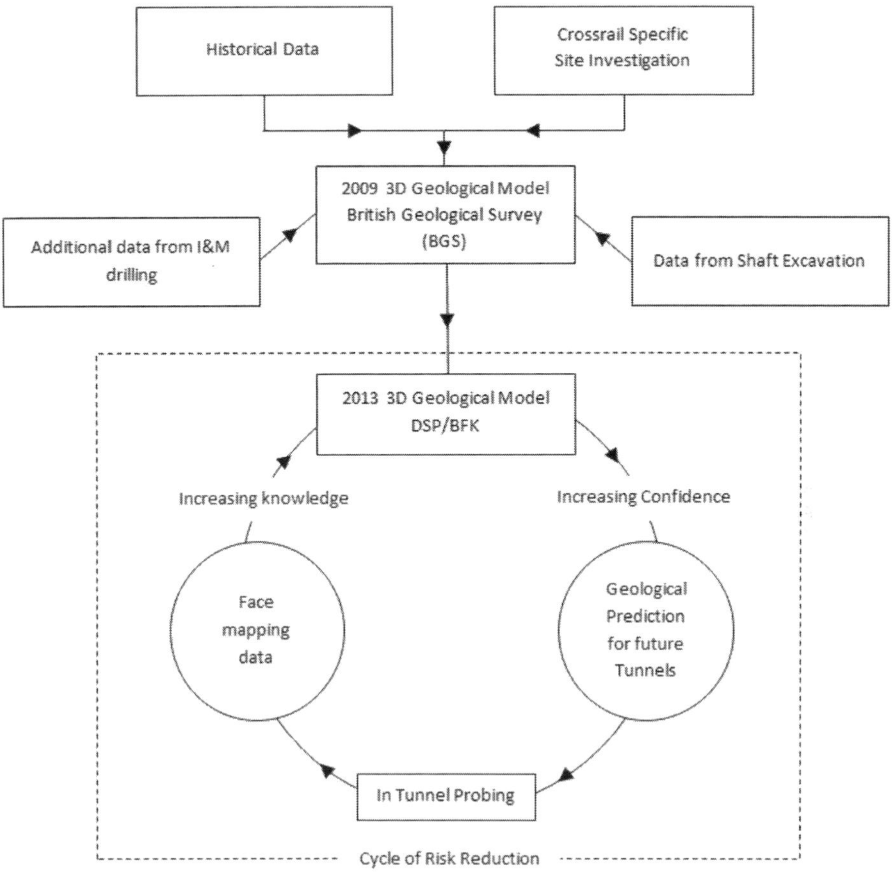

Figure 6. Cycle of risk reduction through the implementation of geotechnical risk management tools

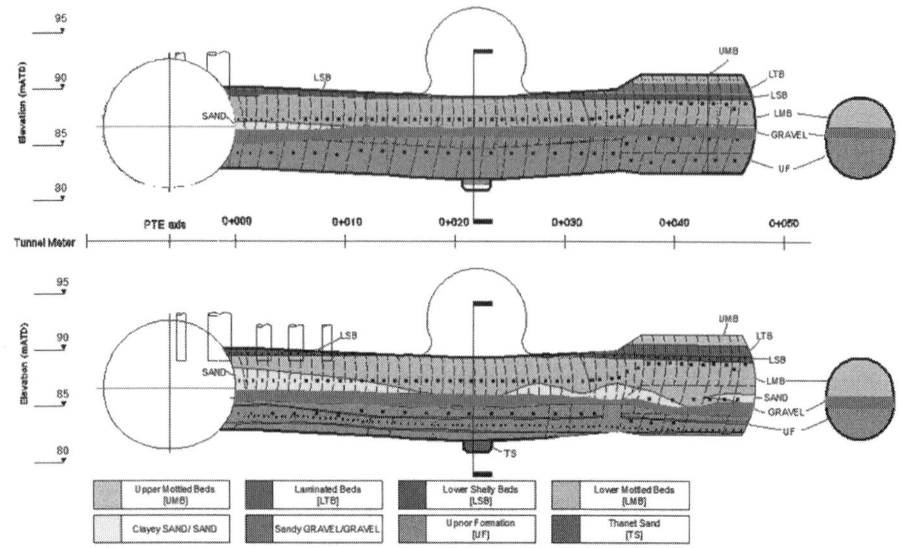

Figure 7. Geological prediction prior to the start of excavation (top) vs. actual (bottom) longitudinal geological section for CP1. The 3D model provided a very good prediction in an area where very little borehole data was available.

Figure 7 presents a comparison of the predicted stratigraphy along CP1 from the 3D model, performed prior to the start of the excavation, against the "real" stratigraphy as recorded through the face mapping process. This provided excellent correlation of results despite the fact that almost no investigation data previously existed in the area of the excavation of CP1.

Strategy for Implementation of Toolbox Items during SCL Tunnel Construction

The chosen tunnelling system for Farringdon Station was a sequential open face SCL tunnelling method that included:

- Prescribed mandatory measures, such as length of excavation steps, sprayed concrete lining thicknesses and in-tunnel probing which were required to be implemented in a prescribed sequence and were considered adequate for the "baseline" ground conditions, and
- Pre-defined contingency measures/toolbox items such as spiles or pocket excavation to be made available for use where localised adverse ground conditions were encountered.

This meant that an element of the design decision making and implementation had to be applied during the construction process. This decision making was managed through the "Shift Review Group" (SRG) and "Required Excavation Support" (RES) process. This ensured that all available relevant data was collated, reviewed and used to inform the next steps for safe tunnel advance, thus managing the risks within a well-defined and robust framework.

The BFK / DSP temporary measures design was built around a depressurisation scheme which included intensive prescriptive in-tunnel probing and pre-defined depressurisation measures ahead of the advancing tunnelling face. This design was supplemented by a set of pre-defined contingency measures/toolbox items which could be applied in the event of encountering localised unfavourable ground conditions. Toolbox Items included pocket excavation/face wedge, rebar spiling, grouted pipe spiling, pipe arches, additional probing and additional depressurisation.

The Senior SCL Engineer and Chief Geotechnical Engineer were employed by DSP and embedded into the BFK Engineering team, their remit being to oversee excavation stability at all times. If needed the toolbox items would be deployed at their discretion via the SRG and RES meeting; this included representatives from Crossrail, the SCL permanent works designer (MMD), the compensation grouting subcontractor, the asset protection engineer together with key members of the BFK and DSP engineering and supervisory teams. This process evaluated of all available date such as geotechnical data, 3D ground model, monitoring results, dewatering/depressurisation measures, compensation grouting data, and SCL early age strength results, etc, and formed the basis for safe advance of the SCL tunnels.

The output from the SRG was a signed RESS, which defined the advance, support, probing, depressurisation and toolbox requirements for the following shift. Five key signatories were required in order to validate the RESS, and no tunnelling was permitted without a valid RESS being displayed at the face. In the event of the RESS becoming invalid, a "Safe Stop Procedure" would need to be implemented.

Conclusions

The Crossrail Farringdon station is located in an area of variable and complex geology with a high degree of risk and uncertainty caused by the presence of multiple faults. Coupled with the presence of critical urban infrastructure and Third Party assets, it is arguably one of the most challenging SCL tunnelling projects on the whole of Crossrail.

The integrated delivery team of DSP and BFK have established a robust and comprehensive geotechnical risk management process to address these risks. This has yielded a greater understanding of the residual risks and enabled to provide a tailor-made pattern of probing and depressurisation measures consistent with the level of risk. Continuous improvements have also been achieved by feeding newly-acquired information into the 3D BGS ground model; this increased knowledge and confidence about the ground has enabled the project team to provide ever-improving predictions about the anticipated conditions.

DSP and BFK have developed an effective strategy for deploying toolbox items to react to adverse conditions encountered, within the wider structure of the SRG and RES process. This has required the full time site presence of DSP supervisory staff into key roles as "Senior SCL Engineer" and "Chief Geotechnical Engineer", complying with onerous mandatory minimum competency requirements specified by CRL, embedded into the BFK organisation, thus enabling the highest possible standards of geotechnical risk management to be delivered on this project.

Acknowledgements

The authors would like to acknowledge the significant contribution from BFK joint venture, Crossrail and Dr Sauer and Partners for providing the necessary resources, guidance and support in the preparation of this paper. We would also like to thank the British Geological Survey that originally developed the 3D geological model of the Farringdon area in 2009 and handed it over to Dr Sauer & Partners in 2013 and especially Dr. Don Aldiss, for his valuable support and input during the development of the model.

References

BTS and ABI (2003) The Joint Code of Practice for Risk Management of Tunnel Works in the UK. London: British Tunnelling Society, 2003.

HSE (1996) Safety of New Austrian Tunnelling Method (NATM) Tunnels: A Review of Sprayed Concrete Lined Tunnels with Particular Reference to London Clay. Sudbury, Suffolk: HSE, 1996.

HSE (1996) The Collapse of NATM Tunnels at Heathrow Airport: A Report on the Investigation by the Health and Safety Executive into the Collapse of New Austrian Tunnelling Method (NATM) Tunnels at the Central Terminal Area of Heathrow Airport on 20/21 October 1994. Sudbury: HSE, 2000.

ICE (1996) Sprayed Concrete Linings (NATM) for Tunnels in Soft Ground. London: T. Telford, 1996. Print.

Crossrail Project: Infrastructure design and construction
ISBN 978-0-7277-6078-4

doi: 10.1680/cpid.60784.363

Stepney Green Cavern – Design Concepts and Performance of SCL Lining

Michal Uhrin MSc, CEng, Mott MacDonald CZ
Thomas Schultheis MSc, CEng, Gall Zeidler Consultants
Nicholas Tucker Eur Ing, BSc, CEng MICE, Mott MacDonald Ltd.

Abstract

Crossrail includes several kilometres of SCL tunnels which were designed by Contract C121, Mott MacDonald supported by Gall Zeidler Consultants. The largest of these SCL tunnels are Stepney Green turnout caverns designed to allow trains from Maidenhead in the west to travel to Shenfield and Woolwich in the east. The caverns represent the largest span underground SCL works constructed below London. This paper concentrates on the primary lining design of the largest Stepney Green sections, excavated with sidewall drift sequences. The eastbound cavern, located approximately 25m below ground surface level, was constructed predominantly within London Clay. The westbound cavern is about 5m deeper. Its upper two thirds were excavated in London Clay while its lower third was excavated in the underlying Lambeth Group stratum, which contains sandy channels with high porewater pressures. This paper discusses the analytical part of the primary lining design which was aimed to simulate the ground–lining interaction considering the ground behaviour, lining parameters and excavation and support sequence and provide predictions to be checked against required standards. The design for the caverns was based on 2D and 3D numerical modelling and the paper outlines the major analytical issues and compares the predicted behaviour against the results of monitoring.

List of notation

2D two-dimensional
3D three-dimensional
SCL sprayed concrete lining
MLGH Mid Lambeth Group Hiatus
p' mean effective stress in the ground
Φ' effective friction angle of the ground
c' effective cohesion of the ground
c_u undrained shear strength of the ground
A^* initial stiffness parameter of the A* small strain stiffness ground model
G shear modulus of the ground

K_0 in situ at rest lateral ground pressure coefficient
E_u undrained elastic modulus of the ground
E_{cm} mean elastic modulus of concrete to Eurocode 2
$E_{c,eff}$ creep reduced elastic modulus of concrete to Eurocode 2

Introduction

Stepney Green Junction, Box and Caverns are located in the London Borough of Tower Hamlets, east of Whitechapel station and at the point of bifurcation of the tracks to Pudding Mill Lane Portal and the Isle of Dogs (see Figure 1.1).

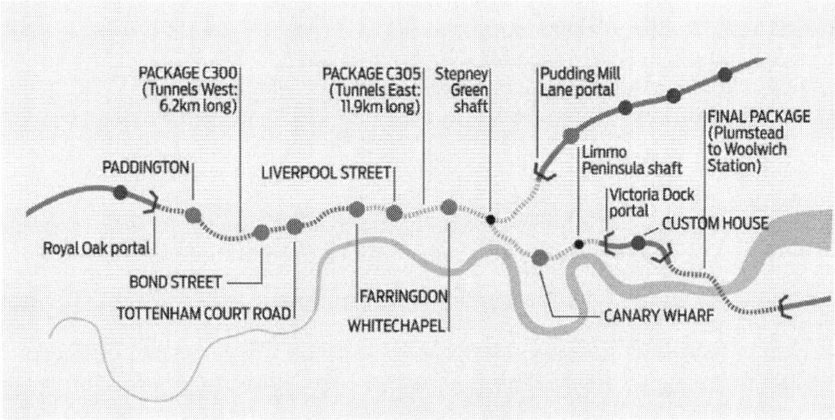

Figure 1.1. Crossrail alignment

Stepney Green is one of the largest work packages on Crossrail and is one of the most complex sections of the new rail link. The caverns will house the crossover tunnels that will allow trains to connect to both eastern terminals. The interchange will allow passengers from the Essex end of Crossrail in Shenfield to connect with the Abbey Wood station in north Kent as well as destinations on the central and west sections of the new rail line.

The two Stepney Green SCL turnout caverns at over 50m long with a quasi-elliptical cross section varying from 8m diameter to 17m wide x 13.m high are the largest caverns ever built using Sprayed Concrete Lining techniques in Central London. At the eastern ends of the caverns structures there are two TBM launch adits, approximately 25m long.

Excavation of the caverns progressed from a rectangular access shaft built across the twin TBM running tunnels to the west of the junctions that will house M&E utility rooms, a ventilation facility and provide an emergency ingress and egress for Crossrail during operation.

The SCL cavern design was carried out by Mott MacDonald supported by Gall Zeidler Consultants. Temporary works were designed by OTB on the Contractor's behalf. The caverns form part of the Eastern Running Tunnels Contract C305 that was awarded to the Dragados/Sisk JV in December 2010 for approximately £500 million.

The SCL excavation commenced on the EB launch tunnel in November 2012 and the EB cavern in January 2013. The EB cavern primary lining was completed in May 2013, followed by the westbound in August 2013. To construct the caverns and launch tunnels, the project team had to excavate approximately 38,000m^3 (bulked) of material and apply over 6,000m^3 sprayed concrete to the temporary and permanent linings.

SCL structures at Stepney Green

Stepney Green SCL construction consists of various structures including:

- two Launch and two reception adit tunnels
- two single side drift caverns
- two double side drift caverns

The total length of the mined tunnels is approximately 120m, containing 50m of adit tunnels, 40m of single side drift tunnels and 25m of the large double side drift caverns. A detailed description of the individual structures is given in the following chapters.

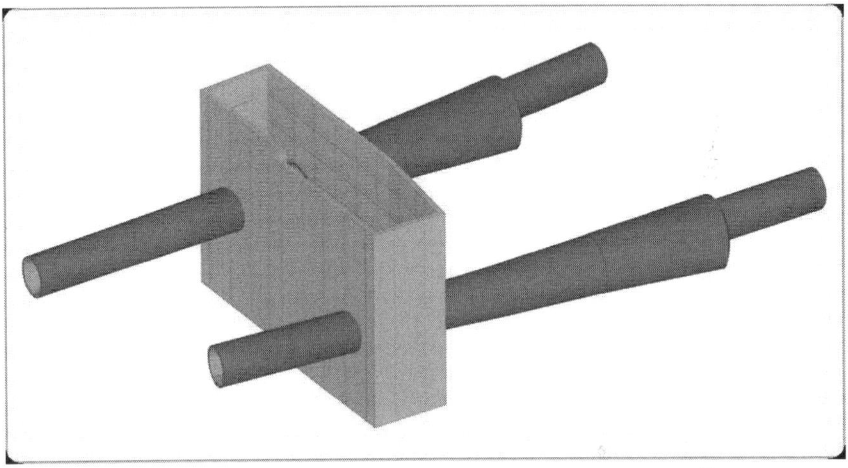

Figure 2.1. SCL structures at Stepney Green

Adit tunnel

The cross section for the adit tunnels is in the range of 8 – 9.4m in width and 8.8m in height. The primary lining consists of steel fibre reinforced shotcrete with a thickness of 350mm including a 75mm sealing layer. The cross section is subdivided in three faces (top heading, bench and invert) with a maximum face area of 32m^2. The resultant excavation cross section is about 55m^2. The secondary lining consists of 250mm sprayed concrete reinforced with steel bars and mesh. A combination of spray applied waterproofing and sheet membrane was designed for SCL structures for Stepney Green.

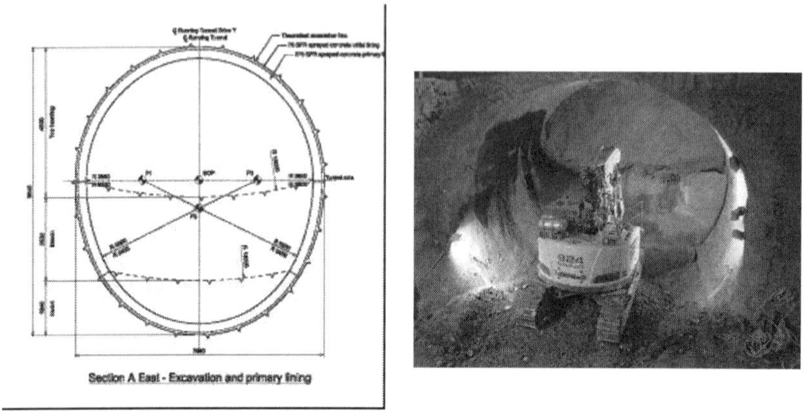

Figure 2.2 Adit tunnel cross section

Single Side Drift

The cross section for the single side drift cavern varies in width from 9.9 to 12.9m and in height from 8.6 to 11.2m.

The primary lining consists of sprayed concrete with a thickness of 400mm including a 75mm sealing layer and 300mm for the temporary sidewalls. The primary lining is divided in two layers consisting of a 300mm primary I layer including steel fibre reinforced shotcrete, 16mm bar reinforcement in the knee areas and a 100mm primary II layer including plain shotcrete and a B503 mesh over the entire circumference and bar reinforcement in the shoulder and invert area. In addition L-bars where installed at the top and bottom connections.

The cross section is subdivided vertically and horizontally in 6 faces with each face area less than 30m^2. The resultant excavation cross section is about 80 to 120m^2.

Whilst analysis does not rely upon the presence of lattice girders for structural capacity in the design cases, it was concluded that lattice girders provide an important contribution to shape control as well as the improving the bending and shear resistance of the linings and therefore the robustness of the lining system.

The secondary lining consists of 350mm sprayed concrete reinforced with steel bars and mesh.

A combination of spray applied waterproofing and sheet membrane was designed for SCL structures for Stepney Green.

Figure 2.3 Single Side drift cross section

Double Side Drift

The cross section for the double side drift cavern varies in width from 13m to 16.5mm and in height from 11.4 to 13.4m.

The primary lining consists of sprayed concrete with a thickness of 500mm including a 75mm sealing layer and 300mm for the temporary sidewalls. The primary lining is divided in two layers consisting of a 375mm primary I layer including steel fibre reinforced shotcrete, 16mm bar reinforcement in the knee areas and a 125mm primary II layer including plain shotcrete and a B503 mesh and bar reinforcement in the shoulder and invert area. In addition L-bars where installed at the top and bottom connections.

The cross section is subdivided vertically and horizontally in up to 10 faces with each face area less than 30m^2. The resultant excavation cross section is about 120 to 180m^2.

Lattice girders were installed for the same reason as mentioned in 2.2.

The secondary lining consists of a 400mm sprayed concrete reinforced with steel bars and mesh.

A combination of spray applied waterproofing and sheet membrane was designed for SCL structures for Stepney Green.

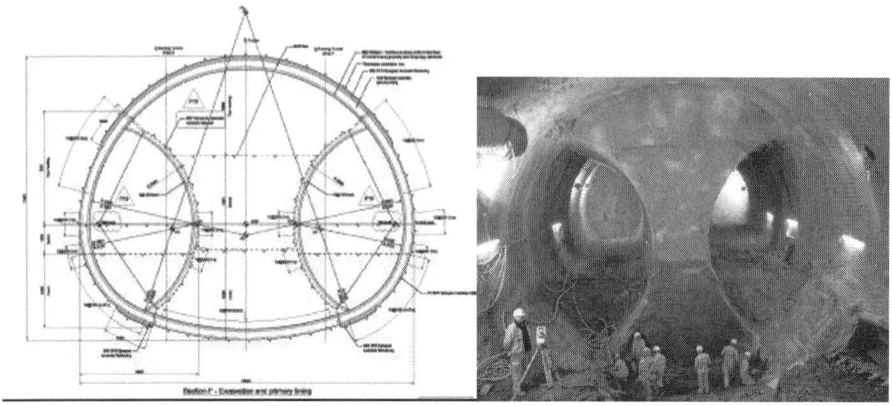

Figure 2.4 Double Side drift cross section

Excavation and Support Method

For the excavation of the tunnels the Method of SCL lining with sequential excavation and support was applied.

The single side drift cavern was constructed by the application of a temporary sidewall and a subdivision of the drifts into top heading, bench and invert. In longitudinal direction a short ring closure with a distance of 6 to 7m was implemented to ensure a rapid stabilisation of the ground.

The double side drift cavern was constructed by the application of sidewall galleries with a subsequent centre drift which was also subdivided into top heading, bench and invert. In longitudinal direction a short ring closure with a distance of 5 to 6m was implemented to ensure an early ring closure.

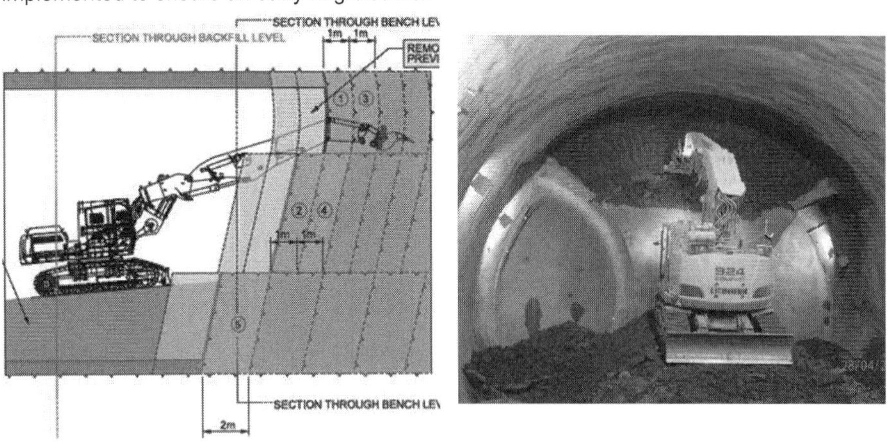

Figure 2.5 Typical Excavation and Support sequence of centre drift.

Ground and groundwater conditions

The ground and groundwater conditions are presented in this paper for the extent and level of detail relevant to the analytical part of the primary lining design for the Stepney Green caverns.

Ground stratigraphy

The ground strata present at the Stepney Green site are as follows (from the ground surface level): Made Ground, River Terrance Deposits, London Clay Formation – Unit B/A3, London Clay Formation – Unit A2, discontinuous Harwich Formation, Lambeth Group Formation, Upnor Formation, Thanet Sand and Chalk bedrock. Made Ground and River Terrace Deposits are relatively thin and have generally low influence on the SCL works. Only a local presence of the Harwich Formation was observed, occurring in a discontinuous band beneath the Stepney Green Shaft footprint. The Lambeth Group Formation includes significant presence of channel sand with high porewater pressures which had to be reacted upon by design of a suitable depressurization scheme.

The depth from the ground surface level to the eastbound cavern axis level is approximately 25m. The westbound cavern is located about 5m deeper. The stratigraphy applicable to the caverns is illustrated by schematic geological longitudinal sections in Figure 3.1 (eastbound cavern) and Figure 3.2 (westbound cavern).

Figure 3.1 indicates that the eastbound cavern was excavated predominantly within London Clay Unit A2. Figure 3.2 shows that the top part (approximately two thirds) of the westbound cavern was also constructed in London Clay. Its bottom part however had to be excavated in the Lambeth Group Formation with significant presence of channel sand. The depressurization scheme was therefore most critical for the westbound excavation. Due to its importance, extent and impact the depressurisation process was simulated in all westbound numerical analyses.

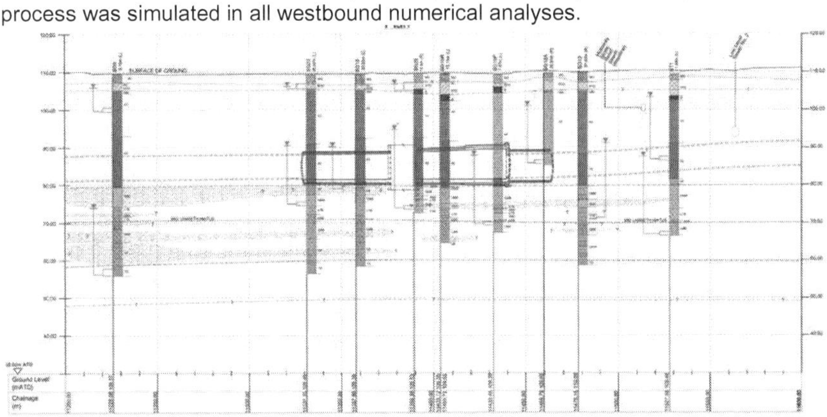

Figure 3.1 Schematic geological longitudinal section eastbound

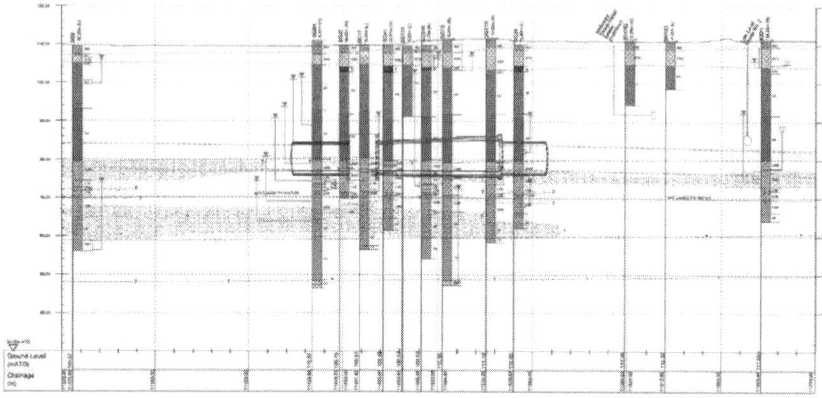

Figure 3.2 Schematic geological longitudinal section westbound

Groundwater conditions

The groundwater conditions were evaluated based on information gathered from multilevel piezometer measurements. The Mott MacDonald geotechnical team reviewed the measurements and rationalised them into series of pore water pressure design lines applicable to various design scenarios. Illustration of the water pressure design lines is provided in Figure 3.3 (a).

The complex in situ piezometric profile includes sub-hydrostatic pressure, changes in pore pressure gradient and pressure drops due to underdrained conditions below London and due to complicated sequence of the clayey and sandy strata. A simulation of the pore water pressure conditions from a numerical model plotted against a theoretical design line is presented in Figure 3.3 (b).

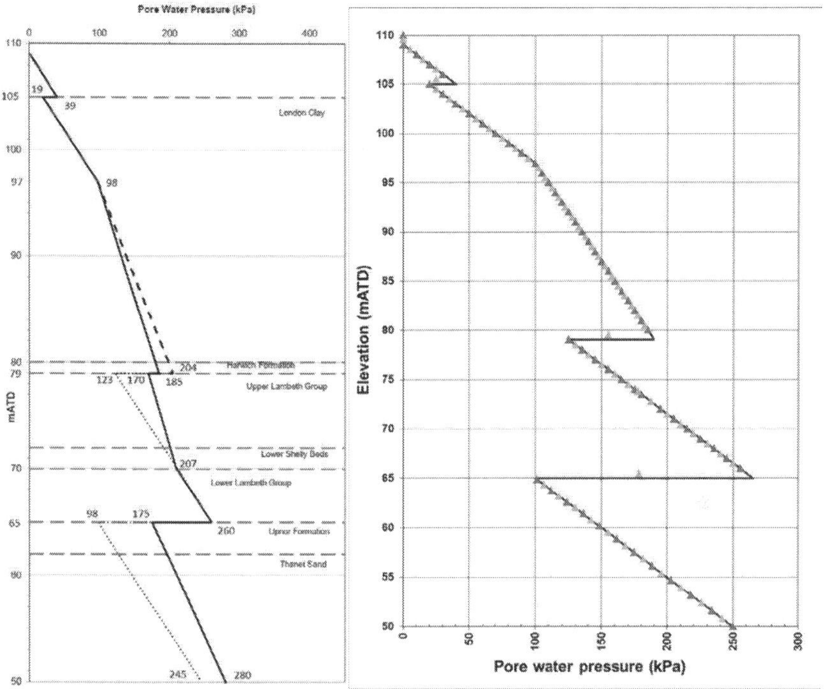

(a) Requirement (b) Numerical simulation

Figure 3.3 Pore water pressure design lines

Numerical analyses of the SCL caverns

Numerical models were carried out to substantiate the design for the SCL caverns. This paper concentrates only on the primary lining analyses. The purpose of the numerical analyses was to simulate the ground-lining interaction and provide model predictions to be checked against required standards. For successful outcome, the numerical models had to deal with the ground behaviour, lining structural behaviour, potential surface surcharge and the excavation and support sequence.

During the course of the design numerical models were undertaken both for the eastbound and the westbound cavern and both with and without the 75kPa surface surcharge required by the project's standards. The site is generally situated under a park area which indicates minimum surface surcharge in the short-term and no need for compensation grouting works. However, the activities of the construction site including heavy storage and plant movement can cause some degree of surcharge. In the long-term the 75kPa surcharge was meant to account for potential future development on the surface.

The numerical analyses were carried out in finite difference program FLAC. Initially, the caverns were modelled in 2D plane strain sections. Later in the design process, 3D numerical models were developed by extruding the 2D sections into the longitudinal direction. One 3D model was built for the largest single side wall drift cavern and one for the largest double sidewall drift cavern.

Drained and undrained ground behaviour

The clayey ground was generally modelled as undrained whereas the sandy ground was generally modelled as drained in the short-term scenarios, i.e. during excavation and SCL primary lining works. In addition, the fully closed and hardened primary lining ring was also checked for full drained load.

The depressurisation of the Lambeth Group sand channels in the eastbound cavern models was simulated in the numerical analyses according to the requirements of its performance specification. Depressurisation carried out from the ground surface was required to reduce and maintain the pore water pressure within the Lambeth Group granular strata below a certain level. Depressurisation carried out from within the tunnel was required to further reduce the water pressure within the actual area of excavation to close to zero.

There was a discussion on the potential effects of partial consolidation of the clayey ground before the ring closure. The debate was specific to the westbound track conditions due to the close proximity to the sand channels in the Lambeth Group Formation, which could act as water re-charge medium and speed up the rate of consolidation (swelling of clay). On the other hand, the depressurization scheme would act against this trend. In order to check sensitivity a fully coupled fluid-mechanical analysis was carried out to investigate the impact of the potential partial consolidation. However, the general reliability of such analysis was questioned due to its extremely complicated nature. The major objections were as follows:

- Ground permeability had to be calibrated to match the complicated in situ pore water pressure profile. The profile tended to get unstable after a certain number of flow steps in the numerical calculation.
- There were several time-related variables in the time-dependant analysis, namely the consolidation time (dissipation of excess pore water pressure), gain in sprayed concrete strength and stiffness over time and the ground relaxation process which could generally occur over time. Simplifying assumptions had to be made.
- It was difficult to determine the effective stress strength parameters appropriate for this short-term situation.

The issue of the appropriate drained strength parameters which should allow for dilation effects had the most significant impact on the outcome of the analysis.

Ground model

The ground was modelled as a non-linear elastic perfectly plastic medium. The strength of the granular ground was simulated using a standard Mohr-Coulomb criterion with effective stress parameters c' and Φ'. The strength of the clayey ground was simulated by a Tresca criterion with undrained shear strength c_u. The undrained strength profile with depth was determined based on data from laboratory testing, in situ testing and empirical relationships. An example plotted from the numerical model against its theoretical design line is presented in Figure 4.1 (a).

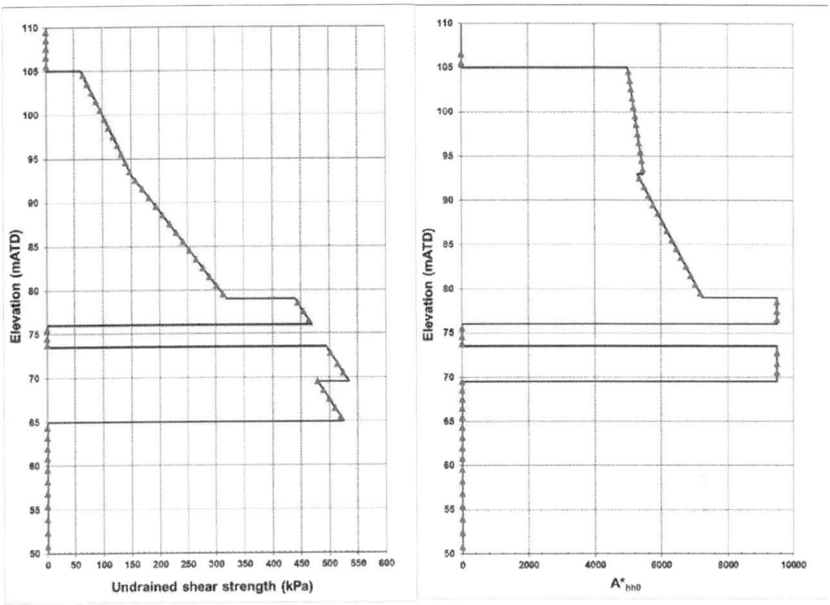

(a) Shear strength profile (b) Initial stiffness profile

Figure 4.1 Ground parameters varying with depth

The stiffness of Made Ground was modelled as linear elastic. The stiffness of all remaining ground strata was modelled as non-linear, dependant on stress and strain including small strain stiffness approach using a Jardine type function (Jardine et al, 1986). The original work by Jardine et al defines a relationship between the undrained ground elastic modulus normalised by the undrained shear strength (E_u/c_u) and the axial strain. For purposes of numerical modelling this approach was modified by Mott MacDonald into a relationship between the shear modulus of the ground normalised by the mean effective stress (G/p') and the shear strain (Eadington and O'Brien, 2011). Also the stiffness parameters were evaluated based on laboratory test results, in situ testing and empirical relationships. Example profile of the initial stiffness parameter varying with depth plotted from the numerical model against its theoretical design line is shown in Figure 4.2 (b).

The in situ at rest lateral ground pressure coefficient was modelled as follows:

- K_0 = 0.5 for Made Ground and River Terrace Deposits.
- K_0 = 1.2 for London Clay and Lambeth Group Clay above MLGH, i.e. in the area most affected by the tunnelling works. This approach represents a simplification to modelling the K_0-profile variable with depth based on the ground investigation measurements and is more representative of the real ground behaviour with respect to simulation of SCL tunnelling works.
- K_0 = 1.0 for ground below MLGH.

The in situ pore water pressure was modelled as described in section 3.2 and shown in Figure 3.3.

Sprayed concrete lining model

The sprayed concrete lining was modelled as linear elastic with perfectly plastic hinges allowing limited redistribution of bending moments. The plastic hinges were allowed only in the "sharp corners" of the perimeter lining – temporary sidewall connection (red dots in Figure 4.2). The rest of the lining was designed for linear elastic stress. The limiting bending moment in the plastic hinges was determined based on the section capacity according to the axial force acting in the lining.

The development of the sprayed concrete strength and stiffness over time was taken into account. It was calculated according to the modified J2 curve in EN 14487-1 for the first 24 hours and according to the Chang and Stille (1993) equations afterwards. The elastic modulus of the sprayed concrete was reduced in order to account for creep effects so that $E_{c,eff}$ used in the analysis was equal to half of E_{cm} (Eurocode 2 notation).

The sprayed concrete primary lining is reinforced by steel fibres, mesh, reinforcement bars and lattice girders. The lattice girders were deemed to be required for profile control only and were not included in the lining structural capacity calculations. The mix sprayed against mesh or re-bar did not include steel fibres. The 75mm initial layer was reinforced only by steel fibres. The mesh and reinforcement bars were installed with 45mm cover by structural concrete to ensure long-term durability because the primary lining (except the 75mm initial layer) is designed as a permanent structure. All these aspects were considered in the lining design.

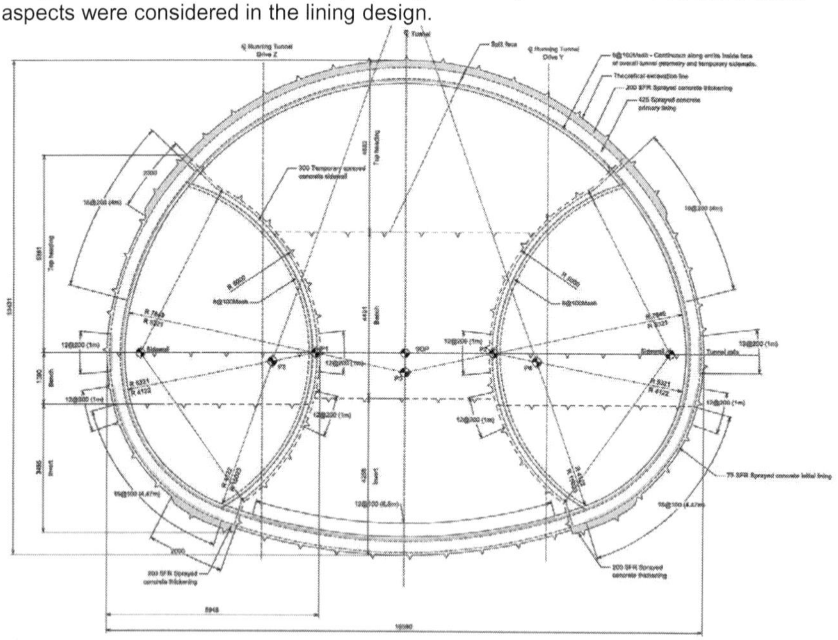

Figure 4.2 Location of plastic hinges

The structural capacity of the sprayed concrete sections was calculated according to Eurocode 2 and RILEM, which provided framework allowing consideration of the influence of the steel fibre reinforcement. A simplification to the RILEM approach was adopted considering only a rectangular tensile stress block in the steel fibre reinforced section, the tensile stress value being equal to the residual strength. The mean residual tensile strength arising from the steel fibre reinforcement was taken at 1.8 MPa at 28 days and 2.0 MPa at 90 days based on the requirements for beam test results in the sprayed concrete specifications.

Limit state design

The ground parameter values adopted in the numerical analyses were moderately conservative, i.e. characteristic according to Eurocode 7. The analysis and design was based on Eurocode 7, Design Approach 2, Note 1. The FLAC models were therefore based on nominal geometry, characteristic ground parameters and mean structural stiffness. The load factor was applied to the resulting lining stress and material factors were applied in the structural design checks.

Plane strain 2D analyses in FLAC

The Stepney Green caverns were analysed first in plane strain 2D sections using the finite difference program FLAC. Generally, 2D models are quicker to build and much quicker to run than 3D models. Also, the results of 2D modelling are easier to process in terms of structural design checks. However, there were some shortcomings of the 2D modelling, applicable especially in the complex case of the Stepney Green cavern excavation sequences.

2D models generally follow the stages of the excavation and lining sequence. These should be related to the ground relaxation process which in turn should correspond to the age and thus strength and stiffness of the SCL lining. C121 carried out a calibration study for the ground relaxation parameters based on the case studies of Heathrow Express, London Bridge (platform tunnels and concourse tunnel) and King's Cross tunnels. Unfortunately, none of these cases was similar enough to the Stepney Green caverns in terms of size, shape and excavation sequence.

Additionally, there was the issue of the appropriate modelling of the temporary side wall removal. In principle, the problem is similar to the 3D effects simulated by ground relaxation. The temporary sidewalls are initially under stress, typically including high axial force. When removed, their supporting effect disappears which should equal to applying point forces equivalent to the former side wall axial stress to the perimeter lining (Figure 4.3). The removal however is a step by step process carried out in short advances and thus the load on the perimeter lining increases gradually with the progress of the construction works.

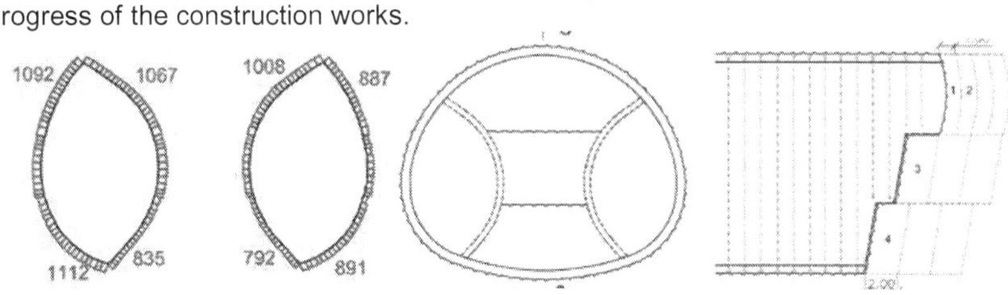

Figure 4.3 Sidewall removal impact schemes

Depending on the details of the modelling approach to the sidewall removal and the capabilities of the analytical software, the outcome of the analysis can vary to a significant extent. In the 2D FLAC models this phenomenon was simulated as follows:

- During excavation of the central core the sidewalls are left in the model supported by relaxation forces which are reduced in the same way as the ground relaxation forces. This leads to a certain drop in the axial force in the sidewall and thus to a certain drop in the support effects.

- The sidewalls are removed with the installation of the central invert and removal of the ground relaxation forces. If the sidewalls were removed from the model earlier the structural impact of the removal would be counteracted by the applied support forces of the simulated ground relaxation process.

3D analyses in FLAC

The FLAC 3D analyses were undertaken for the largest section of the single side wall drift sequence and the largest section for the double sidewall drift sequence extruded in the longitudinal direction. The 3D analyses were carried out for the following main reasons:

- To remove doubts about the representativeness of the 2D ground relaxation parameters and their relation to the SCL strength and stiffness.
- To test the proposed approaches to the temporary sidewall removal sequence and verify the stability of the partially excavated central core during the step-by-step sidewall removal process.
- To test the proposed approach of delayed reinforcement installation on the near face (Figure 5.1)
- To test the impact of TBM transit loads on the primary lining.

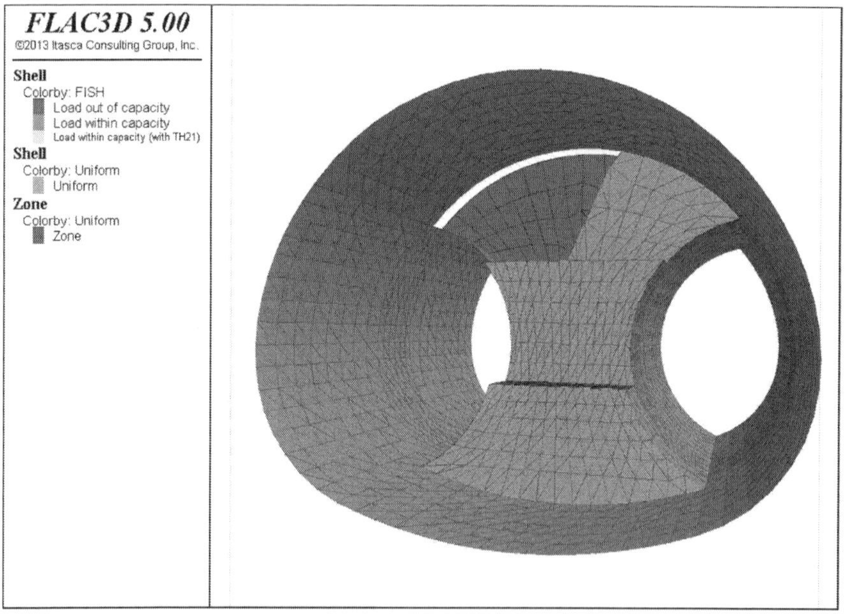

Figure 4.4 Excavation of the central core in the double side wall cavern model in FLAC 3D

3D models do not need to emulate the ground relaxation process. In the Stepney Green analyses the excavation and support sequence was modelled in full detail as illustrated in Figure 4.4. The strength and stiffness of the advances of the SCL shell were updated according to the estimated advance rates and cycle times.

Because the SCL tunnel is sprayed in separate "patches" it could be argued that the primary lining consists of rings separated by "cold" circumferential joints. This is because the patches are sprayed at different times, the steel fibres may not be fully able to bridge across these joints and there is only a minimum splice of the mesh, whereas in the longitudinal joints there are bars provide across (Figure 4.5). The argument was explored with respect to the stability of the partially excavated central core with the proposed step-by-step sidewall removal instead of the more traditional approach to remove the sidewalls after only after the central core excavation and ring closure. Specifically, it was argued in this respect that due to these joints the central core stability should not rely on the cantilever effect in the longitudinal direction. Although the SCL in reality acts as an isotropic shell, several simulations were run in the 3D models with various degrees of separation along the circumferential joints in order to rule out the potential for the central core instability.

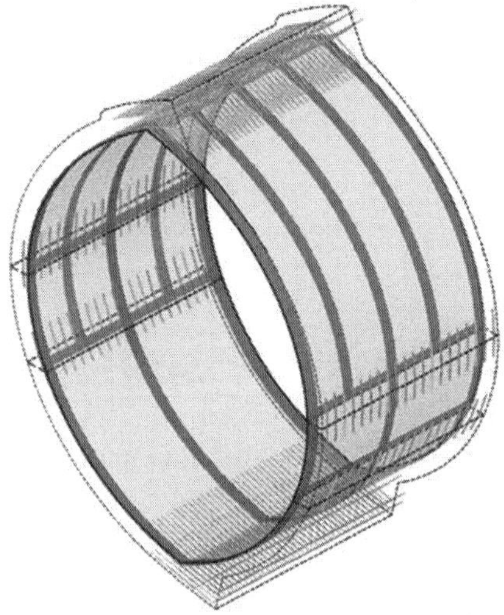

Figure 4.5 Longitudinal and circumferential joints in the SCL shell

Results of numerical analyses compared with monitoring

Structural stability considerations

The numerical analyses have shown that the predicted stress in the primary lining was within the SCL section capacity. The 3D models verified the feasibility of the proposed step-by-step sidewall removal during the excavation of the central core and supported the argument that the partially excavated central core will remain stable during the process. Both concepts proved to be working well during the construction of the caverns. The analytical work therefore substantiated a design leading to easier, faster and safer SCL construction works. Additionally, the numerical modelling results verified the feasibility of the proposed approach of delayed reinforcement installation on the near face (Figure 5.1) which allowed installation of the reinforcement and spraying of the last pass in complete wide rings. The approach reduced the quantity of joints and improved the constructability of the permanent primary lining at high quality. Finally, the 3D modelling allowed assessment of the impact of the TBM transit load on the primary lining supporting the transit frames.

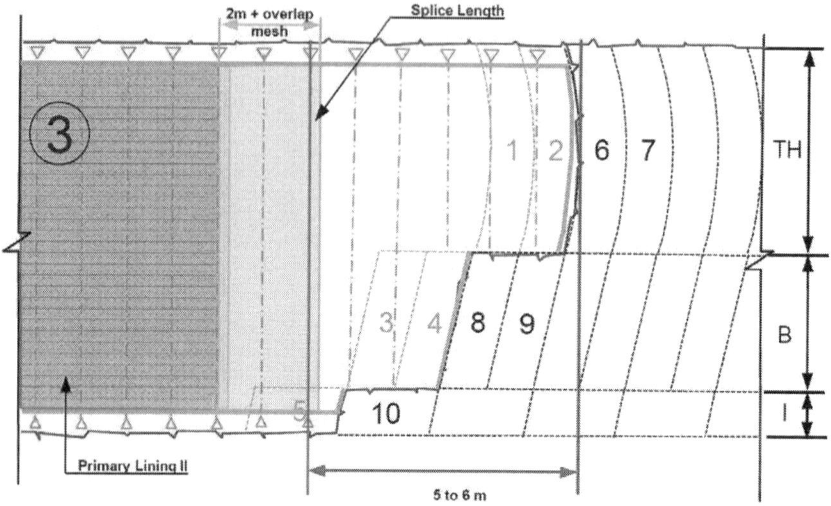

Figure 5.1 Concept of delayed mesh installation on the near face

Ground and lining displacement considerations

From the numerical models, ground movements and volume losses were predicted. At the time of writing this paper there was not enough monitoring information available to the authors to make a valid comparison, although substantial amount of data was available from lining displacement measurements.

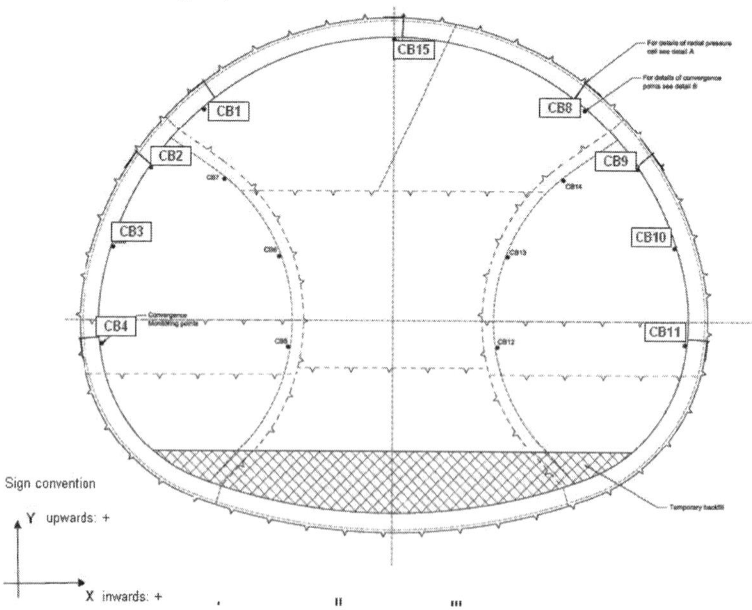

Figure 5.2 Lining displacement monitoring points in double sidewall drift cavern

Tunnel linings are traditionally designed based on stress (to be within section capacity) but lining performance during construction is traditionally monitored based on displacements by means of convergence measurements assessed against trigger levels. Typical arrangement of convergence monitoring points in the double sidewall drift cavern is shown in Figure 5.2. The trigger level philosophy was to base the movement triggers on expected lining displacements which were estimated by means of the numerical analyses. The green trigger was set at 75% of the FLAC predicted displacement considering that approximately 25% of the total lining displacement takes place before the initial reading of the monitoring point. The amber and red triggers were scaled off the green one. During the construction phase the measured displacements mostly occurred within the green zone and converged close to the amber boundary with the amber limit being exceeded in a very limited number of occasions. Therefore, the triggers served their purpose of a conservative benchmark for checking the expected lining behaviour.

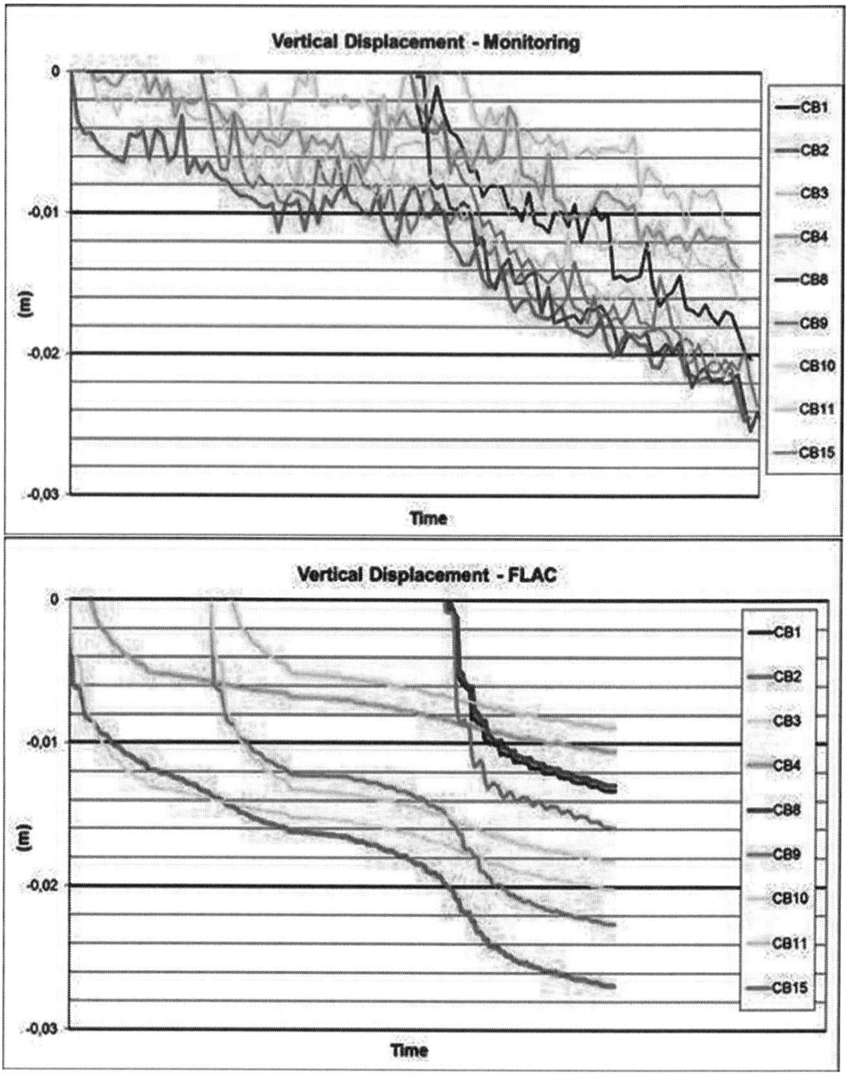

Figure 5.3 Comparison of predicted and measured vertical lining displacement

Figure 5.3 presents a comparison of the development of the vertical displacement of the convergence points (Figure 5.2) with time predicted by 3D analysis and measured by monitoring at the Stepney Green construction site. Figure 5.4 shows the same for horizontal displacements. Figure 5.5 provides the comparison for selected monitoring point CB10 movement in vertical and horizontal co-ordinates. The graphs show that there is a relatively good match in terms of the vertical displacements. There is a similar trend and order of magnitude, some points in reality move slightly more than in the numerical model. The horizontal displacements however seem to be over predicted by the numerical model with respect to measurements. It might indicate less bending load at the sides of the tunnel and perhaps could be adjusted by modified K_0 value or anisotropic stiffness but that is only a speculation at this point.

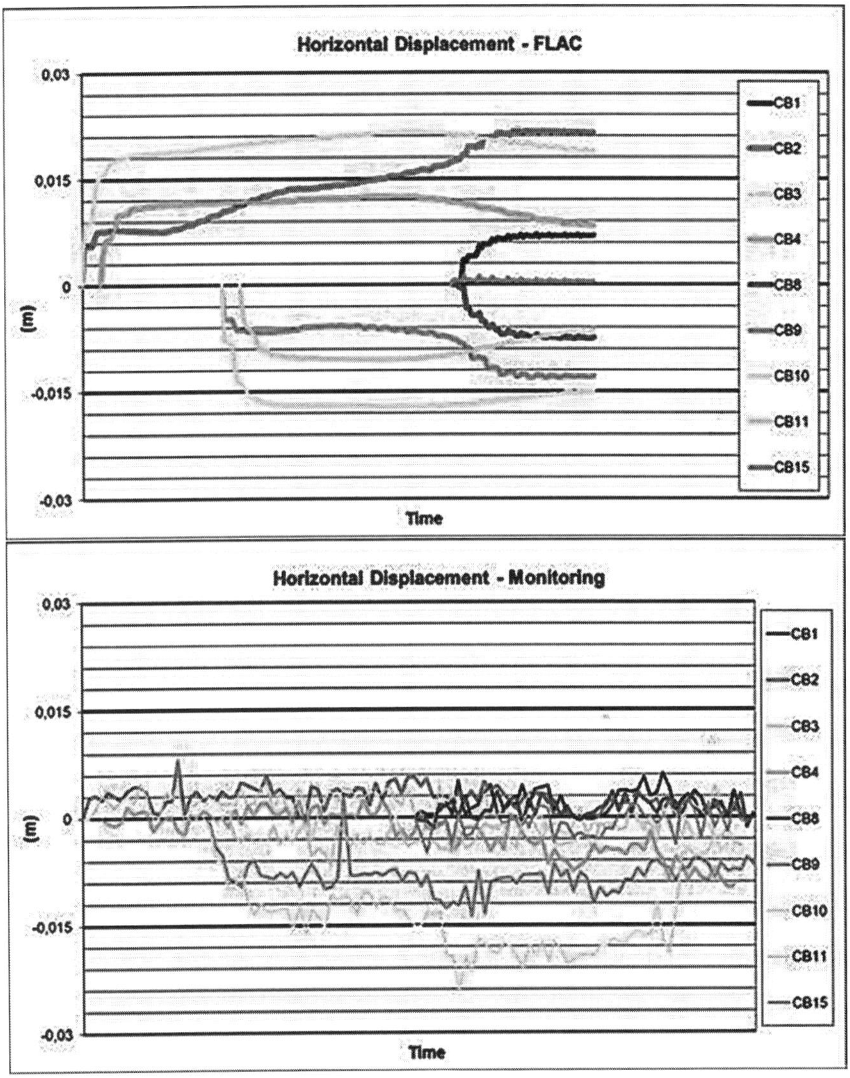

Figure 5.4 Comparison of predicted and measured horizontal lining displacement

In general, the lining movements indicate that after the removal of the temporary side walls the points at the crown move downwards whereas the points at the sides move slightly outwards horizontally, i.e. slightly into the ground. That demonstrates the activation of the lining arch under the ground load after the side walls removal and indicates that the lining is actually under significant stress.

The monitoring measurements so far do not show much difference between the lining displacement of the eastbound and the westbound cavern. That could indicate that the potential partial consolidation during the excavation and primary lining discussed with respect to the westbound cavern either did not happen or did not have a clear measurable impact or was prevented by the dewatering system.

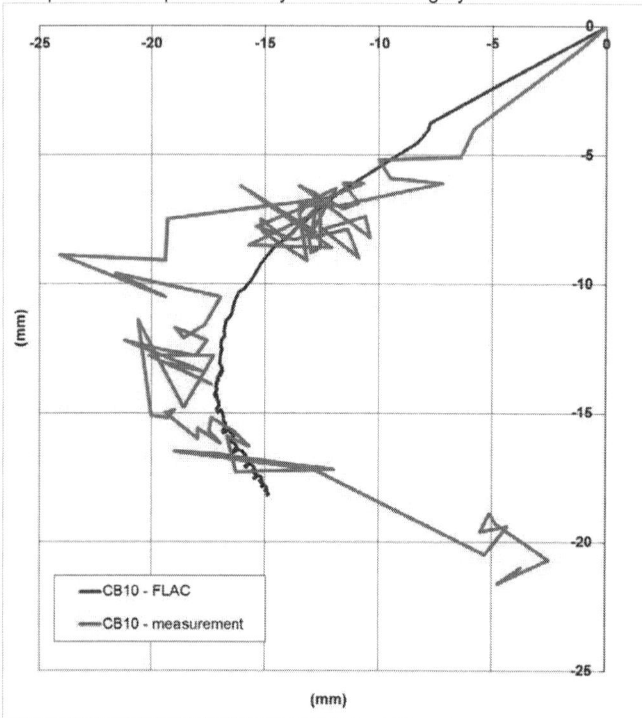

Figure 5.5 Comparison of movement of monitoring point CB10

Conclusions

The Stepney Green caverns represent major SCL works of a large span and complex excavation and support sequence. Sophisticated numerical analyses were undertaken to support the primary lining design, check feasibility of various proposals and sensitivity to potential risks.

From the beginning of the design it was discussed whether the eastbound or the westbound cavern was the more critical in terms of design for the primary lining stress. The eastbound cavern is by 4.5m shallower and excavated predominantly within the London Clay whereas the bottom third of the westbound cavern is excavated within Lambeth Group Formation containing Channel Sands. Depressurisation scheme was designed to address the excavation in the granular strata and proved working during the construction. Potential impact of partial consolidation was discussed with respect to the westbound cavern. The monitoring measurements so far do not indicate a significant difference between the eastbound and westbound lining behaviour. The concerns about the potential partial consolidation during the short-term were acted upon during the design phase using a fully coupled fluid-mechanical numerical analysis however, it was found that such analysis was too complex and too sensitive to inputs to provide reliable predictions. The most important issue was the determination of appropriate effective stress strength parameters because of their major influence on the outcome of the short-term consolidation analysis. Such parameters should include dilation effects.

Traditionally, the temporary sidewalls are removed after full ring closure and full excavation of the central gallery. In such a case the excavation of the central gallery takes place in the confined space between the sidewalls. The numerical modelling work supported the feasibility of an alternative approach with step-by-step removal of the temporary sidewalls and ruled out concerns about instability of the partially excavated central core and closure of the tunnel before installation of invert. That allowed a more efficient and safer construction. The feasibility of the delayed installation of reinforcement on the near face was also verified by numerical analysis. The proposed approach allowed improved constructability and quality of the permanent SCL primary lining. Also the loads from the planned TBM transit on the primary lining were assessed by numerical modelling. Overall, the 3D models served as good means of testing various design concepts in terms of ground-lining interaction and structural impact on the primary lining.

2D numerical models were also carried out but considered less representative of the complex excavation and lining sequence due to lack of calibration against comparable case studies. The key issue in addition to calibration of ground relaxation parameters was found to be the modelling of the effects of removal of temporary structures such as side walls.

Trigger levels for lining convergence monitoring were defined based on displacements predicted by the numerical analyses. The triggers gave a reasonably good agreement between the analytical prediction and the monitored lining behaviour. The numerical models slightly underestimated the vertical movement of some points but generally overestimated the horizontal movement of all points. The lining movement after sidewall removal indicates the activation of the lining arch under the ground load. The crown points move downwards whereas the side points move into the ground.

References

Chang Y and Stille H (1993) Influence of early age properties of shotcrete on tunnel construction sequences. In proc. *Shotcrete for Underground Support VI* (eds. Wood & Morgan), Niagara-on-the-lake, Canada, 110-117. USA: ASCE.

Eadington J and O'Brien AS (2011) Stiffness parameters for a deep tunnel – Developing a robust parameter selection framework. In proc. *15th European Conference on Soil Mechanics and Geotechnical Engineering* (eds. A. Anagnostopulost et al.), Athens, Greece, 531-536.

Jardine RJ, Potts DM, Fourie AB and Burland JB (1986) Studies of the influence of non-linear stress-strain characteristics in soil-structure interaction. *Géotechnique* **36(3)**: 377-396.

RILEM TC 162-TDF (2003) Test and design methods for steel fibre reinforced concrete. *Materials and Structures* **36**: 560-567.

Standards:

BS EN 1992-1-1 - Eurocode 2: Design of concrete structures – Part 1-1: General rules and rules for buildings.

BS EN 1997-1 - Eurocode 7: Geotechnical design - Part 1: General rules.

BS EN 14487-1 – Sprayed concrete – Part 1: Definitions, specifications and conformity

Crossrail Project: Infrastructure design and construction
ISBN 978-0-7277-6078-4

doi: 10.1680/cpid.60784.385

Design and Construction of Crossrail Stepney Green Spayed Concrete Lined Caverns

Nicholas Tucker Eur Ing, BSc, CEng MICE, Mott MacDonald Ltd.
Katherine Scevity BEng, MCQI CQP, Bechtel
John Comins BEng, HND, ACSM, Dragados-Sisk JV
Emilio Linde BSc, MSc, DIC, EurGeol, FGS, OTB Engineering Ltd.

Abstract
At over 50 m long, 17 m wide and 14 m high the Stepney Green Caverns are the largest caverns ever built using Sprayed Concrete Lining techniques in Central London. This paper describes the design, construction and monitoring of the Primary Sprayed Concrete Lining together with lessons learnt that can be applied to other Crossrail structures.

The lining system consists of a fibre reinforced Primary Lining, a waterproofing membrane and a fibre reinforced Secondary Lining. A combination of spray applied and sheet membrane waterproofing was utilised. The permanent Primary Lining was designed to take the full short-term applied ground load and surface surcharges. Additional long-term loads, subsequent to the installation of the Secondary Lining are shared between the two linings.

The excavation and support methodology varies along the length of the cavern. The launch chamber was constructed using top-heading, bench & invert, the smaller section of cavern using single side-wall drift and the largest part of the cavern using double side-wall drifts with a central gallery.

The cavern profile at each advance was controlled using target-less remote laser scanning equipment, set up at a safe distance from the excavation face. The performance of the lining was constantly monitored using deformation monitoring points, pressure cells, inclinometers, extensometers and piezometers and reviewed daily.

Project Description
At over 50 m long, 17 m wide and 14 m high the Stepney Green Caverns are the largest caverns ever built using Sprayed Concrete Lining techniques in Central London.

The Stepney Green Caverns were excavated under the East End of London at Stepney Green to create the underground rail track junctions on the city's Crossrail project that will permit train services to run either north towards Stratford or south towards Canary Wharf. See Figure 1.

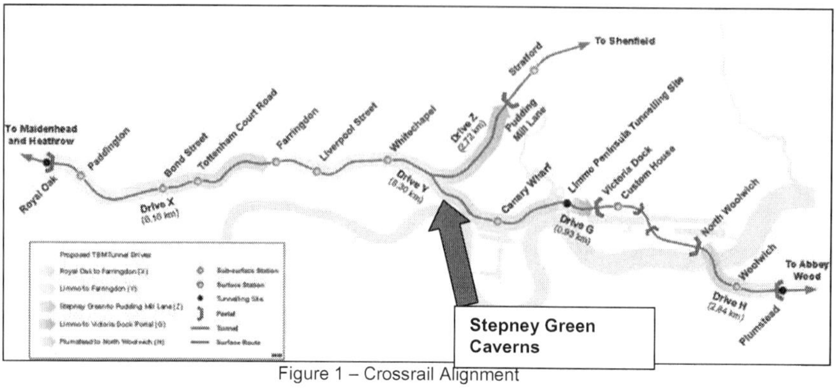

Figure 1 – Crossrail Alignment

The vertical alignments of the caverns are different; with the westbound alignment being approximately 5m deeper. This meant that the eastbound cavern was wholly excavated within London Clay whilst the westbound cavern encountered up to 4m of Lambeth Group below axis level.

Excavation of the caverns progressed from a rectangular access shaft, built across the twin TBM running tunnels to the west of the junctions that will house M&E utility rooms, a ventilation facility and provide an emergency ingress and egress for Crossrail during operation. The SCL cavern design was carried out by Mott MacDonald Ltd. supported by Gall Zeildler Consultants. Temporary works were designed by OTB on the contractor's behalf. The caverns form part of the Eastern Running Tunnels Contract C305 that was awarded to the Dragados/Sisk JV in December 2010 for approximately £500 million.

The contractor took possession of the site in March 2010 to set up and start excavation of the large access and operations shaft. The shaft construction used diaphragm slurry walls of up to 40m deep, 15m wide and 60m long with bar reinforcement cages and over 9,500m^3 of concrete. Shaft construction was completed in January 2013 and SCL excavation commenced on the EB launch tunnel in November 2012 and the EB cavern in January 2013. The EB cavern Primary Lining was completed in May 2013, followed by the westbound in August 2013. To construct the caverns and launch tunnels, the project team had to excavate approximately 38,000m^3 (bulked) of material and apply over 6,000m^3 sprayed concrete to the temporary and permanent walls.

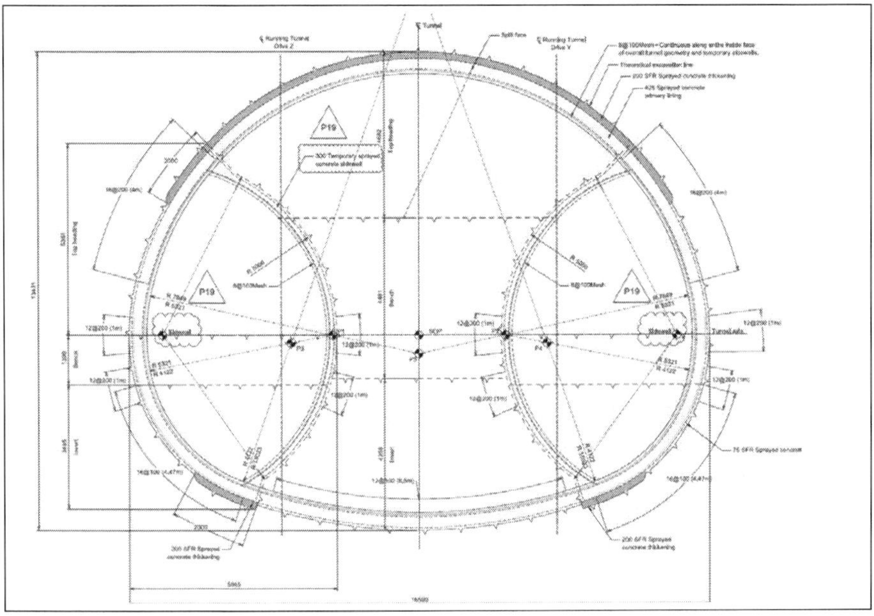

Figure 2 – Stepney Green Cavern – Primary Lining for Largest Section

Risk Management

The construction of the Stepney Green Caverns was one of the highest risks on the CRL programme. The size and location of the caverns are unique; caverns of this size had never been built in the same geological conditions in London. The risks associated with the construction of the caverns needed to be carefully managed in order to meet the Client's and other legal obligations and followed the principles of the Joint Code of Practice for Risk Management of Tunnel Works.

The period of time when the face or crown is left unsupported was when the high risk events were most likely to occur. In accordance with the requirements of CDM legislation, the Designer identified risks that could not be mitigated during design and passed them on to the construction phase.

The main risks included;

- Instability and/or collapse of the tunnel through water ingress due to presence of sand channels.
- Local face instability with block falls or collapse due to highly fissured and slicken-sided zones of clay.
- Potential for isolated pore water pressure in granular layers causing instability of excavation face and inundation.
- Unknown hydrology causing inundation of the tunnel causing drowning or injury.
- Inadequate depressurisation of non-cohesive strata leading to potential tunnel excavation instability, tunnel collapse and impact on third parties and the general public.

In order to manage these risks effectively a suite of documents, review meetings and other tools were utilised.

The SCL Action Plan was the main risk management document used by the Contractor and contained details on how to mitigate risks. The action plan described how the ground and tunnel movements, water inflows and other unexpected situations would be controlled during the works. Specific Action Plans were also developed for high risk areas of the works, e.g. for the Westbound Cavern – Groundwater and Depressurisation.

Due to the high risk nature of these works, key personnel were required to meet strict competency requirements and approved by the Client's Project Manager. To ensure all parties were aware of the works, their responsibilities and the impact of other interfaces a suite of meetings were put in place. Namely the Shift Review Group (SRG), the Contract Technical Committee (CTC) and the Engineering Review Panel (ERP).

The SRG was held daily. Representatives from the Client, the Contractor, the Designer and the Senior SCL Engineer were required to attend the meeting. The role of the group was to review the last 24 hours of work, discuss any observations (health & safety, quality and technical), review the monitoring data, face logs, probe holes etc. Having reviewed all the data, the works for the next 24 hours were then planned and recorded on the Required Excavation and Support Sheet (RESS), acting as the permit to work system.

The CTC met bi-weekly to review the effectiveness of SRG process. Senior representatives from the Client, Contractor, Designer and the Senior SCL Engineer met to review the summary of the previous week's progress. The meeting was also used to discuss technical issues such as excavation methodology and the effectiveness of the depressurisation system. The CTC helped manage high risk operations by highlighting areas which could be improved upon.

The ERP met every four weeks and included members from the CRL Central Engineering Group. The ERP played an overseeing role during the works. They reviewed the output of the SRG and the CTC, gathering and disseminating any lessons learnt. High risk activities undertaken on other parts of Crossrail were discussed at the ERP and any lessons learnt incorporated in to the cavern works.

The CRL Expert Panel provided valuable peer review, allowing the overall team to be questioned on the robustness of the systems in place for controlling the works. They visited the site on a number of occasions, in advance on key risk activities commencing, for which the Contractor and the CRL site team and Designer prepared presentations for discussion.

Primary Lining Analysis & Design

SCL Lining System - The lining system developed for the SCL tunnels at Stepney Green comprised a steel fibre reinforced Primary Lining, a waterproofing membrane and a non-steel fibre reinforced Secondary Lining with bar-reinforcement. The lining is a double-shell lining with both linings considered part of the permanent load bearing structure. A combination of spray applied waterproofing and sheet membrane was designed for SCL structures for Stepney Green.

Material Durability - Key performance criteria that determined the durability of the sprayed concrete material are as follows:

- Concrete mix design and reinforcement cover to meet the exposure class according to BS8500 to achieve 120-year design life.
- Water to cement ratio targeted between the 0.4 and 0.45 range to create low permeability.
- Active pozzolans, namely silica fume slurry, to further decrease the permeability to less than $1x10^{-12}$ m/s enhancing chemical stability of the concrete.
- Structural steel fibres to promote homogenous distribution of micro-cracks.
- Micro-synthetic fibres in the Secondary Lining to provide concrete anti-spalling mechanism in the event of a fire.
- Alkali-free set accelerators at dosages less than 8% by weight of total cementitious materials to promote early and long term compressive strengths.

Performance Requirements – Key performance requirements for the Primary and Secondary SCL are given in Figure 3;

Performance Requirement	**Primary SCL**	**Secondary SCL**
Design life	120 years	120 years
Early age strength development (minimum)	Greater than modified J2 requirements shown in Figure 4.	Greater than J2 requirements shown in Figure 4.
Long term compressive strength	28 days>C28/34 90 days>C32/40	28 days>C28/34 90 days>C32/40
Structural fibre performance	28 days: Class D1 S1.8 ; D3S1.4 90 days: Class D1 S2; D3S1.4 (After BS EN 14487-1)	Residual flexural strengths as for Primary SCL
Concrete shrinkage	Less than 0.03% at 28 days	Less than 0.03% at 28 days
Exposure class	DC-2 as per Table A.11 of BS 8500-1	XC3 as per Table A.5 of BS 8500-1
Water permeability (chemical durability)	< 35mm penetration – See clause c) in sprayed concrete trial section of KT20.2201	Not applicable
Cementitious content	> 380kg/m³ Type: CEM II A-D, CII A-D	> 380kg/m³ Type: CEM I, CEM II A-D, CII A-D
Water-cement ratio	< 0.45	< 0.5
Fire protection	Not required	Fire protection requirements to EUREKA time-temperature curve

Figure 3 – SCL performance Requirements

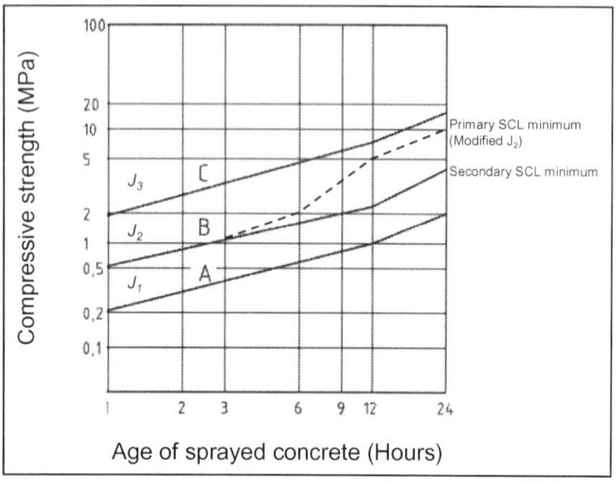

Figure 4: Early Age Strength development for sprayed concrete - modified (based on EN 14487-1:2005)

Primary lining – the Primary Lining was designed to take the full short-term applied ground load and any other loads, such as from compensation grouting and surface surcharges, expected in the two to three years prior to Secondary Lining installation. Additional long-term loads, such as those applied through on-going consolidation of the London Clay, subsequent to the installation of the Secondary lining were shared between the two linings.

The Primary Lining comprised two layers; P1 steel fibre reinforced sprayed concrete and P2 bar reinforced sprayed concrete. Steel fibres were provided in order to increase the ductility of the concrete and provide toughness and post crack resistance in the long term. A minimum of 40kg/m^3 of fibres were specified to reduce crack widths in sprayed concrete to 0.3mm. In addition bar reinforcement was provided to accommodate calculated tensile forces. Lattice girders were provided to ensure shape control and additional robustness to the design although not actually required by structural calculation.

A 75mm thick initial concrete layer, sprayed directly against the ground, was provided. This was included as load bearing in the short term but was considered sacrificial and not considered in load capacity calculations post installation of the Secondary Lining.

Secondary Lining - Taking into account the loads and stresses already taken by the Primary Lining, the Secondary Lining was designed to carry:

- The full long term water pressure,
- Internal loads, such as mechanical and electrical equipment,
- Long term load conditions; e.g. the effects of consolidation,
- The effects of temperature and shrinkage,
- Fire loads,
- The effects of degradation of the 75mm sacrificial layer,
- An allowance for additional loads that might be applied in the future.

The proportion of consolidation loading applied to the Secondary lining was calculated using FLAC numerical modelling. The Secondary Lining was reinforced with conventional bar reinforcement.

Secondary linings were designed to carry sufficient residual capacity to resist ground loading after a EUREKA time/temperature fire curve, as defined in the Technical Specification for

Interoperability – Safety in Railway Tunnels TSI-SRT. The EUREKA curve was developed for the rail industry in Germany. The Secondary lining concrete (cast in-situ or sprayed) contains micro-synthetic fibres in order to limit explosive spalling and maintain structural integrity. The quantity of fibres was determined by pre-construction fire testing.

Waterproofing Systems - Waterproofing membranes were installed between the Primary and Secondary linings. The waterproofing system provided is "un-drained"; i.e. the groundwater is contained by the waterproofing membrane and not managed by an internal tunnel drainage system, the Secondary lining carrying the full hydrostatic loads.

Spray applied waterproofing membranes offer benefits by bonding to both the Primary and Secondary linings. This property offers maintenance and repair benefits in the long term by preventing the movement of water, either behind or, should it be breached, in front of the membrane. Sprayed membranes are not suitable to be applied directly to active water ingress and if this is the case sheet membranes are recommended. The Stepney Green caverns were constructed in both London Clay (sprayed membrane utilised) and Harwich Formation and Lambeth Group (sheet waterproofing membrane utilised).

Due to the presence of structural fibre reinforcement in the Primary concrete linings, a regulating layer (fine aggregate non-fibre reinforced concrete) is applied to the substrate prior to application of the waterproofing layer.

Structural Parameters - The parameters used in the design of the SCL tunnels are presented in Figure 5 below:

Sprayed & Cast In-Situ Concrete	Value
Characteristic compression cylindrical strength of sprayed concrete	32 MPa
Mean Elastic Modulus	33.9 GPa
Coefficient of creep reduction	2
Creep reduced (effective) modulus	16.95 GPa
Steel Bar Reinforcement Parameters	**Value**
Yield Stress	500 MPa
Elastic Modulus	205 GPa

Figure 5: Structural Parameters

In order to model the beneficial effect of steel fibres in concrete sections under bending, the RILEM TC 162–TDF2 guidelines were used to model the behaviour of concrete under tension. The behaviour of concrete under compression followed Eurocode 2. This resulted in the stress strain diagram shown in Figure 6.

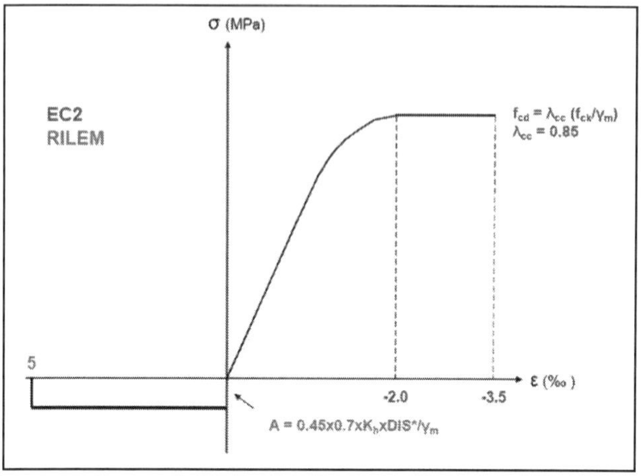

Figure 6 - Stress Strain diagram for SFRS

Where, following RILEM TC162:

- A is flexural tensile stress for ε = 0.0 - 0.5%
- A = 0.45x0.7xKhxD1S*/γm
- Kh (shape factor) = 1.0-0.6x(h-0.125)/0.475
- D1S* is 2.0 MPa at 90 days and 1.8 MPa at 28 days

The following points should be noted:

- The compressive side of the stress strain curve follows Eurocode 2 (Section 3.1.7, Figure 3.3);
- The residual strengths in tension are calculated using a simplification of the RILEM model.
- The residual strength in tension was set as a constant value up to a strain of 0.5%;
- The partial material factor of safety, γm is 1.5 for concrete and fibre-reinforced concrete and 1.15 for steel bars, in accordance with Eurocode 2.

Structural Analysis – The following numerical analyses were undertaken using both FLAC 2-D and 3-D:

- Calibration of 2-D numerical analysis to establish ground models and relaxation parameters.
- Preliminary analyses to confirm the lining shapes, thicknesses and proposed excavation sequences were structurally adequate for the Primary SCL linings.
- Modelling of ground-structure interaction using the finite difference package FLAC was carried out.

A modelling simplification inherent in 2-D modelling is the relaxation of the ground ahead of an excavation face, a largely 3-D effect, and the values used in the 2-D modelling were calibrated against 3-D models and previously constructed tunnels in London Clay, at Heathrow Airport and Kings Cross Station. Time dependent development of sprayed concrete strength and stiffness were included in the models.

In line with London Underground (LU) requirements a surcharge loading of 75kPa was applied to both the Primary and Secondary Lining cases. The train loadings were included in accordance with LU Engineering Standard 1-052 – Civil Engineering – Bridge and Railway Group Standard GC/RT5112.

Excavation Modelling – A FLAC a 3-D extruded model was used to analyse the Stepney Green Caverns. A typical excavation sequence is described below;

The narrower section of the turnout caverns utilised a single sidewall drift construction methodology. The advance lengths were 1 metre for top heading and bench and 2 metres for invert. The following construction sequence was assumed in the analysis;

Sequence Number	**Construction Sequence (assumed in design)**
01	Construct whole length of left hand side
02	Construct whole length of right hand side

Part of lining	**Activity**	**Advance length (m)**	**Duration (hours)**
(1) Left hand side			
TH	Excavation, scanning, mapping	1.0	4.0
	Primary lining		4.0
BCH	Excavation, scanning, mapping	1.0	2.0
	Primary lining		4.0
INV	Excavation, scanning, mapping	2.0	4.0
	Primary lining		6.0
(2) Right hand side			
TH	Excavation	1.0	4.0
	Split 1 initial lining 200mm		2.0
	Sidewall removal		4.0
	Primary lining incl. joint formation		6.0
BCH	Excavation incl. removal of sidewalls	1.0	6.0
INV	Excavation incl. removal of sidewalls	2.0	8.0
	Primary lining incl. joint formation		8.0

Advance rates / cycle times

The excavation sequence is shown diagrammatically in Figure 7 taken from the 3-D model output;

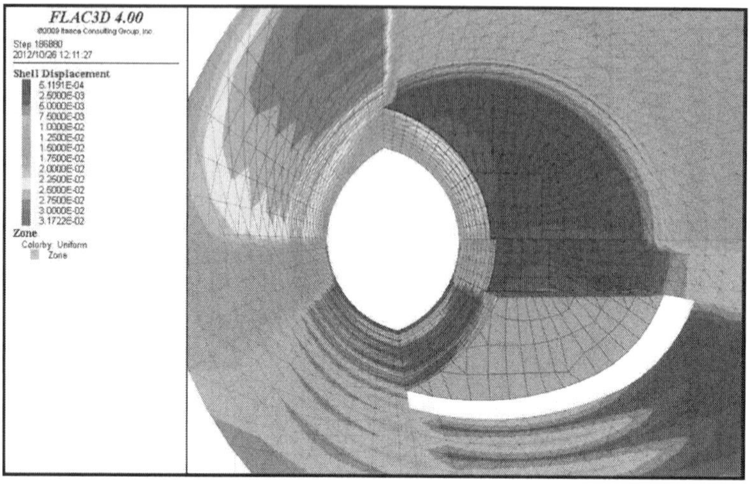

Figure 7 – Excavation Sequence for Single Sidewall Drift – showing Shell Displacement

The calculated volume losses were in the order of 1.0–1.2 %. The maximum predicted displacement for this section was 32mm.

The results of the analysis for long-term condition for the Primary Lining bending moments in the hoop direction are shown in Figure 8.

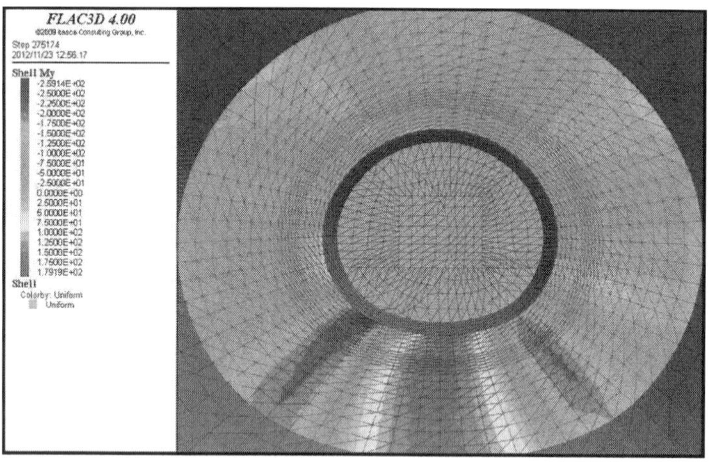

Figure 8 – Bending Moments in Long-term Condition

The red/orange colour in the invert shows tension on the inside face. The green/blue colour shows tension on outside face, with the largest moment typically being at the knees.

The wider section of the turnout caverns utilised a double sidewall drift with central gallery construction methodology. The advance lengths were 1 metre for top heading and bench and

2 metres for invert. The following calculation steps for the central gallery were assumed in the analysis;

Calculation Step	Scheme	Description
Stage 1	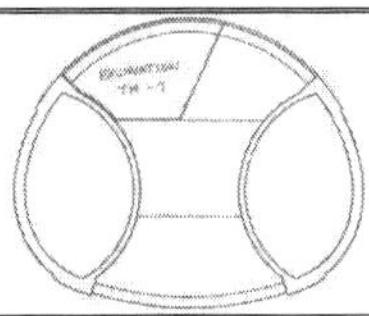	Top Heading – Split 1 – Excavation
Stage 2	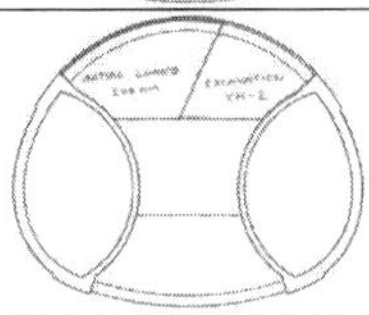	Top Heading – Split 1 – Initial lining 200mm Top Heading – Split 2 – Excavation
Stage 3	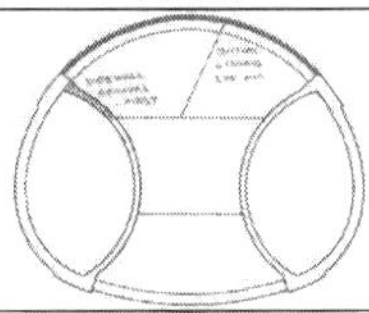	Top Heading – Split 1 – Sidewall removal Top Heading – Split 2 – Initial lining 200mm
Stage 4	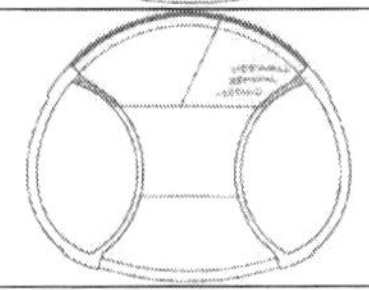	Top Heading – Split 2 – Sidewall removal
Stage 5	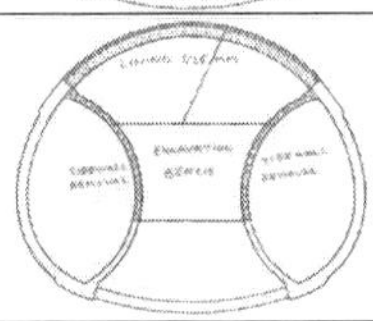	Top Heading – Primary lining 625mm Bench – Excavation Bench – Sidewall removal
Stage 6 until Stage 10	–	As Stage 1 until Stage 5 (second advance)
Stage 11	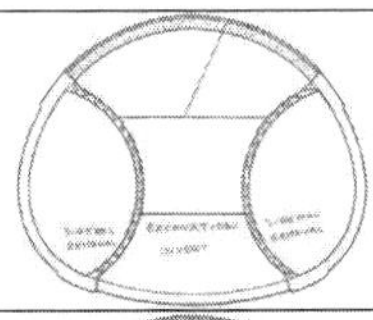	Invert – Excavation and Dewatering
Stage 12	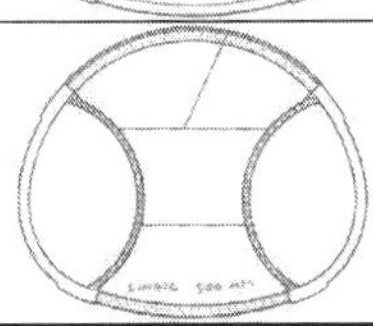	Invert – Primary lining 500 mm and again Stage 1 (third advance)

The calculated volume losses were in the order of 1.3–1.4 % for the sidewall drifts and 1.0 for the central gallery. The maximum predicted vertical displacement at cavern crown was 60mm.

Construction of Cavern Primary Linings

Ground Stabilisation and Depressurisation

As identified during the design stage, the main ground hazard during the construction of the SCL works at Stepney Green was the presence of high pore water pressures in the sandy layers of the upper part of the Lambeth Group. In the Eastbound Cavern, which was fully excavated in the stiff clays of the London Clay and Upper Mottled Beds, this pressure could lead to heave of the base and consequently rupture the impermeable clay layer during the invert stage. In the Westbound Cavern, where sand layers were encountered at bench and invert stages, the pore pressures could cause instability of face and invert due to running liquefied sands.

Based on the results of pumping tests, a surface ejector scheme was developed to dewater the upper part of the Lambeth Group. The strategy was to encapsulate the tunnels within a ring of ejectors. A spacing of approximately 8 metres was initially set by WJ Groundwater during their design based on previous pumping test data and the constraints imposed by surface access. These constraints required the use of inclined ejectors to achieve the drawdown outside the site boundaries. The dewatering layout is shown on Figure 9.

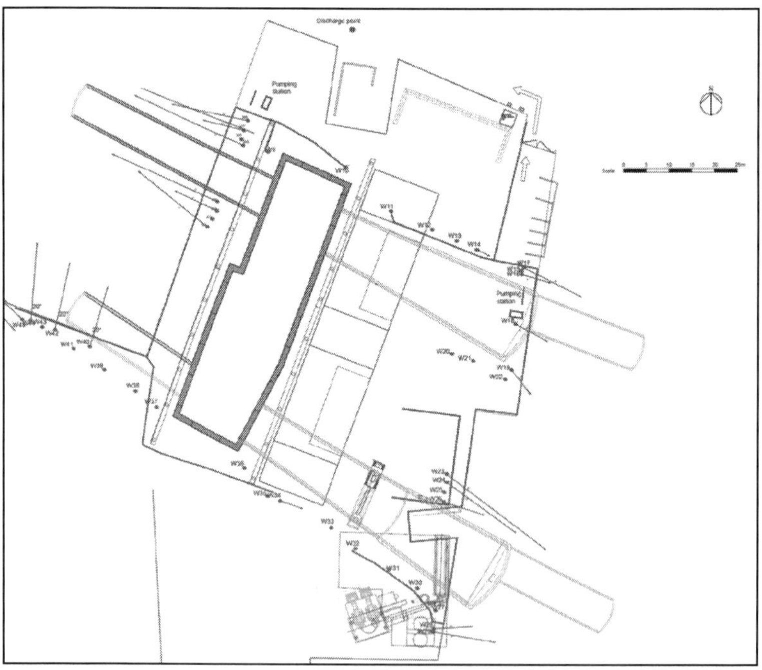

Figure 9 – Layout of Dewatering Scheme

The drawdown achieved by this surface based system was approximately 12m (from 90m to 78m ATD). As the deepest invert excavation level for the Eastbound and Westbound was 79m and 75m ATD respectively it was concluded that an additional in-tunnel dewatering system would be required for the Westbound Cavern excavation.

The nature of the layers to be dewatered (being randomly distributed sand channels with changes in thickness, length and frequency) meant that an intense probing regime was needed to detect these formations. A preliminary design for the in-tunnel dewatering was developed as a starting point, with wells drilled every 4 metres at bench level after the excavation and completion of the primary lining in the invert. The depressurisation worked as a temporary "moving front" of depressurisation with the tunnel wellpoints installed as the excavation progressed. The performance of the tunnel wellpoints was monitored using un-pumped wellpoints as stand pipe piezometers. This allowed the monitoring location and pumped wellpoints to be altered to suite the progress of the works. In addition dedicated piezometers, comprising shallow wellpoints, were installed to just 1 metre below the invert. See Figure 10

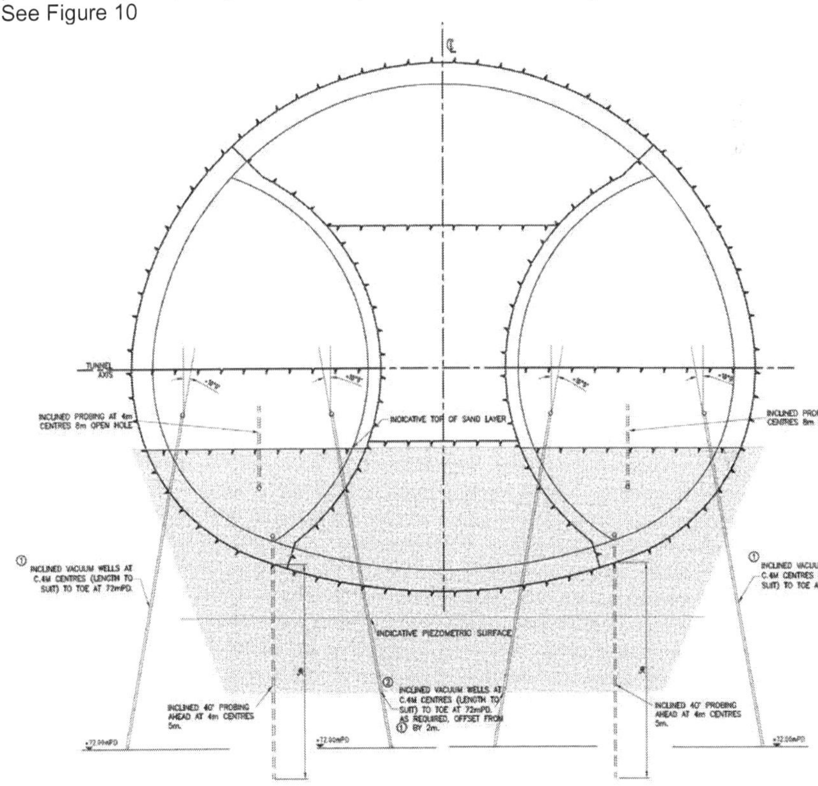

Figure 10 – Layout of Dewatering Wellpoints

Due to the inconsistent distribution of the sand layers, the initial wellpoint layout and frequency were altered following review of information obtained from the probing, face logs and piezometer monitoring. The selection of screen and filters for the wells was modified to accommodate the changes in permeability of the sand lenses with due consideration to the groundwater abstraction flow rate.

The additional drawdown obtained by the surface wells was sufficient for the excavation of most of the tunnels. However, it was not always possible to achieve a full drawdown at the interface boundary between a sand horizon and underlying clay stratum. Flowing water was sporadically encountered leading to challenging construction conditions when invert excavations were "open". A permit to dig system was successfully implemented which ensured certain key checks had to be undertaken before excavation of the invert was permitted. The permit was signed by all of the interested parties and the Senior SCL Engineer indicated the contingency measures that had to be employed depending on the water levels and geology.

Excavation Sequence and Ground Support

Three gangs were mobilised in November 2012 to work the 24/7 shift pattern of 7 days on - 3 days off dayshift and a 7-4 nightshift. At peak times, during concurrent construction of both EB and WB Caverns, over 80 staff and operatives worked at site on each shift.

The EB launch tunnel, situated wholly within the relative "comfort" of the London Clay, was excavated first to familiarise the crews and create more working space at the pit bottom for manoeuvring the large SCL plant before breaking out through the diaphragm wall for the EB cavern. The primary excavation and spraying plant were proved during this juncture and bespoke on-the-job training was given to the crews by the suppliers and manufacturers of the plant.

The invert and crown joint connections for the lattice girders and steel reinforcing bars on the first side drift to the adjacent enlargement proved to be particularly challenging to construct. Despite best efforts during preparation of the joints some remedial works involving the breaking back of previously sprayed concrete proved to be necessary. The acute angle of each joint in section profile proved difficult to excavate and spray accurately in the close confines of the side drift. Lessons learnt from the first side drift in the EB tunnel led to a series of improvements in joint preparation that were quickly implemented in the subsequent double side drift section and the WB Cavern.

The following plant was used to construct the SCL caverns:

- Primary excavation of all tunnel elements by a pair of Liebherr 924 compact excavators c/w various hydraulic tools including Wimmer drilling units, crusher pincers, road-header, breaker, claw & various excavating buckets;
- Pair of Liebherr 928 loading shovels for removal of spoil from face into skips at pit bottom;
- Five Utranazz 5.5m3 car-mixers for conveying fresh concrete from site batch plant to column skip on surface and from pit bottom to the face;
- Primary concrete spraying by three Meyco Potenza spraying mobiles (one for maintenance as spare);
- Two Meyco Suprema shotcrete pumps and two Meyco Oruga spraying robots;
- Meyco Piccola and GM dry bag hand spraying equipment;
- L2C Atlas Copco twin boom drilling jumbo for installation of spile bars, vacuum wellpoints, ground probing and pre-split holes within Drive Y tunnel eyes of headwall;
- Combination of RT 4WD Genie boom-lift MEWPs and two tracked boom-lifts for all working at height activities.

Figure 11 – EB Cavern – Removing temporary sidewall of top heading

A dedicated SCL Plant Manager was appointed to ensure that the equipment was utilised as much as possible. Prestart checks, daily maintenance and cleaning of each item of plant by the responsible trained operative were incentivised by the implementation of a production bonus scheme. Preventative maintenance and reactive repairs as necessary were carried out underground by fitters on each shift. Equipment was removed from service for scheduled maintenance by manufacturers and specialist plant technicians.

Design Development during Construction

In order to improve the constructability of the design the following early changes were agreed with the Contractor, CRL and the Designer and implemented on site:

- Bench excavation rounds for side wall and enlargement increased to 2 metres to improve speed of ring closure;
- Lattice girders were removed from all temporary side walls to minimise handling and erection risks and improve quality of the sprayed concrete lining;
- Alternate lattice girders were removed from Primary SCL main outer shell to provide girders at 2 metre centres;
- The P2 layer (comprising bar reinforcement but no steel fibres) for invert, bench and top-heading was constructed directly after the P1 layer invert in 2 metre long sections to minimise mesh lap joints and improve the quality of the sprayed concrete lining;
- The P2 mesh in main outer SCL shell at the crown and invert of the side drifts was left exposed for 1 metre to enable a better lap connection later during the enlargement construction;
- Split face construction and 200mm capping of the initial SCL layer on the top heading in larger spans of cavern central enlargement (i.e. wider sections of the apple core) was replaced by driven steel spile bars and a 75mm initial layer. This facilitated installation of the lattice girders and reduced further construction joints in the P1 layer

Tunnelling Progress

The average tunnelling rates achieved for the tunnel elements were as follows:

Tunnel Element	Sequence of construction	Eastbound (m/day)	Westbound (m/day)
Launch tunnel	1	1.7	0.9*
	Single side drift section (Cavern 1)		
LH side drift	2	0.9	1.3*
RH enlargement	3	1.3	1.4*
	Double side drift section (Cavern 2)		
LH side drift	4	1.4	0.9*
RH side drift	5	1.1	1.0*
Central enlargement	6	0.6	0.8*
	TBM Reception stub tunnels		
Stub tunnels	7	1.5	1.5*

* includes installation of vacuum well-points for groundwater depressurisation

The average cycle times that were achieved for the single side drift sections were as shown below;

Part of lining	Activity	Advance length (m)	Actual Duration (hours)
	(1) Left hand side drift		
TH	Excavation, survey, ramping	1.0	1.5
	Primary lining (P1 only)		3.2
BCH	Excavation, survey, ramp removal	2.0	2.5
	Primary lining (P1 only)		2.5
INV	Excavation, survey, backfill out	2.0	3.5
	Primary lining (P1 only)		4.0
P2	P2 layer for INV, BCH & TH	2.0	7.5
	(2) Right hand enlargement		
TH	Excavation, survey, ramping & removal of sidewall	1.0	2.0
	Primary lining (P1 only)		2.7
BCH	Excavation, survey, ramp & sidewall removal	2.0	2.8
	Primary lining (P1 only)		2.2
INV	Excavation, survey, backfill & sidewall removal	2.0	3.2
	Primary lining (P1 only)		3.2
P2	P2 layer for INV, BCH & TH	2.0	6.5

Figure 12 – WB Cavern - Installing Lattice Girders in TH of Central Enlargement

Generally the SCL caverns were constructed faster than expected. It is considered that this was attributed mainly to:

- High motivation of work crews (mainly due to bonus pay);
- Design development during construction;
- Synergy and collaborative working between DSJV, OtB, Crossrail, C121 site teams and Designer as evidenced during the effective daily SRG and weekly CTC meetings;
- Good care, maintenance & rotation of equipment;
- Implementation and continual improvement of effective ground dewatering systems;
- Consistent supply of fresh quality concrete from the Hanson on-site batching plant;

Lessons Learnt

The main lessons learnt, relating to health and safety, have been identified as follows:

- Minimise manual hands-on work where personnel must enter the exclusion zone at the face (e.g. for joint preparation and installation of steel continuity bars, lattice girder connections, survey instrumentation);
- Minimise the duration of time that people have to be within the exclusion zone at the face for each round;
- Promote fully mechanised work where possible for joint preparation, excavation & spraying. Selection of plant and tools is an important contributing factor when planning and designing the works. Consider size, weight, power, gradeability, reach, emissions and engine fire suppression. Where possible time should be allowed for trialling each item of plant;
- Provided good shape control can be maintained by other means, remove lattice girders from the design due to the various manual handling, working at height and falling objects risks associated with their transport, handling, hoisting and erection at the face. They slow ring closure, increase potential for voids in the lining at the butt

plate connections and their fixed position pre-determines face size and round length leading to an inflexible design when considering face support. Due to the flaring and changing shape of the cavern each girder had to be uniquely designed, fabricated and installed in pre-determined position.

- Minimise the use of steel bars and mesh by utilising steel fibres where possible. The bars tended to increase potential for shadowing and voids when spraying mechanically using robots;
- Avoid tight corners in the SCL profile (e.g. where the temporary sidewall meets the main outer lining shell in the crown and invert) as they are difficult to excavate and spray accurately using mechanised means;
- Use 2 metre advances for bench and invert as they create better access for excavation and spraying, minimise handling of backfill for ramping and generally enable quicker ground support and reduced ring closure time;
- Minimise construction joints that require preparation, rework and inspection;
- Remove pressure cell instrumentation from the lining as it takes time to install and read, and the results proved to be inconsistent. Other means such as strain monitoring should be considered to supplement the deformation monitoring.

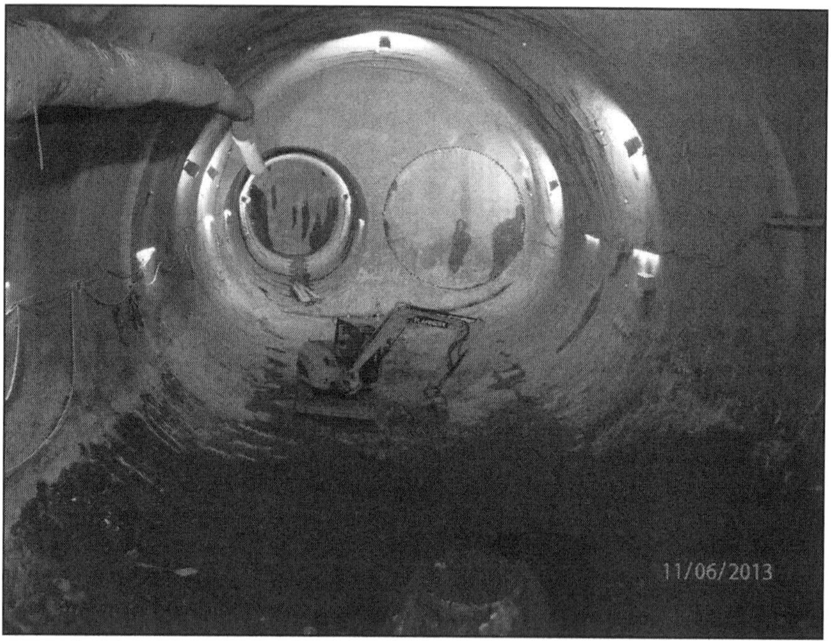

Figure 13 – Removing Backfill from Stepney Green Cavern

SCL Performance

Instrumentation & Monitoring Plan

In line with the CEDS Instrumentation & Monitoring Standard; Green, Amber & Red zones were identified as below;

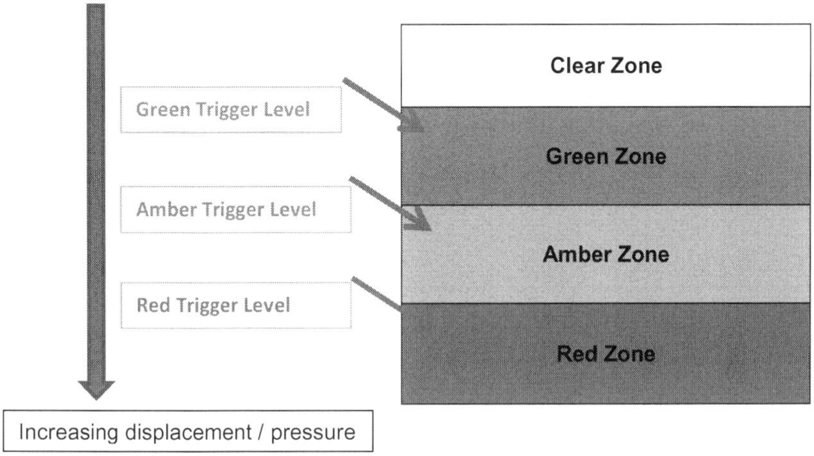

The definition of the zones and trigger levels with associated actions are described below;

- **Clear Zone** – the zone when monitoring was initiated, prior to any triggers being reached. No actions.
- **Green Trigger Level** – this trigger alerted that the Green Zone had been reached. The Green Trigger level was set at 75% of the FLAC calculated lining displacement. This took account of an estimated 25% lining displacement occurring prior to installation and first reading of a monitoring point. The action was to review instrumentation and increase frequency if necessary and agreed at the SRG.
- **Amber Trigger Level** – this trigger alerted that the Amber Zone had been reached. The Amber Trigger level was set at 105% of FLAC calculated lining displacement. The action was to inspect the area of the relevant tunnel and review by the ERP both within 24 hours. The ERP could recommend changes to monitoring and construction methodology.
- **Red Trigger Level** – this trigger alerted that the Red Zone had been reached. The Red Trigger level was set at 150% of FLAC calculated lining displacement. The action was to stop the tunnelling process and make the excavated face safe pending an Emergency Engineering Review within 4 hours.

Predicted vs. Actual Lining Displacements

Given the variable diameter of the cavern, the Designer provided two sections for trigger level monitoring points for the caverns. In each section each monitoring point had its own set of trigger values for both horizontal and vertical deformation.

In general terms, it was observed that the vertical component was similar to the expected one (between 90% and 105% of the amber trigger value), whereas the horizontal component was less than expected (between 30% and 70% of the amber trigger value).

The following tables summarise the green, amber and red trigger values and the average values obtained;

	SINGLE SIDEDRIFT SECTION							
	Actual	Horizontal			Actual	Vertical		
CB1	-0.8	10	14	20	-25.8	-21	-29	-41
CB2	-0.6	11	15	22	-19.4	-19	-26	-38
CB3	-0.5	14	19	27	-15.3	-13	-18	-26
CB4	-1.4	9	13	18	-8.6	-6	-8	-11
CB8	0.6	5	7	10	-18.9	-11	-15	-21
CB9	2.0	7	9	13	-14.1	-9	-13	-18
CB10	3.0	6	8	12	-7.2	-5	-7	-10

	DOUBLE SIDEDRIFT SECTION							
	Actual	Horizontal			Actual	Vertical		
CB15	-				-24.0	-13	-19	-27
CB1	-5.4	5	8	11	-20.3	-11	-15	-21
CB2	-7.3	16	22	31	-24.4	-21	-29	-42
CB3	-4.0	17	23	33	-17.2	-16	-23	-32
CB4	-9.6	9	13	19	-13.8	-9	-13	-18
CB8	2.7	5	7	11	-21.1	-10	-14	-20
CB9	7.4	10	14	20	-23.1	-18	-25	-36
CB10	4.1	14	20	28	-21.0	-15	-21	-30
CB11	4.8	8	12	17	-13.8	-8	-11	-16

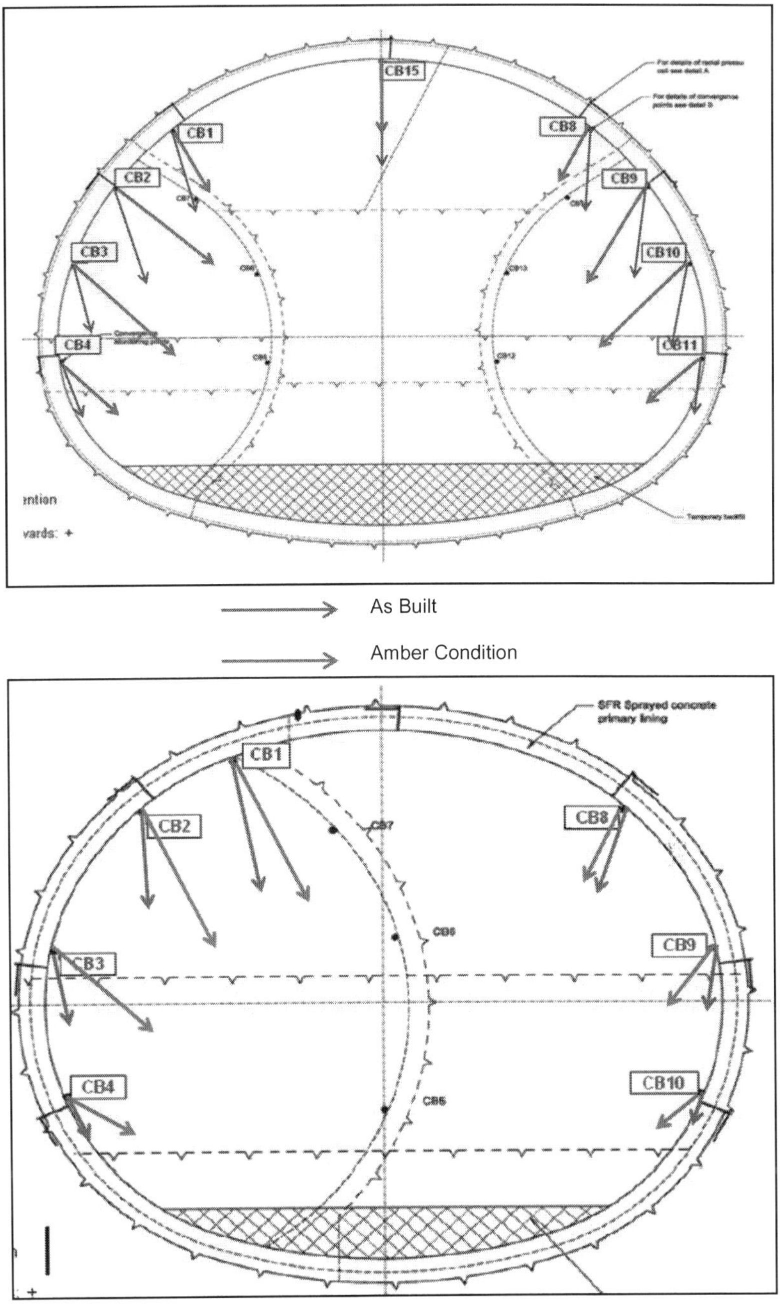

Figure 14 – The expected movement in amber condition and average as-built movement for the single and double side-drift sections.

The discrepancy between vertical and horizontal is considered to be due to the effect of the dewatering that was carried out in the sandy layers beneath the tunnel invert, causing whole-body downward movement of the cavern shell.

Surface settlement

Surface movement was monitored with several arrays of precise levelling points. Readings were taken on a daily basis. Maximum settlement was 60mm above the largest section of the Eastbound Cavern. This value includes the movement due to the SCL works and dewatering. From the analysis of the timeline in settlement, a value of approximately 10mm was determined to be related to the depressurisation and 50mm due to SCL works. The predicted surface settlement, assuming a volume loss of 1.5% was 62mm.

Above the Westbound Cavern the maximum recorded settlement was 51mm, approximately 40mm due to SCL works and 10mm due to depressurisation. The predicted surface settlement for the Westbound Cavern was 30mm.

Conclusions

Design and construction of the Stepney Green SCL Caverns has demonstrated that it is possible to construct large caverns successfully in London Clay and the Lambeth Group, with structure and ground movements controlled to acceptable levels. This was done with a collaborative team effort between Client, Contractor and Designer. There are four further caverns to be constructed on Crossrail and key lessons learnt from Stepney Green can be applied to these structures.

References

1) BTS/ABI 2003 The joint Code of Practice (JCoP) for Risk Management of Tunnel Works in the UK. Joint publication by the Association of British Insurers & the British Tunnelling Society.
2) Technical Specification for Interoperability – Safety in Railway Tunnels TSI-SRT.
3) RILEM TC162-TDF; Test and design methods for steel fibre reinforced concrete. Vol 36, October 2003.
4) LU Engineering Standard 1-052 – Civil Engineering – Bridge and Railway Group Standard GC/RT5112

Crossrail Project: Infrastructure design and construction
ISBN 978-0-7277-6078-4

doi: 10.1680/cpid.60784.407

Back Analysis of Observed Measurements for Optimised SCL Tunnel Design

Angelos Gakis, Dr. Dipl.-Ing, MSc DIC, Chief Geotechnical Engineer, Crossrail Farringdon Station, Dr Sauer & Partners Ltd, London UK
Stephen Flynn, MSc, CEng MIEI, SCL Tunnel Engineer, Crossrail Farringdon Station, Dr Sauer & Partners Ltd, London UK
Ali Nasekhian, Dr.-Ing, FE Design Engineer, Crossrail Farringdon Station, Dr Sauer & Partners Ltd, London UK

Abstract
The Crossrail Farringdon Station is a deep level station with two platform tunnels, 300 metres long at a depth of approximately 30 metres below street level. The running tunnels are primarily TBM bored with the platform tunnels enlarged using sprayed concrete lining (SCL) tunnelling method. The station is located in challenging ground conditions due to the presence of several faults and the heterogeneity of the Lambeth Group, which call for a variety of temporary works contingency measures and in-tunnel depressurisation scheme.

Back analyses for the already constructed cross passage CP1 were performed using finite element (FE) analysis in order to calibrate them against observed deformations, deriving valuable information for future tunnel design. In order to achieve this, the accuracy of assumptions made in the initial design stages had to be checked. Such parameters as ground relaxation factor for construction advance, at rest earth pressure coefficient and soil stiffness are back-calculated and compared with the initial values.

The outcome of this work was refined FE modelling parameters based on measured deformation results, demonstrating a greater accuracy for predicting both in-tunnel and surface deformations and thus allowing for a better understanding of the tunnel lining and ground behaviour to be anticipated in advance of construction.

List of notation

BFK	is BAM Ferrovial Kier Joint Venture, main contractor
CRL	is Crossrail, the Client and Project Manager
DSP	is Dr Sauer & Partners, specialist SCL designer for BFK
FE	is the Finite Element method
k_o	is the at-rest earth pressure coefficient
SCL	is Sprayed Concrete Lining method
λ	is the soil relaxation factor

Crossrail Farringdon Station

The Farringdon Station is one of the 8 Crossrail stations in London that provides an interchange between the Crossrail and existing London Underground networks at the Eastern Ticket Hall/ Barbican Station and an interchange between Crossrail and Thameslink at the western ticket hall/ Farringdon side. The Crossrail Farringdon Station is a deep level station with two platform tunnels stretching between the existing London Underground Farringdon and Barbican Stations at a depth of approximately 30 metres below street level.

The contractor, Bam Ferrovial Kier Joint Venture (BFK), formed between BAM Nuttall Ltd, Ferrovial Agroman (UK) Ltd and Kier Construction Ltd, was awarded the construction contract (C435) for the main construction works at Farringdon station by Crossrail Limited (CRL) in November 2011. The Dr. Sauer & Partners Company Limited (DSP) has subsequently been appointed by BFK to provide sprayed concrete lining (SCL) design and site supervision.

The structures at Farringdon Station are:

- Escalator inclines ES1 and ES2,
- Eastbound Platform Tunnel PTE and Westbound Platform Tunnel PTW (SCL Tunnels enlarged from TBM pilot tunnels),
- Lower concourse tunnels CH1 and CH2,
- Cross passages CP1, CP2a, CP2b, CP3a, CP3b, CP4, CP6a, CP6b, CP7, CP8 and CP9,
- Ventilation adits VA1 and VA2,
- Platform Extension tunnels PL1, PL2, PL3 and PL4,
- Stub tunnels STE1, STW1, and STW2,
- Lift shaft passage (LP1) connection to Thameslink lift shaft and
- Temporary tunnels – PL2RC Wraparound, CP1-CH1 Connection.

At the time that this article was written (October 2013), the completed tunnels in Farringdon station were the "Early Western Tunnels", STW2-PL1, CP1 and TBM reception soft-eyes in shafts SH-W1 and SH-W2, as well as CP1-CH1 connection adit, PL2RC Wraparound, CH1 pilot tunnel and westbound TBM pilot tunnel (shown in light blue colour in figure 1). This reports focuses on CP1, a cross passage of approximately 35m length and cross sectional area of approximately $22m^2$ (7.2m height x 6.2m width).

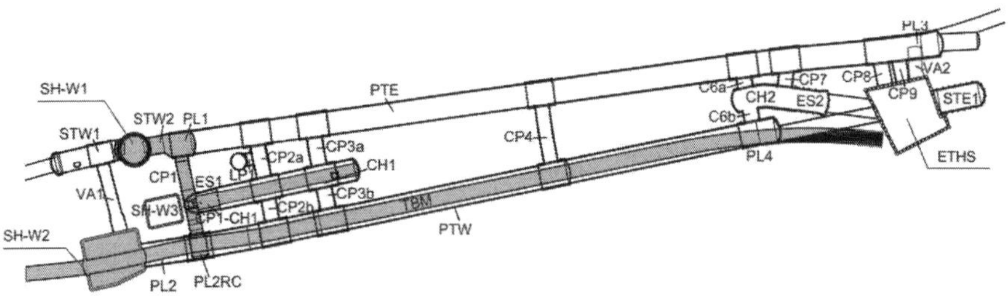

Figure 1. Farringdon Station plan view

Design considerations

The analysis and design of buried structures is one of the most complicated and difficult subjects in tunnel engineering. Construction of urban tunnels requires precise measurements of soil settlement at the surface and in-tunnel deformations. Factors such as nonlinear behaviour of soil, stress history, overburden depth and diameter of tunnel have major influence on the development of the ground deformation.

The main purpose of this research was to investigate the performance of 2D and 3D finite element models with the prescribed design assumptions and to specify constraints for the analysis. By using worst case scenario of faulted ground material properties it was possible to assess to a first degree, how close to the as-built results the preliminary analysis was.

2D & 3D finite element analyses were carried out to determine the in-tunnel deformations associated with the excavation and primary lining construction. This section describes the analysis methodology, soil and sprayed concrete properties, modelling stages and comparison of finite element analysis (FEA) of design with as-built deformations. In an effort to determine a more representative model, a parametric study and a back analysis of significant was performed.

3D Finite Element Analysis

Three dimensional modelling provides an opportunity to consider the 3D effects of tunnelling in the course of construction, ground/structure interaction, face stability and gives information regarding volume loss and ground movement without subjective assumptions. It is necessary to make certain assumptions in 2D analysis to take into account 3D geometrical effects, construction sequences, and stress variations due to previous tunnelling activities. It is common practice to assume a so called relaxation factor in order to consider three dimensional effects in a 2D analysis (presented in 2.2). However all these assumptions are subjective, and depending on the complexity of tunnel geometry and subsoil condition, sometimes it is a difficult task to estimate a reasonable and sound factor. In addition, since the plane strain conditions don't exist at tunnel breakouts or tunnel junctions, therefore 3D analysis is necessary in these situations to analyse tunnel lining and ground stability.

ABAQUS Version 6.12 released 2011 (Dassault Systemes Simulia Company) a general purpose finite element software package was used to perform the numerical analyses. The ground was modelled using linear tetrahedral elements and sprayed concrete linings were modelled using linear triangular shell elements. Soil materials were modelled with the elastic-perfectly plastic Mohr-Coulomb model.

The dimensions of the model boundaries were 145m x 120m x 44m (Figure 2). The model comprised of approximately 600,000 solid elements, 50,000 shell elements and 191 analysis steps. The construction shafts (SH-W1 and SH-W2), were modelled as a wished-in-place structure following a relaxation step in which 50% of soil stiffness was reduced.

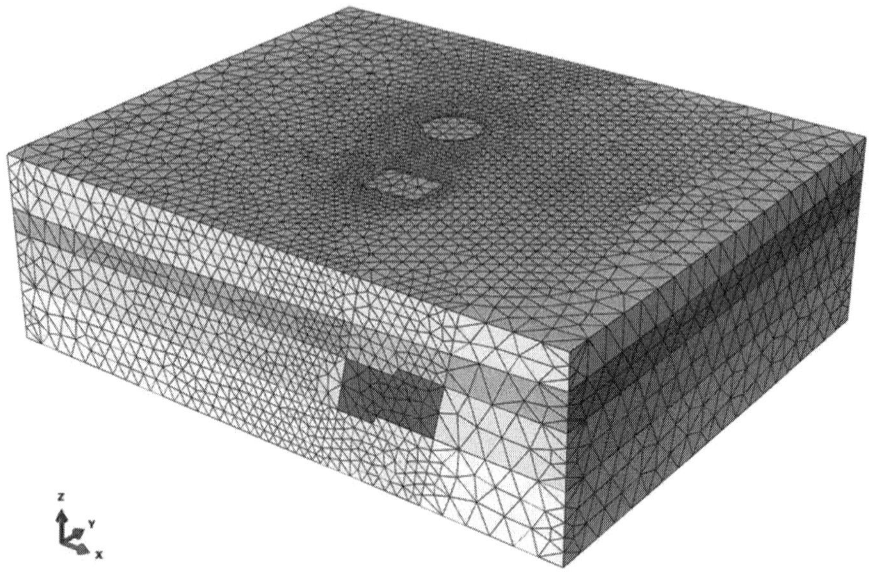

Figure 2. Finite Element mesh and extent of the 3D model

Undrained soil parameters were taken for the analysis in order to account for the "fast" construction in comparison to the time of consolidation in the London Clay. The numerical analyses have been undertaken on the basis of a total stress analysis and no water pressure was produced during the analysis.

Ground Conditions and Model Geometry

SOIL PROPERTIES		Upper Strata	LC (A2)		Lambeth Group		Thanet Sand	
			Unfaulted	Faulted	Unfaulted	Faulted	Unfaulted	Faulted
Unit Weight	[kN/m^3]	17	20	20	21	21	21	21
Young's Modulus	[MPa]	10	40+3.7z1	33+3z1	36+5.9z1	30+4.9z1	209+4.3z2	180+3.9z2
Poisson's Ratio	[-]	0.2	0.495	0.495	0.495	0.495	0.2	0.2
Shear Strength	[kPa]	-	85+6.5z1	40+6.7z1	95+10z1	55+8.8z1	N/A	N/A
Friction Angle	[°]	31	0	0	0	0	39	34
K_0 – Value	[-]	0.5	1.2	1.2	1.2	1.2	1.0	1.0

Table 1: Ground parameters under undrained conditions, except for Upper Strata and Thanet Sand which are drained

The geological formations encountered in the area of Farringdon Station, were the typical of the London basin, i.e. the upper strata comprising Made ground, Alluvium and River Terrace Deposits (mainly in the Eastern part) overlay London Clay, Lambeth Group, Thanet Sand and Chalk. In the FE model Chalk layer was considered as the bed rock and therefore the bottom boundary of the model is top of the Chalk layer. An important feature was that the thickness of the London Clay layer was significantly thinner in the western part relative to the eastern part, varying between 4 and 10m and contains London Clay A2 subdivision only. The layout of subsoil layers used in the FE model in the region around Western Ticket Hall has been illustrated in Figure 3 and soil parameters used in the FE models are given in Table 1.

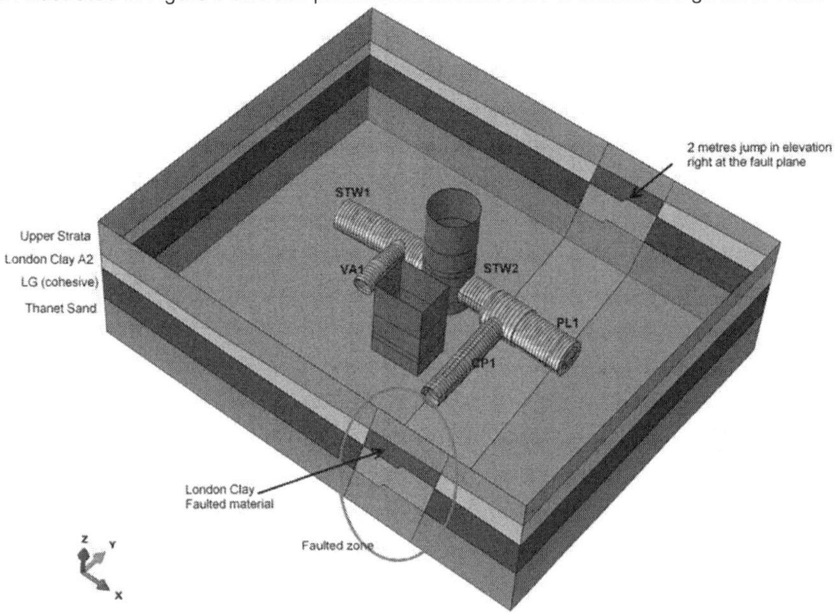

Figure 3. Fault zone and subsoil layers in the 3D FE model

The main characteristics of the geology in Farringdon station around the area subject to this FE model are the following:

- The presence of Farringdon Station Fault which affects the thickness, the elevation and the continuity of the top/bottom of the soil layers.
- The Farringdon Station Fault has been taken into account in the FE model by defining a Fault zone with faulted material properties. Boundaries of the fault zone have been obtained by offsetting ± 10 metres the fault plane identified in the geological section drawings in Geotechnical Interpretative Report. Figure 3 shows the geometry of the Fault zone in the FE model.
- London Clay contains only subdivision layer LC_A2 with a slight inclination towards west. The thickness varies approximately between 4 to 10 metres.

Steel Fibre Reinforced Sprayed Concrete Model

Parameter	Value
Characteristic cylindrical compressive strength of SFRC* (28-days)	28 MPa
Characteristic residual tensile strength for SFRC	0.45 MPa
Elastic Modulus used in FE analysis (Primary Lining only)	13 GPa
Poisson's ratio	0.2

Table 2: Parameters for the steel fibre reinforced Sprayed concrete used in the FE models

The primary SCL consists of a fibre reinforced sprayed concrete lining. The material properties utilised for the sprayed concrete tunnel linings are given in the Table 2. For the primary lining design, the 28-days compressive strength was considered.

Sprayed concrete is modelled as elastic-plastic material. Residual strength parameters are considered in a post failure condition of the material. For instance, tension cut-off 0.4 MPa was considered for the tunnel linings based on Fibre Reinforced Sprayed Concrete material properties. The "concrete damaged plasticity model" was therefore implemented in the model to satisfy this material behaviour. It was a damage model for concrete material which assumes that the main two material failure mechanisms are tensile cracking and compressive crushing.

2D Finite Element Analysis

The two dimensional analysis comprised of staged, plane strain models. Construction and installation of primary lining was modelled following the prescribed predesigned excavation and support sequence. The finite element software package Phase2 Version 8.0 (Rocscience) was used to carry out the numerical analysis. The computer software analysed the ground loading conditions and ground response due to tunnel construction. The finite element mesh used consists of approx. eleven thousand 3-noded triangular solid material elements used to model the ground and beam elements to simulate the concrete lining support. The Mohr-Coulomb constitutive model was used to simulate the elasto-plastic behaviour for the elements forming the ground. The model assumes that failure occurs when the shear stress at any point in a material reaches a critical value that is linearly dependent on the normal stress in the same plane. The sprayed concrete lining in the model follows an elastic material behaviour.

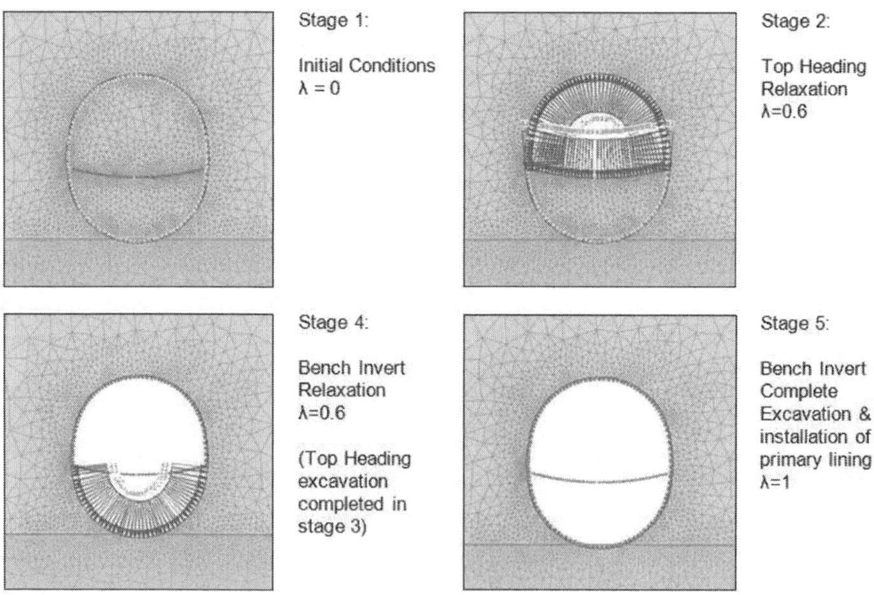

Figure 4. Typical excavation and support stages in the 2D FE analysis

For 2D analyses, a relaxation factor or (confinement loss factor) (λ) was applied in order to simulate the deformation that took place prior to the installation of the support (Figure 5). The approach is referred to as the convergence-confinement method (Potts & Zdravkovic, 2001), to which the proportion of unloading before lining construction is prescribed and volume loss is a predicted value. Hence in the initial condition where no excavation has taken place, a value of λ=0 is applied whereas when the installation of the primary lining is complete, λ=1. The internal pressure reduction method was selected in order to simulate the ground relaxation. Hence, at all stages the internal pressure (p) applied, is a fraction of the in-situ pressure (p_0) that depends on the relaxation factor (λ) and is expressed by the following equation:

$$p = (1 - \lambda) \cdot p_0$$

Equation 1.

Prior to the installation of the primary lining, the pressure, p, was reduced from p_0 (in-situ state) to $(1 - \lambda) \cdot p_0$, with subsequent deformation taking place. When the excavation was complete and the support system had been installed, the internal pressure aws removed $(p = 0)$.

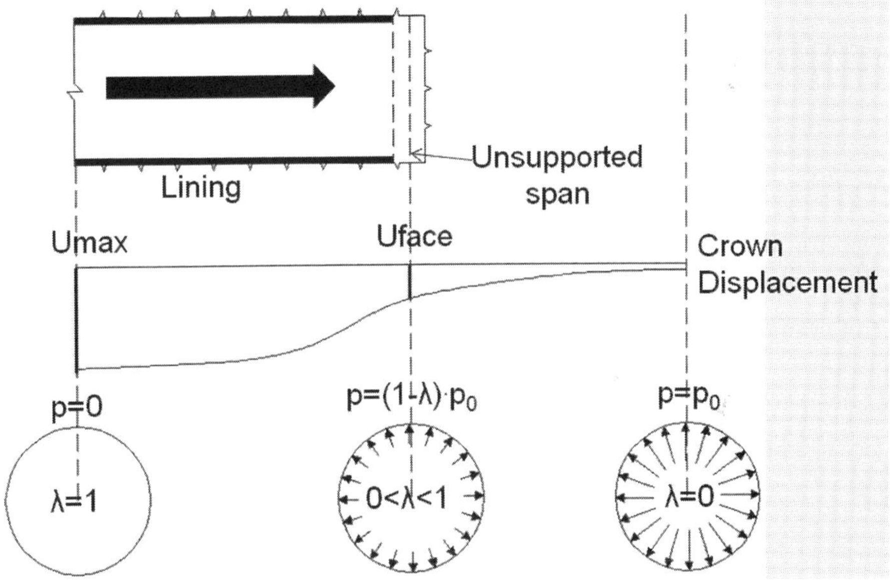

Figure 5. Schematic of the convergence-confinement method (Spyridis, Gakis & Bedi 2012)

A relaxation factor λ = 0.6 was used in the design of CP1, derived from an axisymmetric finite analysis model of equivalent excavation area, performed with Phase2 and later confirmed by the 3D finite element model.

During the parametric study presented in the next sections, the relaxation factor was one of the parameters that proved to have a great influence on the results, hence its value has been refined for the optimised, final analyses.

In-Tunnel Monitoring Data

Monitoring cross sections were installed in several locations along CP1 tunnel, at tunnel meters 1.1, 6.1, 18.1, 22.0, 26.2 and 30.2. For the purpose of this report, the monitoring array at tunnel meter 18.100, installed approximately in the mid- length of the cross passage was selected. Being equally distant from the breakout area (junction with PL1 tunnel) and the headwall (end of CP1), this position was less affected by end effects, hence is more appropriate for plane strain finite element analysis.

Monitoring of this cross section was carried out on a daily basis during the excavation of CP1. It has to be noted that due to the relatively small magnitude of the displacements, the results were affected by the accuracy of the readings. The final monitoring data following the completion of the excavation of CP1 are presented in figure 6. The crown monitoring point, M1, settled by approximately 7mm, as the upper side points M4 and M5. Smaller settlements were observed in the lower side points M6 and M7 (approximately 3 to 4mm). In terms of transverse displacements, points M4 and M5 exhibited approximately 5mm of inward movement, whereas points M6 and M7 moved by approximately 3mm. The 1.8mm transverse displacement at point M1 is attributed to the accuracy of the readings.

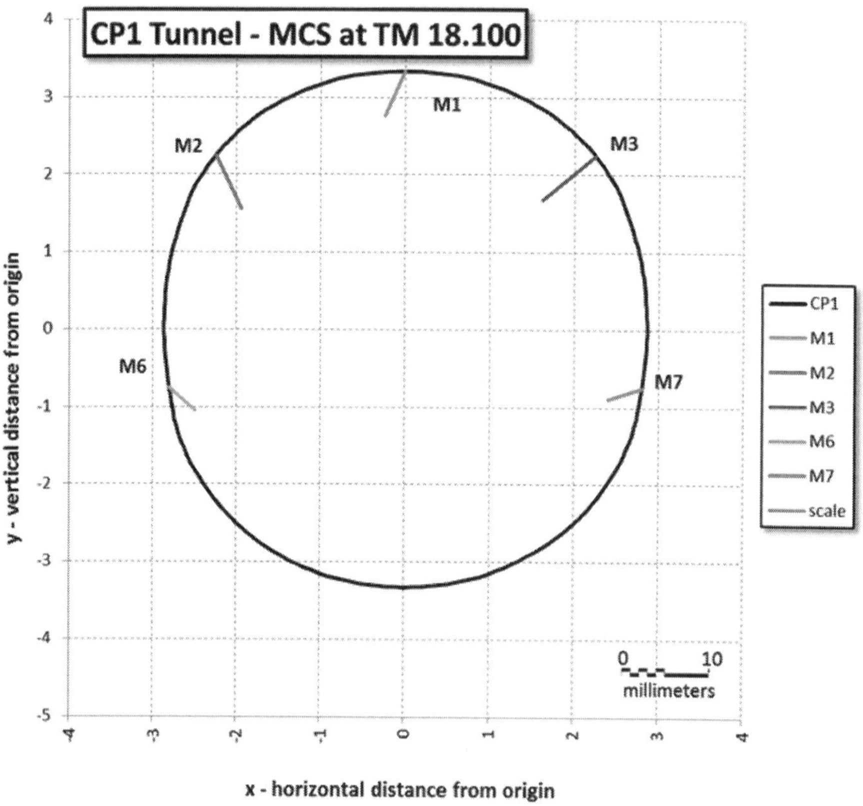

Figure 6. In-tunnel monitoring data for CP1, monitoring cross section at 18.100 tunnel meters

Optimisation of FE Models

Using the 2D and 3D finite element models of the design as a baseline, the influence of various parameters was investigated. The purpose of this study was to assess the effect that these parameters had independently on the displacements of the tunnel section and to try and derive an optimised model that exhibits similar behaviour to the monitoring section at tunnel meter 18.1 of CP1.

Parametric study on the influence of various parameters

In order to investigate the effect that each particular parameter has on the in-tunnel deformations a detailed parametric study performed in the course of the 2D and 3D FE analysis.

The main parameters that were modified were:

- the stratigraphy: (as assumed during the design stage and as observed through the excavation of CP1) – "design" was the stratigraphy used in the design models, based on the original site investigation data, "as-built" is the actual stratigraphy, as derived through the excavation of the tunnels.
- the relaxation factor, λ: a wide range of relaxation factors, varying between 0.3 and 0.8 was tested
- the parameters of the soil: during the design phase, the "faulted" parameters were used, accounting for a potential deterioration of the materials in the proximity of faults. It was however proved that CP1 was not affected by faulting, hence "unfaulted" parameters were used in the majority of the parametric analyses.
- the at-rest earth pressure coefficient, k_0: the effect of k_0 was also investigated, carrying out analyses with values of 1.2 (design value), 1.35 and 0.65.
- the stiffness of the unloaded material: in most cases, the soil material exhibits a stiffer response when unloaded. Using a Mohr-Coulomb constitutive model, does not allow for this behaviour to be simulated. Therefore, an alternative method was used, applying an increased stiffness (3 to 5 times the original Young's modulus – "E x 3" to "E x 5") on a soil wedge, starting from the top-heading/bench interface, extending to the bottom of the model geometry at 45 degrees dip angle.

no.		Analysis	Stratigraphy	λ	Soil paramters	k_o	Unloading Wedge Stifness	In-Tunnel Deformation (mm) Settlements and Horizontal displacements to the right are positive M1 X	M1 Y	M4 X	M4 Y	M5 X	M5 Y	M6 X	M6 Y	M7 X	M7 Y
1	3D	design analysis	Design	-	Mixed	1.2	-	-1	-3	2	-2	-4	-3	4	0	3	0
2	2D	Design	Design	0.6	Unfaulted	1.2	-	-12	-25	16	-18	-16	-18	2	2	-2	2
3	2D	Stratigraphy	Design	0.6	Faulted	1.2	-	-9	-55	31	-60	-31	-60	3	0	-3	0
4	2D		As-Built	0.6	Faulted	1.2	-	-2	-44	28	-64	-28	-64	3	4	-3	4
5	2D	k_o	As-Built	0.6	Unfaulted	1.2	-	0	-12	12	-18	-12	-18	2	2	-2	2
6	2D		As-Built	0.6	Unfaulted	0.65	-	0	-11	6	-14	-6	-14	0	1	0	1
7	2D		As-Built	0.6	Unfaulted	1.35	-	0	-16	14	-24	-14	-24	2	2	-2	2
8	2D	Unloading Wedge	As-Built	0.6	Unfaulted	1.2	E x 3	0	-14	12	-23	-12	-23	2	2	-2	2
9	2D		As-Built	0.6	Unfaulted	1.2	E x 4	0	-14	13	-22	-13	-22	2	2	-2	2
10	2D		As-Built	0.6	Unfaulted	1.2	E x 5	0	-14	13	-24	-13	-24	2	1	-2	1
11	2D	Localized k_o reduction	As-Built	0.6	Unfaulted	0.65	-	0	-11	1	-8	-1	-8	1	2	-1	2
12	2D	Relaxation factor, λ	As-Built	0.3	Unfaulted	1.2	-	-3	-12	12	-17	-12	-17	3	3	-3	3
13	2D		As-Built	0.4	Unfaulted	1.2	-	-3	-12	12	-18	-12	-18	3	1	-3	1
14	2D		As-Built	0.5	Unfaulted	1.2	-	-3	-12	12	-17	-12	-17	2	2	-2	2
15	2D		As-Built	0.6	Unfaulted	1.2	-	0	-12	12	-18	-12	-18	2	2	-2	2
16	2D		As-Built	0.7	Unfaulted	1.2	-	0	-11	12	-18	-12	-18	1	1	-1	1
17	2D		As-Built	0.8	Unfaulted	1.2	-	0	-10	10	-17	-10	-17	1	1	-1	1
19	3D	Optimised	Design	-	Mixed	0.65	-	0	-5	4	-3	-4	-5	3	3	-3	3
20	2D	Optimised	As-Built	0.8	Unfaulted	0.65	-	0	-6	5	-8	-5	-8	0	0	0	0
		Actual displacements						-1.8	-7	5	-7	-5	-7	3	-4	-3	-3

Table 3: List of the analyses performed during the parametric study and the corresponding results

Analyses 1 to 17 in table 3, constituted the framework of this parametric study. These analyses were grouped according to the parameter that was investigated:

- 1 – 2: analyses carried out at the design state using 2D and 3D finite element modelling respectively.
- 3 – 4: effect of the stratigraphy, applying either the stratigraphy used in the design analyses, or the "as-built" stratigraphy.
- 5 – 7: effect of at-rest earth pressure coefficient, k_0.
- 8 – 10: effect of using higher Young's modulus in the unloading wedge (as described above).
- 11: effect of reducing k_0 locally, assuming a zone of reduced k_0 = 0.65, around the tunnel excavation (3 radii from tunnel axis laterally) while the global k_0 remains at the design value of 1.2, after Potts & Zdravkovic (2001).
- 12 – 17: effect of the relaxation factor

Back analysed model results

The results of the analyses preformed in the context of the parametric study (analyses 3 to 17 of table 3), were compared to the actual displacements measured in the monitoring cross section of tunnel meter 18.1 (refer to table 3 and figure 6) and the weight of each parameter was originally assessed.

The outcome of this comparison, was that k_0, the relaxation factor and the use of faulted/unfaulted parameters were playing the principle role.
Therefore, after a selection of the optimal parameters for the 3D and 2D finite element models was made, the optimised model results are presented in table 3 (analyses 19 and 20 respectively).

For the 3D model, a very good match was achieved by reducing k_0 from 1.2 to 0.65. For the 2D model, the optimised analysis used a relaxation factor 80%, k_0 of 0.65 and unfaulted soil parameters.

Conclusions

Obtained field data for the in-tunnel monitoring of the case study, cross passage CP1, were evaluated against 2D and 3D finite element models, after an extensive parametric study on the influence of various parameters.

The main conclusions of this research suggest that:

- The soil in this geological environment exhibits a significantly high relaxation prior to the installation of the primary lining. Additionally, the value of k_0=1.2 that was used in the design, overestimated slightly the magnitude of the displacements, with a value of k_0=0.65 providing results closer to the actual displacements.
- The use of faulted parameters for CP1 was slightly conservative, the application of un-faulted parameters proved to be more realistic in this study.

The 2D and 3D multi-parametric back-analysis above show good agreement with field data. The outcomes are very promising for future optimisation of two dimensional plain strain geotechnical models. However, its application is problem specific. The utilisation of measurements from field data and FE models reveal important aspects of the ground-support interaction applicable to the derived softening factor. The derived softening factor estimated from the axisymmetric model should require further investigation. Researching trends show a lower k_0=0.65, this assessment is more representative for Lambeth Group and could be adopted for future design assumptions.

Acknowledgements

The authors would like to acknowledge the contribution of BFK joint venture for the support and the guidance towards the completion of this study, Crossrail Ltd. for the motivation and for granting the permission for this paper to get published and last but not least, Dr Sauer and Partners Ltd. for providing resources and technical support that made this paper possible.

References

Potts, D. M., and Zdravković, L (2001) Finite Element Analysis in Geotechnical Engineering: Theory and Application. London: Thomas Telford, 2001

Simulia (1992) ABAQUS: User's Manual. [Providence, R.I.]: Hibbitt, Karlsson & Sorensen, 1992.

Spyridis, P., Gakis, A., Bedi, A. (2012). Numerical Modelling. Presentation to the BTS-YM, London, October 2012.

Crossrail Project: Infrastructure design and construction
ISBN 978-0-7277-6078-4

doi: 10.1680/cpid.60784.419

Stabilisation of soil pillar adjacent to Eastern Ticket Hall at Bond Street Crossrail Station

Hilary Skinner MA (Cantab) FICE Donaldson Associates Ltd, London, UK
Vicky Potts, MA MSc PhD Donaldson Associates Ltd, London, UK
Adrian St.John, BEng (Hons) Eur Ing MICE CEng BAM Ferrovial Kier (BFK) Joint Venture, London, UK
Damian McGirr, BEng (Hons) MICE CEng Donaldson Associates Ltd, London, UK

Abstract
Sprayed concrete lining (SCL) tunnelling alongside the secant pile walls forming the Eastern Ticket Hall (ETH) box at Bond Street station left a narrow pillar of clay, less than 1m wide in places, between the piles and the tunnel lining. Since the tunnelling was to occur after excavation of the box, the stability of this soil pillar during excavation was identified as a concern. Analysis undertaken as part of a due diligence review of the lining design indicated that there was indeed a high risk of overstressing in this area with potential for a collapse mechanism to develop during SCL excavation.

This paper describes this identification and modelling of the ground stability risk, and also the selection and design process that was employed in the development of a detailed temporary works scheme to mitigate that risk. The difficulties associated with the implementation of that design in a highly constrained location are also discussed. A significant degree of collaboration between the construction and design teams enabled amendments to be made to alleviate some construction difficulties without compromising the effectiveness of the soil stabilisation scheme.

Ultimately it was decided to use a grouted pipe arch at the breakout from the box and, where a pipe arch was no longer feasible due to the tunnel geometry, an innovative solution to create a mattress of reinforced soil above the tunnel was developed using overlapping arrays of grouted spiles installed from the shaft.

The works undertaken to stabilise the soil pillar adjacent to the ETH at Bond Street exemplify the successful use of an integrated approach to design and construction to mitigate potential ground instability during SCL excavation.

List of Notations

BFK	Bam Ferrovial Kier JV (Contractor)
DAL	Donaldson Associates Ltd (Contractor's Temporary Works Designer)
DDR	Due Diligence Review
ETH	Eastern Ticket Hall
SCL	Sprayed Concrete Lining

Introduction

Crossrail is Europe's largest construction project, and will transform rail transport in London, increasing capacity by 10%, supporting regeneration and cutting journey times across the city. The Crossrail route will run over 100km from Maidenhead and Heathrow in the west, through new tunnels under central London to Shenfield and Abbey Wood in the east. There will be 38 Crossrail stations including 9 new stations at Paddington, Bond Street, Tottenham Court Road, Farringdon, Liverpool Street, Whitechapel, Canary Wharf, Custom House and Woolwich.

A joint venture of BAM Ferrovial Kier (BFK) was awarded the Crossrail contract C300/C410 "Western Running Tunnels and Station Caverns" in January 2011. This comprised the construction of two 6.2km tunnel drives between Royal Oak and Farringdon and the construction of sprayed concrete lining (SCL) tunnels for Bond Street and Tottenham Court Road Station tunnels. Donaldson Associates (DAL) was appointed as BFK's SCL designers for these works. These two parties had a clear responsibility for excavation stability. DAL's scope of work comprised a Due Diligence Review (DDR) on the primary lining design, which formed part of the permanent works, and the design of "Toolbox items" and any other necessary measures required wherever an unacceptable risk of instability during excavation was identified. Further to this, a 24/7 team of Senior SCL Engineers and SCL Shift Superintendents from Donaldson Associates was embedded into the BFK Engineering team.

This paper relates to a specific element of these works at Bond Street Station - the construction of LCE1 and LCE2 tunnels. A plan of the affected area is shown in Figure 1.

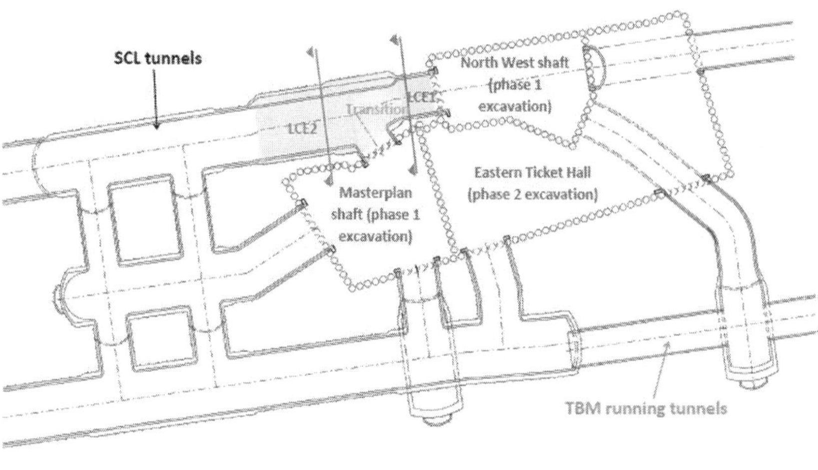

Figure 1: Layout of Eastern Ticket Hall at Bond Street Station

The Eastern Ticket Hall (ETH) was formed using a secant piled wall by another Crossrail contract, with pile depths of up to 45m below ground level. Female piles terminated just above the tunnel horizon, meaning that at tunnel level the piles were contiguous. Temporary internal piled walls divide the ETH into three parts; the north-west shaft, the masterplan shaft, and the rest of the ETH, see Figure 1. This was done to allow early access for SCL works from some parts of the box. As part of the advance works (prior to TBM arrival at Bond Street) SCL tunnelling was to commence working westwards from the north-west shaft.

Tunnels LCE1 and LCE2 were to be constructed parallel to the ETH wall, mostly alongside the recently-constructed masterplan shaft. At some locations, the extrados of the primary lining for LCE1 and LCE2 was less than 1m from the piles forming the ETH wall.

It was realised at an early stage in the DDR that this would leave a narrow pillar of clay between the tunnel and the piles, and that this would potentially be vulnerable to overstressing during SCL tunnel excavation. The installation of the piles in the ETH and subsequent excavation of the north-west and masterplan shafts may have caused disturbance of the ground and lowered the in-situ stresses, both of which increased the risk of overstressing during excavation. This concern was compounded by LCE2 being the largest tunnel profile to be excavated at Bond Street, with an excavated horizontal span of over 12.8m, see Figure 2. The risks associated with construction of these tunnels and possible solutions were discussed at a series of workshops between MM, BFK, Crossrail and DAL.

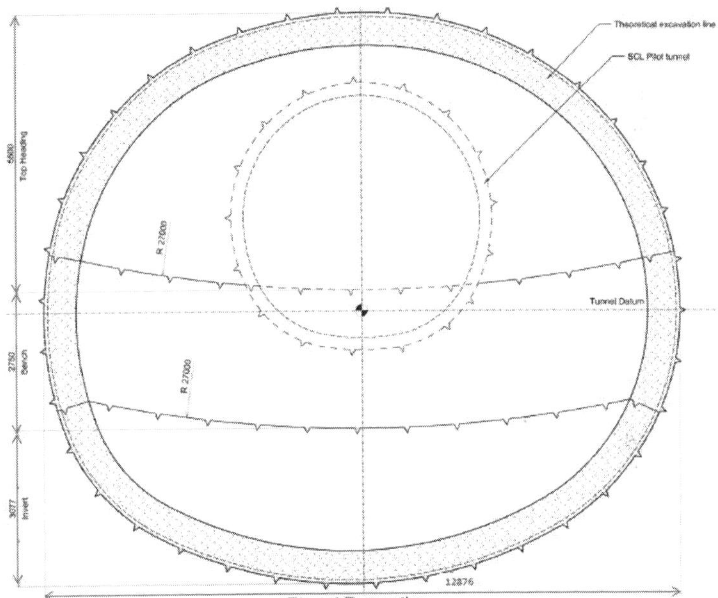

Figure 2: LCE2 cross section

This paper describes the assessment of the risks at this location, and describes the development of a safe, practical and effective scheme by DAL and BFK to mitigate these risks.

Assessment of Risk

Having identified a potential concern over stability at this location, a 2D finite element model was developed using Zsoil (2011) to investigate the stress conditions in the soil pillar between the tunnel and the piles at various construction stages. Two sections, across LCE1 and LCE2, were chosen for the analysis as shown in Figure 1. The analysis and the results are described in the following sections.

Ground Conditions

The stratigraphy at this location is indicated in Figure 3, along with the pore water pressure profile in the London Clay, which is underdrained by the lower chalk aquifer.

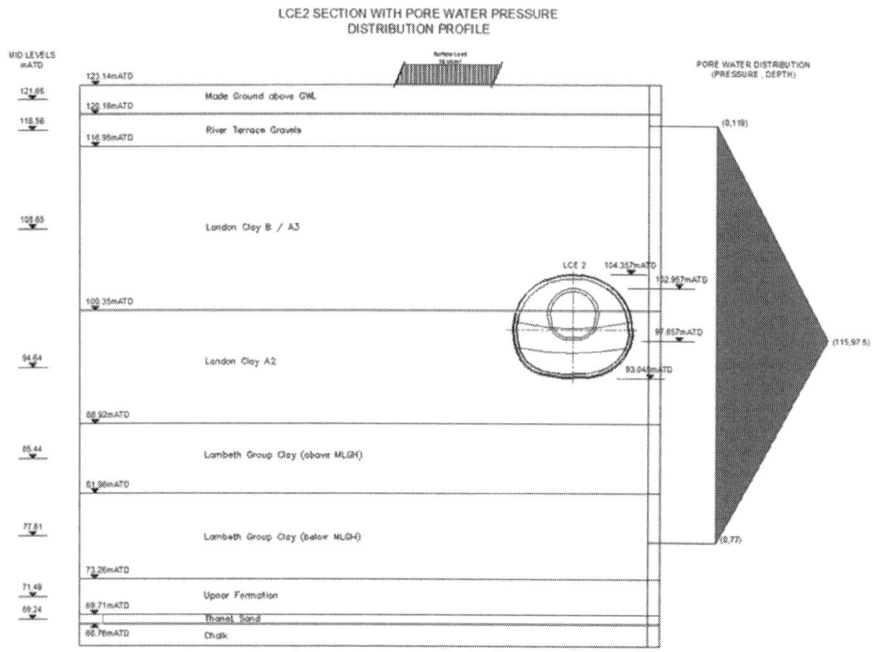

Figure 3: Modelled ground conditions for LCE1 and LCE2 (including location of LCE2)

A coupled consolidation analysis was undertaken using a hardening soil model for the clay materials at or near the tunnel horizon to replicate reasonable small strain stiffnesses during excavation. For the superficial soils, a Mohr-Coulomb constitutive model was considered adequate. The parameters were chosen to give a reasonable prediction of the behaviour, and were not factored. The geotechnical parameters used in the finite element analysis are summarised in Table 1. Permeability was selected to replicate the known insitu water pressure distribution.

No.	Description	Layer Thickness	Level of Top	Element Type/ Plasticity Model	Bulk Unit Weight	Dry Unit Weight	Initial Void Ratio	Permeability vertically	Permeability horizontally	Residual Saturation	Seepage Parameter	Lateral Earth Pressure Coeff. at Rest	Young's Modulus	Poisson's Ratio	Friction Angle	Cohesion	Dilatancy Angle
					γ_{bulk}	γ_{dry}	e	$k_{s,v}$	$k_{s,h}$	S_r	α	K_0	E'	ν	ϕ'	c'	ψ
		[m]	[mATD]		[kN/m³]	[kN/m³]	[-]	[m/s]	[m/s]	[-]	[m⁻¹]	[-]	[MN/m²]	[-]	[°]	[kN/m²]	[°]
1	Made Ground	3.0	123.1	Continuum (MC)	20.0	16.0	0.50	-	-	-	-	0.58	5.00	0.20	25	0	0
19	River Terrace Gravels (above GWL)	1.2	120.2	Continuum (MC)	20.0	19.0	0.50	-	-	-	-	0.50	25.00	0.25	38	0	0
2	River Terrace Gravels (below GWL)	2.1	119.0	Continuum (MC)	20.0	19.0	0.50	1.00E-07	1.00E-06	0.50	0.25	0.50	25.00	0.25	38	0	0
3	London Clay B/A3	16.6	117.0	Continuum (HSS)	20.8	15.3	0.70	7.00E-11	7.00E-10	0.80	0.05	1.20	-	0.20	25	5	3
4	London Clay A2	11.4	100.4	Continuum (HSS)	21.6	16.7	0.65	2.00E-11	2.00E-10	0.90	0.05	1.20	-	0.20	25	5	3
5	Lambeth Group Clay (above MLGH)	7.0	88.9	Continuum (HSS)	21.7	17.6	0.55	1.00E-11	1.00E-10	0.90	0.05	1.20	-	0.20	25	5	3
6	Lambeth Group Clay (below MLGH)	8.7	82.0	Continuum (MC)	21.0	17.2	0.70	-	-	-	-	1.20	123.84	0.20	25	5	0
7	Upnor Formation	3.6	73.3	Continuum (MC)	21.0	18.3	0.54	-	-	-	-	1.20	90.00	0.20	30	5	0
8	Thanet Sand	0.9	69.7	Continuum (MC)	19	14.4	0.55	-	-	-	-	1.00	130.00	0.20	38	0	0
9	Chalk	-	68.8	Continuum (MC)	20	15.2	0.67	-	-	-	-	0.80	5000.00	0.20	35	5	0

Table 1: Geotechnical parameters used in FE analysis

Additional analyses were undertaken varying the strength and stiffness parameters of the soil to assess the sensitivity of the lining response to changes in ground conditions.

Sequence of Construction

The FE model replicated the following sequence of construction:

1. Generation of initial stresses and water pressure distribution (based on K_0 and K_w values in Table 1)
2. Installation of the ETH piles – these were wished-in-place
3. Excavation of the north-west shaft and masterplan shaft. A sequence of successive excavation to 0.5m beyond propping levels shown on the design drawings, followed by installation of the prop, was adopted for the excavation of all 5 levels within the shaft down to the base slab level
4. Casting of the base slab
5. Excavation and spraying of LCE1/2, according to the design sequence (top heading and invert for LCE1, pilot, top heading, bench and invert for LCE2)

Modelling of SCL Tunnel Construction

In order to model tunnel excavation steps such that the 2D model was representative of 3D tunnel excavation, ground stress redistribution was modelled by allowing 30% of the internal stress within the elements to be excavated for the pilot and top headings, to be carried outside the model and 70% of the stress redistributed to the model itself. For the bench and invert these figures were adjusted to 20% and 80% respectively to take account of the prior excavation steps. The time over which the redistribution was allowed to occur coincided with approximately 10m of tunnel construction. The sprayed lining gained stiffness and strength according to the modified J2 curve.

Results of FE Modelling

Figures 4 and 5 show the stress level contours at the end of tunnel excavation for LCE1 and LCE2 respectively.

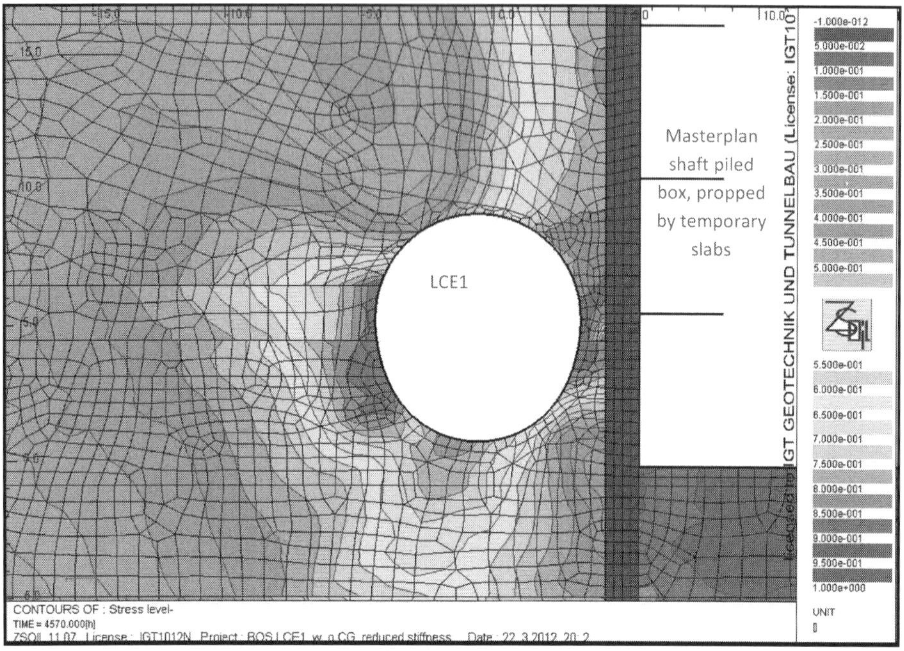

Figure 4: Stress level contours at the end of excavation of LCE1

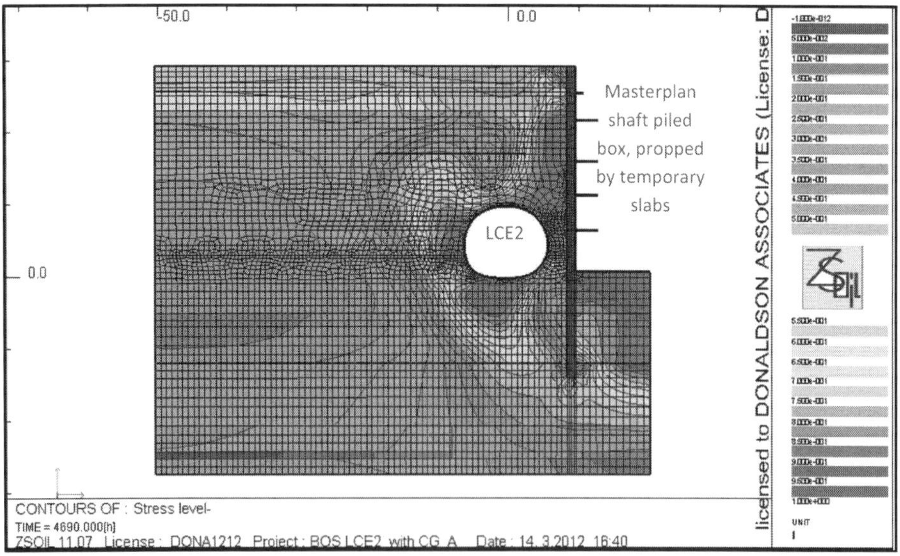

Figure 5: Stress level contours at the end of excavation of LCE2

The results suggested that in both cases more than 85-90% of the available soil strength in the London Clay soil pillar between the tunnel and the wall would be mobilised at the end of construction. This was highest for LCE2, where the results suggested that 96% of the available shear strength was utilised (Figure 5). At 100% utilisation, the soil would begin to yield and in a situation such as this significant deformations would be likely to develop. The extent of the highly stressed soil in Figure 5 was also of concern.

These results indicated that – in an unfactored analysis – the soil pillar was highly stressed and approaching full mobilisation of the soil mass. This was a result of the relaxation of horizontal stress locally following the shaft excavation and then the tunnelling work. The modelling assumed that the London Clay was homogenous throughout its mass, and did not take account of any fissuring. The possible presence of adversely-oriented fissures presented an unacceptable risk of instability of the soil column during excavation.

Mitigation of Risk

Having identified a potential risk to the stability of the excavation a solution was required to mitigate this risk. Any solution needed to take account both of physical constraints, and due to the proximity to adjacent works there was potential for numerous interfaces with other designers and contractors. These are discussed in detail in this section.

For the SCL tunnels constructed under contract C300/C410, the primary lining was designed to form part of the permanent works as a composite SCL lining and therefore any solution could not compromise the integrity or performance of the SCL primary lining. The complex geometry of this part of the SCL tunnelling further limited the options for installing support measures.

Installation of pre-support measures from within the ETH was also complicated due to the box being only partially excavated at the time of construction of LCE1 and LCE2. The north-west shaft would clearly need to have been excavated to allow tunnelling for LCE1 and LCE2 to commence; in contrast, the masterplan shaft (being constructed by an adjoining Crossrail contractor) was programmed to be excavated concurrently with the construction of LCE1. The remainder of the ETH would be excavated at a later stage. Furthermore, it was not possible to impose any load on the ETH piles as these had not been designed to allow for this. Any solution therefore needed to be independent of the piled wall structure.

The constraints for the three parts of the tunnel (LCE1, flared transition from LCE1 to LCE2 and LCE2) are summarised in Table 2.

Part of tunnel	Constraints
LCE1	Accessible from the already-excavated north west shaft.
Transition	Flare commenced at approximately 8m from the north-west shaft. Support measures to be installed from within tunnel or masterplan shafts (NB. ETH not excavated)
LCE2	Support measures to be installed from within tunnel (or masterplan shaft, if constructed).

Table 2: Constraints on installation of support measures

In practice the support measures were selected to both address the risk of soil failure and to address the constraints for installation. Support measures were preferred that could be installed ahead of tunnel excavation, as delay in achieving ring closure was seen as increasing the risk to excavation stability and potentially increasing settlement.

For LCE1, installation of a grouted pipe arch from the north-west shaft would provide protection against both local and global potential instability (risk of falling blocks and failure of the soil pillar respectively). This could be safely installed from the north-west shaft, and a portal arch was designed to sit inside the piles of the north-west shaft to support the pipes independently of the piles.

Provision of support measures for the transition and LCE2 was particularly difficult. The complex geometry of the transition (see Figure 6) was not well suited to installing pipes from the ETH or spiles from within the tunnel. Besides which, spiling would only provide protection against potential instability local to the tunnels; it would not be able to support the vulnerable soil pillar.

Significant amendments to the permanent works design were ruled out due to spaceproofing requirements and programme constraints. Soil stabilisation from the surface was practically impossible due to logistical constraints and the presence of a temporary shaft with Tube à Manchette (TaM) arrays for the compensation grouting. Reducing the length of the excavation rounds was discounted as it was considered that these would make excavation and spraying more difficult without accelerating ring closure and therefore any benefit would be limited.

Consideration was given to using tensioned cables between the pilot tunnel and the piled wall to raise the horizontal stresses in the soil pillar and improve its strength. However this inevitably involved an interface with both the primary lining and the piles would need assessment and agreement from the designers of both of these elements. The spacing involved in installing the cables between piles did not suit the round lengths proposed for the SCL tunnelling. To prevent the primary lining from being adversely affected would have meant releasing the tension in the cables at the critical moment of excavation and this solution was therefore not considered appropriate.

The preferred solution was to create a reinforced earth mattress above the tunnel to intercept the vulnerable soil column and enhance the capacity of the soil to spread loading over a wider area. This was to be achieved using an array of overlapping grouted pipes installed near-horizontally from the masterplan shaft. This approach had numerous advantages:

- The complex geometry of LCE2 and the transition did not interfere with the installation
- As a pre-support system the grouted spiles could be installed offline
- No delay to achieving ring closure
- No impact on adjacent permanent works
- Additional “Toolbox items” (e.g. spiles, pocket excavation) were still feasible if problematic ground conditions were encountered locally

Design Development

The design of the pipe arch followed conventional methods, with loads determined from the FE model and based on an agreed excavation sequence. The design was able to allow for the installation of grouted spiles above the transition, if needed for local ground stability (e.g. if ground found to be highly fissured). This is shown indicatively in Figure 6. As the pipes were to be installed in close proximity to the excavation, the key issue was to ensure the pipes would not protrude into the transition.

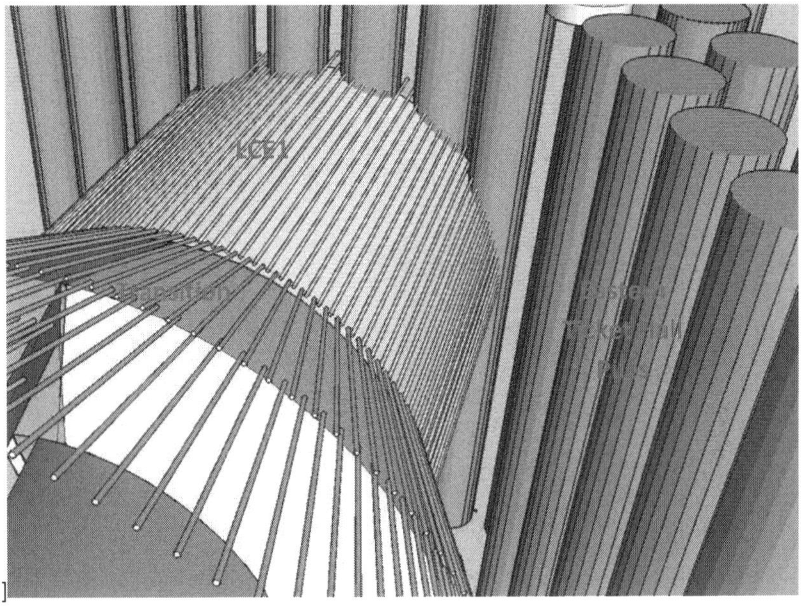

Figure 6: Indicative sketch of grouted pipe arch above LCE1 allowing for installation of spiles above transition if needed

The reinforced soil mattress was designed in two ways. Firstly, the degree of improvement in soil parameters and the depth over which this would be required was investigated using the FE model described in Section 3. A modest improvement in soil parameters (around 10%) was required over a 1m band above the tunnel to disrupt the formation of the potential failure mechanism. The potential for local failure would remain but could be addressed by the pre-designed toolbox measures. A mattress was judged to be created provided that a system of reinforcing elements could be inserted that crossed each other more than once over the required depth, as indicated in Figure 7. A layout for grouted pipes was devised, which would not be required to carry structural load but to enable the soil to act as a more coherent mass. The pipes would be installed from the masterplan shaft, from between the contiguous piles, sufficiently above the transition and LCE2 that spiles could be installed if necessary. The pipe size was selected such that it was large enough to enable drilling of around 12m, but sufficiently small to facilitate drilling at angles in between the existing piles with a free space of around 150mm. Self-drilling pipes of 76mm diameter were judged to fulfil both requirements.

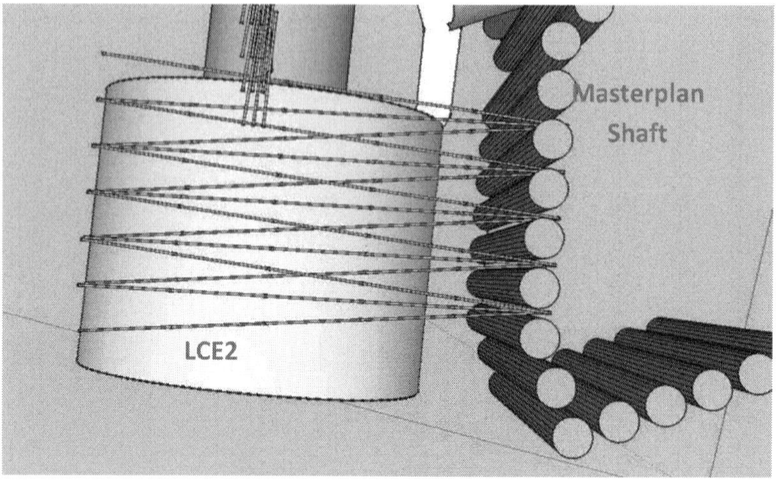

Figure 7: Indicative sketch showing overlapping array of grouted spiles above LCE2

In order to install these safely and to avoid the masterplan prop slabs and/or the requirement for additional temporary working platforms, it was proposed that the pipes were installed at an angle to the horizontal. The lengths and angle of the pipes was intended to avoid the TAMs already in place 4m above the tunnels.

Figure 8 gives an indicative elevation showing the relative location of the temporary props in the masterplan shaft, the drilling rig, the position of the TaM's and how these factors combined to constrain the installation of the grouted spile array, along with the proposed solution. Figure 9 shows the design layout of the grouted spile array above LCE2.

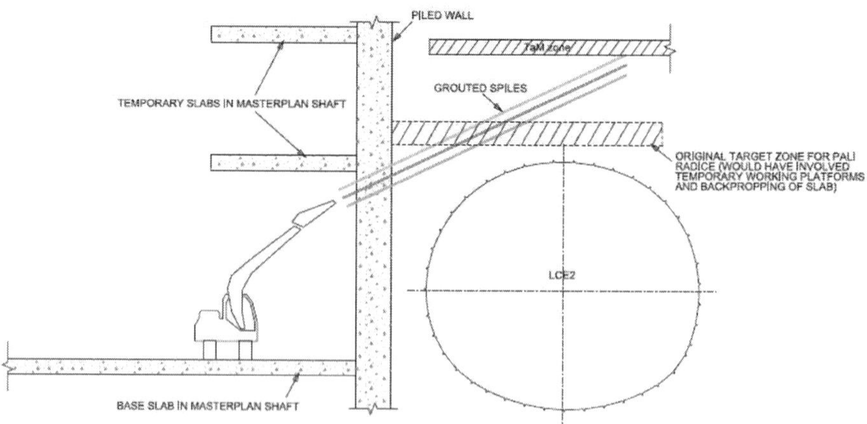

Figure 8: Elevation showing installation constraints for grouted spile array

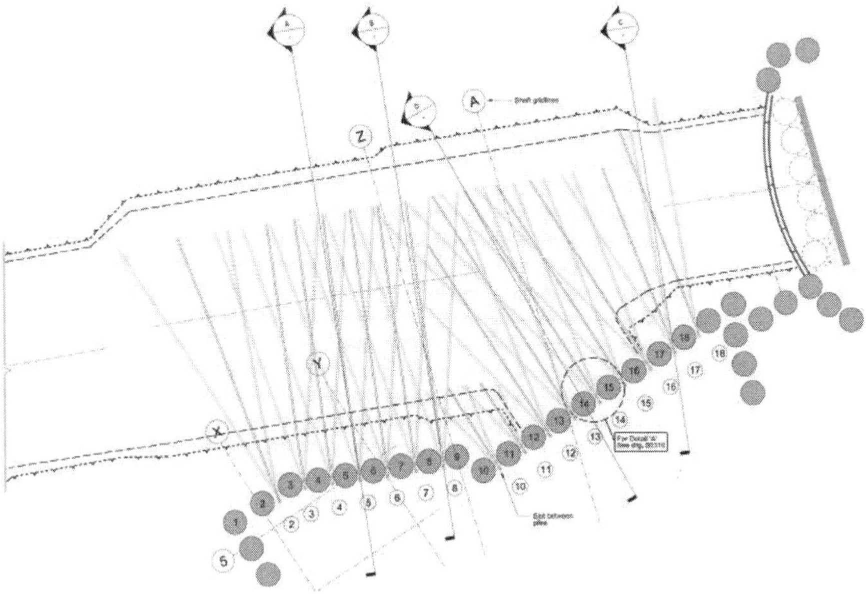

Figure 9: Design Layout of Grouted Spile Array above LCE2

Residual Risk

The construction team had been consulted extensively during the design development period, yet there remained concerns that, on such a large project with multiple shifts, the design philosophy and residual design risks associated with a novel scheme such as this may not have been fully appreciated by the wider construction and supervisory teams. Designer-led toolbox talks were considered to be the most effective means of communicating these issues, and this innovation was sufficiently successful to be rolled out elsewhere on the project.

Construction

In spite of the best efforts of all parties during the design process, some difficulties were still encountered during construction of the grouted spile array above LCE2. In particular, during installation of the first high level grouted spiles, grout ingress was observed from one of the nearby TaMs. A further review was undertaken to assess the as-built TaM positions, the design locations of the spiles and the drilling tolerances to assess their proximity and the likelihood of clashes. As a precaution the separation between the two was increased to reduce the likelihood of recurrence. The length reduction of some pipes was anticipated to reduce the effectiveness of the mattress, but not in the key area between tunnel crown and masterplan shaft. The full time site presence of Donaldson staff within the BFK Engineering team enabled rapid and effective communication of issues and decision-making without delaying site progress.

Once drilling from the masterplan shaft had commenced it also became evident that the limitations on the location of the drilling rig during operation were more onerous than anticipated, due to the working room requirements for the drilling rig being used and the awkward nature of the space in the masterplan shaft. Not all of the specified pattern of spiles could be installed and therefore a rapid re-assessment was needed. This was the case in the north-west corner of the masterplan shaft, where working room was particularly limited, and meant that spile coverage above the final few metres of LCE2 would be particularly difficult to achieve. Figure 10 illustrates the awkward and congested nature of the site at the time these works were being undertaken.

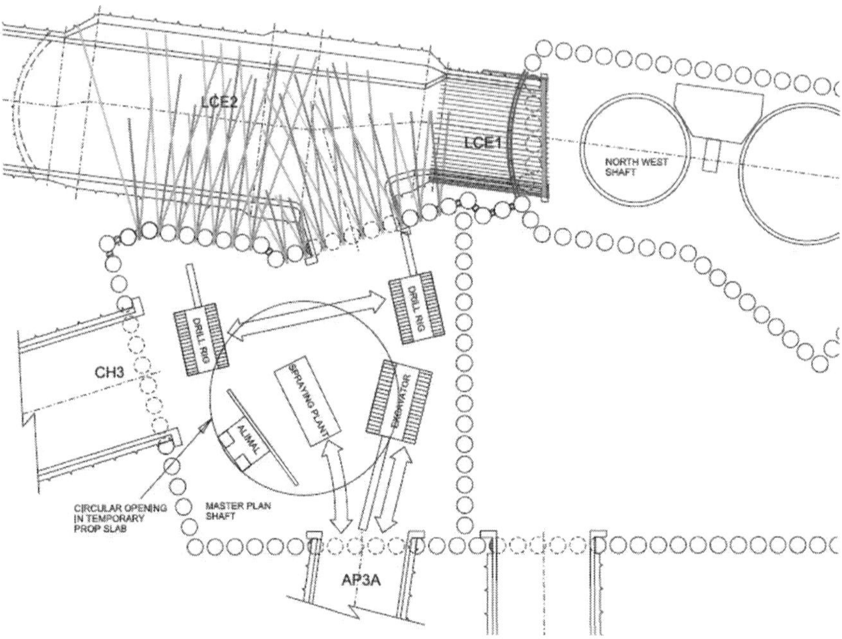

Figure 10: Site Layout at the time of installation of the grouted spile array

The geotechnical records obtained from the enlargement of LCE1 and the start of LCE2 (above which it was feasible to install the grouted spile array) indicated that the ground conditions were likely to be better than allowed for in the design. This was confirmed by further site visits to observe the geotechnical conditions during enlargement of early parts of LCE2 and logging of the excavation of the pilot tunnel through the affected areas. In view of this, the design was reviewed to confirm the requirements if the ground conditions remained favourable. It was found that the coverage of the array could be reduced where the distance between the masterplan shaft and LCE2 increased provided that the ground conditions remained favourable.

As more information became available following excavation of the pilot tunnel through LCE1 and LCE2, and as excavation of LCE2 progressed under the reinforced soil mattress, confidence in the ground conditions grew and the design was adapted accordingly. The final plan layout of the grouted spile array once the various constraints had been taken into account is shown in Figure 10.

Conclusion

Both tunnels LCE1 and LCE2 were constructed successfully with no block failures or excessive movement either within the tunnel or of the surrounding assets. Whilst the pipe arch solution over LCE1 was a tried and tested method, the grouted pipe 'mattress' over the transition and LCE2 was a novel approach and driven out of necessity as a result of the challenging geometry and significant physical and logistical constraints at this location.

The combination of a pipe arch and an innovative grouted spile array as a pre-support system at this complex location had numerous benefits to the project, but the principal advantages were that there was no adverse impact to any of the adjacent permanent works, and that no further in-tunnel measures were required during excavation, leading to more rapid ring closure. Furthermore these works could be carried out offline, without having any impact on the SCL tunnelling.

Whilst the design of measures could be carried out readily, the installation constraints drove the choice of solutions, and the installation issues required close consultation between installer and designer together with careful re-evaluation of the design. The benefit of embedding key members of the design team onto site and carrying out careful observation and recording allowed rapid re-analysis of the design based on site observations and a reduction in the requirement for installation to the benefit of the project. This highly effective communication between the design and construction teams resulted in the successful and timely re-evaluation of the design wherever this was required by the challenging conditions encountered on site.

Crossrail Project: Infrastructure design and construction
ISBN 978-0-7277-6078-4

doi: 10.1680/cpid.60784.433

Design of Crossrail's Precast Tunnel Linings for Fire

Jessica Ryburn BEng (*Hons*) MSocSci, CEng, MICE, Atkins
Harnaik Mann MEng, Arup
Stephen Doran BSc, CEng, FICE, Tunnel Consult Ltd
Karen Cheung BEng, MSc, CEng, MICE, Arup

Abstract

This paper describes the approach taken to design Crossrail's precast concrete running tunnel linings for structural stability during a severe fire incident.

In recent decades there have been a number of fire incidents in road and rail tunnels which have caused severe damage to the lining structures. These incidents have demonstrated that fires in tunnels can heat up much more rapidly than assumed by the standard cellulose design fire curve and that concrete in tunnel linings is prone to explosive spalling. As these issues are not as prevalent for the design of other types of structures they are not (yet) explicitly covered by prescriptive design rules in design codes and standards.

The approach taken for the design of Crossrail's precast concrete tunnel linings involved an initial design phase assessment followed by pre-construction phase verification testing. The design phase assessment used thermal material properties obtained from historic testing of the Channel Tunnel Rail Link (CTRL) precast lining segments. These properties were used to model the temperature distribution in the Crossrail lining segments when exposed to a RABT-ZTV fire design curve. A simplified calculation method was then used to estimate the capacity of a reduced structural cross-section, which had been estimated from the temperature distribution and included a nominal allowance for spalling. This was compared against an accidental loading case.

Prior to construction, prototype Crossrail lining segments were subjected to full scale fire testing under load. Crossrail's main running tunnels are split into three separate construction contracts, each of which is independently manufacturing precast concrete segments using different concrete mix designs. Therefore three separate test programmes were undertaken. The results from these test programmes are compared with the design phase predictions and show good agreement.

Introduction

Crossrail is a new high frequency, high capacity railway, involving the construction of 21 kms of twin 6.2 m diameter precast concrete segmentally lined running tunnels beneath central London.

An important design requirement for the Crossrail tunnel linings is that in the event of a fire the linings must continue to carry the ground loads. This is not only for passengers and staff in the tunnel escaping and the emergency services responding, but also to ensure that such an accident does not result in a collapse that endangers populated areas of the city on the surface above.

In recent decades there have been a number of fire incidents in road and rail tunnels which have caused severe damage to the lining structures. These incidents have demonstrated that fires in tunnels can be much more intense than assumed by a standard design fire curve and that, especially in such conditions, the high strength concrete used in tunnel linings is prone to explosive spalling.

The design for such severe fires is not a prevalent requirement for other types of structures and therefore design for these fires is not (yet) explicitly covered by prescriptive design rules in design codes and standards. The approach taken by Crossrail therefore involved not only a design phase assessment based on the principles of BS EN 1992-1-2 but also pre-construction testing in order to verify the design.

The main running tunnels were split into three construction contracts, each of which is independently manufacturing the precast segments used in their section of the project. Therefore three sets of verification testing were carried out for three different concrete mixes and are presented for comparison with the design phase assessment.

Specified requirements

Applicable Time-Temperature Curve

Structural fire resistance is usually assessed against an assumed “design fire” time-temperature curve, which defines the temperature at the surface of the concrete throughout the duration of the fire.

The design fire for building structures is usually taken to be the ISO 834 (BS 476) standard fire curve, which is a “cellulosic” fire based a fuel load comprised of wood, paper, furniture and fabrics. Fuel loads in tunnels however, can be greater than this and burn much more rapidly with greater intensity. Therefore project standards required that the tunnel lining meet the more severe RABT-ZTV (EUREKA) curve, reflecting the recent adoption of EU Directive 2008/163/EC –L64 – TSI on Safety in Railways Tunnels in the trans-European conventional and high-speed rail system.

The RABT curve assumes an intense fire of short duration. A maximum temperature of 1200°C is reached within 5 minutes, and the fire duration is 60 minutes. A cooling period between 60 minutes and 170 minutes is defined. By contrast the standard curve only reaches a temperature of 576°C after 5 minutes and a temperature of 945°C after 60 minutes. No cooling period is defined. The significance of the much higher rate of heating (thermal shock) of the RABT curve fire is that the risk of explosive spalling of the concrete lining is greatly increased.

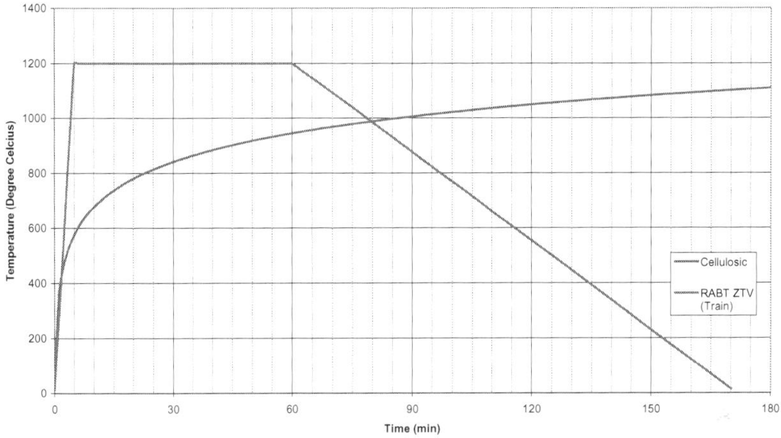

Figure 1: RABT EUREKA Time –Temperature Curve and Standard ISO 834 Time Temperature Curve

Mitigation against spalling

Crossrail's precast tunnel linings comprise of Steel Fibre Reinforced Concrete (SFRC) and except in a few specific locations, do not contain conventional reinforcement bar. A minimum concrete compressive strength of C50/60 was specified although the actual concrete strengths used in the works tended to exceed this requirement in order to satisfy other aspects of the specification.

Concrete is usually considered to perform well in fire scenarios as it does not combust and has good insulating properties. Concrete begins to loose strength around 300°C and suffers considerable reduction in strength if subject to temperatures above 550-600°C, however if there is no spalling, the over-heated concrete is usually limited to the surface facing the fire and this heat damaged concrete surface layer continues to act as insulation for the inner concrete which, being cooler, retains strength and continues to carry load. If spalling occurs, that is if pieces on the surface of the concrete break-off, a new layer of concrete will be exposed to elevated temperatures. This layer may in-turn overheat and spall exposing the next layer. If such progressive spalling is allowed to occur the section size could be reduced to the point where it is no longer adequate to sustain the loads. Hence minimising the potential for concrete to spall is a key design requirement.

Following industry practice, Crossrail's approach to mitigate against the risk of spalling was to specify the use of polypropylene fibres, which melt at relatively lower temperature (140°C) creating a more porous structure which allows vapours to escape thereby mitigating potential spalling due to excess pore vapour pressure. In order to reduce the severity of potential spalling due to internal thermal stresses, the choice of aggregate was also limited to

Limestone or Granite, these materials have a coefficient of thermal expansion closer to the cement paste compared with other aggregates.

However the phenomenon of concrete spalling is not well enough understood to be sure that a prescriptive approach to concrete mix design will guarantee that spalling will not occur. Therefore the specification additionally required that the chosen mix design, when tested under a load which representing the maximum ground load, did not exhibit any surface spalling exceeding 25 mm depth.

Structural Design Assessment

Thermal Analysis

As the design fire curve varied from the standard fire curve, the general thermal profiles given in Eurocode 2 were not applicable to this design and so another source of information was required to estimate the temperature profile through the lining section under the design fire conditions. The approach adopted for Crossrail's assessment was to investigate the data from physical fire tests that were carried out on an earlier project, the Channel Tunnel Rail Link (CTRL). This was based on the reasoning that CTRL had used similar Steel Fibre Reinforced Concrete to that specified for the Crossrail project and would therefore be applicable data.

However a straight correlation was not possible as the CTRL tests had been carried out against the standard fire curve rather than RABT data. Therefore instead the temperature data from the CTRL tests was back analysed using a software package called TEMP/W. This information had been obtained with thermocouples that had been placed at different depths within the test panel. The aim of the back analysis was to establish appropriate values for the thermal parameters: thermal conductivity and volumetric heat capacity.

The final calibrated temperature profile showed good agreement with the CTRL data. Contrary to expectations for a higher strength concrete, the lower limit of thermal conductivity given in BS EN 1992-1-2:2004 3.3.3 provided the best match. Volumetric heat capacity was taken to be 2.5×10^6 J/m^3.

These thermal properties were then used to model the thermal response of the Crossrail lining against the specified RABT curve.

The results of Temperature vs. Time for different thermocouples are presented in Figure 2.

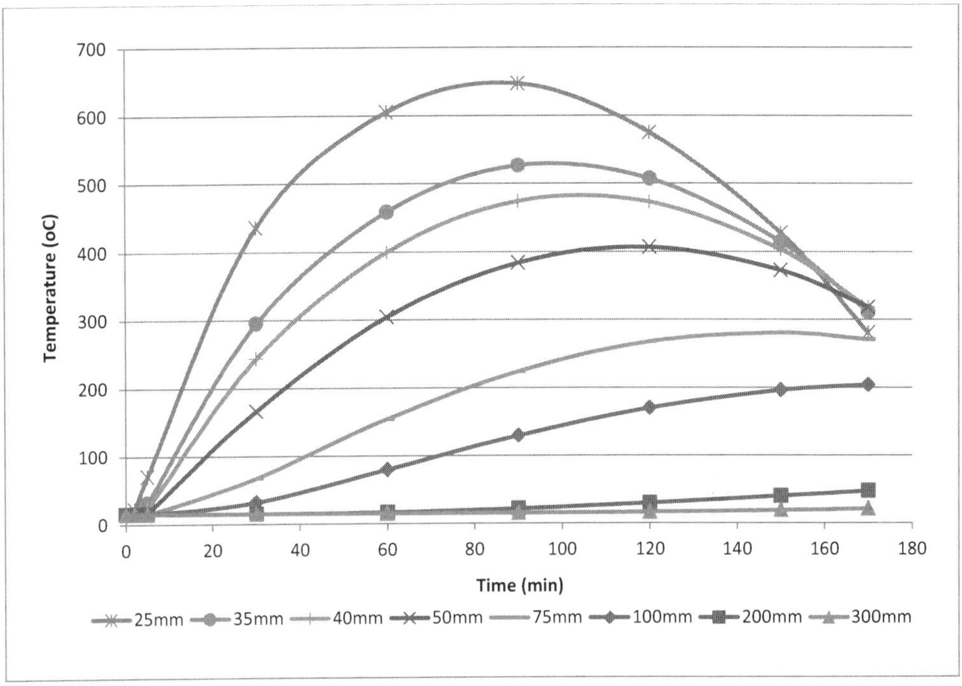

Figure 2 Modelled thermal response of Crossrail lining under RABT-ZTV fire curve (Cheung and Doran, 2010)

Structural capacity calculation

Eurocode 2 broadly categorises three different methods for verifying the structural capacity of concrete structures exposed to fire. These are:

- Tabulated Data
- Simplified Calculation Methods
  - 500 ^{0}C Isotherm method
  - Zone Method
- Advanced Calculation Models

Both the tabulated data and 500°C Isotherm Method calculation method are based on standard cellulosic fire exposure and were therefore not strictly applicable the Crossrail segment design. The zone method and advanced calculation methods require knowledge of the material properties of the SFRC for the relevant temperature range and more significantly, for the relevant rate of heating. This information was not known. Hence the designers had no strictly complying codified design methodology for this situation. However the 500°C Isotherm Method, described Annex B of EN1992-1-2, did provide a codified principal for basing structural capacity calculations on a reduced of the cross-section thickness, ignoring the heat damaged zones at the concrete surface. Therefore the 500 ^{0}C

Isotherm method was adopted because in the absence of relevant data it was considered that a more onerous calculation method would not have guaranteed a more accurate result and also because pre-construction testing would provide the opportunity to verify the design assessment.

Annex B of EN1992-1-2 describes the 500°C Isotherm Method. The heat damaged zone is defined as the depth of the 500°C isotherm. The strength and stiffness of the heat damaged zone is neglected, whilst the remaining section retains its initial values of strength and elasticity.

Based on the TEMP/W modelling, the 500°C Isotherm curve was found to be located 40mm from the heated side. An additional loss of section of 25-35 mm due to spalling was allowed. Therefore section capacity calculations and segment joint capacity calculations were based on a reduced 225mm thick section. These calculations assumed an accidental load case with partial load factors taken as 1.0 and material factor of 1.2 in accordance with BS EN 1990 and BS EN 1992-1-1 respectively.

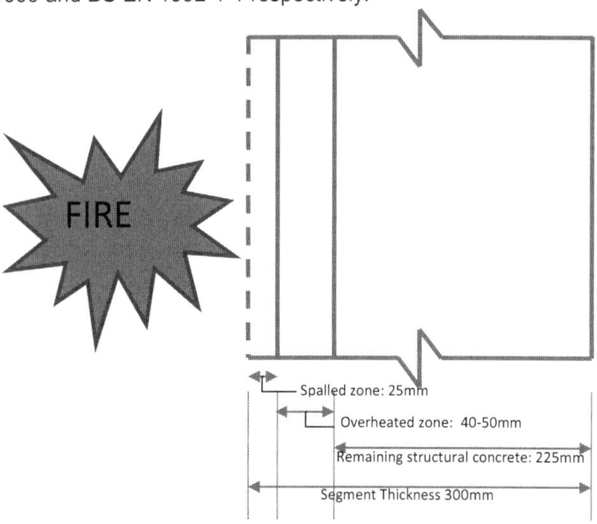

Figure 3: Section model adopted for structural calculations.

Testing of Prototype Segments

Loading during testing.

For physical fire tests to be relevant they need to mimic in the in-service conditions in terms of the magnitude and duration of the fire, the ground loading and restraint of the element tested (refer Figure 4). Owing to a circular shape and embedment in the ground, segmental tunnel linings will be prevented from thermal expansion during a fire and therefore be subject to additional stress due to thermal effects. These thermal stresses can be quite significant especially in the hotter surface layers.

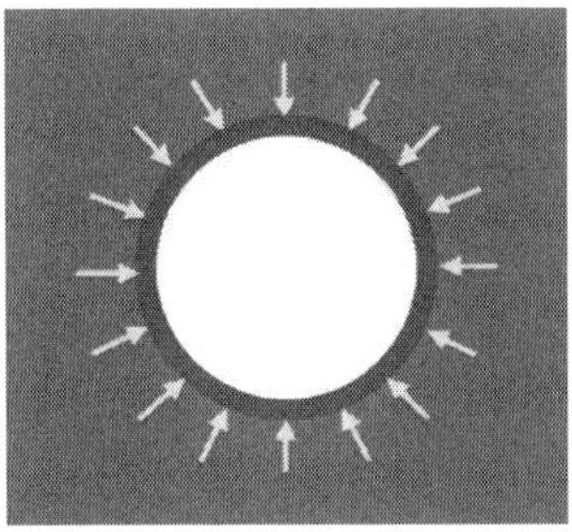

Figure 4: Typical in service loads on a tunnel.

In the last few decades there have been a number of research projects in Europe to investigate the effects of fire on tunnel linings which have used varying test set-ups, including the use of post tensioning cables to mimic in-service hoop loads. For Crossrail a similar test set-up that was developed for the EUREKA project was adopted. The test set-up seeks to mimic the ground loads via application of force via horizontal and vertical jacks as shown in Figure 5.

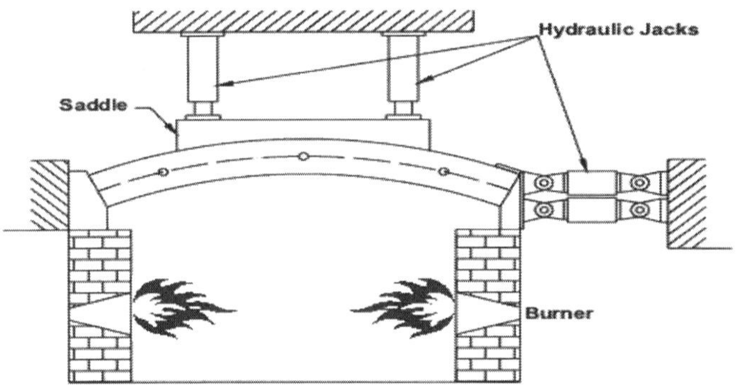

Figure 5: Fire test set-up arrangement used at Crossrail, which was adopted from the EUREKA-Project EU499.

The initial intention of the fires test specification was to preload a test segment using hydraulic jacks to a load equivalent to the ground load and then restrain the segment so that it could not deform during the fire test. However it would have been difficult to restrain the segments in such a manner with the available test equipment, so instead the segment was preloaded to a load equivalent to the worst case in-service ground loading plus an additional load calculated to be equivalent to the thermal stress that would be experienced by the restrained fibres on the intrados of the segment during a fire. During the fire test the segment was permitted to deform while the load was held constant.

Test Configurations

Fire Test for two of the Construction Contracts (C300 and C305) were carried out by Efectis Nederland BV Centre for Fire Safety and the third (C310) at MFPA Leipzig GmbH, Leipzig Institute for Materials Research and Testing. The two test facilities employed slightly different test set-up arrangements to suit their existing facilities and the individual requirements of each particular contract as shown in Figures 6a and 6b and 7a to 7c.

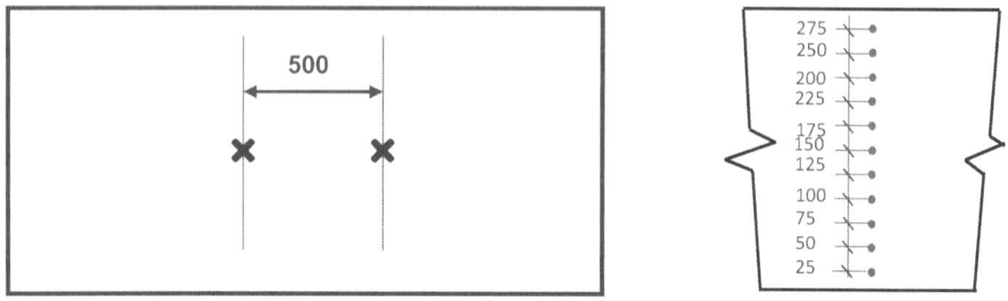

Figure 6a: Arrangement of Thermocouples in Test Segments tested at the Efectis Nederland facility (thermocoupes at 125mm – 275mm depth were omitted in the C305 samples). (Vermeer and Kortekaas, 2012), (Vermeer and Noordijk 2012).

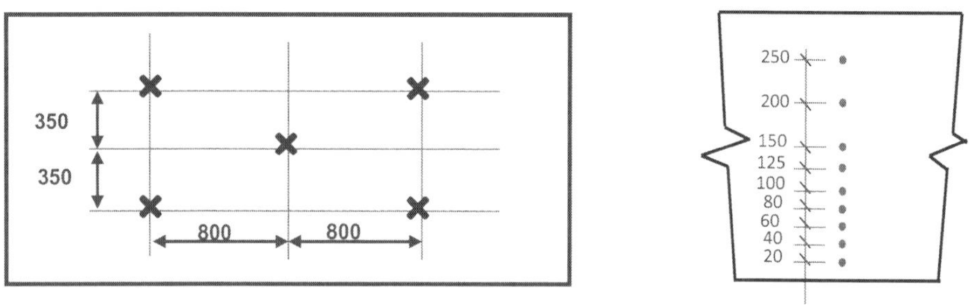

Figure 6b: Arrangement of Thermocouples in test segments tested at the Liepzig facility (Dehn and Juknat, 2012)

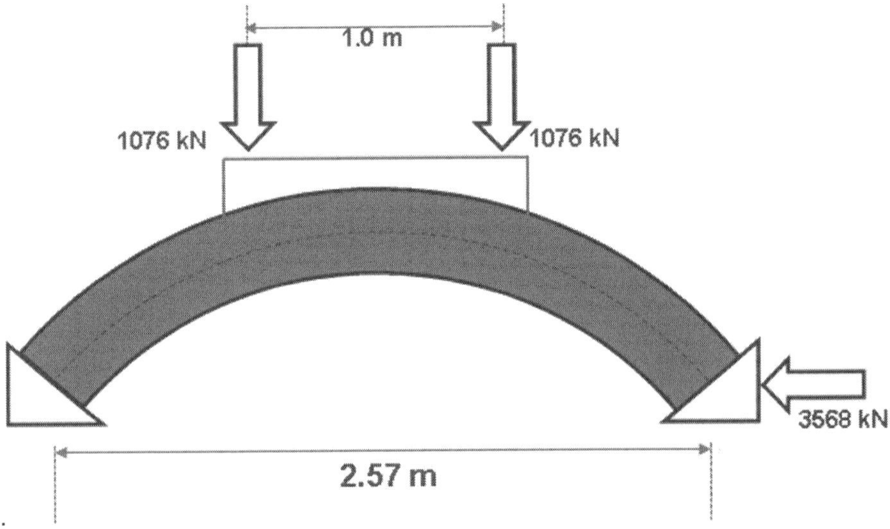

Figure 7a: Test load arrangement for C300, Efectis Facility (Vermeer and Kortekaas, 2012).

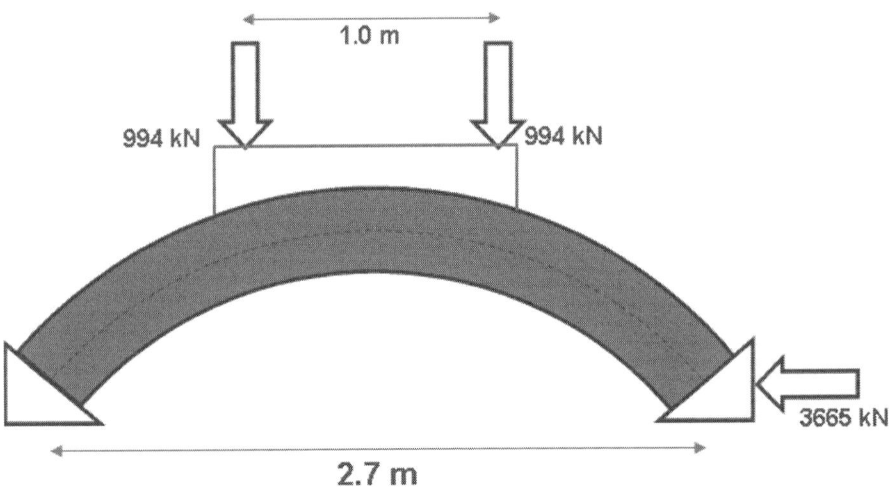

Figure 7b: Test load arrangement for C305 Efectis Facility (Vermeer and Noordijk 2012).

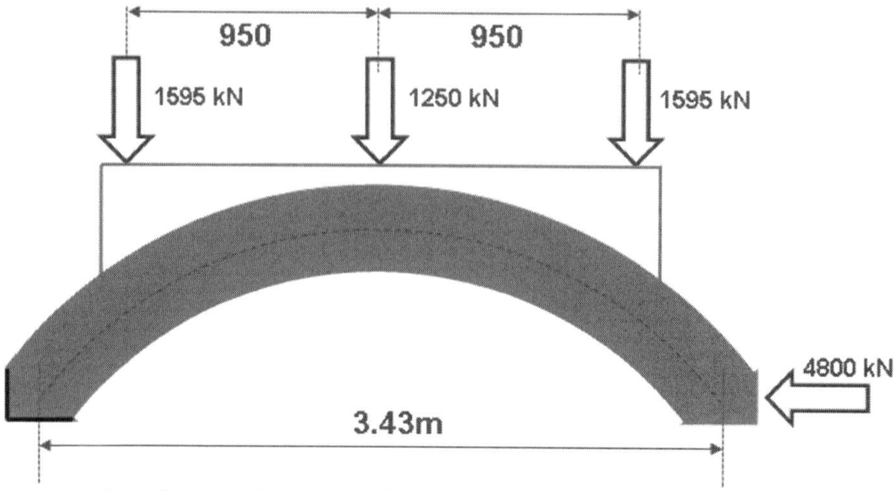

Figure 7c: Test load arrangement for C310 Leipzig Facility (Dehn and Juknat, 2012).

Mix Design

	C300 BFK-13 (mix B10)	C305 Specimen 1	C310 CRL8
Test Laboratory	Efectis	Efectis	Leipzig
Age at testing (days)	68	65	36
PP-fibre content (kg/m^3)	1.0[1]	2.0	1.5
Steel fibre content (kg/m^3)	35	30	30
Powder content (kg/m^3)	410 (287 CEM ; 123 PFA)	450 (337 CEM I, 113 PFA)	460 (258 CEM I, 172 GGBS, 30 filler)
Aggregate type	Limestone	Limestone	86-93% Limestone[2]
Moisture content	Not known	Not known	3.75%
28 day cube compressive strength (MPa)	69.1-88.3	68.5-75.2	87.9 – 94.7

Table 1: Passing test sample mix design

[1] The steel fibre content was increased to 40kg/m^3 for the actual segment production in order to satisfy other specification requirements.

[2] Permission granted to use an aggregate source local to the precast factory, which contains 6-14% other minerals.

All three Contracts tested more than one sample with differing amounts of polypropylene fibres. The data of the passing samples are presented in Table 1. There were a number of other samples with lesser poly-propylene fibre contents that were tested and did not pass.

Results and discussion

Passing Samples

The measured thermal distribution through the segment during the fire test are shown in Figure 10a to 10c below and summarised in Table 2. It can be seen that these results are comparable to the modelled design temperature distribution in Figure 2.

For further comparison, Figure 8 from BS EN 1992-1-2 shows the design thermal response under standard fire conditions for a 200 mm thick slab and is also summarised in Table 2. Figure 9 shows the results from the CTRL physical fire tests. The maximum temperatures obtained from the BS EN 1992-1-2 standard fire curve temperature distribution are in a comparable range to maximum temperatures observed in physical tests again RABT curves at depths of 25-100 mm, although the standard values are not always conservative compared to the physical test results. However as BS EN 1992 allows the use of 500°C Isotherm Method with any time heat regime which causes similar temperature fields in the fire exposed member, these results gives confidence in the applicability of the structural calculation method used in the design assessment.

Scenario	Average spalling depth (mm)	Max Spalling depth (mm)	Temp at approx. 50mm depth[1] °C	Temp at approx. 75mm[2] depth °C)
Design prediction, RABT fire	0	0[3]	405 (after 120 minutes)	260-280 (after 120 - 170minutes)
C300 passing test sample, RABT fire	13.5	35	350 - 380 (at 100-120 minutes)	250-280 (after 120 - 170minutes)
C305 passing test sample, RABT fire	8.4	13 mm	300 - 325 (at 120 minutes)	230 (after 120 - 180minutes)
C310 passing test sample, RABT fire	3	14 mm	375 - 420 (at 100-120 minutes) 40 mm depth 275 -315 (at 100-120 minutes) 60 mm depth	220-250(after 120 - 175 minutes) 80mm depth
Standard fire, 200 mm thick slab, EC 1992-1-2:2004 Figure A.2	0	0	390 (after 120minutes)	240-330 (after 120-180 minutes)

Table 2: Summary of passing Crossrail fire tests

[1]For the C310 test, no thermocouple was present at 50 mm depth – 40 mm & 60 mm is data presented in table instead.

[2]For the C310 test, no thermocouple was present at 75 mm depth – 80 mm is data presented in table instead.

[3]25mm allowance for spalling was taken into account in the structural capacity calculations, however thermal modelling was based on 300mm section.

It can be seen from Table 2 that the C300 sample with 1% poly-propylene content did not strictly comply with the specification for maximum spalling depth. However as the depth of the 500°C isotherm was less than the design expectations and at the 75mm depth the temperature was less than 300°C, it was accepted to use 1% polypropylene fibres with that concrete mix.

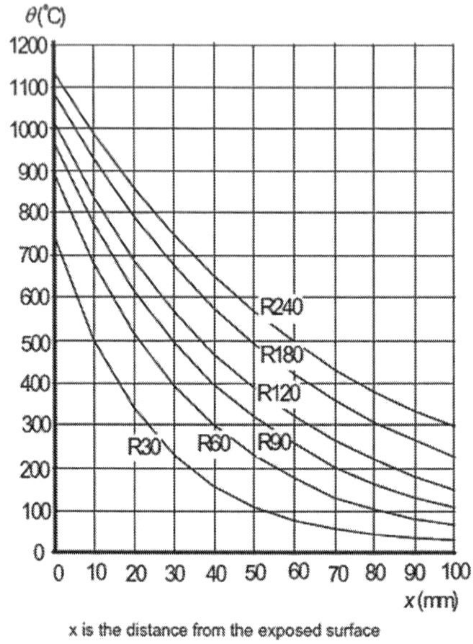

Figure A.2: Temperature profiles for slabs (height h = 200) for R60 - R240

Figure 8: Figure A.2 from BS EN 1992-1-2:2004 showing design thermal response for a 200mm thick slab exposed to a standard fire.

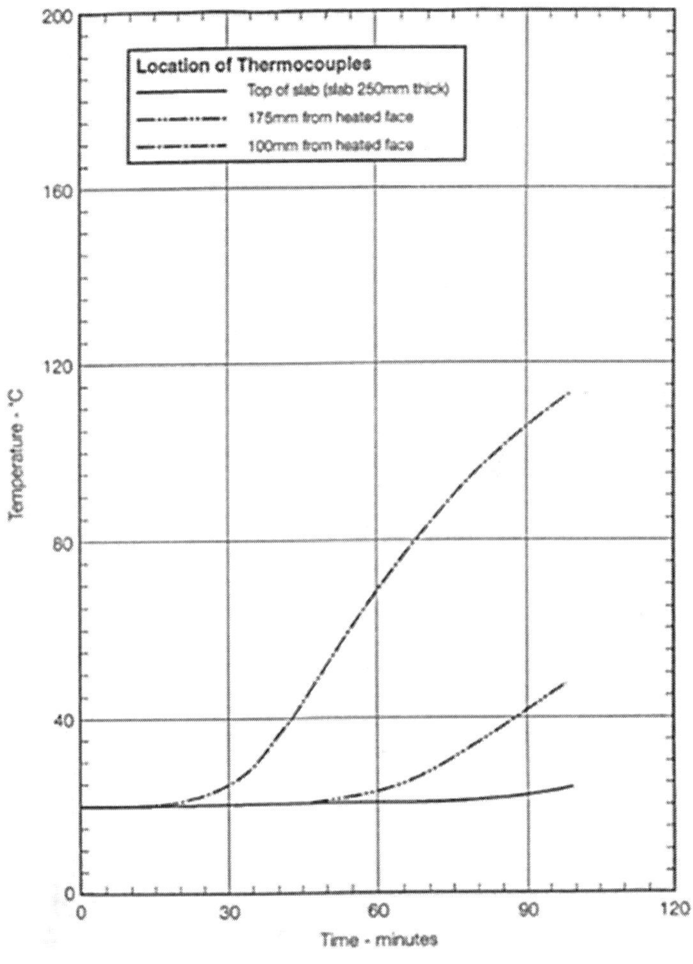

Figure 9: Results from CTRL physical fire tests of
test panels expose to a standard fire (Union Railways, 1997)

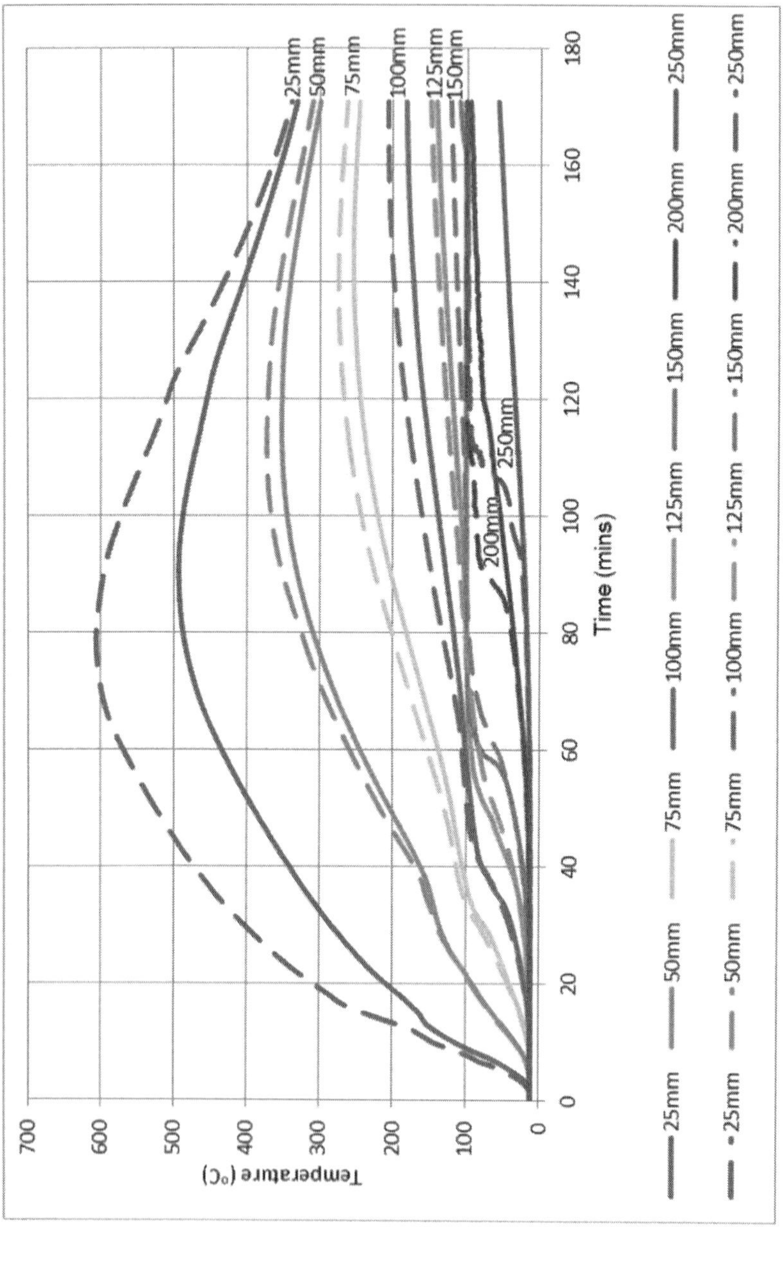

Figure 10a: C300 measured thermal response of passing test sample (Vermeer and Kortekaas, 2012).

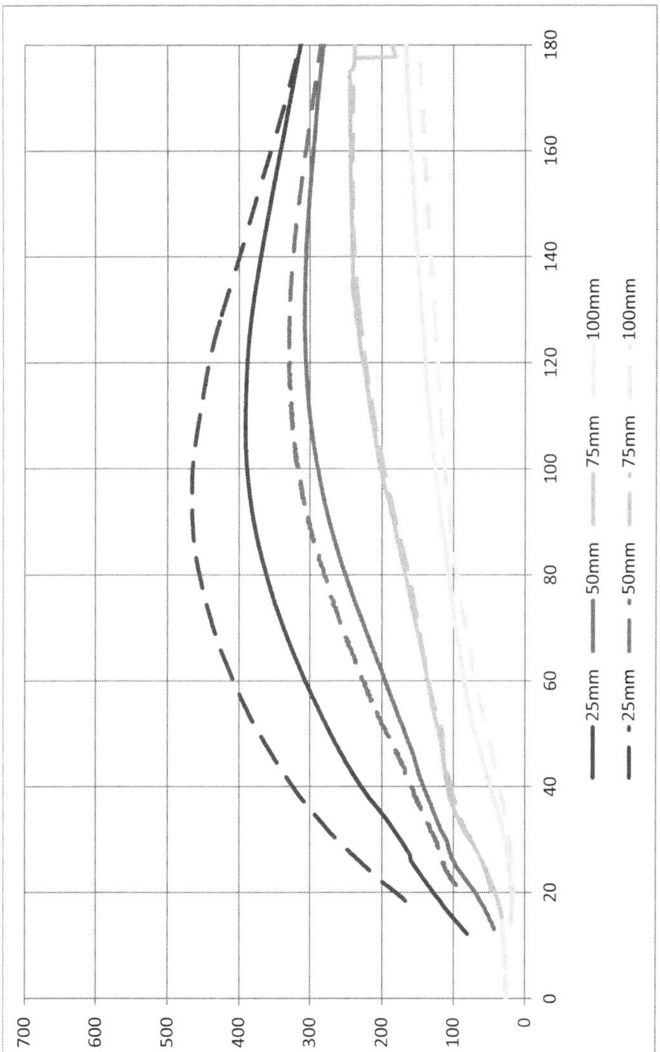

Figure 10b: C305 measured thermal response of passing test sample (Vermeer and Noordijk, 2012)

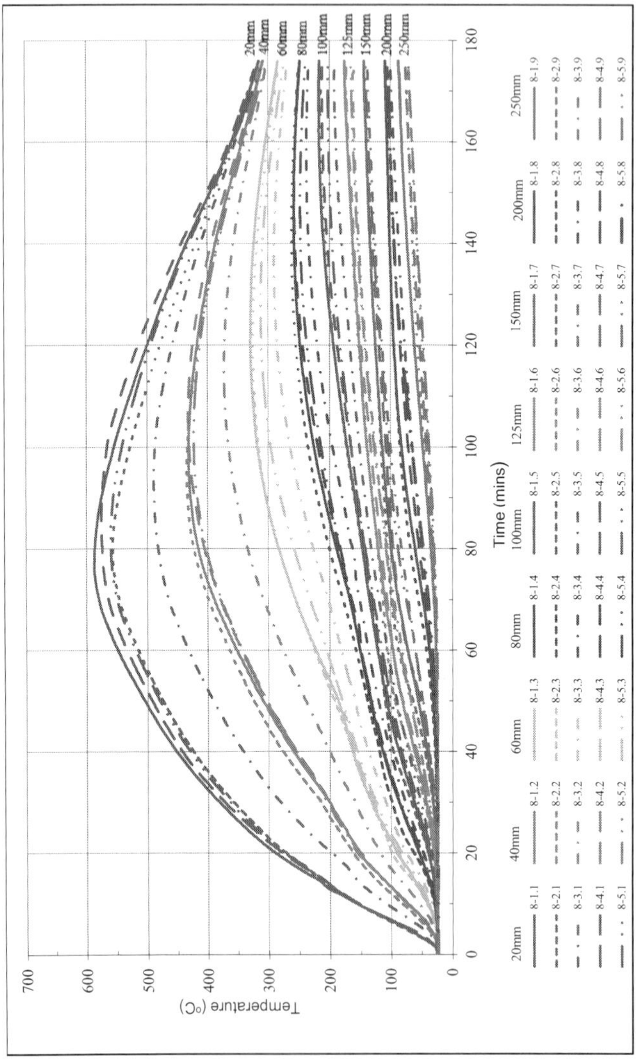

Figure 10c: C310 measured thermal response of passing test sample (Dehn and Juknat, 2012)

Non-passing samples

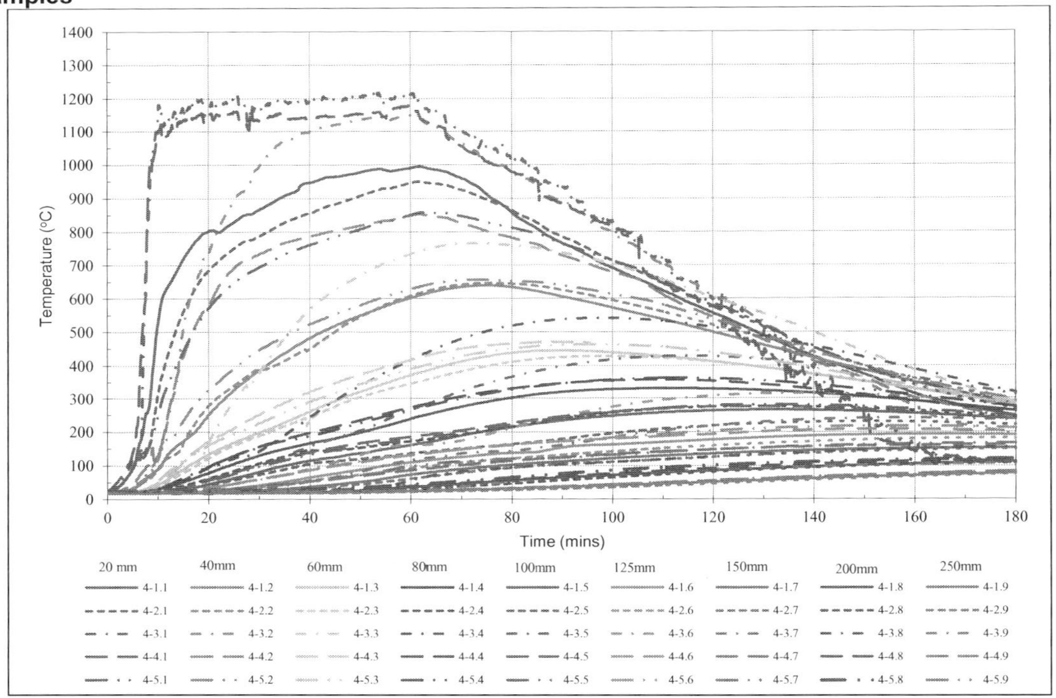

Figure 11: Non passing test sample (for C310) using 0.9 kg/m^3 polypropylene fibres (Dehn and Juknat, 2012)

For comparison an example from a non-passing sample is shown in Figure 11 below. This sample exhibited average spalling 60 mm and a maximum spalled depth of 155 mm, with the 500 oC isotherm essentially extending to a depth of 100 mm. It should be noted however, that despite being preloaded with a compressive load representing the maximum depth ground load plus restrained thermal expansion, none of the segments, not even the non-passing segments, exhibited a structural failure during the fire test.

Conclusion

The design phase thermal analysis results, which were based on thermal properties obtained from back analysed data from testing carried out on the CTRL project, show agreement or conservatism compared with the thermocouple data from the Crossrail physical test results. Additionally none of the physical fire test samples failed structurally while tested under loads representing the deepest cross section of the route. Also the temperature profile observed in the segment provides verifying evidence to support the validity of structural design calculations against a RABT-ZTV fire curve using the 500 °C Isotherm Method.

The non-passing fire test samples demonstrate that the most important factor in designing for fire is to ensure the potential for spalling is limited. This was achieved in the three concrete mix designs for Crossrail Segments by the inclusion of between 1.0 and 2 kg/m^3 polypropylene fibres and use of limestone aggregate.

Although Crossrail intend to actively minimise the risk of a major fire incident from occurring in the tunnels through such steps as controlling the potential fuel loads and ignition sources within the tunnel and employing appropriate emergency strategies, over a long anticipated service life of 120 years and beyond the risk of a major fire incident can never be completely ruled-out. However, by taking the steps described above, and in particular by verifying that excessive spalling does not occur during a RABT-ZTV fire exposure, the designers have ensured that the potential for a major fire incident to escalate into a collapse disaster is implausible.

Acknowledgements

Figure 10a and 10b reproduced by permission of Efectis Nederland BV.

Figure 10c reproduced by permission of MFPA Leipzig GmbH.

Reference

Cheung, K. and Doran, S. (2010) “Bored Tunnel PCC Segmental Linings – Fire Calculation” C122-OVE-C4-XCL-CRG01-00008, Crossrail limited
Dehn, F. and Juknat, M. (2012) “Fire Test for Crossrail C310” Test Report No. PB 3.2/11-247-1 MFPA Leipzig GmbH
Union Railways (1997) “Fire Performance of Concrete for Tunnel Linings” Channel Tunnel Rail Link Technical Report No 000-RUG-RLEEX-00005-AB, Union Railways Ltd
Vermeer, M. and Kortekaas, P. (2012) “Fire resistance tests for Crossrail C300 tunnel” Efectis Nederland report 2012-Efectis-R0129 Efectis Nederland BV
Vermeer, M. and Noordijk, L.M. (2012) “Fire resistance tests for Crossrail C305 tunnel” Efectis Nederland report 2012-Efectis-R0698 Efectis Nederland BV

Crossrail Project: Infrastructure design and construction
ISBN 978-0-7277-6078-4

doi: 10.1680/cpid.60784.451

Ground Settlement Behaviour in Chalk due to TBM Excavations

Mark Cheng Asset Protection Engineer
Patryk Mikulski Asset Protection Engineer

Abstract
In Crossrail project contract C310, two new tunnels – the Thames Tunnel (westbound and eastbound) will be constructed between Plumstead portal and North Woolwich portal. The Thames Tunnel is 2.6 km long and about 15 m below the existing Thames river bed. Construction of Plumstead portal commenced in 2011 and completed in 2012 for the westbound TBM launching in January 2013 and eastbound TBM launching in May 2013. The construction of North Woolwich portal has begun in early 2012 in order to receive the first TBM in 2014. The running tunnels will be constructed predominantly through chalk. Two slurry TBM machines, Sophia and Mary named after the wives of Marc Isambard Brunel and Isambard Kingdom Brunel will be adopted for the sake of the existing ground condition – majority in chalk. The degree of settlement control is one of the prerequisites for tunnelling design for the consequent risk of damage to third-party assets. To obtain the surface and sub-surface settlement profile and corresponding volume loss due to the tunnel constructions against the predictions, various instrumentation and monitoring systems, namely from typical manual levelling studs to advanced automated precise hydrostatic water cell system, are used. This paper is to compare the observed tunnelling-induced ground settlements with the predicted by empirical methods. This study will be a valuable addition to the available literature on ground settlements in chalk, particularly since there are so few case histories available and the predictions of volume loss are often conservative as a result.

Introduction
Crossrail is a new high-frequency, high-capacity railway with route running over 100 km from Maidenhead and Heathrow in the west, through central London to Shenfield and Abbey Wood in the east (Figure 1).

The £200 million contract C310 of Crossrail is part of the Crossrail project which will link the east and west London rail routes. The C310 contract is unique for whole Crossrail project as it is the only one using mix-shield tunnelling boring machines (TBMs) with slurry face support. The construction works comprise the construction of the 430m long Plumstead portal, 390m North Woolwich portal and the 2.6 km twin-bore Thames Tunnel between Plumstead and North Woolwich, also known as Tunnel Drive H.

The Thames Tunnel, which will be predominantly constructed through chalk, will use Slurry TBMs. The main tunnelling works started at the beginning of 2013 with the launch of the first slurry TBM Sophia with a diameter of 7.12 m at westbound tunnel on 14th Jan 2013, and the second TBM Mary launched at eastbound tunnel on 22nd May 2013. Westbound and eastbound TBMs arrived at Woolwich box respectively on 15th May 2013 and 12th August 2013. It is anticipated that the tunnelling works will be completed in October 2014.

Assessing tunnelling-induced ground loss is a major issue when predicting ground movements and their effects on adjacent structures. This paper is to study the settlement observed from the monitoring instrumentations between Plumstead portals and Woolwich station, and compared with the settlement values predicted by the Xdisp software, which calculates the tunnelling-induced settlement of assumed volume loss (VL) and trough width parameter (K), primarily under a normal probability (Gaussian) distribution theory.

Tunnel Alignment

C310 contract relates to Drive H Thames Tunnel located at the east end of the Crossrail alignment (Figure 1). The Thames Tunnel covers the two running tunnels from North Woolwich portal to Plumstead portal, including four cross passages. The total length of two tunnels is about 6 km including the cut and cover sections of North Woolwich and Plumstead portals. Maximum ground overburden along route is 22.8 m and deepest point is 24.1 m below sea level, beneath the river Thames. The alignment is passing through river Thames, Woolwich station box, Southern Outfall Sewer, Dockland Light Railway tunnels and below 12 listed buildings within zone of influence. This paper covers only half of the route from Plumstead portal to Woolwich station (at the time of the paper writing, the TBMs are at Woolwich station box.

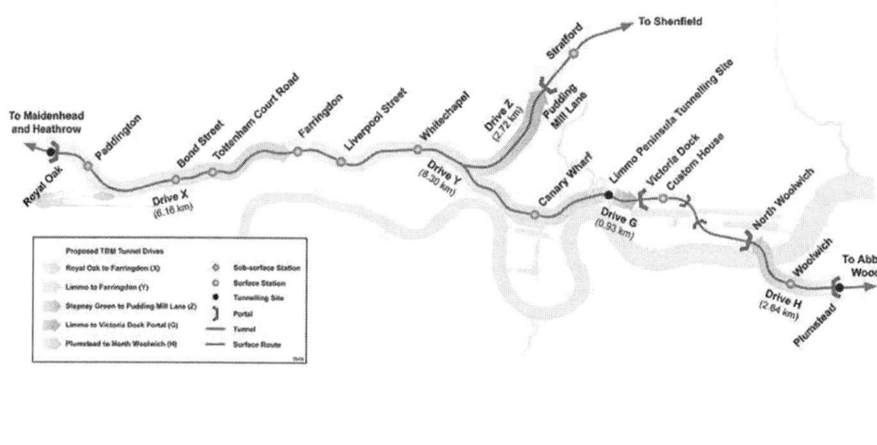

Figure 1 Crossrail Running Tunnel Alignment and Drives

Tunnel Characteristics

The tunnels are constructed by two identical mixed-shield Slurry TBMs with a cutter head extrados of 7.12 m. The TBMs with slurry face support can partly or completely replace the slurry by compressed air for maintenance work, and it has an outside diameter of 7.12 m and forms a tunnel with an inner diameter of 6.20 m. The slurry, which consists of water and bentonite or clay, is selected depending on the ground conditions. The bentonite-soil mix is hydraulically transported out of the tunnel by pipes to be separated into liquid and solid parts (Maidl et al, 2011).

The tunnels are designed as segmentally lined with universal ring. One ring consists of eight segments 1.6 m long, 300 mm thick, reinforced with steel fiber concrete. A total of 1,636 rings were installed for two tunnels between Plumstead portal and Woolwich station. Due to technical constraints on site, only one TBM is under operation during the tunnel constructions. The westbound drive has been built first before the eastbound drive. The average advance rate for first drive was 6 to 7 rings per day and for second drive the average was 10 rings per day. The difference between averages is probably due to the learning curve of the first drive where the TBM team gained experience in site specific soil conditions.

Ground Condition

Ground investigations along the route alignment between the Isle of Dogs and Plumstead have been undertaken by Crossrail and the data are summarised in the Crossrail Geotechnical Sectional Interpretative (May 2007). This report is based on borehole investigations, historic boreholes, piezometers analysis, laboratory tests, etc. which were undertaken between 2003 and 2007.

The longitudinal geological ground profile along the running tunnels alignment of Drive H is shown in Figure 2. The general geological stratum sequence from the ground level between Plumstead portal and Woolwich station is: made ground, alluvium, river terrace deposits (RTDs), Thanet sand (TS) formation and chalk.

Tunnelling has started from Plumstead portal at chainage 22556 in the RTDs and the underlying Thanet sands. The chalk horizon will be met after 100 m. The mixed stratum conditions end at chainage 22300. The tunnel is situated completely in the chalk until chainage 21600 where Thanet sands are encountered again. Between this chainage and Woolwich box at chainage 21250, mixed stratum conditions are present. Approximately 100 m before reaching the Woolwich box the tunnel is situated completely in the overlying Thanet sands.

Brief descriptions of the encountered stratum are given below. RTDs are well graded sand and gravel with clayey/silty sand sediments occurring locally as discontinuous beds, and cobble sized gravel may be present. Thanet sand maximum thickness is 15 m. Generally comprises of two main beds: homogenous silty fine to medium sand with increasing clay/silt content in lower part of the formation and bullhead beds, discontinuous thin layer of rounded course flint gravels with cobbles in sandy clay matrix. Chalk is the major strata present along whole C310 route. The chalk group in England is divided into two sub-groups, grey and white chalk, which in turn are divided into a number of formations. Along the drives, the chalk belongs to the Seaford chalk formation of the white chalk. The Seaford chalk formation comprises three distinct beds: the Tout beds, the Cuckmere beds and the Haven Brow beds. All three beds were encountered within boreholes. However the Tout beds are not along tunnel alignment. Encountered chalk is described as white, off-white or cream coloured, medium density, weak to moderately weak fractured chalk with flint bands, occasional marl seams.

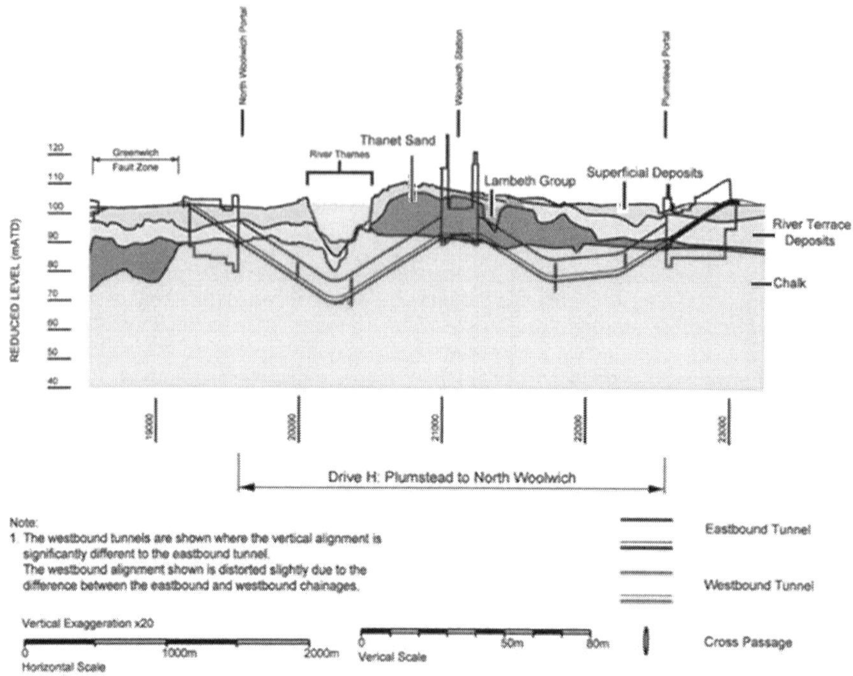

Figure 2 Longitudinal Geological Section along the C310 Tunnel Drives

Empirical Ground Movement Prediction

For the prediction of the tunnelling-induced ground vertical movements, the methodology described in Crossrail Civil Engineering Design Standard (CEDS) Part 7 Ground Movement Prediction (2010) is adopted. The report states two empirical methods for the ground vertical settlements:

- O'Reilly and New (1982) appropriate for predicting the surface movements where the ground overburden is greater than one tunnel diameter above the crown of the tunnel;
- New and Bowers (1994) appropriate for predicting surface and sub-surface movements where the ground overburden is within one diameter of the crown of the tunnel.

It is noted that the locations where the tunnel with ground cover less than one diameter of the tunnel are close to the portal and station, and the ground movements have been highly affected by the TBM launching and landing at these locations. The first method is therefore adopted and discussed in this paper. Based on the O'Reilly and New (1982) method, the shape of the surface vertical settlement can be represented by a Gaussian error equation:

$$S = S_{max} exp^{(-y^2/2i^2)} \quad (1)$$

where

S is the surface settlement at a distance y from the tunnel centre line,

S_{max} is the maximum settlement at y=0,

i is the standard deviation of the curve.

The value of i defines the trough width and corresponds to the value of y at the point of inflexion of the settlement curve. For practical reasons it can be defined as product of trough width parameter (K) and depth of the tunnel.

In this project, Oasys software Xdisp V19.1 has been used to analyse the ground movement induced by tunnelling and portal/box excavations, complying with the empirical method by O'Reilly and New (1982). Ground modelling in Xdisp has been developed in accordance with the requirements of CEDS part 7 and 8. The user is required to define the estimated volume loss (VL) above the tunnel due to deformation, as well as the trough width parameter and tunnel geometry. Xdisp will then use this to define the settlement profile at the surface or specified depth. The input parameters for the generic settlement assessments of the Drive H are:

- Running tunnel (excavated) diameter = 7.18 m
- Trough width parameter, K = 0.5
- Volume loss for running tunnels, VL = 1%
- Homogenous ground conditions

Volume loss is the ratio between volume of excavated material and theoretical tunnel volume. It causes settlements at the ground surface with the same volume. For the Crossrail project, CEDS Part 7 states that the default moderately conservative assumption of volume loss for running tunnels (ii) is 1% along the entire route of Drive H. For the areas designated as 'controlled tunnel drive zones', the target volume loss is specified as at a reduced rate of 0.5%. There are two areas along H drive defined as controlled zone. The first area is located between chainage 22600 and 22500, where the ground cover to tunnel crown is less than one tunnel diameter, and critical assets - Cathedral Substation and White Hart Road Bridge are located. The second area is between chainage 21520 and 21450, where critical assets - Middlegate House (supported by combination of shallow and piled foundations), Southern Outfall Sewer and Docklands Light Railway tunnels. It is noted that both Cathedral Substation and Middlegate House have been strengthened by grouting compensation before and during the TBM arrivals, and they are not studied in this paper as their settlement behaviour has been affected by the grouting.

The results of the predicted tunnel included surface movement (with 1% VL and 0.5% VL at the controlled zone) over the H drive alignment are shown in Figures 4 and 5. The maximum predicted surface settlement contour are up to 20 mm for tunnelling excavations and 50 mm for both tunnelling and portal excavations.

Observed Ground Movements and Back Analyses of Volume Loss

For measuring the tunnelling-induced ground vertical movement, and addressing corresponding volume loss, numerous geotechnical monitoring instrumentations have been installed on/under the ground and at the structures (e.g. buildings and bridges).

For monitoring the ground surface and shallow sewer drainage pipes, manual levelling studs with an accuracy of 1-2 mm are adopted and installed typically in 5 m spacing crossing about the two tunnels. For the buildings and bridges, automated hydrostatic levelling cells (HLCs) with an accuracy of 0.1 mm are used. As for the sub-surface monitoring, extensometers are installed and adopted. Figures 6-8 show the locations and arrangements of the monitoring instrumentation at 17 cross sections (across two tunnels) over the tunnel alignment up to Woolwich station.

Readings were taken daily when the areas within TBM monitoring zone of influence (ZOI). When outside the ZOI the monitoring would be carried out weekly for 4 weeks, and subsequently monthly for 3 months. Long-term monitoring will be taken quarterly until the settlement rate is equal or less than 2 mm per annum. Figure 3 illustrates the definition of TBM ZOI, which is defined as one total depth (D) from the TBM base forwards and 2D backward.

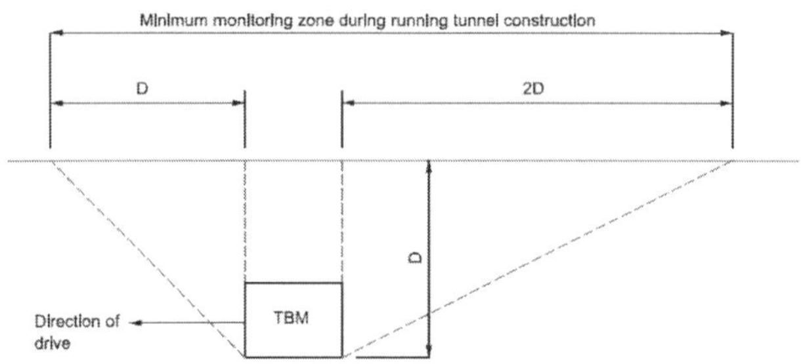

Figure 3 Zone of Influence of the TBM

17 cross sections are selected to examine the observed actual ground surface settlement and re-assess the actual volume loss due to the tunnel excavation. The locations of 17 cross sections are shown in Figures 6-8. The re-assessment (back analyses) of the actual volume loss is performed by Xdisp as discussed earlier. K value of 0.5 is adopted. Figures 9-25 graphically present the monitoring data and the best-fit settlement curves generated by Xdisp for the 17 sections.

Table 1 summaries the two-tunnel geometry of westbound (WB) and eastbound (EB), maximum observed actual settlement, the re-assessed VL of the sections. The maximum ground surface settlement of 10 mm is recorded for the tunnelling at the mixed stratum of RTD/TS, and 9 mm at chalk. The maximum back-calculated volume loss is 0.35% at either RTD/TS or chalk, which is much less than the required volume loss of 1% in CEDS Part 7, and below 0.5% specified for the controlled tunnel drive zones.

Section	Tunnel Depth [m]	Distance between Two Tunnels along the Sections [m]	Maximum Settlement [mm]	Volume Loss (VL) [%]		Stratum Encountered
				WB	WB & EB	
1	-12.0	10	-5	0.125	0.125	RTD
2*	-12.5	11	-5	0.125	0.125	RTD
3	-16.0	16	-6	0.125	0.150	TS/Chalk
4	-18.0	18	-6	0.125	0.125	Chalk
5*	-18.0	20	-7	0.250	0.350	Chalk
6*	-18.5	32.5	-9	0.250	0.300	Chalk
7*	-24.0	30	-9	0.250	0.300	Chalk
8*	-19.6	37	-4	0.125	0.125	Chalk
9	-20.8	19	-6	0.125	0.125	Chalk
10	-21.3	19	-2	0.125	0.125	Chalk
11	-22.0	18	-6	0.250	0.300	Chalk
12	-23.0	14	-5	0.125	0.250	Chalk
13	-18.0	15.5	-3	0.100	0.125	TS/Chalk
14*	-17.0	14	-5	0.100	0.250	TS/Chalk
15	-13.0	17	-7	0.250	0.250	TS
16	-13.0	14	-10	0.250	0.350	RTD/TS
17	-12.0	15	-3	0.100	0.100	TS

Note: * - sections within the controlled tunnel drive zones where VL of 0.5% is specified.

Table 1 Summary table for the actual tunnelling-induced settlements and volume loss

Discussion

Settlements and Volume Loss in Chalk

The observed settlements induced by two tunnels are in the range between 3 mm and 10 mm, whereas the predicted maximum settlement is of 20 mm. The actual volume losses derived from the observed settlements in Table 1 range from 0.1% to 0.35% for the tunnelling in different strata. The actual volume losses are calculated by the back analyses using the settlement curve generated by Xdisp showing a good agreement with the actual monitoring data. They are all below the required volume losses of 1% and 0.5% for the controlled zone, with the same trough width parameter (K) of 0.5.

Settlements occurred in the chalk, where is defined as weak to moderately weak fractured chalk in this project, is less than 10 mm and the actual volume loss associated with this settlement is 0.35%. Available study in chalk tunnelling is limited. The actual volume loss obtained in this project could be considered to be relatively small, compared with the volume loss obtained from other published case studies for tunnelling in soft soils.

The small observed settlement and volume loss is likely to be attributed to both the application of the slurry TBMs and the discrepancy on the assumption of the strength and stiffness of the chalk encountered in tunnelling.

Trough Width Parameter K

There is no suggested design value for the trough width parameter for the chalk in the CEDS Part 7 (2010). For the design purpose, the chalk is considered to have the same K value as cohesive soils (0.5). Based on various projects Jones (2010) has proposed the logarithmic relationship between the parameter K and the overburden height above the tunnel in clays. However there is no such relationship developed for the chalk. The tunnel depth in the chalk on C310 drive is within a narrow zone between 18 m to 24 m. By applying Jones' (2010) logarithmic relationship K values of 0.43-0.5 are found. Based on the back analysis results discussed earlier, K value of 0.5 has been obtained with best-fit curves generated by Xdisp to match with the actual settlement data. This implies that, due to the rock nature of the chalk, the width of the surface settlement trough over the chalk is slightly narrower than those over the clays in the published studies.

Interferences of Twin Tunnels

In the twin tunnels, pillar width (distance between two tunnels) is the most decisive factor for both tunnel stability and ground movement, because of the interference between them. With two parallel identical tunnels in this project, due to construction of the second EB tunnel, the first WB tunnel and the surrounding soil may move as a rigid body causing interference. Fang et al. (1994) considered the displacements and strains associated with twin-tunnelling. Their study shows that if the second tunnel construction generates stresses within the zone created by the first construction a 'large and irregular' volume loss could be expected.

In C310 drive, two tunnels are apart from the distance in the range of 2 to 5 tunnel diameter. Clear interferences between two tunnels are established for the sections with the pillar width up to 3 tunnel diameter (i.e., all sections except Sections 6-8), as shown in Figures 9-13 and 17-25. Sections 6-8 (in Figures 14 to 16) where two tunnel distances are more than 4.5 tunnel diameter show the no twin tunnel interference effect.

Conclusion

Total length of 2.6 km of twin bore tunnels was constructed by the slurry TBM in chalk at C310 contract between Plumstead portal and Woolwich station. The comparison of the observed ground surface settlements with the predicted values has been addressed. Very low values of tunnelling-induced settlements have indicated a successful slurry TBM operation, which delivered safe tunnelling experience for the future tunnelling projects. Volume loss values achieved are far below required 1%. The purpose of this paper is to share this case study of the slurry TBM excavations in the chalk and provide a reference for designers and contractors what the ground is behaving and what amount of volume loss can be expected during tunnelling in the chalk. TBMs are on the way to build the twin tunnels between Woolwich station and North Woolwich portal, and the ground movement behaviour would be worth studying to provide more valuable information on the tunnelling in chalk.

Figures

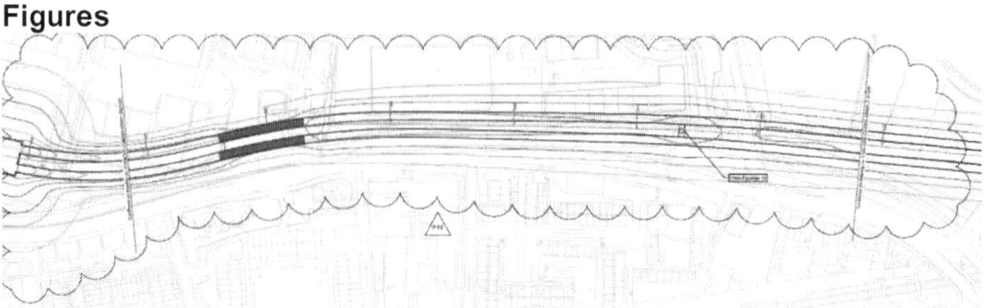

Figure 4 C122-OVE-C2-DDA-CR001_Z-21332 Settlement contours, final stage. Routewide Plan Sheet 32 of 38

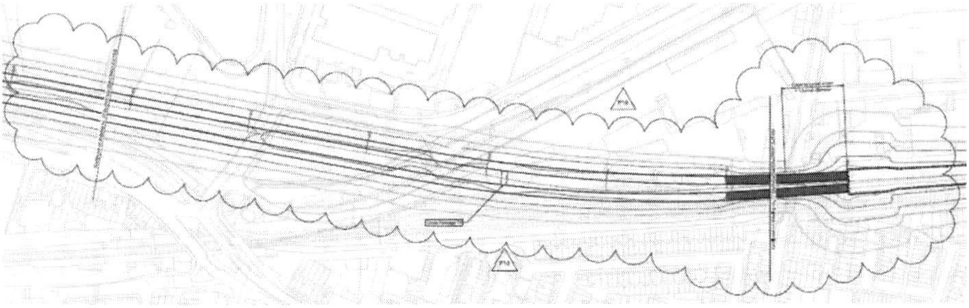

Figure 5 C122-OVE-C2-DDA-CR001_Z-21333 Settlement contours, final stage. Routewide Plan Sheet 33 of 38

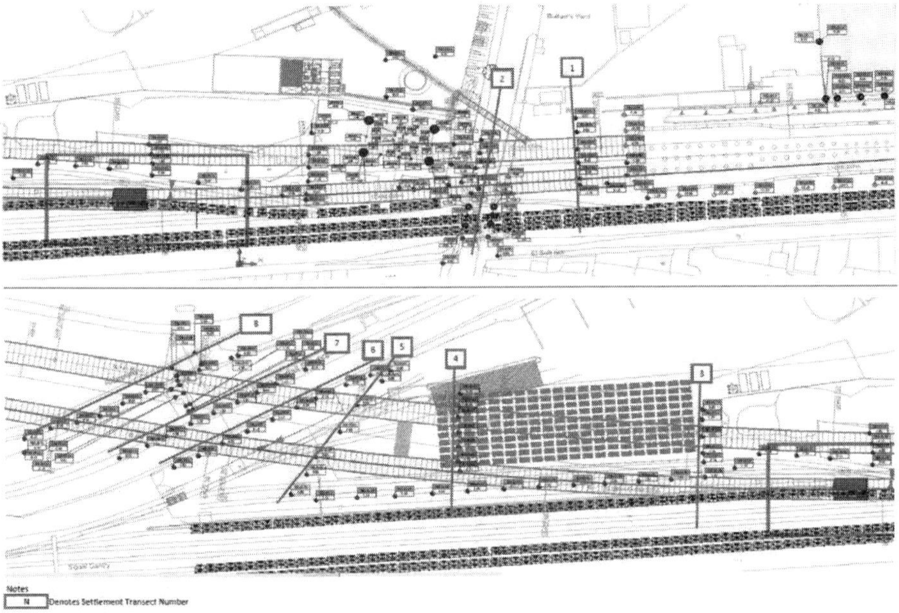

Figures 6 C310 Instrumentation and monitoring sensor layout 1

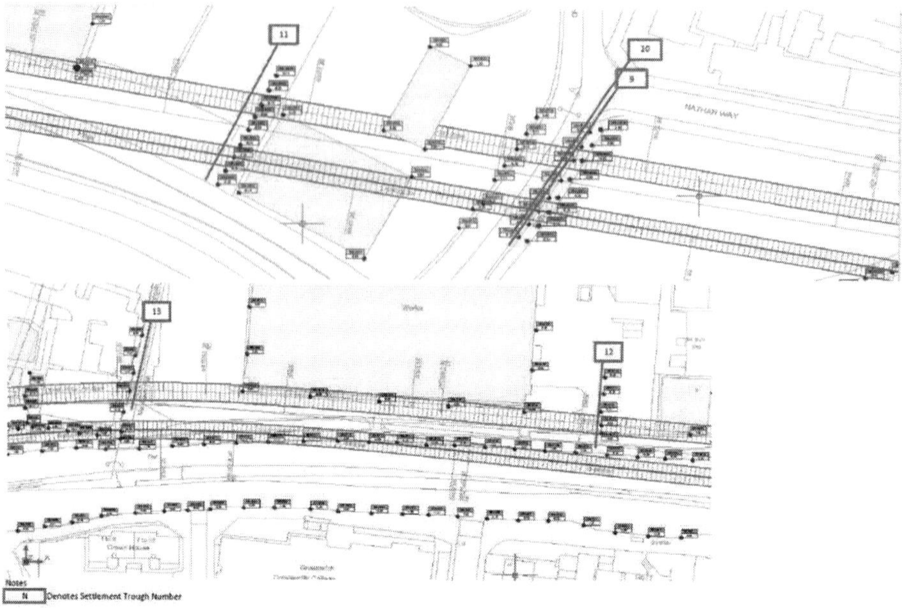

Figures 7 C310 Instrumentation and monitoring sensor layout 2

Figures 8 C310 Instrumentation and monitoring sensor layout 3

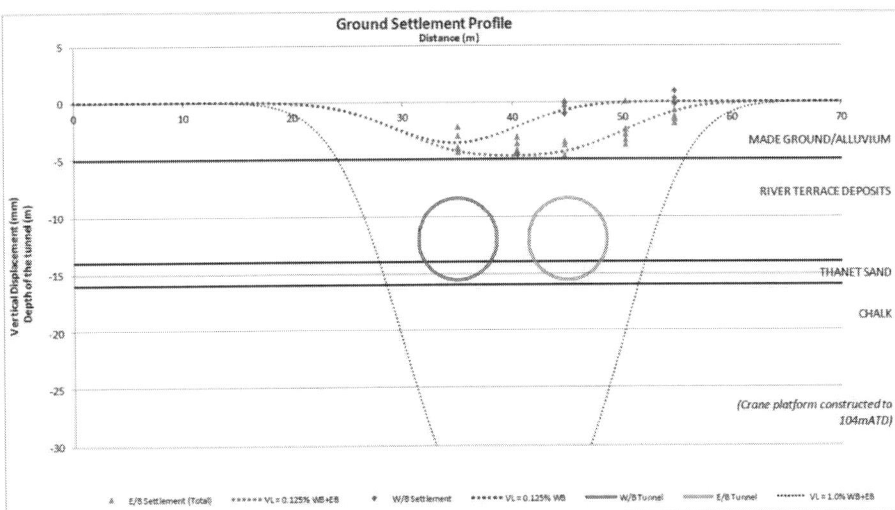

Figure 9 Section 1 Plumstead portal – crane slab 2

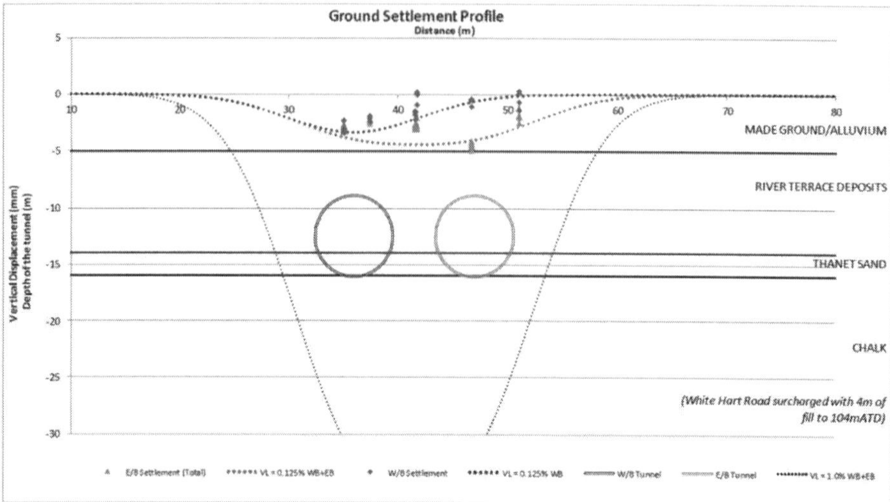

Figure 10 Section 2 White Hart Road

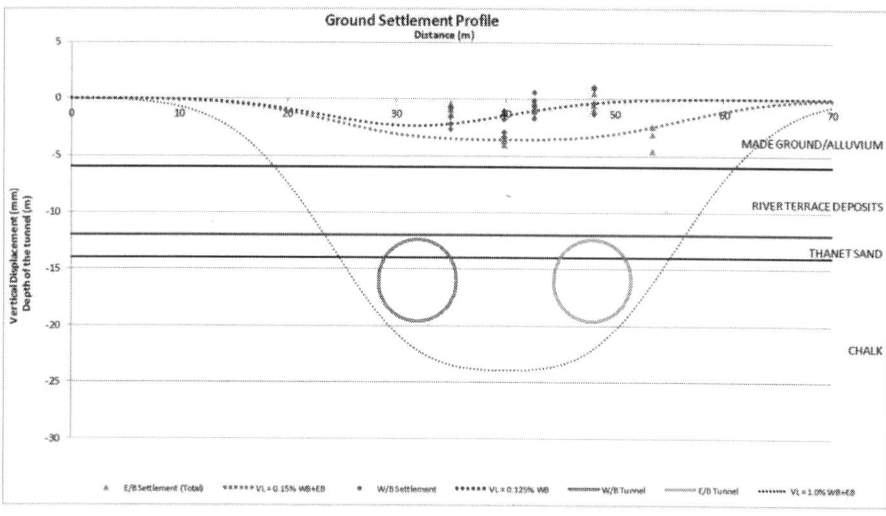

Figure 11 Section 3 STP 02

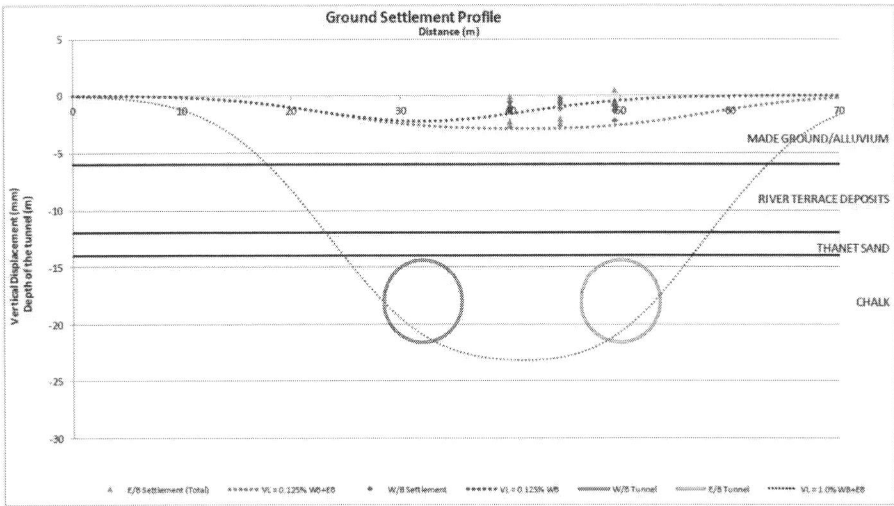

Figure 12 Section 4

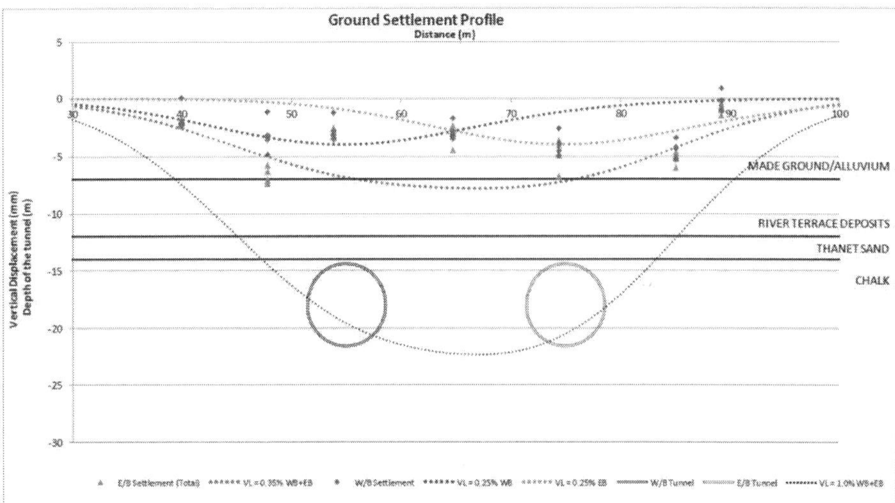

Figure 13 Section 5 Geotechnical cross section 1

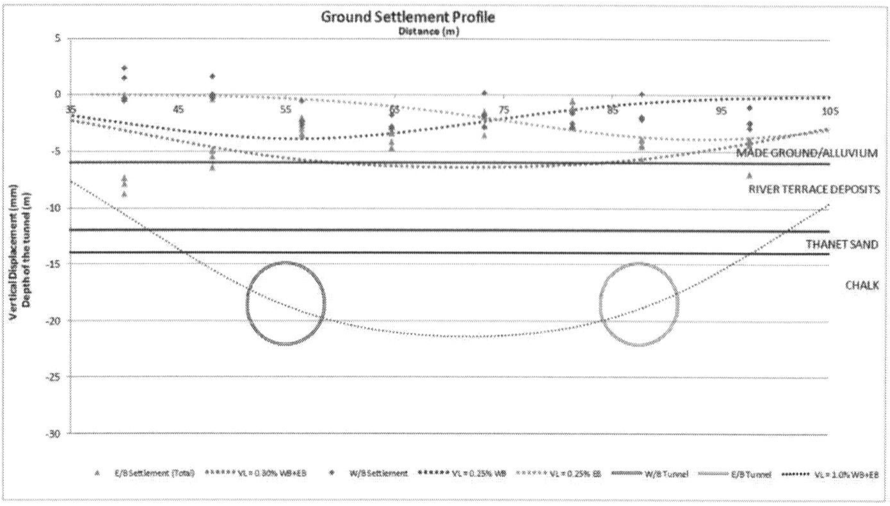

Figure 14 Section 6 Southern Outfall Sewer 1

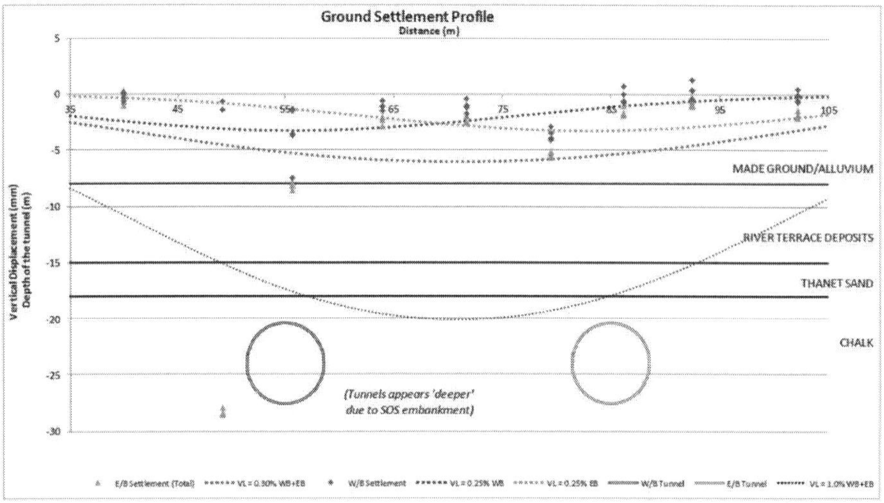

Figure 15 Section 7 Southern Outfall Sewer 2

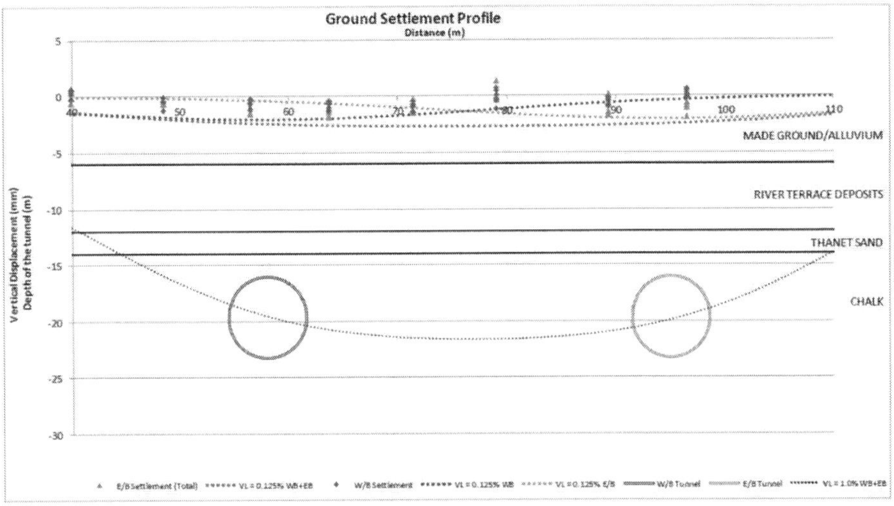

Figure 16 Section 8 Southern Outfall Sewer 3

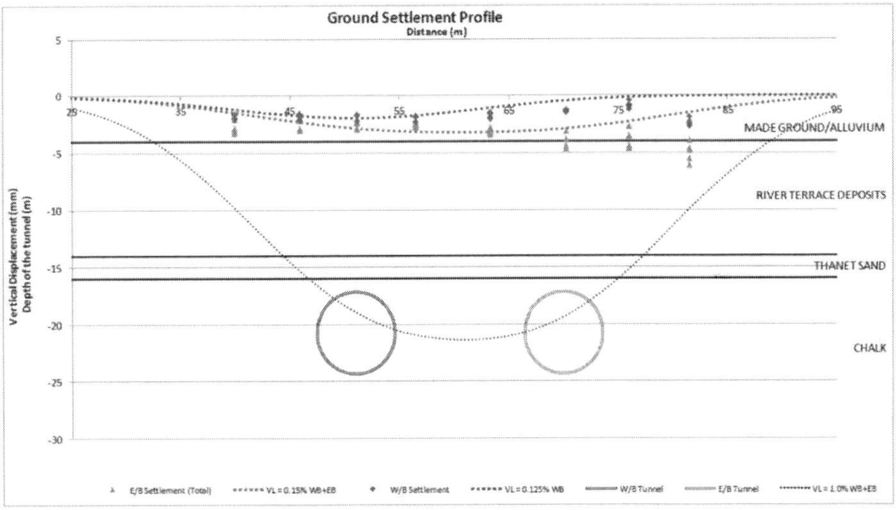

Figure 17 Section 9 Pettman Crescent bridge

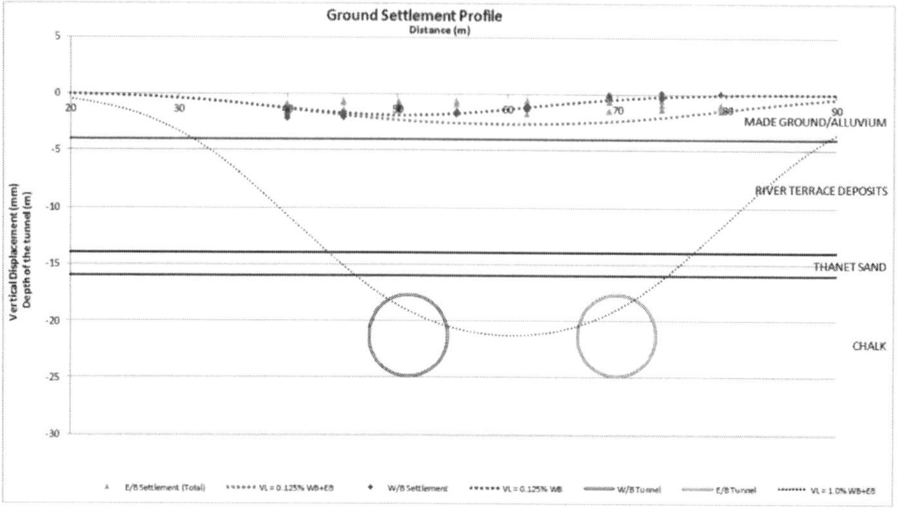

Figure 18 Section 10 Pettman Crescent HLC (E)

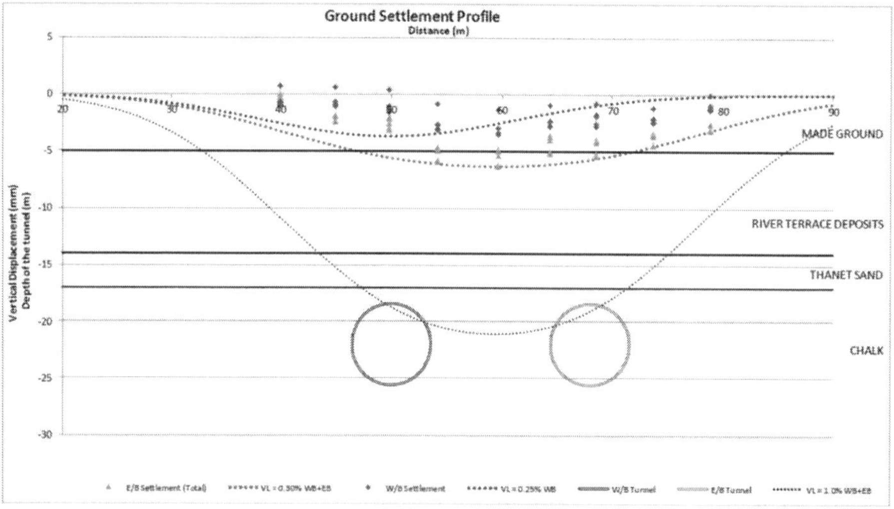

Figure 19 Section 11 Bus garage

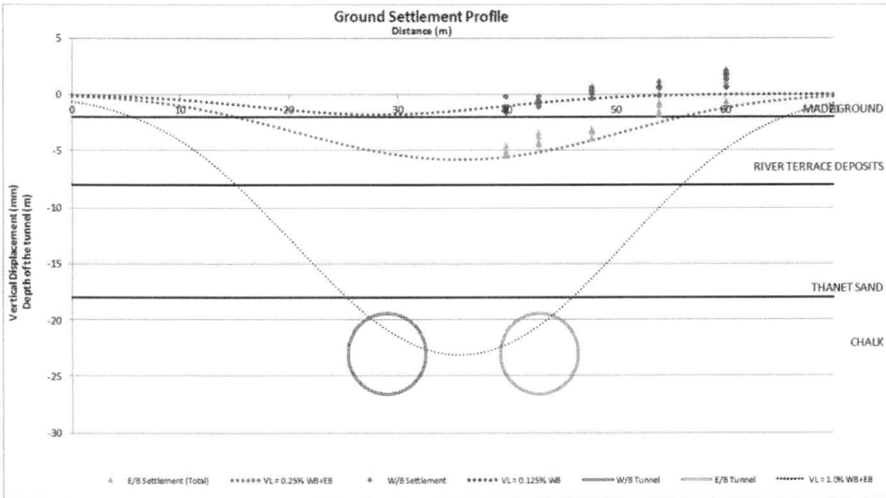

Figure 20 Section 12 Signature industries

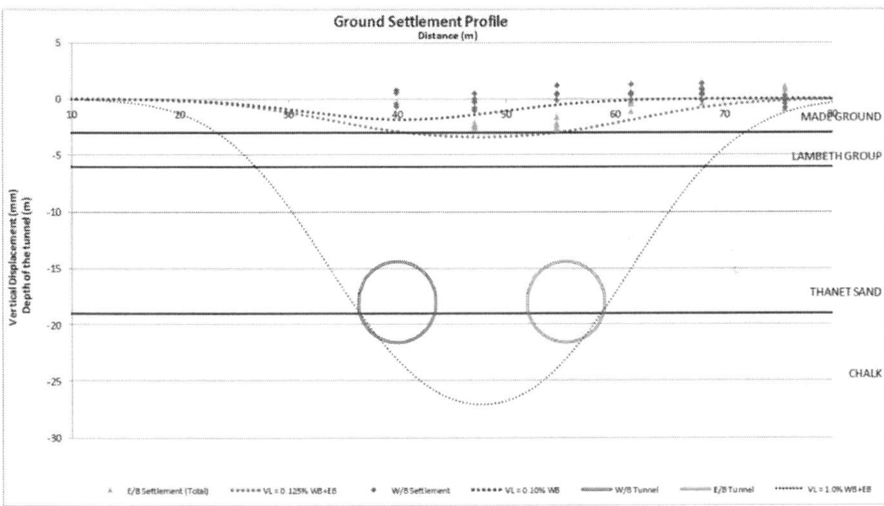

Figure 21 Section 13 Marshgate path

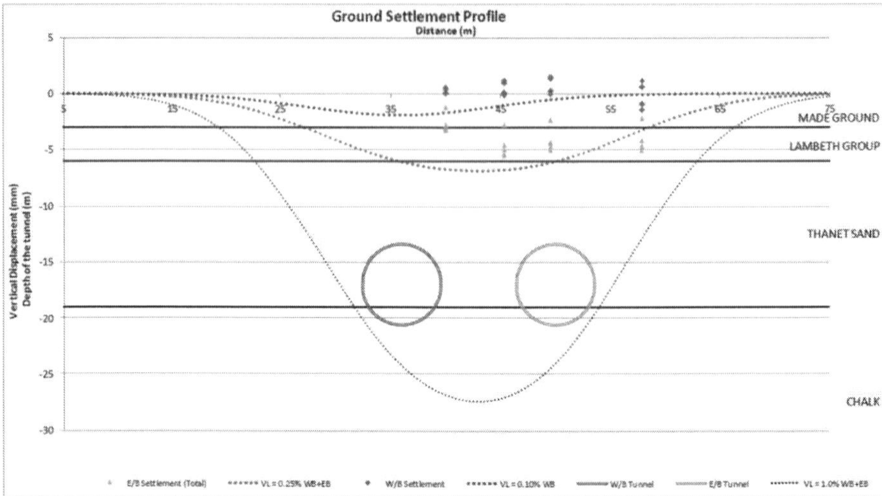

Figure 22 Section 14 MGH wall (E)

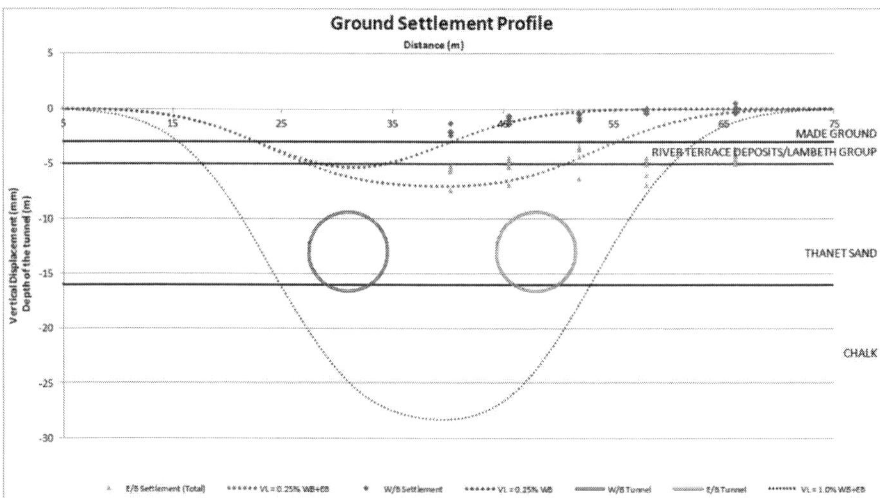

Figure 23 Section 15 MGH footpath

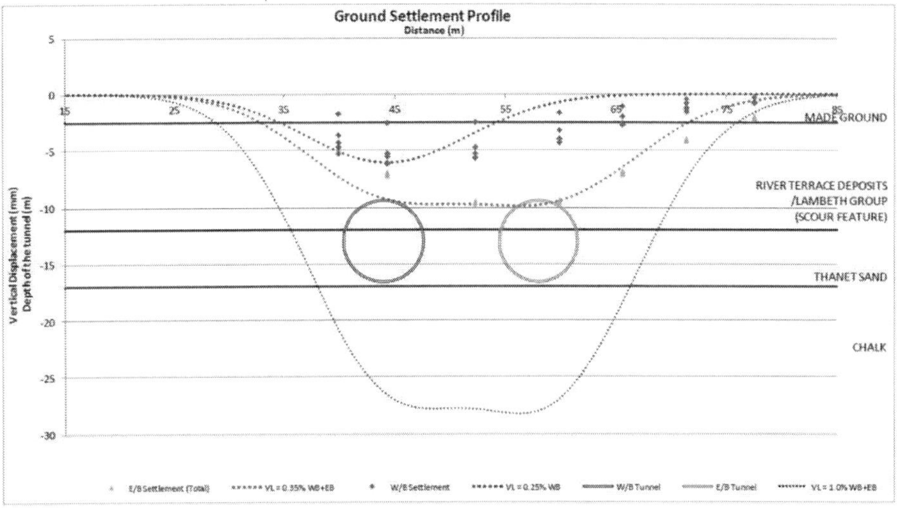

Figure 24 Section 16 IO centre

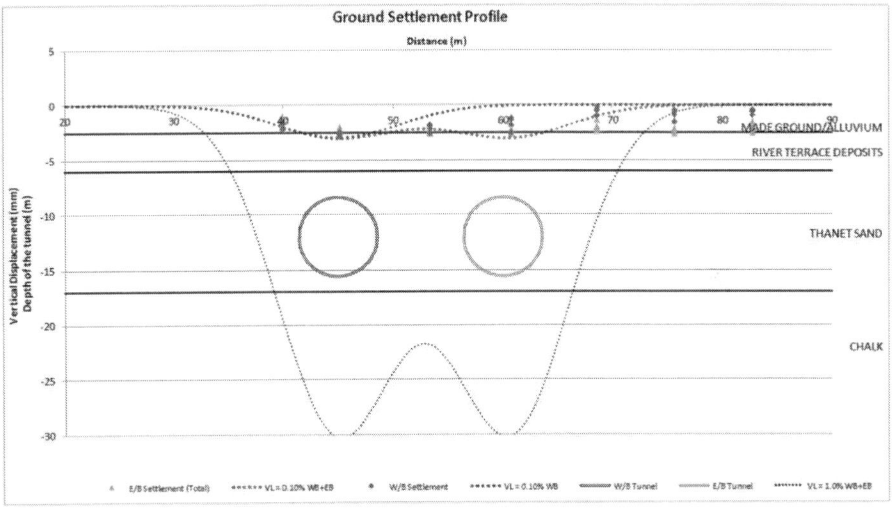

Figure 25 Section 17 Cornwallis Road

References

Crossrail CEDS - Civil Engineering Design Standard Part 7 (2010). *Ground movement prediction.*

Crossrail Geotechnical Sectional Interpretative Report (May 2007). *Isle of Dogs to Plumstead, Volume 1*: Text, Report No. 1D0101-G0G00-00520, Rev.B.

Fang, Y. S., Lin, L. S. and Su, C. S. (1994). *An estimation of ground settlement due to shield tunnelling by the Peck-Fujita method.* Canadian Geotechnical Journal, Vol. 31, No. 3, pp. 431-443.

Jones, B. D. (2010) *Low-volume-loss tunnelling for London Ring Main Extension.* Proc. Instn. Civ. Engr. Geotechnical Engineering 163, June, Issue GE3, 167-185.

Maidl B, Herrenknecht M, Anheuser L (2011) *Maschineller Tunnelbau im Schildvortrieb.*
New BM and Bowers KH (1994) *Ground movement model validation at the Heathrow Express trial tunnel*, 301-329.

O'Reilly MP and New BM (1982) *Settlements above tunnels in the United Kingdom – their magnitude and prediction*, 173-180.

Crossrail Project: Infrastructure design and construction
ISBN 978-0-7277-6078-4

doi: 10.1680/cpid.60784.471

The Prediction of Ground Movements Adjacent to Box Excavations

Jonathan Gatter BSc (Hons), MSc (Eng), CGeol, FGS, Hyder Consulting (UK) Limited
James Latham MEng (Hons) MSc, Water Environment Ltd
Kevin Raven BSc (Hons), MSc, CEng, FGS, IOM^3, Hyder Consulting (UK) Limited

Abstract
Thames Water's Crossrail Project team required a simple method of predicting ground movements around complex shaped station excavations as a verification tool for the Crossrail 3D FEA based damage assessments.

Existing commercial software did not take into account the corner stiffening effects of box excavations where the deflection of the wall will be greatly reduced.

The London Underground (LU) technical advice note for the prediction of movements around 'short wall' excavations provided a method of predicting ground movements around the Crossrail stations. However, the use of the Advice Note was time consuming and extrapolation of the data provided gave rise to irregular displacement profiles. Horizontal movements can only be predicted perpendicular to the box excavation wall.

Development of the LU methodology was undertaken and incorporated into an Excel based software package. Innovative use of the LU horizontal movement predictions also allowed x and y vectors of movement to be developed along the asset.

This output of the digital LU based method for the prediction of ground movements compared very closely with the 3D FEA generated predictions and was a valuable tool in the agreement of the final mitigation strategy for Thames Water assets.

List of notations

a	Relative distance from corner
b	Relative distance from corner
C_1	Corner 1
C_2	Corner 2
CF_a	Correction Factor for distance a
CF_b	Correction Factor for distance b
CF	Total Correction Factor
D	Distance along asset

R Resultant horizontal ground movement
Δ_{horiz} Factored horizontal ground movement
Δx Longitudinal ground movement
Δy Transverse ground movement
Δz Vertical ground movement
H_b Elevation at base of excavation
H_g Elevation at Ground Level
H Depth of Wall
S_h Assumed "long wall" horizontal displacement
S_v Assumed "long wall" vertical displacement
WL Width of Wall
x_0 Offset from wall to asset

Introduction

A simple empirical based methodology has been developed by the authors to predict ground movements due to deep excavations, for the purpose of assessing whether these movements will have the potential to cause damage to existing underground utility assets.

The methodology, referred hence as the Gatter-Latham system (GLS) is a simplified approach to the assessment of ground movements and has been adapted from the published London Underground (LU) technical standard for ground movement (*London Underground: Civil Engineering (2009) – Technical Advice Note G-058 A10*).

Thames Water required a method of rapidly assessing the Crossrail ground movement predictions, particularly around the proposed station boxes, where it was anticipated the greatest impacts on Thames Water assets would occur. This tool was required to verify Crossrail's methodology and for agreement of levels of damage and mitigation works required due to the impacts of tunnel and excavation induced ground movements on a representative number of the approximate 10,000 Thames Water assets potentially affected.

Existing commercially available software used on Crossrail for prediction of ground movements over tunnels did not, at the date of this phase of the damage assessment process, model ground movements around box excavations with sufficient accuracy for the purposes of calculation of damage to utility assets.

The GLS methodology creates a simple ground movement profile from a relatively small amount of input data for both the asset being assessed and the corresponding excavation. Ground movements are predicted in the vertical (z), transverse (y) and longitudinal (x) directions along a specified asset with the resulting output being used in a separate analytical process to assess the potential damage to the asset.

The paper introduces the concept of ground movements close to excavations and provides information on the currently available methodologies which have been used to predict these movements and following an appraisal of the LU G-058 methodology, the adaptations for use in damage assessment calculations are discussed.

Ground movement prediction

Ground Movements due to deep excavations

Introduction

The complexity and scale of ground movements around the deep box excavations proposed at seven of the new Crossrail stations (excluding Canary Wharf) within the tunnelled section of the scheme (See Figure 1) were likely to pose the greatest risk of damage to existing assets in close proximity to these works, particularly when combined with the tunnelling and sprayed concrete lining (SCL) works proposed as part of the station complex.

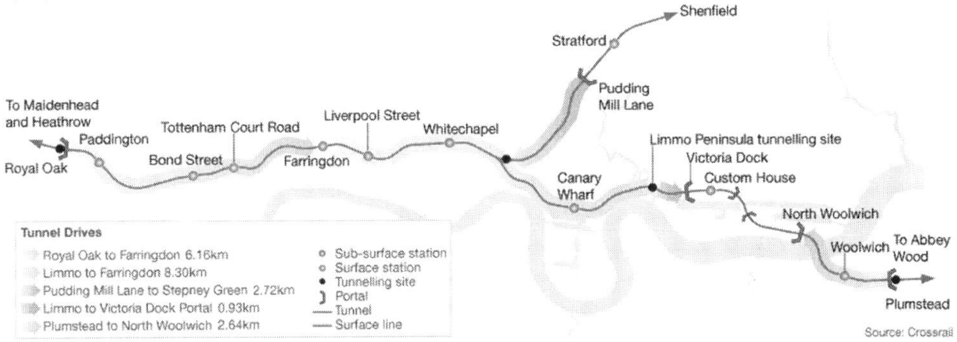

Figure 1: Simplified plan of Crossrail underground stations and tunnel alignment

The piling and excavation works for the station boxes were planned to be one of the early activities at the station locations, hence early agreement on the impact of these works on Thames Water owned assets was essential to allow planning and implementation of protective/mitigation measures in advance of these proposed works.

Theory

Ground movements associated with excavations act in three vectors. Firstly, vertical ground movement will occur around the entire excavation. Generally, the greatest vertical settlement will be located close to the excavation itself and reduces with distance from the wall, see Figure 2a. Directly at the face of the wall, there will be other factors that affect the distribution of ground movement, such as the installation of the wall itself. The methodology developed however was to incorporate the ground movement impacts of the retaining wall installation and the subsequent excavation works.

Horizontal ground movements will also occur, and in general will act towards the excavation with the greatest movement occurring at the centre of the wall, see Figure 2b. This movement is made up of displacements in the *x* and *y* vectors.

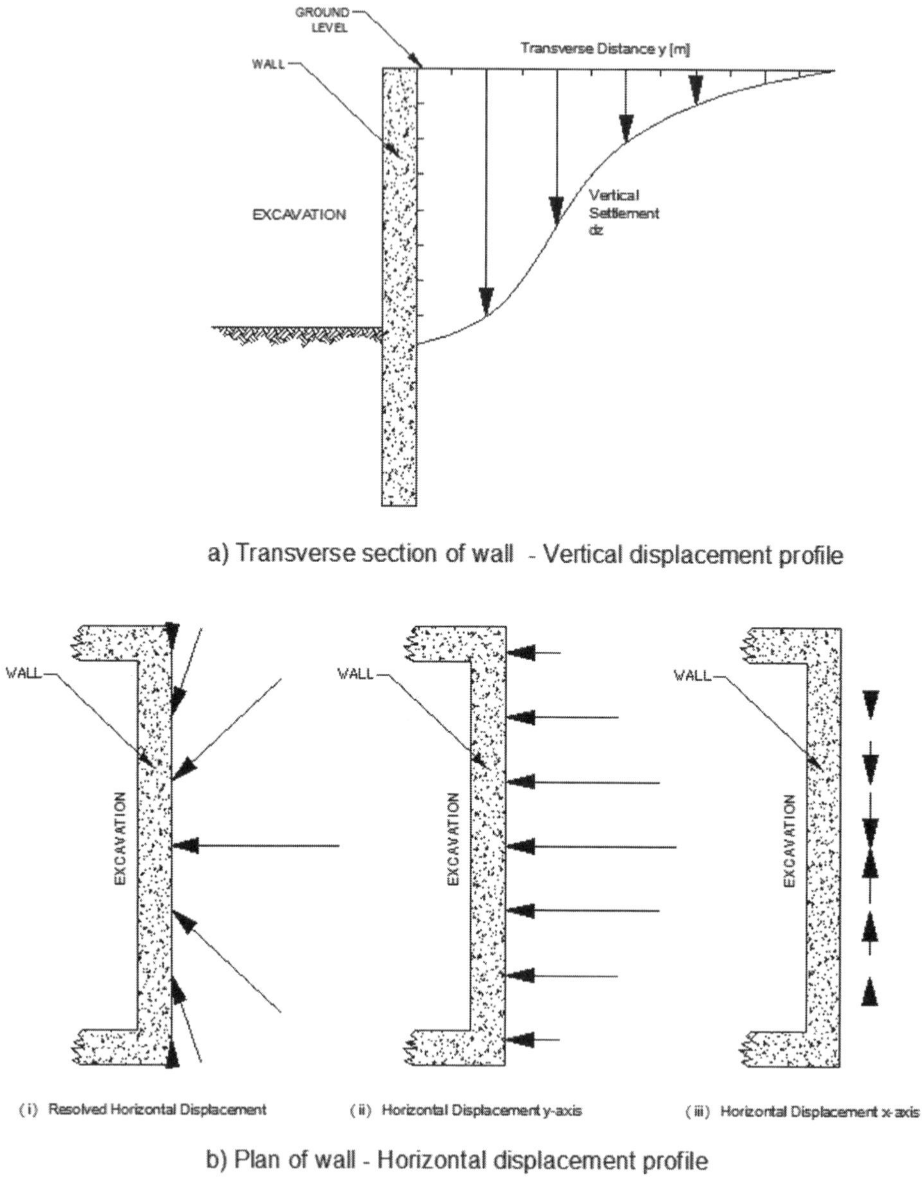

Figure 2: Schematic section and plan indicating vertical settlement and horizontal displacement profiles.

Current methodologies

For the prediction of ground movements around box excavations, the general practice is to adopt the Ciria C580 (*Gaba A.R., Simpson B., Powrie W., Beadman D.R. (2003): Embedded Retaining Walls, Guidance for Economic Design*) ground behaviour model. This data is primarily based on 'long wall' retaining structures defined as having a length of greater than 2.5 times its depth, and the data tends to represent plane strain at the centre of the 'long wall' remote from the corners of the excavation.

The majority of the station boxes on Crossrail could be defined as 'short wall' excavations, where the length of the wall is similar to its depth. Use of the C580 data, which is based on an upper-bound envelope of data for longs walls, which tend to be less stiff than short walls, is considered to be over-conservative for the modelling of ground movements around 'short wall' shafts. The C580 approach also does not take into account the stiffening effects of the corners to the short walls, which significantly modifies the ground movement pattern around the shafts, although reference is made to the effects of stiff corners in the Ciria document.

The commercially available ground movement software at the date of the start of the damage assessment process for Thames Water assets used the Ciria C580 ground movement prediction model and assumed that the entire length of the wall displaces by the same degree of movement, regardless of its length. This assumption is known to be incorrect for box excavations where, as noted in the LU document, because many of the walls are relatively short compared to their depth, the total movement of the wall and deflection of the wall at the corner of the excavation will be greatly reduced in reality.

The combination of short walls in a shaft excavation creates a more rigid structure, as the wall perpendicular to the one being assessed acts as a prop and limits its deflection in the direction of the excavation. The greatest deflection will therefore be located at the centre of the wall, furthest away from the corners.

A typical plot from the commercial software showing displacements in the *x* (axial along the length of the asset), *y* (transverse towards the excavation) and *z* (vertical) directions are shown as dashed lines in Figure 3 and settlement contours are given in Figure 4. It can therefore be seen that along the length of the wall, between chainages -15 and +15, displacements towards the excavation are equal and therefore induce no displacement in the *x* direction, which is incorrect.

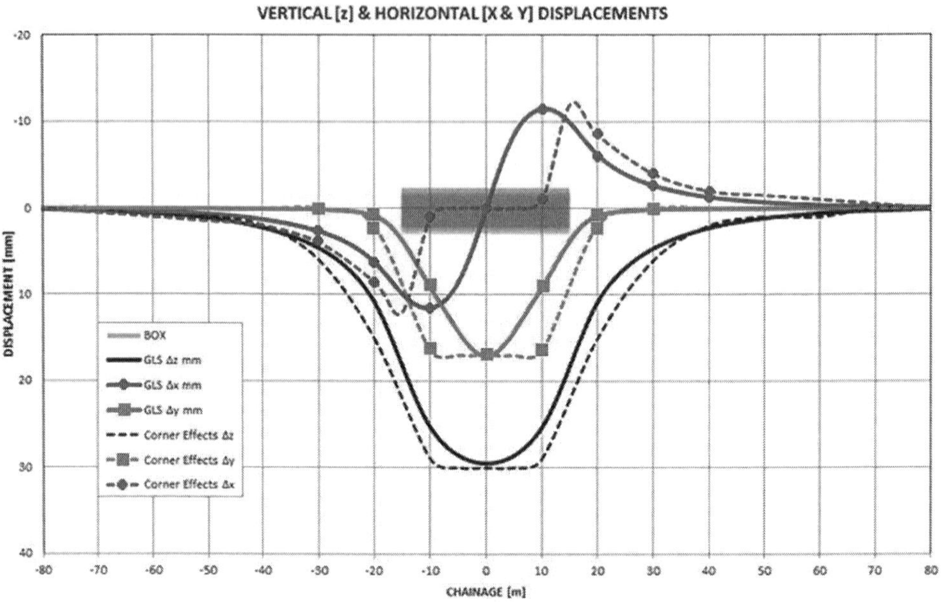

Note: *Dashed lines indicate corner effects. Solid lines indicate GLS corner stiffening effects.*

Figure 3: Typical output of ground displacements along line of asset

As a consequence, high values of differential ground movement are calculated at the corners of the excavation, due to abrupt changes in displacement, resulting in artificially elevated magnitudes of strain when assessing the potential damage to assets in proximity to the ends of these excavations. The model also over-predicts total ground movements around the excavation. Hence, an alternative method of the prediction of ground movements around the short wall excavations typical of the Crossrail station geometries was required to be developed.

In comparison, a more representative profile of ground displacements taking into account corner stiffening using the GLS method is shown as solid lines in Figure 3. Maximum displacements are in the same order as the typical displacement plot, however there are no sudden changes in the deflection, as anticipated, a smoother transition is demonstrated at the limits of the excavation.

Typical contours for the standard C580 method and the proposed GLS method are given in Figure 4. Ground displacements are shown in Figure 4a and contours follow equally spaced, concentric rings around the corner of the excavation based on the C580 model.
Figure 4b indicates the proposed GLS method and shows limited displacements due to the corner stiffening effects at the corner of the excavtion.

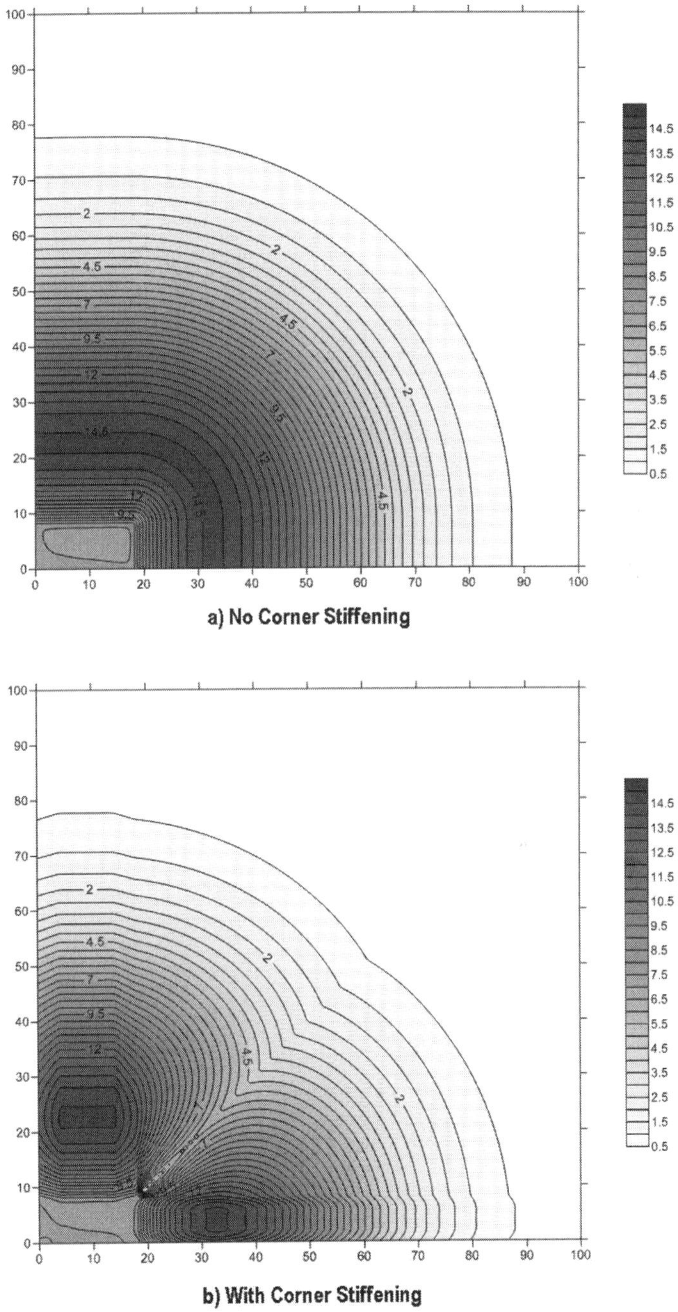

Figure 4: Typical contour output of ground displacements adjacent to corner of asset

Generic/parametric numerical modelling

Selected publications were reviewed in order to ascertain guidance on developing a simplistic approach for verification of the 3D FEA models including Wu et Al.(*Choa-Hui Wu, Chang-Yu Ou, Ningchien Tung (2010): Corner effects in Deep Excavations, Establishment of a Forecast Model for Taipei Basin T2 Zone, Journal of Marine Science and Technology, Vol 18, No 1, pp 1 -11*) and Crossrail, where a number of finite element analysis (FEA) parametric studies (*Crossrail Ltd. (2009): Sub-Surface Movements from Deep Excavations 1D0101-G0G00-01025 Rev. A. and Crossrail Ltd. (2010): Movements due to Box Excavations, Phase 3/Detailed Assessment Methodology, C122-OVE-C2-RAN-CRG01-00059*) were undertaken to investigate an improved approach to the modelling of ground movements around the proposed station excavations.

Crossrail developed analyses included the XR2 model, which included a wrap-around of contours similar to the current empirical methods and was considered to be an over-conservative approach.

The XR3 method, based upon the elliptical ground displacement theory, was developed and compared against a series of 3D finite element analyses which suggested a less conservative approach for the estimation of ground movements than comparable empirical methods. Review of case history data however indicated that the XR3 parametric study did not represent the displacement correctly along the long wall of the excavation.

In long excavations, ground displacements will become constant further from the corners, towards the central the part of the excavation, as it behaves predominantly as a retaining wall due to the diminishing effects of 'stiffening' from the corners. The XR3 method calculates displacements in an arc pattern between the corners along the length of a long wall as opposed to the spandrel displacement profile generated by field observations. A ground movement contour output from the XR3 method is included as Figure 5.

In addition, calculations of vertical and horizontal displacement using this parametric study were carried out on an irregular grid pattern which necessitates surface 'contouring' using specialist software to provide displacement data specific to the location of the asset for damage assessment calculations.

The limitations and practical applicability of these techniques therefore precluded them from being adopted as a simple method for modifying computer based empirical predictions for use in the damage assessment process.

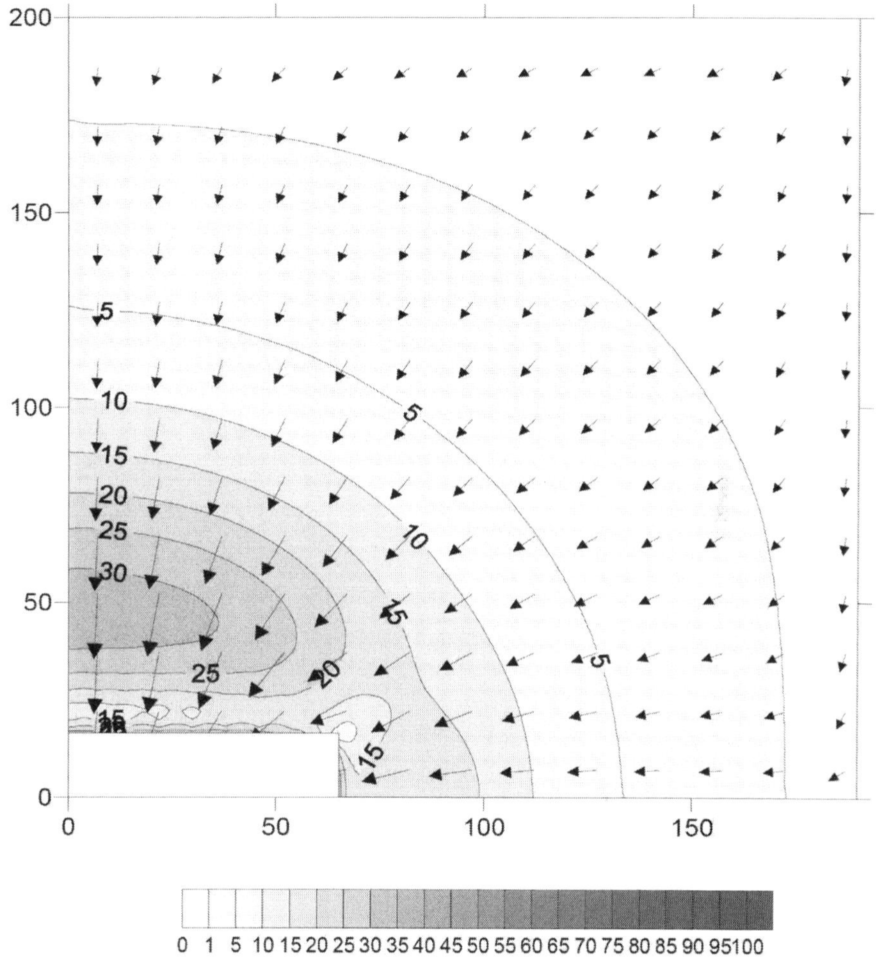

Figure 5: Typical XR3 contour output of ground displacements

From the study of various empirical and FEA based ground movement models for box excavations, it was concluded that due the large number of variables associated with each particular station site, the use of site specific 3D FEA analysis would be adopted by Crossrail, modified to provide a 'best estimate' rather than a moderately conservative approach to the prediction of ground movements and associated damage to 3rd party assets around station boxes.

Thames Water approach

The use of site specific 3D FEA to model ground movements around the station excavations on Crossrail was not considered appropriate for use by Thames Water Crossrail for a number of reasons, particularly as the calculations were to be used as a verification tool only. These were;

- Insufficient time within the damage assessment programme to generate independent 3D FEA models for each site, particularly as it was known that the stations designs were undergoing development as part of the ECI process.
- A budget to construct duplicate 3D FEA models to those developed by Crossrail could not be justified.
- A method was required that could easily be validated against previous case history or monitoring data.
- A method that would allow rapid review and agreement of the Crossrail ground movement predictions.

Following a detailed literature review of available method of ground movements around non-circular excavations, the LU Technical Note G-058 was identified for further study and analysis.

London Underground G-058 technical advice note

Following extensive experience in deep excavations in London, London Underground have developed a technical advice note which sets out a methodology for analysing various types of excavations appropriate to the ground conditions with the City , including short walled shafts, circular shafts and long wall excavations.

A simplified approach for the assessment of short walls is set out in the LU technical advice note and uses correction factor percentages to vary the estimated ground movements close to the corner of an excavation.

The methodology firstly calculates the assumed maximum vertical settlement at an offset *x* from the wall for each point to be assessed, using Equation 1:

$$\frac{Sv}{H} = 0.15\%\, e^{\frac{\left[1-\left(1+\frac{2x}{3H}\right)^{2}\right]}{2}}$$

Equation 1

A correction factor is applied to the equivalent vertical settlement which is determined based on the position of the assessment point along the wall. The relative distances from the assessment point to each corner of the wall are calculated and divided by the depth of excavation. These values, and the value of the offset from the wall divided by the depth of the excavation, are then matched on a lookup table to give the corresponding correction factors. These factors are either summed or subtracted depending on the position of the point along the wall to obtain a corrected value for settlement.

Horizontal movements are calculated in a similar way with the equivalent horizontal settlement determined using the expression:

$$\frac{Sh}{H}(\%) = 0.016\left(\frac{x}{H}\right) + 0.08$$

Equation 2

The corresponding correction factors are applied to the horizontal settlements in the same way as the vertical settlements to give the value of the total horizontal displacement at each assessment point. The predictions of horizontal movement assume however that they only act in a direction perpendicular to the wall being analysed.

Limitations

The standard LU methodology provides a simplified approach to the assessment of ground movements adjacent to excavations. However the level of detail required for a viable damage assessment along a linear asset such as a sewer or water main is much greater than the data provided via the look-up tables within G-058, and therefore there are a number of limitations to using the standard methodology.

Firstly, the interval between the values in the lookup table is fairly large and so when assessing points at close intervals along an asset, interpolation is required. As a result, steps are generated in the ground movement profile which gives rise to abnormal peaks in strains in the damage assessment, which are unrealistic, and the process is very time consuming.

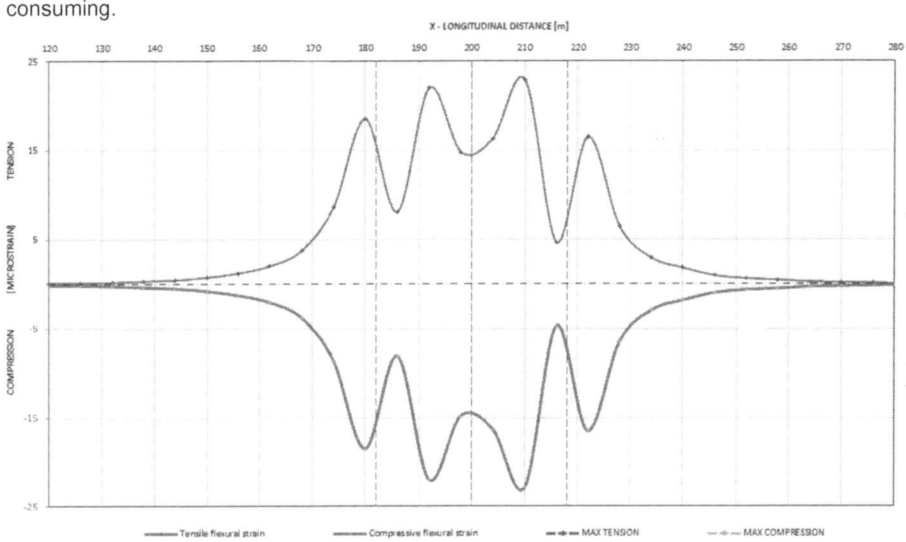

Figure 6: Example of GLS strain analysis using original lookup table

Secondly, the LU methodology is predominantly set out to solely consider vertical ground movements due to excavations. Horizontal ground movements have only been briefly considered in the LU technical advice note and can only be calculated using this method perpendicular to the line of the excavation. For the purpose of a detailed damage assessment, horizontal ground movements in both the *x* and *y* vectors as defined by Figure 2, but related to the orientation of the asset under analysis, need to be calculated in addition to the vertical settlement, for multiple points set at relatively small intervals along the asset

Adaptations to LU G-058 methodology

In order to assess the ground movements along Thames Water assets at specified intervals, calculations as described above need to be undertaken and corrected for their position relative to the excavation wall using predefined values from the lookup table as set out in G-

058. The damage assessment analysis, based on the ground movement predictions, require ground movement data at between 0.5 m to 2.0 m intervals along the line of the asset.

However, the original lookup table, provided within the LU G-058 report, is too coarse and was not developed for the detail of analysis required, and therefore had to be enhanced to a much finer grid of values in order to provide correction factors at the required intervals.

A three dimensional grid surface was generated from the original correction factor table data using a Kriging contouring process, which allowed the interpolation of values to a much finer grid of values. This grid was then developed as a refined lookup table in a spread sheet calculation by enabling a significantly finer selection of correction factors to be determined for specific input values using an advanced lookup table. Further high level smoothing algorithms were then developed in the spread sheet, to provide seamless interpolated correction factors for any required calculation interval, thus ensuring that differential movements between each assessment point follow a continuous curve to represent a realistic ground movement profile.

A manual check and validation process was undertaken to confirm that this refined process did not impact on the magnitude and distribution of ground movements as predicted by the G-058.

Review of horizontal movements

As the damage assessment analysis based on the ground movement predictions requires ground movements in the *x*, *y* and *z* vectors related to the orientation of the asset, and based on an understanding that the current LU methodology provides only vertical settlement and total horizontal displacement perpendicular to the excavation walls, it was necessary to provide a solution to separate the total horizontal displacement into its corresponding *x* and *y* component vectors. To overcome this limitation, a process was developed which included the following steps:

Step 1: Firstly, as previously described, the formula for the trend line in Figure 3 of the LU technical advice note can be used to calculate the total equivalent long wall horizontal ground movement at a particular assessment point. The same correction factors can also be applied to the horizontal movements that were applied to the vertical movements, creating a similar profile.

For each assessment point, the offset from the centre of the wall to the location of a point with an equal horizontal displacement can be established.

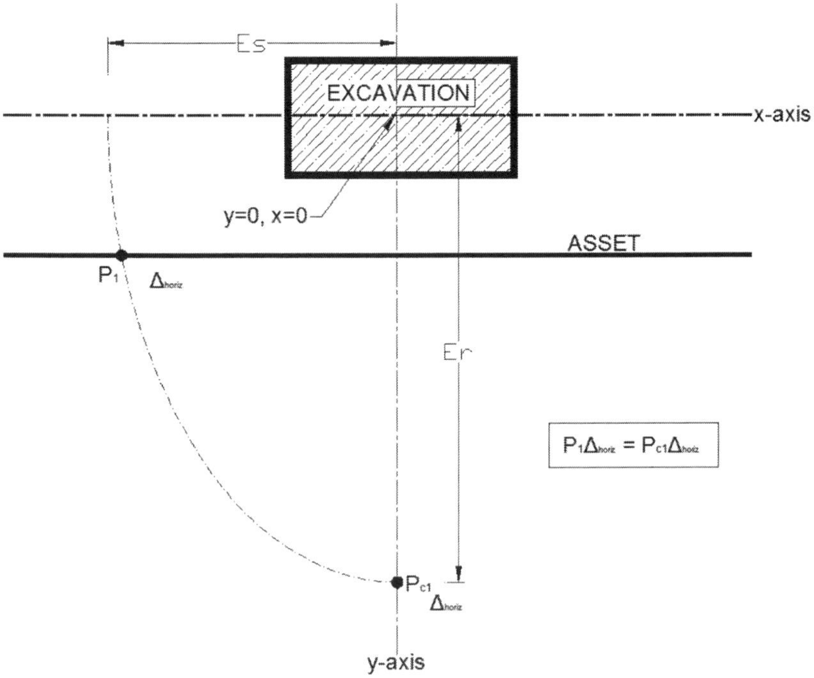

Figure 7: Model of assessment points of equal horizontal displacement and elliptical contour

Step 2: It can be assumed, based on the study of case history data and parametric studies undertaken, that the points of equal horizontal ground movement around an excavation are elliptical in distribution. If the central coordinates of the ellipse are at 0,0 the calculation of the major and minor radii for this ellipse is straightforward.

$$\frac{x^2}{{E_s}^2} + \frac{y^2}{{E_r}^2} = 1$$

Equation 3

Horizontal displacement will act in a direction normal to the curve of the ellipse, and the direction or gradient of this normal can be calculated to give the angle at which the displacement acts. The total horizontal ground movement can therefore then be resolved into the *x* and *y* axes.

The resultant gradient is given by:

$$\frac{dx}{dy} = \frac{{E_r}^2 \times x}{{E_s}^2 \times y}$$

Equation 4 and

$$\mathrm{R} = \frac{\Delta y}{\cos\theta}$$

Equation 5

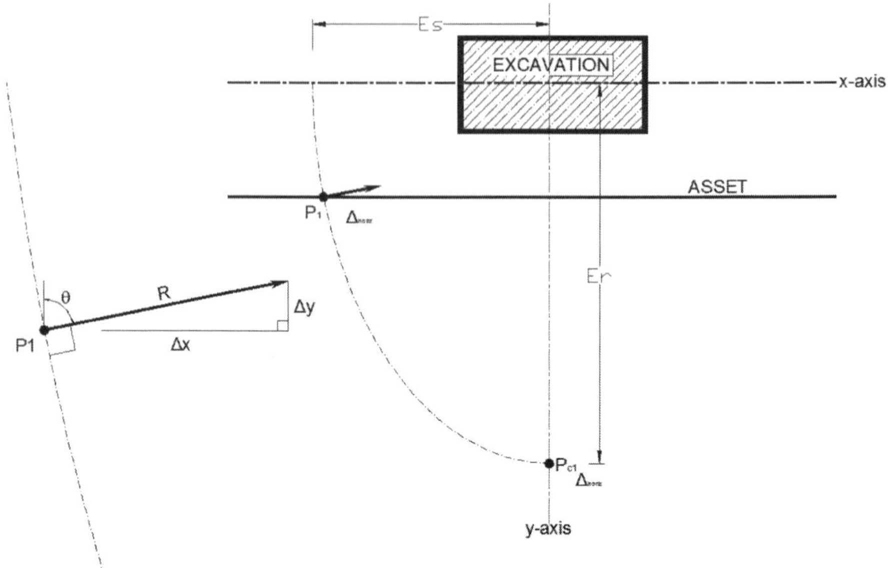

a) Resolving horizontal displacement in x and y vectors

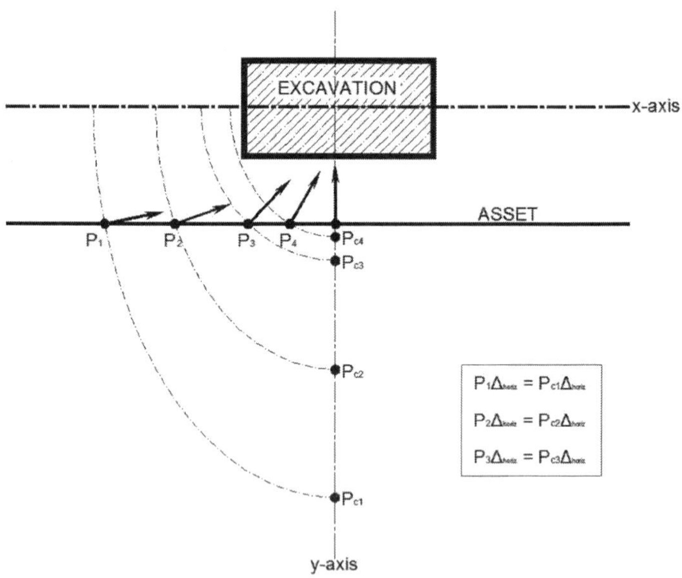

b) Horizontal displacement x and y vectors along asset

Figure 8: Resolving horizontal displacement in x and y vectors

Step 3: The previous stage is repeated for all locations along the asset. It should be noted that the angle of incidence θ of the horizontal ground movements towards the excavation varies according to the position along the asset as would be expected.

Using the GLS

The GLS calculation has been set out so that the ground movements are assessed at set points along the length of an asset impacted by a box excavation. The interval of these points is determined by simple divisions of an individual pipe length within the asset or an arbitrary value of between 0.5 m and 2 m for continuous structures such as a brick sewer. For each assessment point along the asset, a number of calculation steps are carried out to determine the vertical (*z*), transverse (*y*) and longitudinal (*x*) ground movements.

Coordinate system and symbols

Details of both the asset and the dimensions of the excavation adjacent to the asset are required for the assessment. The Thames Water GIS database provided the plan location of the asset and information on the proposed excavation works was extracted from the current scheme drawings. A CAD system was used to locate the coordinates of the asset and the walls of the excavation.

A coordinate system was set up so that a simplified analysis can be carried out for any asset. Looking towards the excavation, corners 1 and 2 of the wall are located on the left hand and right hand side respectively. The central axis centre is located at the centre of the assumed excavation (i.e. at the centre of the wall/box excavation in the *x* direction and half of the assumed breadth of the excavation behind the wall in the *y* direction).

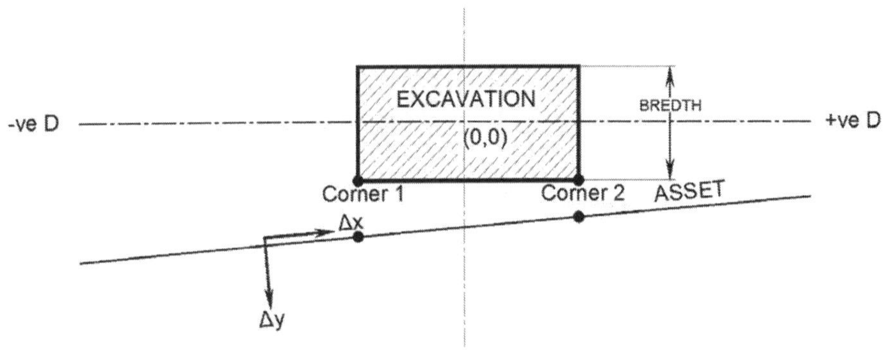

Figure 9: Diagrammatic representation of coordinate system

The coordinate system is used for the purpose of simplifying the location of the assessment points and for calculation purposes. Values for the location of the wall and the asset were entered using the local grid system.

The only additional data required to be entered is;

- Breadth of the excavation (distance between the wall and the other face of the excavation)
- Elevation at the base of the excavation and at ground level

The coordinates of the asset and excavation works input into the programme allows automatic calculation of the relative distances of the asset to the wall, width and depth of the excavation, and the predictions of ground movement based on the methodology as set out in LU G-058, with the refinements as set out in this paper, to allow a fine grid of assessment points to be used and the calculation of horizontal movements in the *x* and *y* direction along the asset being assessed.

The GLS programme generates a plot of ground movements in the *x*, *y* and *z* directions along the length of the asset being assessed as shown in Figure 3. The digital data the GLS uses to generate the ground movement profile in a graphical format was also designed to be directly input into a separate software package to undertake damage assessment calculations, and to be combined with other ground movement data produced for adjacent tunnels and SCL works to provide a composite ground movement model by superposition for the whole station complex.

By the principal of superposition, complex station geometries other than square shaped excavations can be modelled using GLS, in addition the assessment of stages of construction can be analysed for multiple excavations.

Validation and comparison with Crossrail FEA data

Extensive validation using the manual calculations from the G-058 technical advice note was undertaken and independent calculation of horizontal movements using the elliptical pattern of ground movements around box excavations were performed before use of the software on Crossrail.

The first station where the Crossrail FEA ground movement models were compared for validation purposes against the GLS was Bond Street Station, where calculations additional to normal verification testing was undertaken to test the software.

As Figure 10 demonstrates, there is an extremely close agreement between the 3D FEA and the empirical GLS software, in terms of ground movement predictions, which resulted in the calculation of very similar levels and location of damage to the Thames Water assets based on the respective ground movement prediction methods. This close agreement was replicated across all the stations subsequently analysed. For most validation checks performed, the 3D FEA and the GLS output was within +/- 10% of the calculated tensile strain/joint rotation/joint pull-out values calculated, a variation which was acceptable to all parties involved in the damage assessment process.

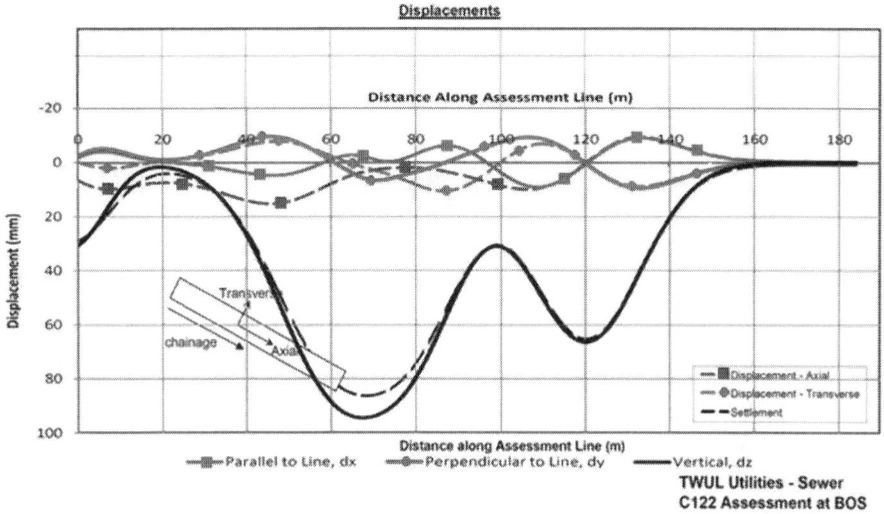

Note: *Solid lines – GLS, dashed lines - FEA*

Figure 10: Comparison of 3D FEA and GLS generated ground movement profile along a TW sewer asset

Conclusions

Refinement of the LU G-058 technical advice note by developing a spread sheet based analysis of the methodology, and further development of the analytical process to allow validation of Crossrail generated ground movements and damage assessments to be undertaken for utility assets, met all the objectives of the brief specified by Thames Water. It provided;

- An empirical method based on case history data to validate the sophisticated 3D FEA model developed by Crossrail as part of their damage assessment and asset protection process for all 3rd party assets
- A highly efficient method of undertaking validation checks on the Crossrail calculations, with multiple analyses being capable of being run within a single working day
- The grid referencing system and output data was compatible with the scheme wide ground movement prediction model produced using commercially available software for the tunnel and SCL works, allowing composite ground movements models to be easily generated for each asset assessed using GLS for input into the damage assessment calculations
- The simple input data and automatically generated ground movement plots allowed the software to be used with minimal training, which provided flexibility of resources within the Thames Water geotechnical team working with Crossrail on the damage assessment and asset protection process
- Supported an efficient process for validation of the predicted levels of damage to Thames Water assets as a result of the highly complex ground movements around the station complexes and agreement of the levels of mitigation/protection required

The one limitation of the GLS software is that the ground movement predictions are limited to those at the ground surface, which needs to be taken into account when analysing deep sewer assets. Because of the significant depth of the majority of the Crossrail station boxes, there was minimal difference in predictions between the Crossrail 3D FEA output which modelled sewers at the correct depths, and GLS, for sewers down to depths of the order of 5 m to 7 m depth. For deeper sewers, because the patterns of movement were similar, although there were some slight variations in the scale of movements, engineering judgement could be used to agree the results of the damage assessment output and required level of mitigation.

Acknowledgements

The Authors would like to acknowledge the contribution and support received from Steve Wilkinson of Thames Water, John Lamont of CH2MHILL, Chris Drew of CH2MHILL, Victoria Crewe of Hyder Consulting (UK) Limited and Dr Barry New of GCG. We would also like to thank Keith Bowers of London Underground for permission to publish this article for the 2013 Crossrail Paper Competition.

References

1. London Underground: Civil Engineering (2009) – Technical Advice Note G-058 A10.
2. Gaba A.R., Simpson B., Powrie W., Beadman D.R. (2003): CIRIA C580 – Embedded Retaining Walls, Guidance for Economic Design.
3. Crossrail Ltd. (2009): Sub-Surface Movements from Deep Excavations 1D0101-G0G00-01025 Rev. A.
4. Choa-Hui Wu, Chang-Yu Ou, Ningchien Tung (2010): Corner effects in Deep Excavations, Establishment of a Forecast Model for Taipei Basin T2 Zone, Journal of Marine Science and Technology, Vol 18, No 1, pp 1 -11.
5. Crossrail Ltd. (2010): Movements due to Box Excavations, Phase 3/Detailed Assessment Methodology, C122-OVE-C2-RAN-CRG01-00059.

Crossrail Project: Infrastructure design and construction
ISBN 978-0-7277-6078-4

doi: 10.1680/cpid.60784.489

Installation of Optical Fibre based Sensors within Sprayed Concrete Lined Tunnels to measure Strain and Temperature

Joe O'Donnell Graduate Design Engineer, Gary Gabriel Associates, formally of BAM Ferrovial Kier JV, C300/410.
Dr Giorgio Nosenzo Technical Director, Monitor Optics Systems,
James Preston Business Development Manager, Monitor Optics Systems,
Tom Sturman C410 Field Engineer, Crossrail,

Abstract
The purpose of this paper is to summarise the practical applications and benefits of optical fibre based sensors over traditional monitoring systems installed within a sprayed concrete lined tunnel. The installation was undertaken as part of the Crossrail project at Bond Street station East ticket hall. The tunnel chosen for installation of the sensors was a concourse tunnel, which is one of the largest sprayed concrete lined tunnels at the station.

The optical fibre based sensor installed is based on the Fibre Bragg Grating strain sensing technique. These cables are designed to survive the installation in harsh environments with aggressive construction techniques like the use of sprayed concrete; whereas standard optical fibres would not survive this process. The cables provide real time structural monitoring with continuous and accurate measurement of strain and temperature values. This provided valuable insights into the mechanical behaviour of the tunnel lining during its early construction and post construction phases. The cables were fully encapsulated within the sprayed concrete lining of a concourse tunnel within the primary lining and at a thickened primary lining location.

The lessons learnt throughout the process have been recorded to aid future installations, as well as suggestions for improvements and further investigations.

Requirements for Monitoring
Monitoring is a vital requirement throughout the whole life of an underground structure to assess its performance in resisting the forces applied to it from the surrounding ground and other structures. This monitoring takes place throughout the construction, and serviceability phases, and has been seen as essential by the British Tunnelling Society (BTS) and the Health and Safety Executive (HSE) report into the 1994 tunnel collapse at Heathrow.

From the findings of the HSE, the Heathrow tunnel collapse occurred for a number of reasons, one being that the tunnel monitoring instrumentation that had been specified by the designer was inadequate for monitoring crucial movements in the tunnel and the designer, as part of their design, had not set rates of change for allowable tunnel movement. According to the report, monitoring is vital in determining the behaviour of the tunnel lining and the surrounding ground. Unsatisfactory trends should be identified sufficiently early to enable corrective steps to be taken. Careful consideration has to be given to the number and location of monitoring points and direct measurement of the movement on a tunnel's cross-section is the most vital and provides the most easily interpretable data.

Traditional methods of tunnel strain measurements involve the measurement of points around the perimeter of tunnel profile with the use of geodetic prisms and surveying equipment. The position of the prisms is measured at various time intervals and a displacement ascertained from subsequent readings. This displacement can be compared with the designer's expected values, with certain limits requiring action to be taken to prevent further movement and the risk of potential structural failure. This method however requires many assumptions to be adopted and in the back calculation from displacement to strain. Another form of monitoring uses pressure cells that are fixed to the extrados of the tunnel and measure the force exerted upon the plate of known area. To be effective and provide reliable results however these cells (earth pressure cells) must be installed to the excavated face of the tunnel. This raises issues regarding the health and safety of operatives during installation due to working under non-supported ground. The reliability of information attained from pressure cells is also questionable due to the wide scattering of data and factors that affect their performance such as installation method.

Once a tunnel has been constructed and is in its serviceability phase, there are often different methods that are used to measure the performance and condition of the structure. These can include further measurement of geodetic prisms and analysis of movement and visual inspections.

Fibre Bragg Grating Sensors

Optical Fibres (OF) are flexible optically transparent fibres through which light can be transmitted by successive internal reflections. This type of transmission allows the propagation of a light signal along a fibre for very long distances with minimal attenuation. This ability to carry data over long distances together with the OFs capability of carrying a high number of different signals along a single fibre, are the reasons for their widespread use in telecommunication.

By exploiting the capability of external actions on a fibre to induce variations of properties to the light travelling within the fibre itself, a number of sensing techniques have been developed. Among the various techniques, Fibre Bragg Gratings (FBG) were selected as the technique that is best suited to providing accurate strain and temperature measurements. This technique is based on the use of selectively modified areas of special telecom-grade optical fibres that have the property of reflecting a narrow bandwidth signal centred on a certain wavelength. This is a function of a number of design parameters and shows the property of varying linearly when strain or temperature variations are applied to the FBG. This wavelength can be measured very accurately using a dedicated demodulation unit capable of converting the wavelength shift in a digital signal. By measuring the wavelength signal through an FBG, it is possible to turn an OF into an accurate strain or temperature transducer.

The use of FBG based sensors provides a number of benefits over other types that make them ideal for use in an underground environment. FBG based sensors are self-referencing because the sensing principle is based on a reversible physical modification of the FBG structure, a structure that does not degrade with the time. It is therefore possible to disconnect and reconnect the cables without having to recalibrate the sensors as they are self referencing and do not suffer from drift. Due to their operating principles and no moving parts, the operation of optical fibre sensors does not cause wear to the sensors or other parts of the sensing network making them highly resistant to the elements, allowing them to survive indefinitely in harsh environments. Sensing cables based on FBG technology typically have a diameter of 1-2 mm and therefore their inclusion in concrete structures does not affect the structural properties.

Regarding measurement performance FBG based optical sensing offers both high resolution and accuracy, for strain typically in the order of 1-2 micro strain and for temperature 0.1 degrees Centigrade, whilst being able to successfully perform dynamic measurements up to several kHz. Due to the nature of OF there is no interference caused from electromagnetic noise, and there is no effect on other systems. As they are transmitters of light there is also no risk of a spark, and no requirement for a power supply to each sensor location. The interrogator and data logger can be installed long distances away from the sensors, such as in an office above ground.

Each FBG reflects a signal that is a function of its design parameters; therefore FBGs can be employed in high numbers, e.g. up to 40 sensors on a single optical fibre (in-series multiplexing), a capability that is a unique feature of the FBG. The use of standard OF couplers and switches and the use of multi-channel interrogation equipment allow a number of different fibres to be connected to a single interrogation unit. This allows the creation of long term, large and sophisticated networks that can be used to provide real time monitoring of a structure.

Installation

The installation was undertaken at two locations within a concourse tunnel at Bond Street Station, one of largest sprayed concrete lined (SCL) tunnels at the station. The successful installation of FBG sensors within an SCL tunnel requires the availability of a sensor capable of surviving what is a very aggressive installation process, capable of seriously damaging most types of sensors. Many FBG sensors that are commonly available are too fragile to be used in such an environment. It was therefore necessary to use reinforced cables that had been previously proven to survive the spraying process.

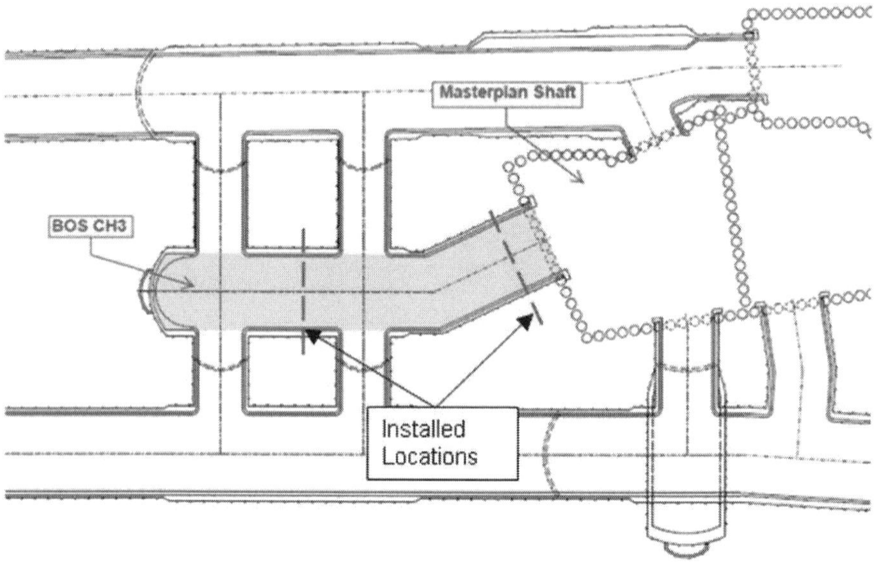

Figure 1 – Installation Location within Bond Street Station, Plan

Reinforced FBG sensing cables are based on the use of FBG arrays encapsulated in a small pultruded fibreglass rod 2mm in diameter. Even with the fibreglass reinforcing and external coatings the sensors still have near perfect strain transfer. This manufacturing technique allows the production of sensors within fibres that whilst being slender are also very robust, easy to handle and install. The robustness of these sensing fibres was further enhanced by the addition of an external nylon coating. This nylon coating makes the sensor more capable of surviving the mechanical effect of concrete being sprayed onto it. The overall diameter of the nylon reinforced sensing cables is 2mm.

The strain sensing cables used were characterised by the presence of 5 FBGs, each of them corresponding to one strain sensing positions within the initial heading of the tunnel. The location of the FBGs on each cable was designed so that each sensing point would be in the desired position around the tunnel profile. Each sensing cable was supplied with two connectors, one at each end. As FBG based sensors are connected to the interrogation units from one end only, this configuration introduces a degree of redundancy. If a cable or connector is damaged or broken it is possible to simply connect the cable from the other end, and acquire all the sensors along the full length of the cable up to the position of the break in the cable.

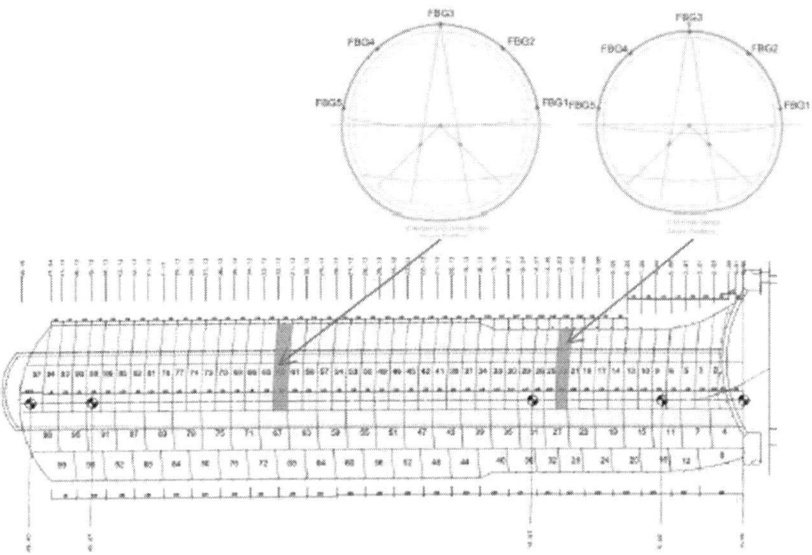

Figure 2 – Long Section through BOS CH3 tunnel showing installation locations.

Temperature sensing cables were also used with a dual purpose: to monitor the temperature variation during the concrete curing phase and also compensating any temperature effect on the strain sensing cables. Each installation was instrumented with a single FBG temperature sensing cable located near the tunnel axis of that advance. As a result of this configuration, temperature compensation for all strain sensing points in a section was supplied from the single temperature sensing point in the same advance.

In addition to the standard cable features, particular care was taken in providing reinforcements to the connectors of both strain and temperature sensing cables to guarantee the robustness of the sensing network.

The FBG sensing cables were interrogated using a dedicated FBG interrogation unit controlled by a local control computer. The interrogator selected was a Micron Optics Sm130 that was selected due to its dynamic interrogation capabilities, allowing the unit to interrogate up to over 500 sensing points at a frequency of 1 kHz. A wireless 3G connection dongle was installed on the control computer to allow the transmission of data to a remote server. This data was then viewable via an internet connection for any number of users with a logon to view in real time using DaMinS software.

Signals from the sensors were carried through optical fibre cables connected to the sensors via standard telecom grade optical fibre adaptors, with cables fixed along the length of the tunnel at the shoulder, and run to the surface via already installed service trunking.

The interrogation equipment and dedicated control computer were installed in the site office located at the top of the main shaft. The equipment was installed within a secure server rack designed to host internet & network components in order to provide protection against dust and dirt.

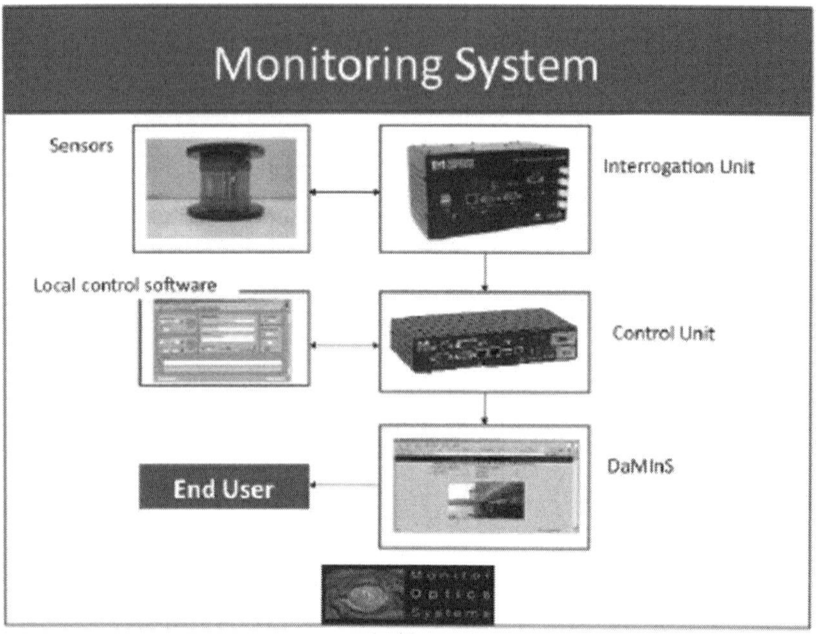

Figure 3 - Monitoring Installation Equipment

Signal cables were run through the shaft and to the predetermined location of the sensors connectors. The cables were inserted within conduits in the sections where they were most exposed to the risk of damage, such as along the length of the tunnel adjacent to construction activities, but were left unprotected in other areas where there was minimal risk of damage. The intrinsically passive nature of FBG signals made possible laying the cables in conduits together with electric cables without any risk of interference.

The sensors were installed onto the intrados of the initial 75mm lining once the minimum safe strength had been achieved for the safe entry of operatives. The installation of the strain sensors was achieved by positioning the sensing cables so that each sensing point corresponded to the level of the intended location, designated by a laser from surveying equipment. The sensing cable was then pinned in place using 30mm galvanised staples. These prevented any undesired movement of the cable during the concrete spraying process while at the same time not exerting forces on the sensors capable of altering the strain measurements. The survivability of the temporary fixings had previously been proven during a pilot tunnel installation.

The installation procedure for the temperature sensing cables was similar, with the only difference that the cables were fixed in a more limited number of points due to their shorter length. The installation of the sensing cables required the use of a Mobile Elevated Working Platform (MEWP) with a crew of two operatives in the basket in order to reach the tunnel crown. At the same time, a third person could complete the connection of the sensing cables to the pre-installed signal cables.

Once the sensing cables were pinned in the correct position and connections confirmed by verifying the sensors provided data, spraying of the primary lining commenced that fully encapsulated the sensing cable within the concrete lining.

Data recording started immediately after the connection of the sensors to the interrogation unit through the sensing cables. This allowed the measurement of strain and temperature variations on the sensors during the concrete spraying and the early age strength development. The acquisition frequency was set at 1 acquisition per second although data was saved with a frequency of 1 data set per minute.

Results

The results that were attained from the installations within advance 22 and advance 62 are shown in chart 1 and chart 2 respectively. Results have been referenced to zero when the spraying activity within that advance ended. There is also a timeline of events shown along the bottom of chart 1.

Chart 3 shows the temperature readings of the first installation, however is not calibrated to the correct temperature and so a relative temperature reading is only available.

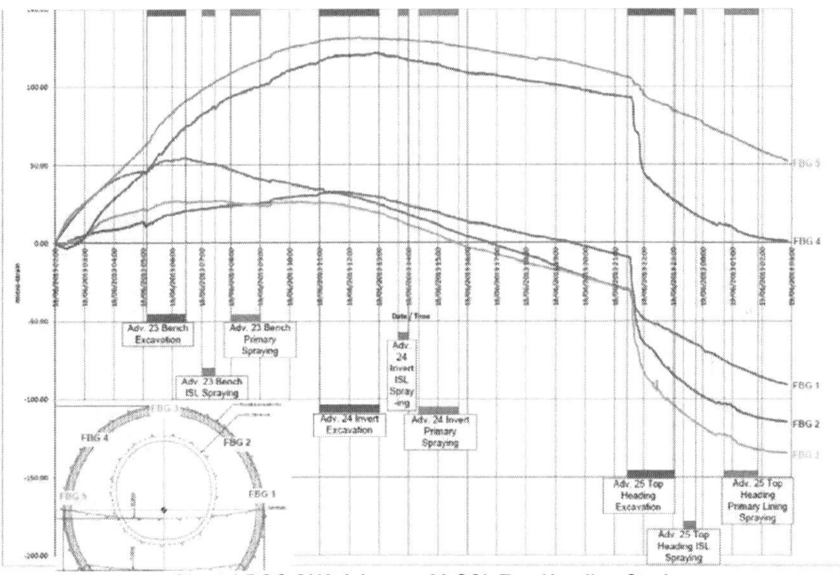

Chart 1 BOS CH3 Advance 22 SCL Top Heading Strain

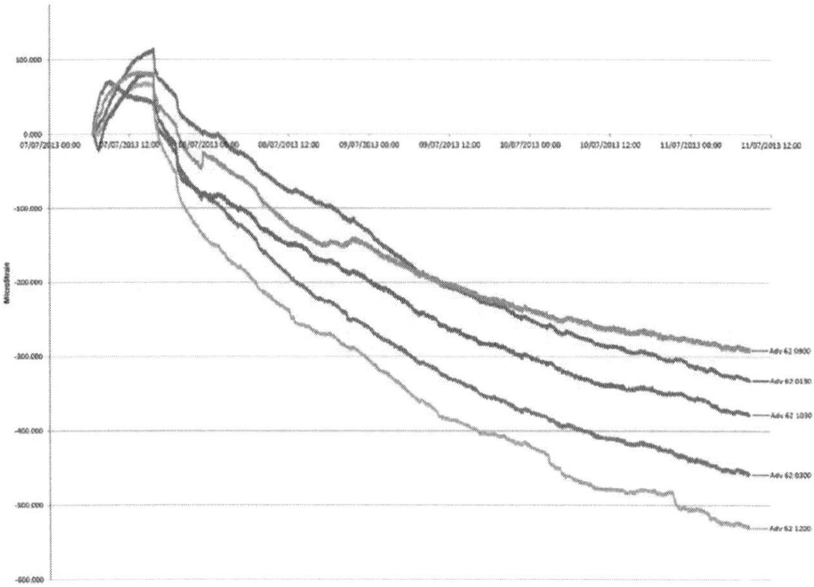

Chart 2 BOS CH3 Ch32 Baselined from Convergence monitoring reading (07/07/13 06:30)

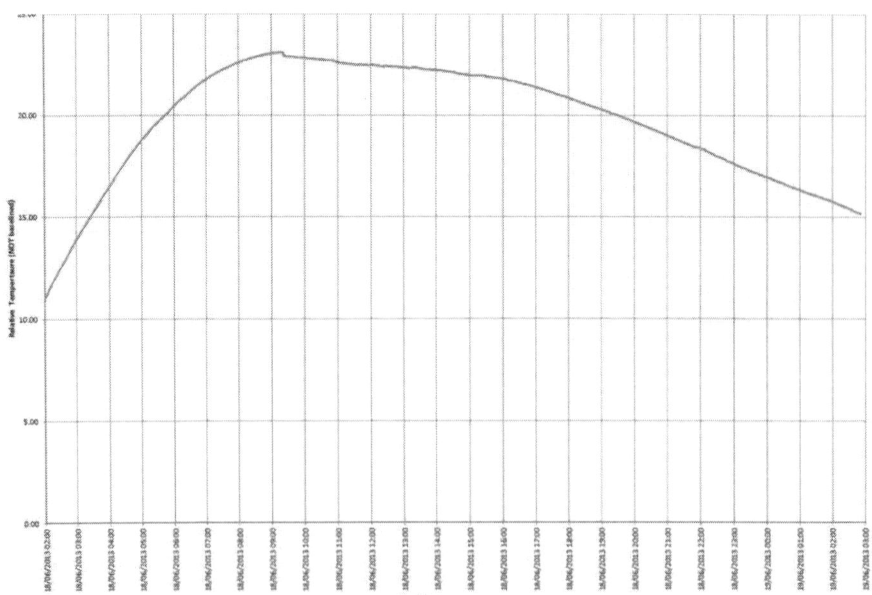

Chart 3 BOS CH3 Advance 22 SCL Top Heading Temperature

Conclusions

It can be seen from both charts 1 and 2 that there was an initial contraction of the tunnel lining as the concrete initially cured, followed by an increase in micro strain, showing tension within the lining. After both installations, within approximately 24 hours all the sensors shown compression when compared to the base lined value once spraying had been stopped. There was also an increase seen in the compression of the lining at various stages during the subsequent stages of tunnelling. This was generally seen during the excavation stages, and was most pronounced during the first top heading excavation after the installations had taken place. This slowed once the exposed faces of the tunnel have been sprayed.

The increase in compressive strain can be attributed to the tunnel lining deforming as it takes loading from the surrounding London Clay. These deformations are mostly seen during nearby breaks within the tunnel lining due to unsealed faces, and are less pronounced once the tunnel ring profile has been completely sprayed up.

It was seen from chart 3 that the maximum temperature is reached approximately 7:30 hours after the encapsulation of the sensing cable.

Further Recommendations

It was noted that the temperature compensation sensor was installed near the left hand side of the tunnel above the axis. During the spraying process the lining commences from the axis and works upwards from both sides, before spraying the crown. Depending upon which side is sprayed first has an effect on the temperature compensation of the sensors within the lining during the early age strength development. Long term however, this would be negligible as it can be assumed that the tunnel would reach a consistent temperature around its profile at each chainage.

Further to the results that were provided from the two installations within BOS CH3 it would be advisable to expand the trial and install further sets of monitoring sensors within tunnels of varying profile. This would aid the comparison of data that can reviewed for a localised area with the assumption that similar loadings would be applied to the tunnel from the ground, and the effects of varying tunnel thicknesses and geometric profiles can be compared. To gain a further understanding of the temperature variation, and as such an estimate of the rate of the concrete curing, it would be necessary to instrument a profile with several temperature sensors. This would allow the temperature variations around the profile to be ascertained, as well as provide a more accurate temperature compensation for strain sensors within the lining. It would also be able to see if there are any local variations in temperature around the ring due to construction activities.

A further investigation would be the installation of sensors around the complete tunnel profile to see the behaviour of the lining around the temporary backfilled invert when it is being tracked upon by tunnelling plant, the installation of mechanical and electrical services, and tunnel finishes.

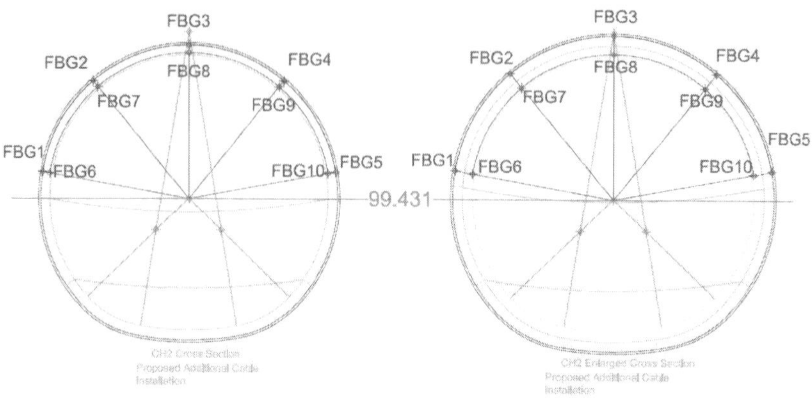

Figure 4 - Proposed Profile and Monitoring Sensor Locations

A further investigation is an installation of two sensor cables within the same advance, one near the intrados, and one neat the extrados. This would allow the bending moments within the tunnel lining to be calculated.

Acknowledgements

The authors would like to thank Monitor Optics Systems for providing the equipment and materials to undertake these trials, as well as the all of the staff from both BFK and Crossrail for their continued support, perseverance and advice throughout both the installation and monitoring of the tunnel.

References

Health and Safety Executive 2000 “The collapse of NATM tunnel at Heathrow Airport” HSE Books

British tunnelling Society 2011 “Monitoring Underground Construction – A best practice guide” ICE Publishing

Crossrail Project: Infrastructure design and construction
ISBN 978-0-7277-6078-4

doi: 10.1680/cpid.60784.499

Tunnelling Induced Strains and Deformations at Central Line

Jessica Yu BE MSc DIC MICE, PhD researcher, Imperial College London & Geotechnical Engineer, Crossrail Ltd
Jamie Standing BSc MSc PhD DIC MICE, Senior lecturer, Imperial College London
Robert Vollum MSc PhD DIC CEng MIStructE, Reader in Concrete Structures, Imperial College London
David Potts Phd DSc FREng CEng FICE FCGI, Professor of Geotechnical Engineering, Imperial College London

Abstract

In October and November 2012, Imperial College installed electrical resistance strain gauge rosettes, potentiometric displacement transducers and tape extensometer eye bolts at selected locations in the Central Line eastbound running tunnel between Lancaster Gate and Marble Arch. The installations were completed before the two Crossrail tunnel boring machines (TBMs) passed below the Central Line running tunnel at this interface.

Automatic continuous measurements of strain in the selected tunnel segments were made before, during and after the passage of the TBMs at this interface. Similarly, automatic continuous measurements were made of any opening and/or closing of horizontal joints of the tunnel rings during this time period. Manual measurements of tunnel convergence and diameter/chord length changes were made during engineering hours.

This paper describes the installation process and presents the monitoring data for each TBM passage as well as the combined effect of both Crossrail tunnels on Central Line. The interpretation of the strain measurements to provide information on the mode of deformation of the instrumented tunnel segments is explained. This information is assessed in conjunction with the manual tape extensometer measurements to ascertain the deformed shape of the instrumented tunnel rings as a result of the Crossrail tunnelling works.

The paper also discusses the use of the field data in guiding the research experiments currently being conducted on a half-scale grey cast iron tunnel ring at Imperial College. Finally, recommendations for future in-tunnel monitoring are proposed based on the wisdom gleaned from this experience.

Introduction

In October and November 2012, Imperial College installed electrical resistance strain gauges, potentiometric displacement transducers and tape extensometer eye bolts at selected locations in the Central Line eastbound running tunnel between Lancaster Gate and Marble Arch. The instrumentation was installed shortly before the first of the Crossrail tunnel boring machines (TBM) passed below the Central Line tunnel at this interface.

The Crossrail twin tunnels begin at Royal Oak Portal in Westbourne Park, West London, and traverse westwards towards Farringdon. The Crossrail tunnels had a clearance of 4.2m to 4.9m under the Central Line twin tunnels east of Lancaster Gate station running at approximately 40 degrees skew (Figure 1). The Crossrail westbound tunnel passed under the Central Line eastbound tunnel at Crossrail chainage 2085m. The Crossrail eastbound tunnel passed under the Central Line eastbound tunnel at Crossrail chainage 2111m.

According to the daily progress update provided by Crossrail, the westbound tunnel TBM (TBM1) cutter head passed under Central Line eastbound tunnel at around 2:00pm on the 19th November, 2012. The cutter head of the eastbound tunnel TBM, (TBM2), passed under Central Line eastbound tunnel at around 19:30pm on the 5th February, 2013.

Continuous measurements of the strain in selected tunnel segments were made before, during and after the passage of the TBM at this interface. Similarly, continuous measurements were made of any opening and/or closing of horizontal joints of the tunnel rings before, during and after the passage of the TBM at this interface. Finally, manual measurements were made of tunnel convergence and diameter/chord length changes at several rings of the eastbound Central Line tunnel before, during and after the passage of the TBM at this interface.

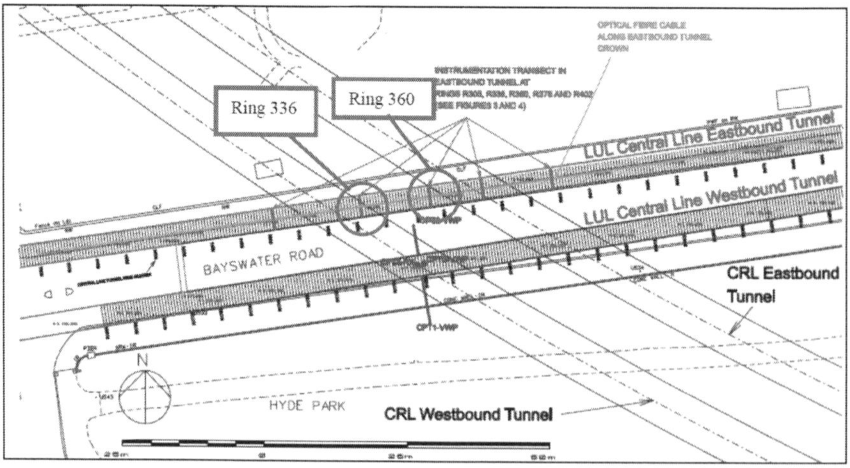

Figure 1. Plan view of Crossrail interface with Central Line east of Lancaster Gate Station

As-built locations

Table 1 summarises the locations of the instrumentation installed by Imperial College in the Central Line eastbound running tunnel.

Instrument	Location
Five no. strain gauges per segment	Axis, and two shoulder segments on R336 and R360
Displacement transducers at 3 horizontal joints	At the crown and at two shoulder level joints on R336 and R360
Seven no. temperature sensors	R336 and R360 - within each datalogger unit in the tunnel
Tape extensometer eye bolts at shoulder and knee levels.	R305, R336, R360, R378 and R402

Table 1. Location of installed instrumentation.

Figure 1 shows the locations of the instrumented tunnel rings in plan. Figure 2 shows the location of the instrumented segments and joints. Figure 3 shows the locations of the strain gauges on Rings 336 and 360.

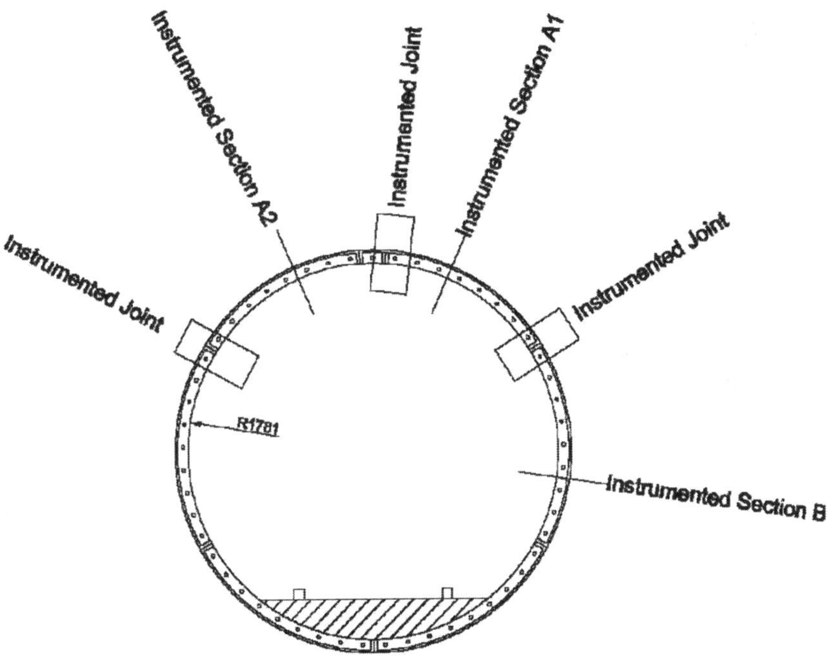

Figure 2. Location of instrumented segments and joints on ring 336 and 360

Segment A2	
Ring 336 Distance from Intrados: Gauge 1 = 13 mm Gauge 2 = 73 mm Gauge 4 = 73 mm Gauge 5 = 12 mm	**Ring 360** Distance from Intrados: Gauge 1 = 12 mm Gauge 2 = 72 mm Gauge 4 = 76 mm Gauge 5 = 11 mm
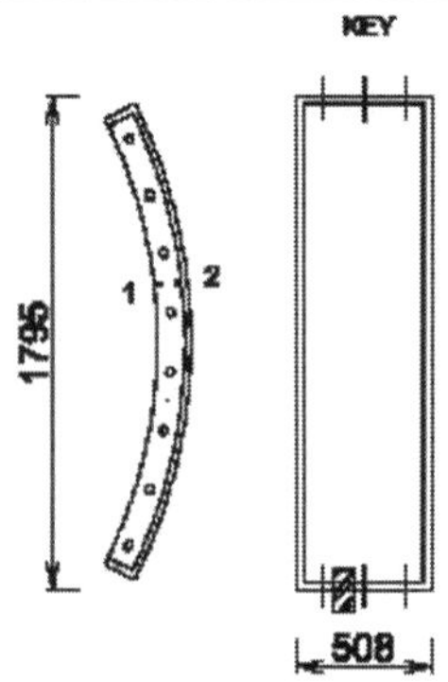	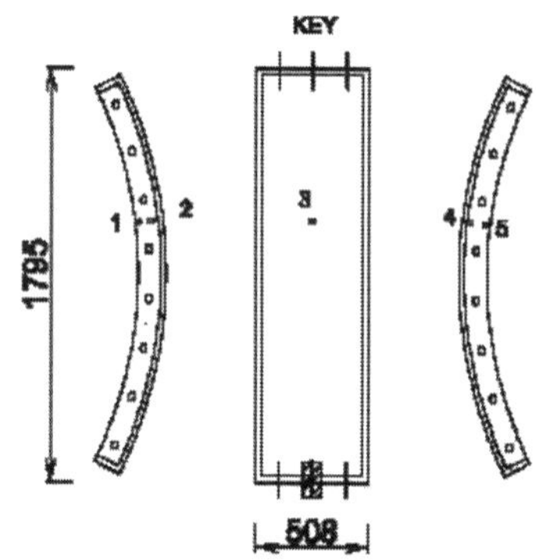
Segment A1	
Ring 336 Distance from Intrados: Gauge 1 = 14 mm Gauge 2 = 73 mm Gauge 4 = 75 mm Gauge 5 = 15 mm	**Ring 360** Distance from Intrados: Gauge 1 = 15 mm Gauge 2 = 75 mm Gauge 4 = 73 mm Gauge 5 = 14 mm

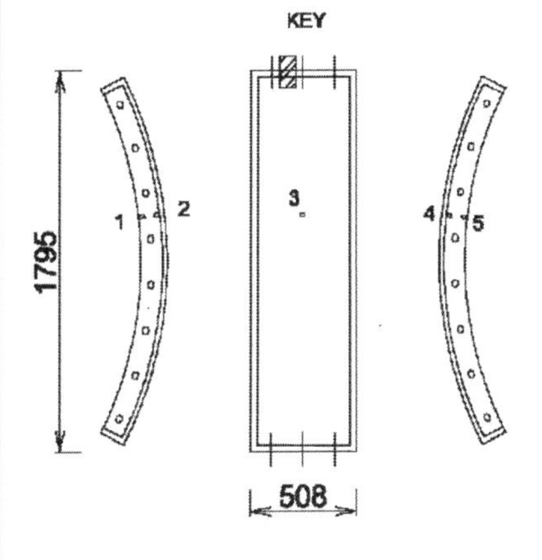

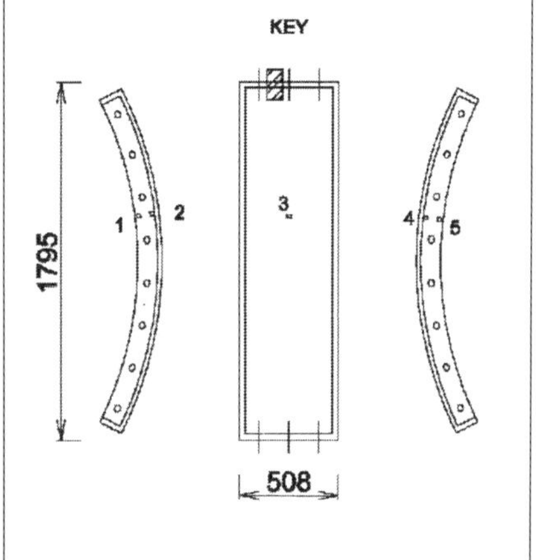

Segment B	
Ring 336 Distance from Intrados: Gauge 1 = 12 mm Gauge 2 = 75 mm Gauge 4 = 74 mm Gauge 5 = 15 mm	**Ring 360** Distance from Intrados: Gauge 1 = 15 mm Gauge 2 = 75 mm Gauge 4 = 73 mm Gauge 5 = 15 mm

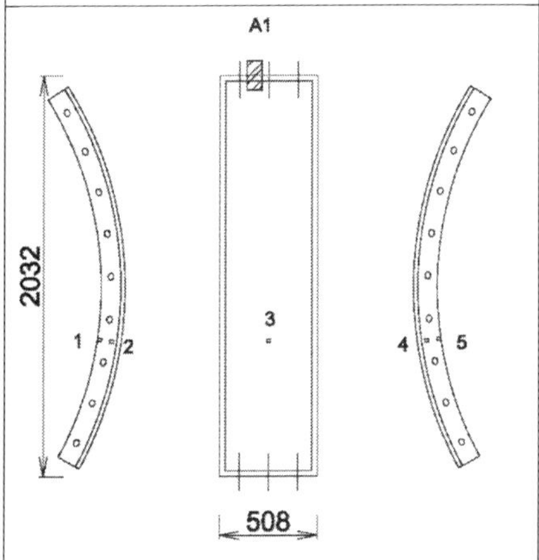

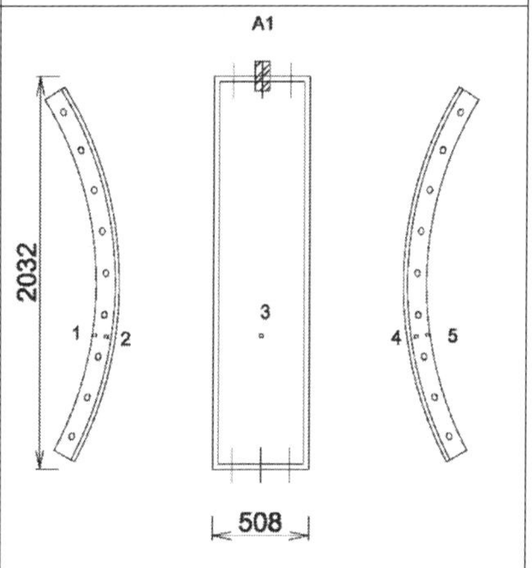

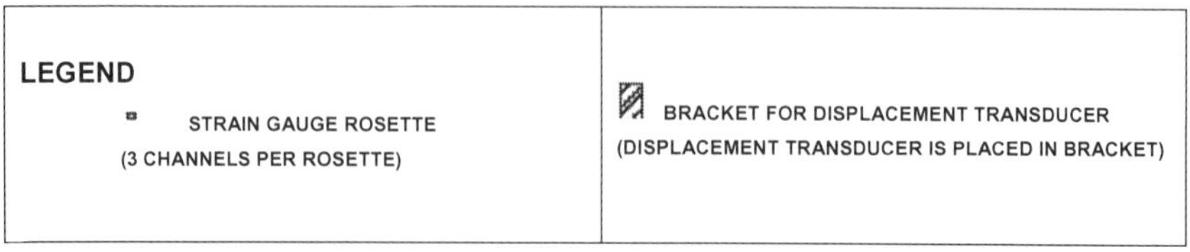

Figure 3. Location of strain gauges and displacement transducers on Ring 336 and Ring 360

The Instrumentation

Electrical resistance strain gauges

Universal general-purpose strain gauges consisting of large three-element 45° single-plane rosettes were used. Each gauge has a resistance of 350Ω ±0.4% and a strain range of ±5%. The Self-Temperature-Compensation (STC) of the gauges was 06 (in ppm/0F) to match the thermal expansion coefficient of grey cast iron which is around 10 to 12 ppm/K, i.e., 5.5 to 6.6 ppm/0F (Gilbert, 1977). The rosette matrix dimension is 16.5 by 20.3 mm.
Each gauge in the rosette was pre-wired with a 330-DFV leadwire. The leadwire was a flat cable of tinned stranded copper, and insulated with vinyl.

Displacement transducers

Potentiometric displacement transducers were used to measure any opening/closing of the horizontal joints. The transducers were supplied by Construction Monitoring Control Systems Ltd (CMCS) who were also responsible for the installation of the displacement transducers.

Temperature gauges

There was one temperature sensor placed within each datalogger unit in the tunnel. There were seven temperature sensors in total and the temperature was logged continuously.

Tape extensometer

An Ealey tape extensometer was used for taking manual measurements. The digital extensometer was able to provide a tenth of a millimetre accuracy. All the readings were taken by the same operative to ensure consistency.

The digital extensometer was hooked into eyebolts secured to the tunnel using the U-clamp system to measure the distance between two eyebolts. This allowed the tunnel convergence and diameter/chord length changes to be monitored. The tension sensor ensured the same tension was applied to the tape each time.

When the correct tension was reached, a green light switched on, and a red light indicated too much tension was applied.

Datalogger for strain gauge, displacement transducer and temperature gauge

The downloading unit on the platform consists of a power supply and an RS485 to RS232 converter through which the data from the in-tunnel loggers is transmitted.

The installation

Electrical resistance strain gauges

For the installation of the rosette type electrical resistance strain gauges, the surface of the cast iron was prepared by using a grinder to remove the 'black crust' until a relatively polished surface was obtained. The area to be gauged was then degreased with Isopropyl Alcohol and further polished by emery paper.

The strain gauges were bonded directly onto the cast iron surface using cyanoacrylate.

A coating consisting of butyl, rubber and foil tape was used to protect the gauges.

The leadwires were bundled together and brought down to the dataloggers by running the bundles through cable ties which were looped around and anchor glued onto the circumferential bolts.

A total of 29 rosettes, i.e. 87 strain gauges, were installed.

Displacement transducers

The displacement transducers were clamped onto the segments. The clamping bracket weighed about 1kg. A total of 4 springs was used to hold the bracket in place. Each spring could apply 2.1N/mm of force giving a total range of 100N-150N for all 4 springs. These springs pushed pointed rods onto the flange to hold it in place securely. The installation was done by CMCS. Figure 4 shows the installation by CMCS for Imperial College in Central Line.

The cable for each displacement transducer was brought down to the dataloggers by running the cables though cable ties which were looped around and anchor glued onto the circumferential bolts.

A total of six displacement transducers were installed.

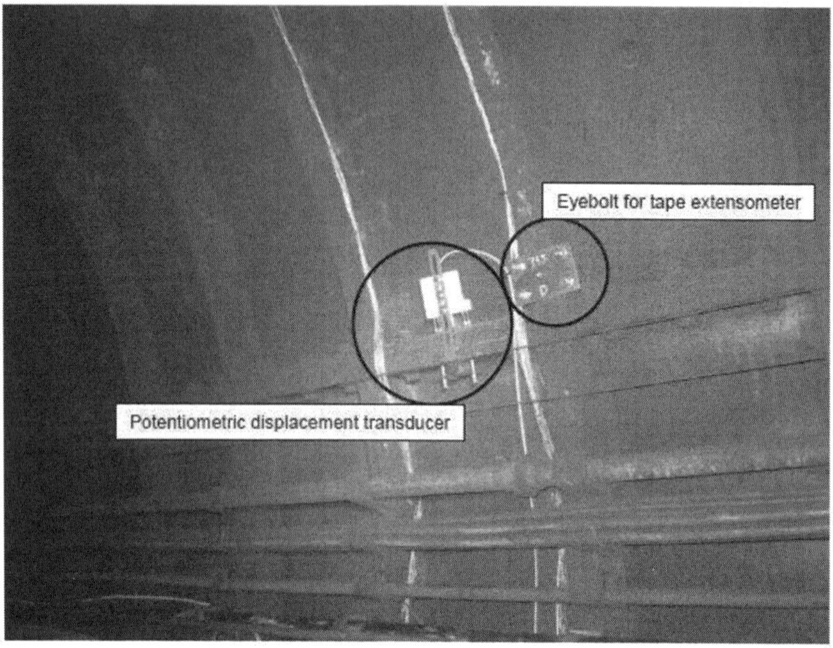

Figure 4. Photo of potentiometric displacement transducer installation by CMCS and tape extensometer eyebolt installation by Imperial College in Central Line.

Dataloggers

Datalogging units for the strain gauges, displacement transducers and temperature gauges were secured onto rings 336, 337 and rings 360, 361. The units were fixed to the segment using a bracket as shown in Figure 5. The logger box had counter sunk screws fed through the back and M6 nuts were used to secure the box on to the bracket. The weight of the logger with batteries is 3.93kg.

Initially the dataloggers were powered via two wires in a six core cable. However, because of the distance between the power supply and the dataloggers, there was significant voltage drop and resulted in data corruption. Therefore batteries were placed in the dataloggers to provide a stable power supply during the passage of the first TBM. Subsequently, a second power cable was installed to power the loggers via six wires, thus reducing the voltage drop. The batteries were no longer necessary and were removed. The voltage requirement for the loggers is 9V.

An additional datalogger unit was secured onto the headwall at the east end of the eastbound platform at Lancaster Gate (Figure 6). This unit allowed the readings stored on the in-tunnel dataloggers to be downloaded onto a laptop during non-engineering hours.

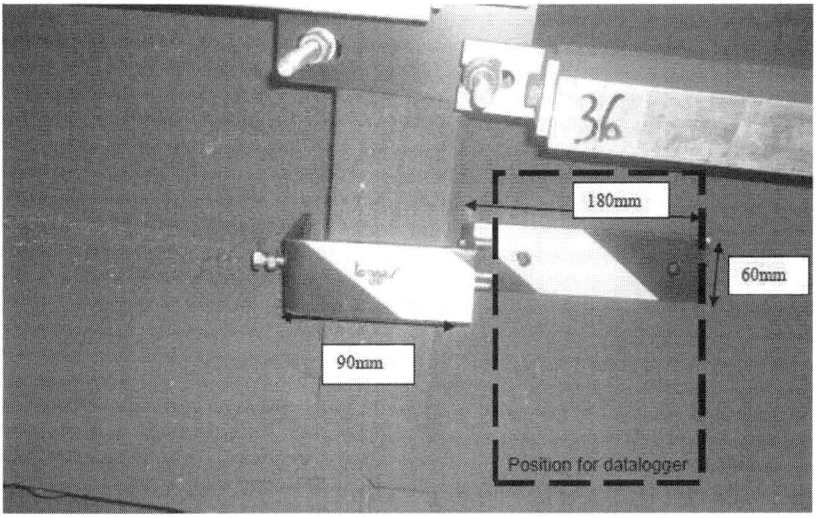

Figure 5. Bracket for securing dataloggers to tunnel segments

Figure 6. Additional datalogging unit on the eastbound platform headwall.

Measurements and calculations

Electrical resistance strain gauge

Prior to the arrival of TBM1, 24 hours of stable baseline monitoring was obtained for all but five gauges. The data are presented as change in strain since the 18th November 2010 after 0100 hours in Figure 7. The measurements were initially unstable due to inadequate power supply. This was corrected by installing batteries at around 1am on the 18th November.

The vertical lines on the graphs mark the time when the cutting face of the TBM was directly beneath the Central Line tunnel. For subsequent discussions and calculations, the change in strain between the initial baseline and the final stabilised readings will be termed 'final change'.

Figure 7 gives an example of each of the three responses that was captured by the strain gauges. The title on the graph gives the location of the strain gauge. R336 or R360 refer to Central Line ring number 336 or 360. A1, A2 and B refer to the location of the segment in the ring as specified in Figure 2. 'Outer flange' refers to the strain gauge located near the intrados of the circumferential flange and 'Inner flange' refers to the strain gauge located near the extrados of the circumferential flange. R1, R2 and R3 refer to the strain gauge locations 1, 2 or 3 in the rosette (as seen in the inset in Figure 7). Finally, the 'E' or 'W' for the series on each graph indicates whether the gauge is on the west circumferential flange or the east circumferential flange of the tunnel ring.

The top graph in Figure 7 gives an example of the gauges on the east and west flanges measuring the same 'mode' of change, i.e. gauges at the same location but on opposite circumferential flanges would both measure either a tensile change or a compressive change.

The middle graph in Figure 7 gives an example of the gauges on the east and west flanges measuring opposite 'modes' of change. Three pairs of gauges displayed this behaviour. The bottom graph in Figure 7 gives an example of erratic measurement from a strain gauge (L341W). The unstable gauges are located on Ring 336 and their results are not processed further.

Figure 7 also shows that the strain gauges were detecting changes in strain in the segments prior to the TBM cutter face being directly below Central Line.

Figure 8 gives an example of the strain measurement over the period when TBM2 passed under Central Line. Unfortunately the power supply to the dataloggers was accidentally switched off for approximately 24 hours, resulting in no data when the TBM was passing under the Central Line. However, using the 48 hours of data prior to the power outage and the stabilised data collected after the TBM had passed; a 'final change' in strain in the segments as a result of TBM2 could still be obtained.

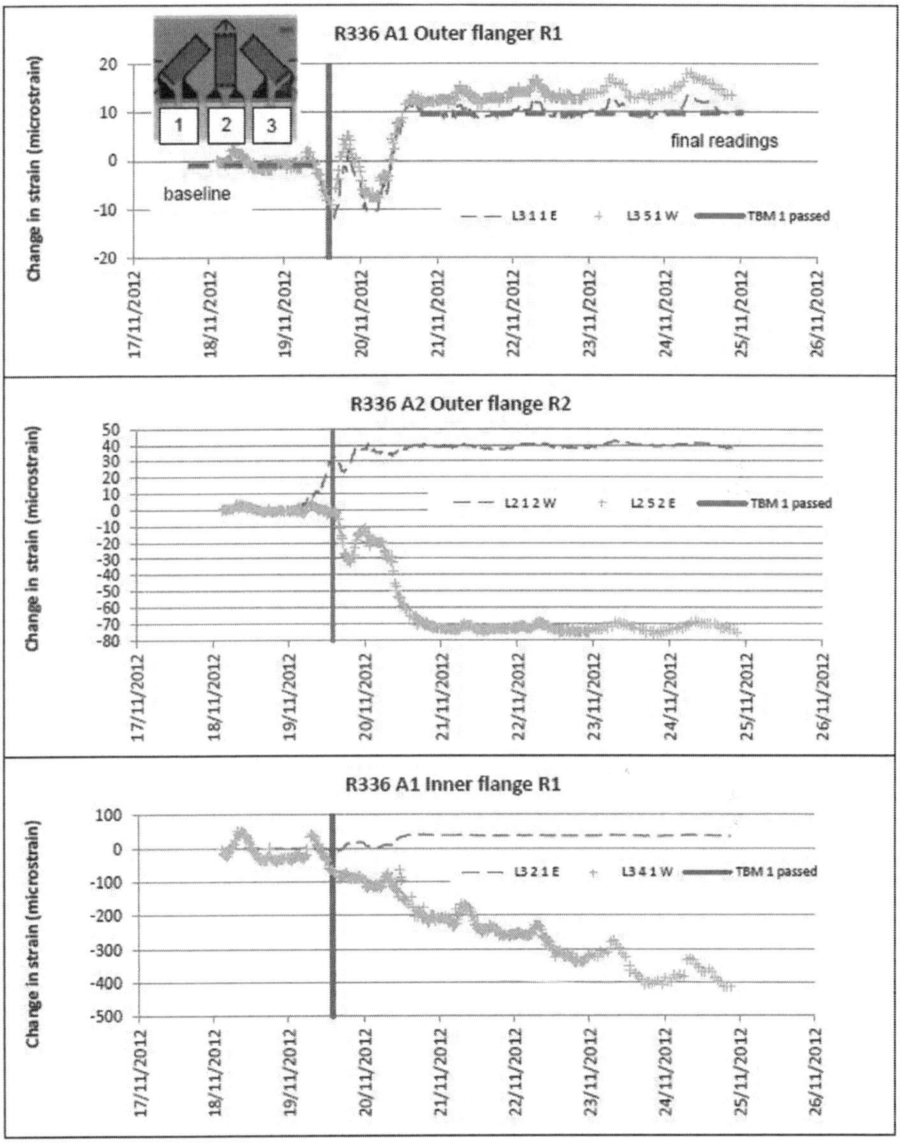

Figure 7. Typical strain gauge response for TBM1 passage showing segment bending (top), segment twisting (middle), and erratic gauge readings (bottom)

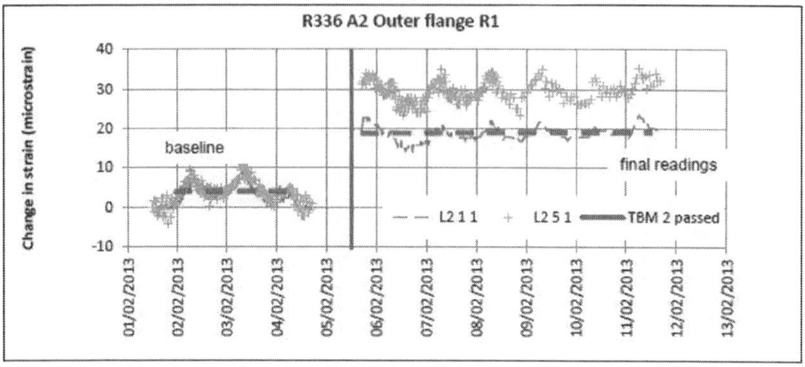

Figure 8. Typical strain gauge responses for TBM2 passage

Using the 'final change' strain measurements from each strain gauge in a rosette, the principal major and minor strains and the principal strain directions were calculated for each rosette location. The calculations included corrections for strain gauge thermal output and gauge factor variation with temperature. Due to the erratic gauges described above, it was not possible to calculate the principal strains at several locations on Ring 336. These locations are listed in Table 2.

Refer to Figure 3 for the locations of the rosettes. Circumferential strain and longitudinal strain directions are defined as shown in Figure 9. Positive bending refers to the mode of bending having a straightening effect on the segment, and negative bending increases the curvature of the segment as illustrated in Figure 10.

Once the magnitude and direction of the principal direct strains on the surface of the cast iron segment was determined, the direct strains on the circumferential flange along the circumferential direction could also be calculated. The circumferential strains were used to derive the bending moment induced in the segment at the instrumented locations. The calculations assumed a linear variation in strain from the circumferential flange intrados to extrados (i.e. between inner and outer measurements), and a constant modulus of 100 GPa. Due to the erratic measurements on ring 336, only the results for ring 360 are presented in Figure 11. The curved arrows indicate the mode of bending in the segment. In all cases the calculated bending moments in the east and west circumferential flanges are in the same direction and have similar magnitudes. It should be noted that the change in strains in the segments was more severe from the passage of the second TBM than from the combined effects of both TBM1 and TBM2.

Gauge designation	Location
L222	Ring 336 Segment A2 inner flange west
L242, L243	Ring 336 Segment A2 inner flange east
L341	Ring 336 Segment A1 inner flange west

Table 2. Locations where principal strains could not be calculated due to erratic gauge measurements

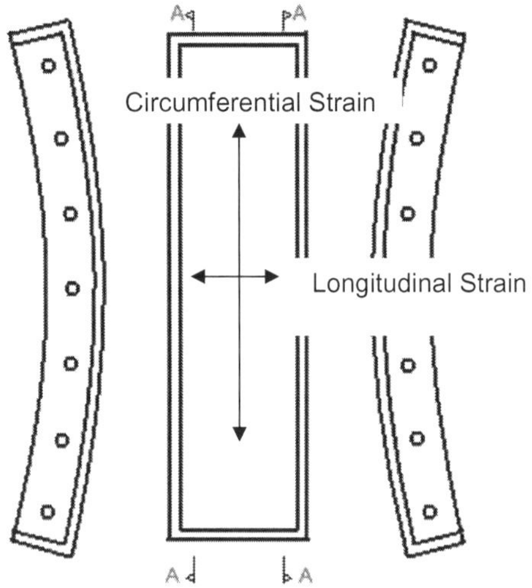

Figure 9 Directions of circumferential and longitudinal strains

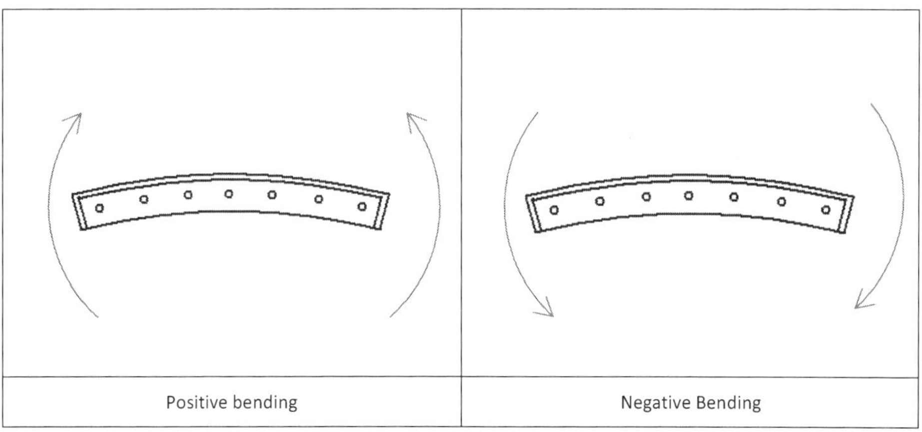

Figure 10 Positive bending and negative bending of a lining segment

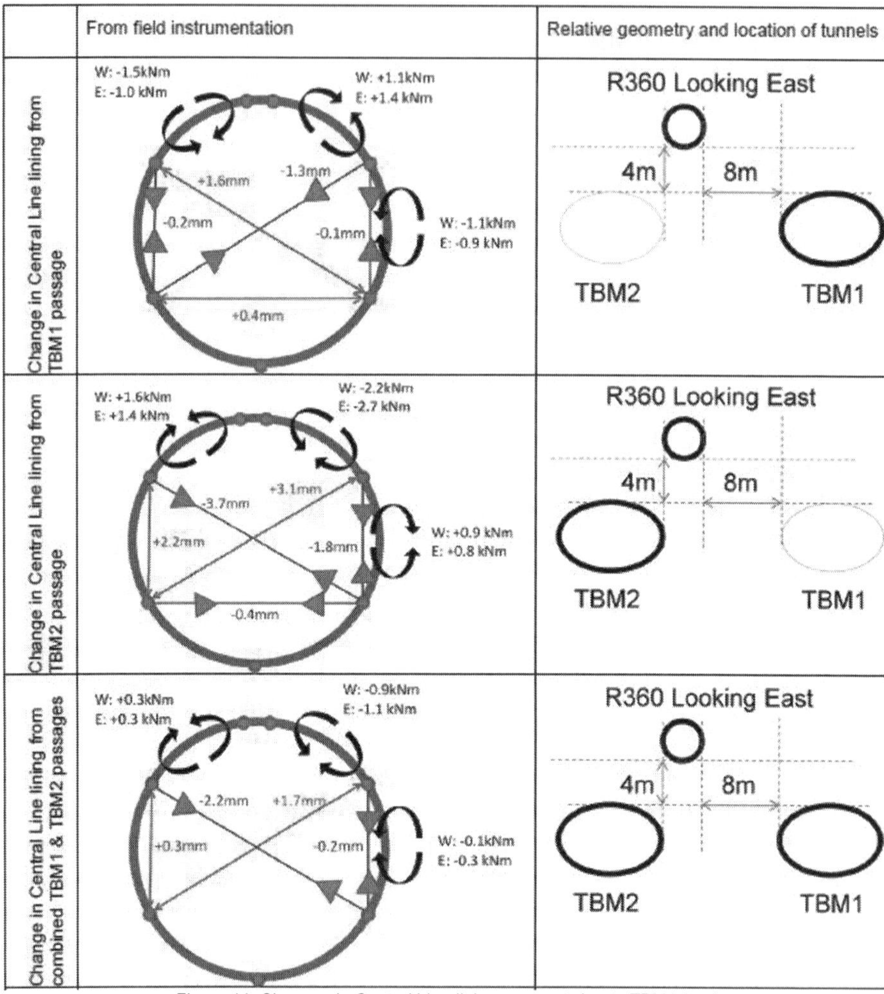

Figure 11. Changes in Central Line lining segment due to TBM1 (top)
TBM2 (middle) and combined TBM 1 and TBM 2 (bottom)

Displacement transducer

The locations of the displacement transducers are shown on Figure 2. There were three displacement transducers on each ring. On ring 336 the displacement transducers were installed midway between the edge bolt and the middle bolt on the horizontal flange. On ring 360 one displacement transducer was installed between the edge bolt and middle bolt and two displacement transducers were installed at the middle bolt location. Displacements were recorded every fifteen minutes.

The displacement readings are given in Figure 12. The vertical axis shows the change in displacement. An increase in reading indicated the joint closing. The red vertical line indicates when TBM1 passed under Central Line and the black vertical line indicates when TBM2 passed under Central Line.

The displacement transducers on ring 336, located above the first Crossrail tunnel, registered joint opening magnitudes of less than 5 microns from the first Crossrail tunnel excavation beneath. Negligible change in displacement was measured from the second Crossrail tunnel excavation. The displacement transducers on Ring 360, located approximately 12m in front of the interface between the Crossrail tunnel and the Central Line tunnel, registered negligible changes due to the TBMs passing.

The gap in the data between 17/12/2012 and 06/01/2013 for five out of the six displacement transducers was due to data corruption during the data retrieval process. After the passage of TBM1, the logging interval was set to 4 hours and attempts were made to download the data once every few days. It was during this time that the data came back corrupted. At the beginning of February 2013, the logging interval was reset to 20 minutes and the data were retrieved once a day.

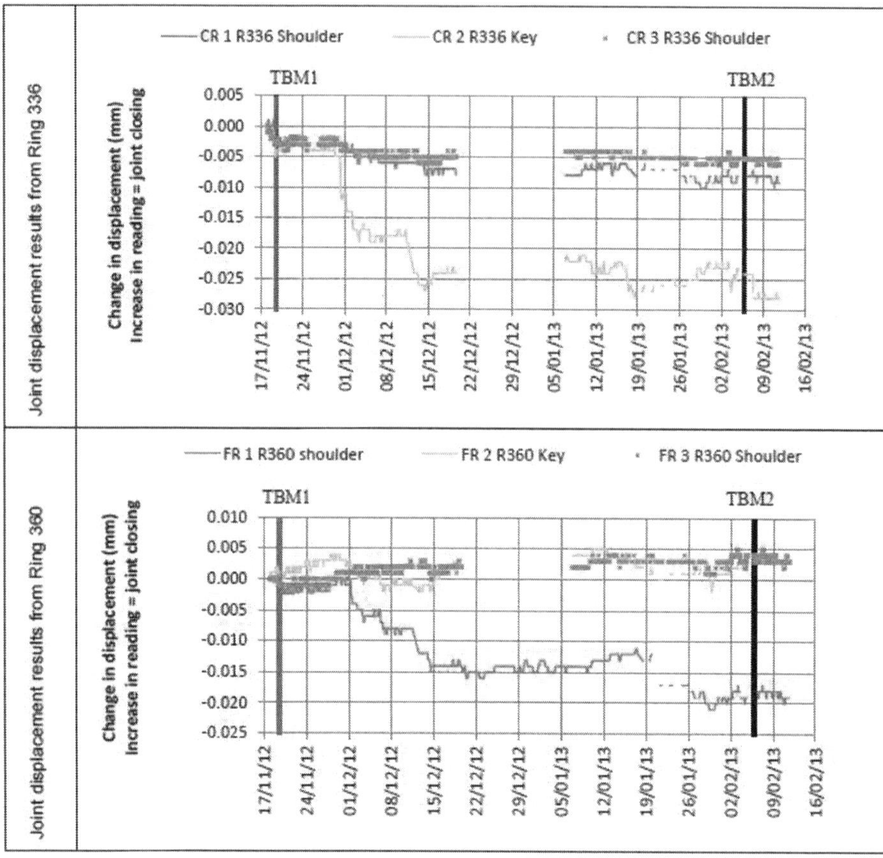

Figure 12. Results from potentiometer displacement transducers at the joints for Ring 336 (top) and Ring 360 (bottom)

Tape extensometer

Tape extensometer measurements were corrected for temperature and the results are shown in Figure 13 for the passage of TBM1 and Figure 14 for the passage of TBM2. The tape extensometer eyebolts for ring 336 were secured across the circumferential flanges of ring 335 and 336. Similarly, the tape extensometer eyebolts for ring 360 were secured across the circumferential flanges of ring 360 and 361. The designations A, B, C and D shown on the graphs are for looking east along the tunnel. The final change in span for each case is shown in Figure 11 along with the results from the strain gauges.

Figure 13 shows that the tape extensometer readings stabilised quickly after the passage of TBM1. The baseline value for calculating the change in span as a result of TBM2 was taken as the measurement recorded on the 28/11/2012. As seen in Figure 14, ring 360 underwent more deformation as a result of the second Crossrail tunnel excavation because it lies closer to the second Crossrail tunnel than ring 336.

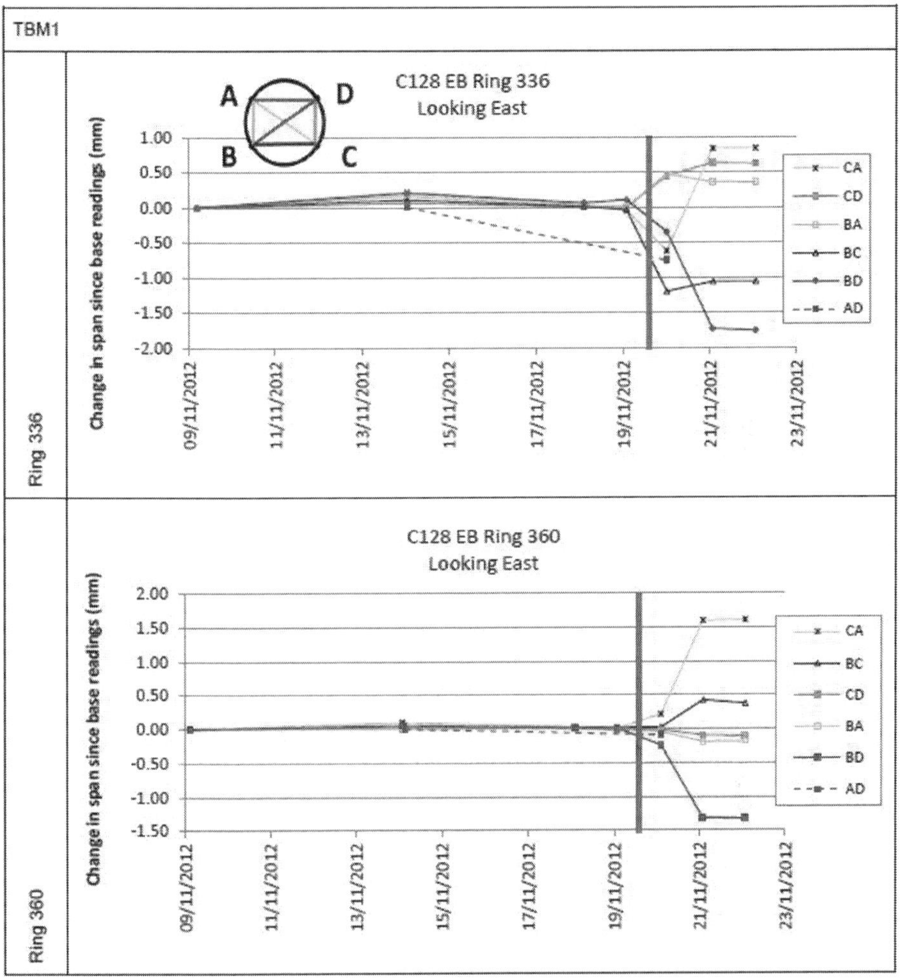

Figure 13. Tape extensometer readings covering the passage of TBM1 for Ring 336 (top) and Ring 360 (bottom)

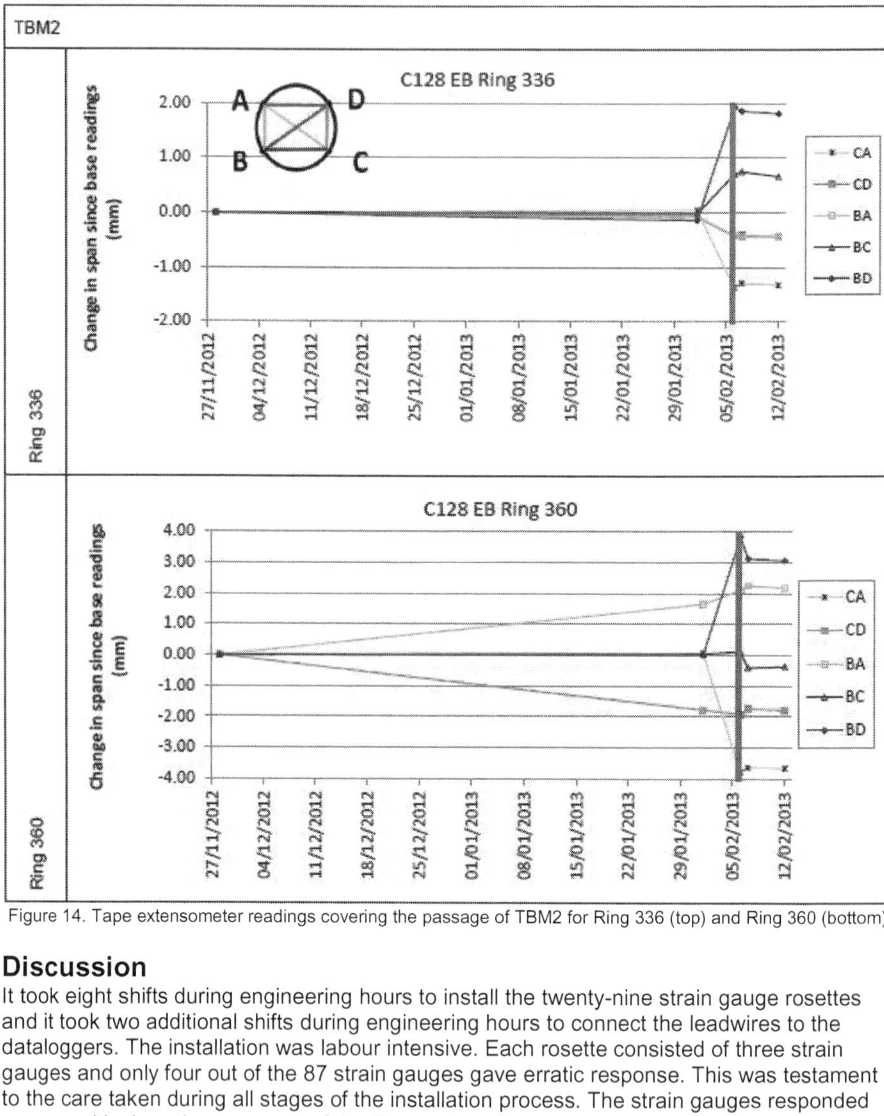

Figure 14. Tape extensometer readings covering the passage of TBM2 for Ring 336 (top) and Ring 360 (bottom)

Discussion

It took eight shifts during engineering hours to install the twenty-nine strain gauge rosettes and it took two additional shifts during engineering hours to connect the leadwires to the dataloggers. The installation was labour intensive. Each rosette consisted of three strain gauges and only four out of the 87 strain gauges gave erratic response. This was testament to the care taken during all stages of the installation process. The strain gauges responded very sensitively to the passages of the TBMs. The readings began to deviate from the baseline levels approximately 12 hours before the TBM cutter head was beneath Central Line. The readings reached a final stable value no more than 48 hours after the TBM had passed. The most useful information provided by the strain measurement was to enable the calculation of the change in bending moment and to ascertain the mode of bending. The main limitation was that because the external loading on the tunnel lining is unknown, it is not possible to calculate the bending moment distribution around the lining as a result of the excavations nearby.

The tape extensometer measurements were invaluable in providing straightforward information on the overall deformed shape of the tunnel lining. The procedure for the installation of the eyebolts was efficient. Even though the discrete, manual tape extensometer readings could only be taken during engineering hours, nightly readings around the time of the TBM arrival were able to capture the elongation of the Central Line tunnel ring towards the new tunnel excavation. As seen in Figure 11, the deformed shape of the tunnel lining determined by the tape extensometer measurements corroborated well with the mode of bending in the lining segments captured by the strain gauge measurements. It is strongly recommended to include tape extensometer readings for future in-tunnel monitoring. Given the relative ease of installation and confidence in data interpretation of the tape extensometer relative to the strain gauges, it is suggested that where tape extensometer readings are taken, strain gauge installation would not provide significant additional useful information.

The main advantage of the potentiometric displacement transducers was that they were relatively easy to install. It took two shifts during engineering hours to install all six displacement transducers and have the associated cabling in place. However, the design of the bracket for the displacement transducer limited where the instrument could be placed on the flange. Unfortunately it was not possible to install the instrument in line with the edge bolt or to install the instrument close to the circumferential flange. In terms of quality of data, as evident in Figure 12, the lack of long-term stability of the potentiometric displacement transducer rendered it unsuitable in determining the magnitude of joint movement. The drift in the readings was several magnitudes higher than any potential joint movement. As mentioned previously, when the logging rate was set to a lower frequency and the data retrieval attempted once every few days, the data would be corrupted. The reason is unknown but it appeared that daily data retrieval was necessary to ensure reliable data collection. Unless the stability of the system could be proven, and the bracket design modified to allow more flexibility in terms of installation location, it is not recommended to use this system for in-tunnel joint movement monitoring.

The field measurements made in the Central Line will be used to guide the laboratory tests conducted at Imperial College on a half-scale cast iron tunnel lining. The test ring will be deformed to match the magnitudes measured in the field. In the laboratory setup, the joint movements will be measured and the bending moments will be calculated from strain gauge measurements. The aim is for the laboratory measurements to provide greater insight into the behaviour of the segmented tunnel lining, thus allowing more informed planning of in-tunnel instrumentation and monitoring for future tunnelling projects.

Conclusions

Imperial College successfully planned, executed and supervised the installation of electrical resistance strain gauges, tape extensometer eyebolts and potentiometric displacement transducers in Central Line before the first of the Crossrail TBMs passed below the Central Line tunnel in November 2012.

The installation procedures and the instrumentation details have been presented in the paper.

The continuous strain measurements have been analysed and the calculated bending moments have been presented in the paper. The continuous measurements of joint movements have been assessed and negligible movement was registered. The manual tape extensometer measurements have been corrected for temperature effects and presented in conjunction with the strain gauge readings to provide a more complete picture of the behaviour of the Central Ling lining as a result of the Crossrail tunnel excavations. The manual data consistently showed the elongation of the instrumented Central Line tunnel rings towards the new tunnel excavation. The tape extensometer readings and the strain gauge measurements supported each other.

Recommendations for future in-tunnel monitoring have been suggested based on the experience gained from this instrumentation and monitoring exercise.

Acknowledgements

The authors wish to thank Tim Morrison, Michael Maloney and Anurada Balasubramanian from London Underground for making this field work possible.

References

Gilbert, G. N. J. (1977) *Engineering Data on Grey Cast Iron – SI Units.* Alvechurch, BCIRA. Report number: Report 1285.

Crossrail Project: Infrastructure design and construction
ISBN 978-0-7277-6078-4

doi: 10.1680/cpid.60784.519

Field Measurement by Fully Grouted Vibrating Wire Piezometers

Michael S.P. Wan – BEng, DIC, MSc, CEng, MICE, Geotechnical Engineer at Crossrail (Postgraduate Researcher at Imperial College London)
Jamie R. Standing – BEng, DIC, MSc, PhD, CEng, MICE, Senior Lecturer at Imperial College London

Abstract

Subsurface instrumentation was installed at a field monitoring site in Hyde Park and bordering Bayswater Road for measuring the ground responses to Crossrail tunnelling near Lancaster Gate. Prior to the passage of the tunnel boring machines, pore water pressures were measured, both in the "greenfield" ground and the ground in the vicinity of existing London Underground running tunnels, by three multi-level vibrating-wire (MLVW) piezometers installed in boreholes fully backfilled with cement-bentonite grout. The pore water pressures in the "greenfield" ground were at the same time measured by conventional standpipe piezometers and four pushed-in pressure spade cells with built-in VW piezometers.

This paper investigates the performance of the MLVW piezometers by comparing the post-installation and steady-state pore water pressure measurements made using these devices with those from the other piezometer types. Results from falling head permeability tests in the standpipe piezometers are presented and discussed. The measurements indicate drainage of groundwater into the existing century-old segmental cast iron lined tunnels. This measured drainage effect is compared with predictions made using a simplified finite difference model. It is concluded that the MLVW piezometers performed well and provide an efficient means of measuring pore water pressures at several depths within one borehole.

Introduction

Research is being undertaken by Imperial College London to investigate the effects of tunnelling on existing tunnels in London Clay. The research opportunity arose with the Crossrail construction of underground twin-bore railway tunnels through central London where the new tunnels interface with existing networks of London Underground Limited (LUL) tunnels and other utility tunnels. The Crossrail western tunnel section involves construction of twin tunnels using 7.2m diameter earth-pressure-balance tunnel boring machines (TBMs) mostly in London Clay. Comprehensive field monitoring was performed at an instrumented site in Hyde Park to assess the effect of new tunnel construction on both the "greenfield" ground and the ground in the near vicinity of existing segmental cast iron lined tunnels. During the year prior to the passage of TBMs, baseline monitoring work on the installed instruments was carried out. Wan and Standing (2014) describe the instrumentation scheme which includes various types of borehole instruments such as rod extensometers, in-place inclinometers and piezometers.

Recent advances in vibrating-wire instrumentation technology have led to an increasing popular application of VW piezometers in geotechnical projects. Dunnicliff (2008) and Contreras et al (2008) report successful installation and monitoring of VW piezometers installed by the fully-grouted method at various sites around the world. The VW piezometers, being a diaphragm type, require only a very small amount of water flow into the sensor to operate, making it possible for them to be installed in a borehole fully grouted with cement-bentonite grout of a suitable permeability. The fully-grouted method has a major advantage in that the straightforward backfilling procedure allows multi-level vibrating wire (MLVW) piezometer sensors to be installed in one single borehole. That makes it possible to obtain a pore water pressure profile with depth at one borehole location, significantly reducing the number of boreholes required. This current paper presents and discusses the post-installation and the subsequent baseline measurements from the different types of piezometers, demonstrating the successful performance of the MLVW piezometers installed in single boreholes using the fully-grouted method. The measurements also establish a steady-state pore water pressure profile at the site as a baseline reference to compare with the future piezometric response from the Crossrail tunnelling operations.

Instrumented site

The field research site is located near Victoria Gate, at the northwest corner of Hyde Park, where the Crossrail west-bound and east-bound tunnels pass beneath the LUL Central Line running tunnels just east of Lancaster Gate. The actual crossing is directly beneath Bayswater Road.

Instrumentation plan

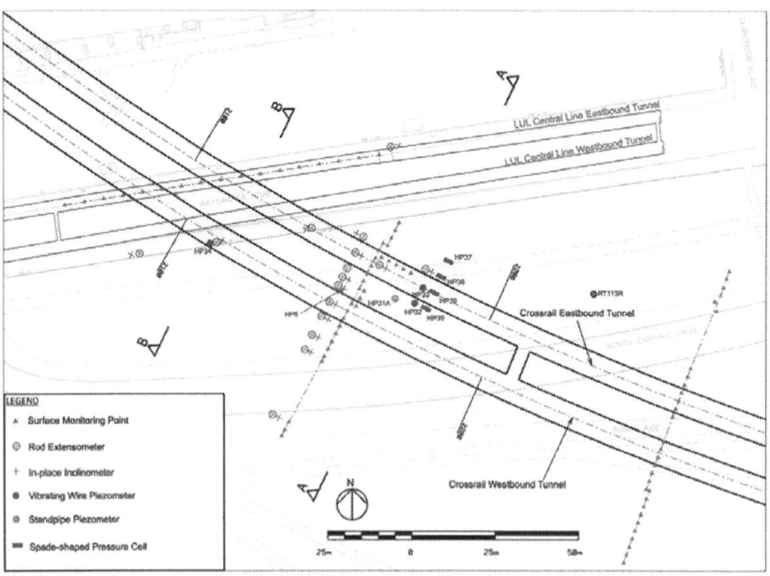

Figure 1 – Plan of instrumentation in Hyde Park and Bayswater Road.

A plan of the instrumentation layout is shown in Figure 1. There are in total 38 boreholes, each accommodating one or more instruments. Among other instruments, piezometers and spade-shaped pressure cells (or spade cells) have been installed around the Crossrail tunnel alignments to measure the pore pressure response resulting from the new tunnel construction. Within Hyde Park, the piezometers and spade cells were installed, as shown in Figure 2a, with the pressure sensors located at various distances vertically and horizontally around the east-bound tunnel to monitor changes in pore water pressure and total horizontal earth pressure as the TBMs approach and pass. There are two MLVW piezometer boreholes (HP32 and HP33) within Hyde Park, with six piezometer sensors at different depths in each borehole. A conventional standpipe piezometer (HP31A) was also installed at 6m away from one MLVW piezometer borehole to verify long-term pore water pressures measured by the VW piezometers. One spade cell (HP39) and three spade cells (HP35, HP36 and HP37) were installed at the level 1.5m above the tunnel crown and at the tunnel axis level respectively where the total horizontal earth pressures and pore water pressures are measured. There is one MLVW piezometer borehole (HP34), also containing six piezometer sensors at depths down to the Crossrail west-bound tunnel crown level, on the south pavement of Bayswater Road, being sunk at a distance of 5m from the LUL Central Line west-bound tunnel wall (Figure 2b).

Site geology and lithological units

Hyde Park is in part of London Basin where marine sedimentation was affected by sea level changes, and has deposits from the London Clay Formation, Lambeth Group Formation and Thanet Sand Formation on the Chalk bedrock – a geology typically found in Central London. During instrument installation, a 68m deep borehole (HP6) was sunk by rotary core drilling through the whole thickness of London Clay Formation and the top few metres of the Lambeth Group Formation. Continuous triple tube core barrel samples were retrieved from the borehole and carefully preserved on site and transported to Imperial College for visual inspection and detailed logging. The lithological units of the London Clay Formation, as characterised by King, (1981), were identified by visual inspection and moisture content measurement (Standing and Burland, 2006).

In the instrumented site, the Chalk bedrock is believed to be at about 80m below ground level, overlain by a thin layer of Thanet Sand and about a 15m thickness of Lambeth Group deposits. At about 60m below ground level is the base of London Clay Formation which has an overall thickness of about 55m. There is a 3m thick layer of Terrace gravel above the London Clay, with a thin layer of made ground at the surface. Most of the 7.2m diameter Crossrail tunnel excavation is within the lower level of the London Clay unit B (B1 and B2), while about one metre of the base of the excavation (invert) is within the more permeable London Clay unit A3ii. The borehole instruments were positioned at different depths in the London Clay lithological units. Figure 3 shows the detailed soil descriptions of the ground determined from the borehole HP6. The boundaries of these lithological units are also shown in Figure 2.

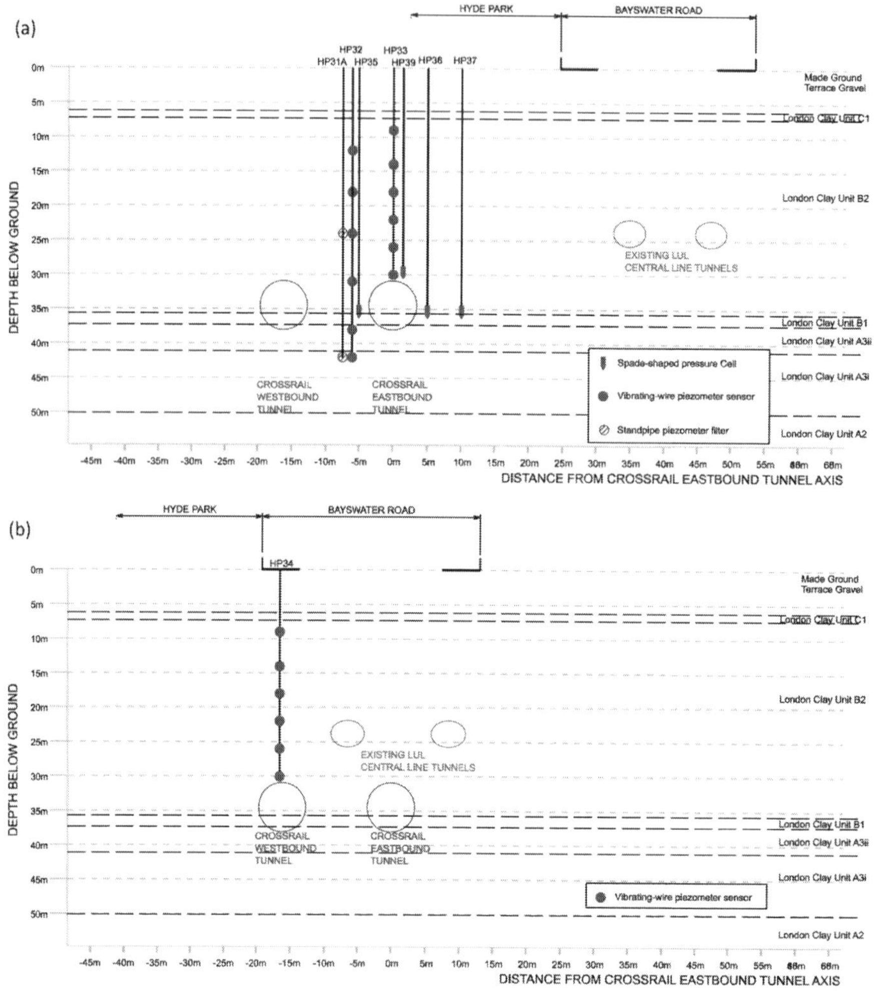

Figure 2 – Cross-section of piezometers (a) Cross-section A-A: piezometers in "greenfield" ground. (b) Cross-section B-B: piezometers in the vicinity of the existing LUL running tunnel.

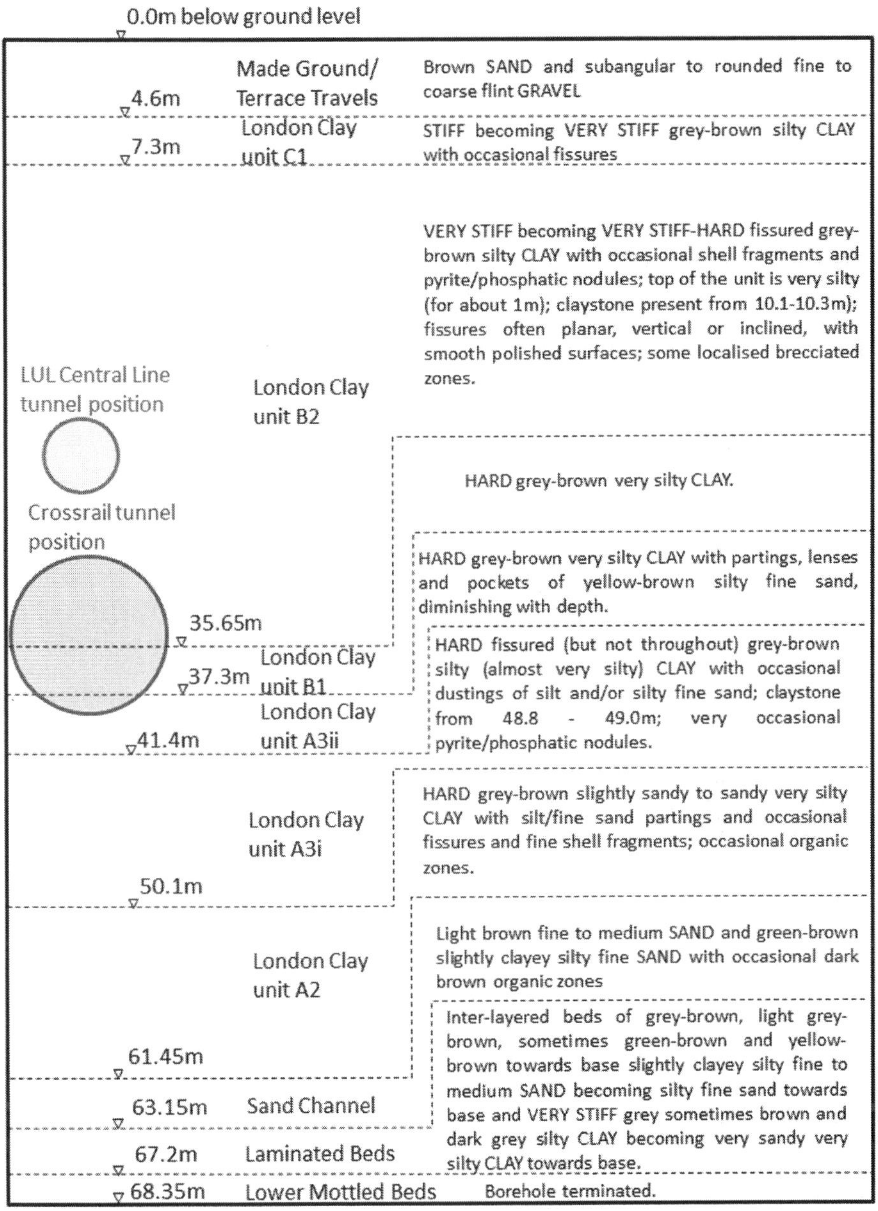

Figure 3 – Detailed soil descriptions of the ground at the instrumented site.

Site hydrogeology

Two aquifers are present at the instrumented site, as reported by many authors (e.g. Simpson et al., 1989). The Terrace Gravels constitute the upper aquifer and the lower granular units of the Lambeth Group Formation and the Thanet Sand Formation the lower aquifer. These aquifers are separated by the London Clay Formation and, if any, the top clayey Lambeth Group Formation. During borehole drilling for installation of instrumentation, the upper water table was established to be about 4-5m below ground level. The deep water table was historically artesian but is now much lower due to excessive abstraction activities in early 1900s. Following the GARDIT strategy the water table is maintained below its natural level by controlled abstraction (Jones, 2007). As a result, the London Clay formation is under-drained, which is confirmed by the steady-state pore water pressure profile measured by the MLVW piezometers and the standpipe piezometer (See Section 6).

Piezometer types and installation

The pore water pressures in the ground near the Crossrail tunnel alignments are measured by three types of piezometer: (a) MLVW piezometers in fully grouted boreholes (HP32, HP33 and HP34), (b) VW piezometer sensors incorporated within spade cells (HP35, HP36, HP37 and HP39), and (c) a conventional standpipe piezometer (HP31A). These three types of piezometers are shown schematically in Figure 4. Table 1 shows the installation depths and the London Clay divisions where the instruments are positioned. All boreholes were formed by cable percussion drilling.

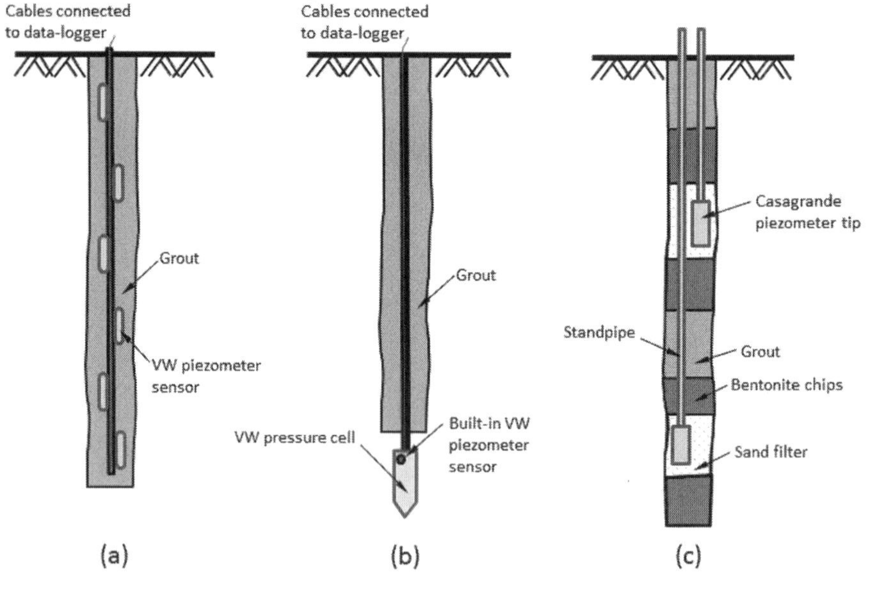

Figure 4 – Schematic presentation of three types of piezometers installed at the instrumented site: (a) Multi-level VW piezometers in fully grouted borehole; (b) VW piezometer sensor built in spade-shaped pressure cell; (c) Conventional standpipe piezometer.

Borehole	Piezometer type	Piezometer sensor/ response zone depth below ground level (m)	London Clay unit	Remarks
HP32 (Hyde Park)	Multi-level VW piezometer	12.0	B2	
		18.0	B2	
		24.5	B2	Reading erratic after 3 months
		31.0	B2	
		37.0	B1/A3ii	
		42.0	A3i	
HP33 (Hyde Park)	Multi-level VW piezometer	9.0	B2	
		14.0	B2	
		17.0	B2	
		22.0	B2	
		26.0	B2	
		29.5	B2	
HP34 (Bayswater Road)	Multi-level VW piezometer	9.0	B2	
		14.0	B2	
		17.0	B2	
		22.0	B2	
		26.0	B2	Reading erratic after 3 months
		29.0	B2	
HP35 (Hyde Park)	Spade cell	34.0	B2/B1	
HP36 (Hyde Park)	Spade cell	34.7	B2/B1	
HP37 (Hyde Park)	Spade cell	34.6	B2/B1	
HP39 (Hyde Park)	Spade cell	28.9	B2/B1	
HP31A (Hyde Park)	Standpipe piezometer	24.5	B2	
		37.0	B1/A3ii	
RT113R (Hyde Park)	Standpipe piezometer	15.8	B2	Borehole installed by Crossrail's contractor prior to current instrumentation fieldwork
		47.0	A3i/A2	

Table 1 – Details of piezometer installations at the Hyde Park site.

Multi-level VW piezometers

The VW piezometer is of the diaphragm type which requires only a very small amount of water flow into the piezometer to operate, eliminating the need for granular filters and meaning that the piezometer sensors can be fully backfilled by a low permeability grout. This has a major advantage in that the straightforward backfilling procedure allows multi-level piezometer sensors to be installed in one single borehole, thus making it possible to obtain a vertical pore water pressure profile at just one single borehole location, significantly reducing the number of boreholes required.

The installation was achieved by lowering the VW piezometer sensors into a borehole at pre-determined depths, before backfilling the borehole with a suitable cement-bentonite grout. The cables leading from the piezometer sensors were then connected to a data-logger housed within a headworks. The detailed installation method and selection and preparation of suitable cement-bentonite grout is described by Wan and Standing (2014). The backfill grout serves as a filter and also as a seal between piezometer sensors, prohibiting the vertical flow of water between the sensors within the borehole and allowing the pore water pressure at different depths to be measured independently.

Spade-shaped pressure cells

A spade cell consists of a pointed rectangular spade-shaped oil chamber, formed by two steel plates, connected to a VW pressure transducer. It measures total stresses acting on the pressure cell in the direction perpendicular to the spade surface. In-built at the top side of the spade is a porous filter which is connected to a second VW pressure transducer, so that the pore water pressure is also measured at the same time.

Each spade cell was installed by first drilling a borehole to a pre-determined depth (about 1m above the design instrument depth). The spade cell was then lowered into the borehole using hollow steel rods connected by couplers, until it reached the borehole base. The cables connected to the two VW transducers were fed through the middle of the steel rods and held at ground surface. The orientation of the spade was then adjusted so that the spade surface was facing perpendicular to the alignment of the Crossrail tunnel. A CPT rig was then mobilised to push the pressure cell vertically into the ground to the design depth (usually about 1m below the borehole base). Subsequently the steel rods were detached from the spade cell, and retrieved one by one, before the borehole was backfilled with cement-bentonite grout. The cables were connected to a data-logger housed within a headworks.

Unlike the multi-level piezometers, only one instrument can be installed within each borehole.

Standpipe piezometers

A two-level standpipe piezometer was installed by placing two conventional Casagrande type piezometer tips connected with PVC standpipes into two response zones at specified depths. The response zones were backfilled with sand to from granular filters 1m long which were sealed at the top and base by at least 1.5m thick layers of bentonite (installed using bentonite chips). The remaining length of the borehole is backfilled with cement grout.

About two months before the TBM arrival at the site, a pressure transducer was inserted into the base of each of the two 19mm diameter standpipes. Each of the transducers was connected via a cable to a data-logger housed within a headworks. This enabled the water pressure, and hence the water level in the standpipes, to be measured and recorded automatically. Manual measurements using a small dipmeter were also made from time to time, without the need of removing the transducers from the standpipes, to calibrate the automatically logged result. The cable connecting the transducers to the top of the borehole has a diameter of 7.5mm and occupies about 15% of the cross-section area of the 19mm standpipe. This was taken into account when interpreting the results of the falling head tests performed in the standpipe piezometers.

Falling head permeability test from standpipe piezometer

The pressure transducers were installed in the standpipe piezometer HP31A, with centre of the two response zones at 24m (London Clay unit B2) and 37m (London Clay unit B2/A3ii) below ground level, almost one year after the standpipe installation. By then the bentonite seals isolating the granular filters at the two response zones should be fully effective. A pressure transducer was also installed within a 47m deep standpipe (pieozometer tip near boundary between London Clay unit A3i and unit A2) RT113R constructed by Crossrail's contractor prior to the current instrumentation fieldwork (it should be noted that this borehole is about 200m from the borehole HP6 used to compile Figure 3). Figure 5 shows the measured water levels in the standpipes in borehole HP31A for the 3 months leading up to the first TBM arrival. Before the transducer installation, the steady-state water levels in the 24m, 37m and 47m standpipes were measured to be 7.1m, 8.7m and 8.4m deep below ground surface respectively – an indication of slight under-drainage within the London Clay formation. The placement of the transducer assembly immediately displaced the water and led to a water level rise within the standpipe, before it dropped back to the original level. The process in itself can be treated as a falling head field permeability test. For the 37m standpipe in borehole HP31A, the excess water head took just about 3 days to dissipate, with the measured water level returning to its original value before the placement of the transducer (Figure 5a). On the other hand, the water level in the 24m standpipe dropped at a much slower rate, and did not return to the level before the transducer placement even after 3 weeks. It was later found that the transducer in the 24m standpipe was faulty and measuring erratic values and therefore only the dipmeter measurements are presented and used for data interpretation (Figure 5b). Three weeks after installation the transducers, the standpipes were topped up with water to ground level to perform another set of falling head permeability tests. The calculated permeability values from the falling head tests for both boreholes HP31A and RT113R are presented in Table 2. The coefficient of horizontal permeability was determined to be of the order of 10^{-9}m/s for London Clay unit A3ii and unit A2, and 10^{-11}m/s for London Clay unit B2. These values compare well with field test results from other sites in Central London and Heathrow (Standing and Burland, 2006; Hight et al, 2007).

Borehole	Depth below ground level (m)	London Clay unit	Falling head test no.	Coefficient of horizontal permeability (m/s)
HP31A	24.0	B2	1	1.6×10^{-11}
			2	4.9×10^{-11}
	37.0	B1/A3ii	1	1.2×10^{-9}
			2	1.2×10^{-9}
RT113R	47.0	A3i/A2	1	2.5×10^{-9}
			2	4.1×10^{-9}

Table 2 – Results of falling head permeability tests in standpipe piezometer boreholes HP31A and RT113R.

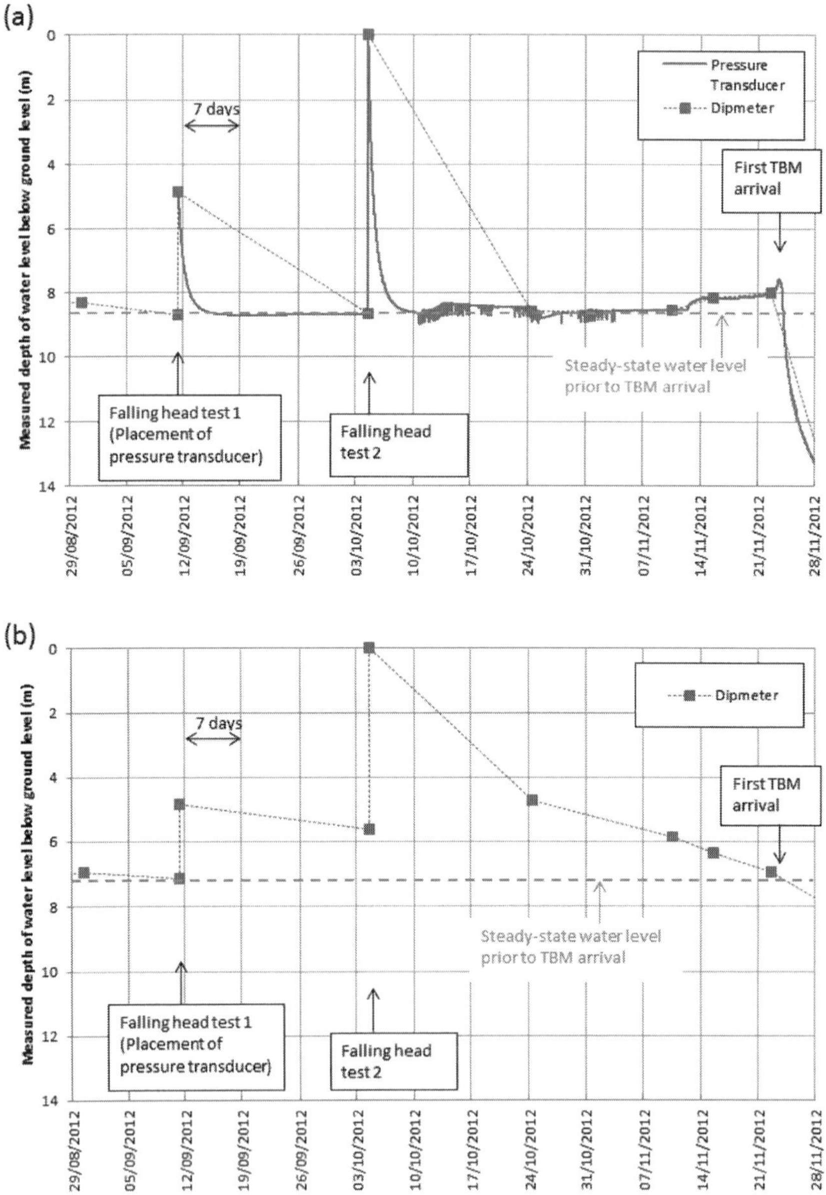

Figure 5 – Measurements of water level in standpipes during the 3 months leading up to the first TBM arrival (a) 37m standpipe of borehole HP31A (b) 24m standpipe of borehole HP31A

Post-installation measurements from VW piezometers

During the instrument installation, negative excess pore water pressures were generated in the stiff clayey ground induced by stress relief and undrained shearing during the cable percussion borehole drilling. The magnitude of the excess pore water pressures depends on the existing ground stresses, the soil's over-consolidation ratio, and degree of borehole support after drilling. Depending on the permeability (predominantly in the horizontal direction) of the ground and the spatial distribution of the initial excess pore water pressure, the excess pore water pressures at different depths would dissipate with time at different rates, until they balance with the far-field steady-state values.

Initial measurements of the pore water pressure were started within the first few days after the installation of the VW piezometers in the fully grouted boreholes (HP32, HP33 and HP34). The initial measurements in borehole HP32 shows that the measured pore water pressures came to a steady-state within about 1 month (Figure 6).

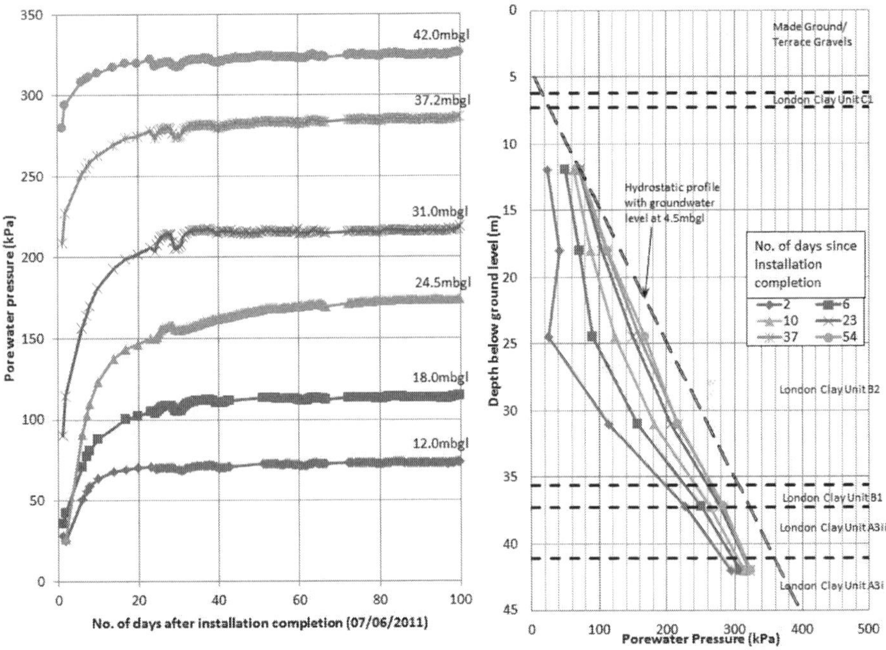

Figure 6 – Post-installation measurement of VW piezometers in fully grouted borehole HP32.

The measured equilibration of pore water pressures can provide an indication of differences in the field permeability, at least qualitatively, of the soil at the respective depths of the piezometer sensors. For piezometer borehole HP32, the first readings were taken 24 hours after the installation was complete and therefore the initial pore water pressures immediately after the installation are not known and need to be determined. In order to obtain the initial excess pore water pressures, first the steady-state pore water pressures are assumed to correspond to the values at the end of the equilibration period. Then the measured excess pore water pressures are plotted against the square root of the time elapsed after the installation completion (Figure 7a). The initial excess pore water pressures for all piezometer sensors are obtained by extrapolating the early straight-line portions of the curves back to time equal to zero (broken lines in Figure 7a).It is then possible to normalise the measured excess pore water pressures with the corresponding initial excess pore water pressures and to plot the normalised values against logarithm of time (Figure 7b). By comparing the time for 50% dissipation (i.e. t_{50}), for example, of total excess pore water pressures for different sensors, the coefficients of horizontal consolidation (c_h) for the soils at respective piezometer depths can be ranked. In fact, if the initial spatial distribution of the initial excess pore water pressure is accurately estimated (e.g. by analytical solutions for a cavity contraction problem), the exact c_h values can be determined from the t_{50} values using Terzaghi's 1-D consolidation theory, assuming all ground water flow is horizontal and radial to or from the borehole wall.

The t_{50} values determined from Figure 7 are tabulated in Table 3, and suggest that the ground within London Clay unit A3i (42m below ground level) and B1/A3ii (37mbgl) has a higher coefficient of horizontal consolidation than the shallower ground within London Clay unit B2 (24mbgl and 18mbgl). This agrees with the trend observed from the results of the falling head permeability tests performed in the nearby standpipe piezometer (HP31A).

The observation that different magnitudes and rates of dissipation of the initial negative excess pressures were measured at different depths within the boreholes indicates that the cement-bentonite grout backfill formed a good seal between the piezometer sensors, otherwise similar magnitudes and dissipation rates of negative excess pressure would have been expected.

Depth below ground level (m)	London Clay unit	t_{50} (hours)
12.0	B2	50
18.0	B2	89
24.5	B2	80
31.0	B2	51
37.2	A3ii	32
42.0	A3i	23

Table 3 – t50 values determined from initial excess pore water pressure dissipation curves in borehole HP32.

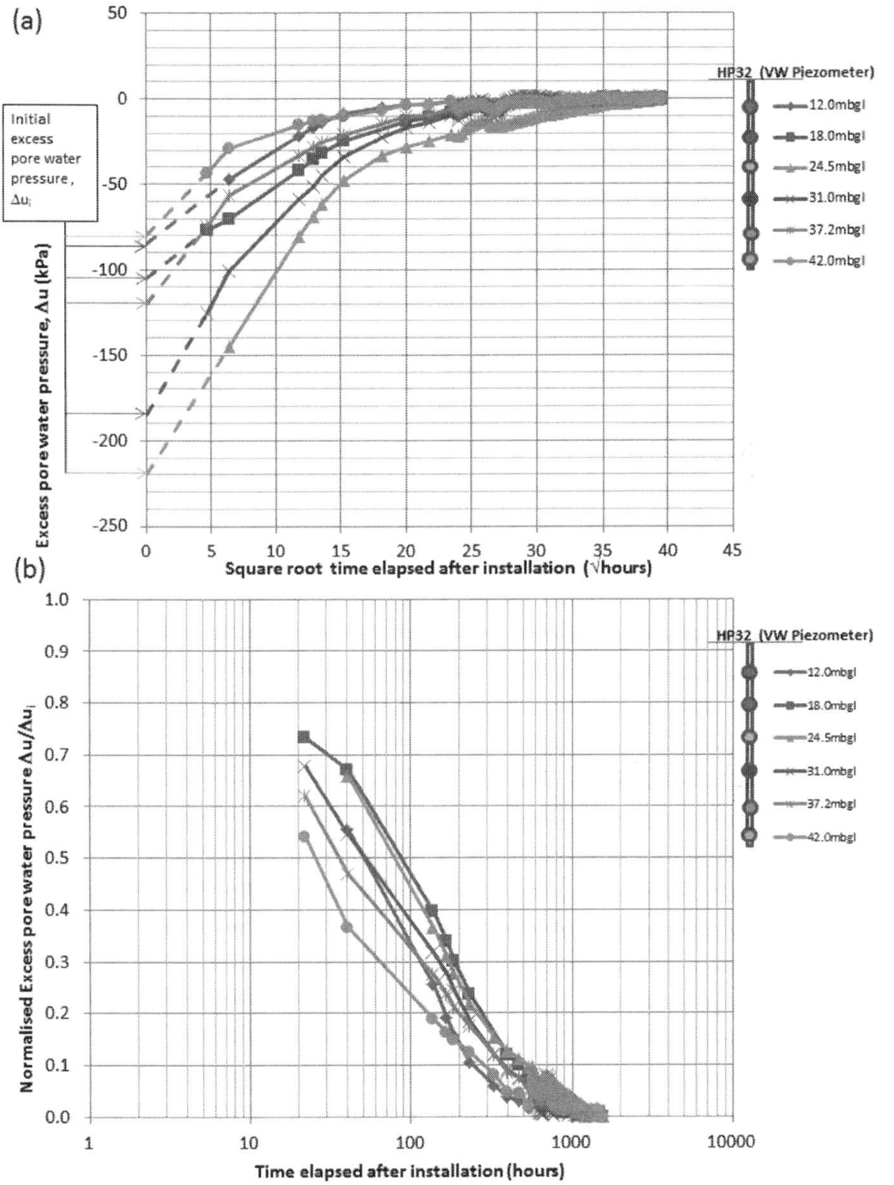

Figure 7 – Dissipation of initial excess pore water pressures in different depths of VW piezometer sensors after installation. (a) Determination of initial excess pore water pressures immediately after installation; (b) Normalised excess pore water pressure dissipation with time.

Steady-state pore water pressures

One of the main aims of the piezometer instrumentation is to establish the steady-state pore water pressure profile prior to any tunnel construction work beneath the site. This would be used to verify the input to the numerical model for simulating the tunnel excavation at the instrumented site. The profile is also used as a reference baseline for determining the subsequent changes induced by the TBM passages. Two profiles of steady-state pore water pressure were established at the site, one in the "greenfield" ground (Figure 2a) and the other in the near vicinity to the existing LUL running tunnels (Figure 2b).

"Greenfield" ground

The steady-state pore water pressures measured by the MLVW piezometers (HP32 and HP33) about two months before the first TBM arrival are plotted against depth in Figure 8. The measured pore water pressures at HP33 are almost consistently about 20kPa higher at all depths than those at HP32, despite the two piezometer boreholes being only 5m apart. This could be due to the error associated with translating the measured voltages from the VW transducers to pressure values through calibration factors. Nevertheless, both show the same trend that the increase of pressure with depth is smaller than a hydrostatic pressure profile, indicating the slight under-drainage of the London Clay formation. The near surface London Clay is influenced by the pore water pressure associated with the upper water table within the Terrace Gravels, whereas the deeper London Clay is under the influence of the lowered pore water pressure within the underlying deep aquifer.

In addition to performing falling head permeability tests, the other purpose of installing the standpipe piezometer near the VW piezometers is to provide an alternative manual measurement of the pore water pressure to confirm the steady-state measurements from the VW piezometers. Also plotted in Figure 8 are the steady-state pore water pressures measured at the two response zones of the piezometer borehole HP31A, which is 6m from HP32. These manual measurements coincide nicely with the profile measured by the VW piezometers in HP32, suggesting the profile measured from HP32, instead of HP33, is more likely to be the true steady-state pore water pressure profile in the site.

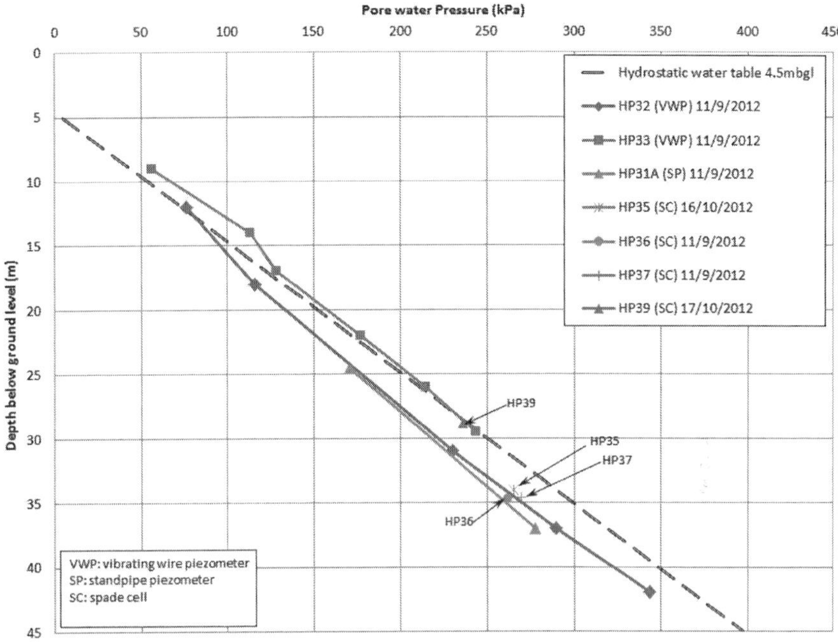

Figure 8 – Steady-state pore water pressures measured in "greenfield" ground by standpipe piezometer (HP31), VW piezometers (HP32 and HP33), and spade cells (HP35, HP36, HP37 and HP39).

Further steady-state pore water pressure measurements from the three spade cells (HP35, HP36 and HP37) also confirm the profile of HP32, although measurements from the remaining spade cell (HP39) agree more with the profile of HP33 (Figure 8).

Ground in the vicinity of existing cast iron lined tunnel

There is one MLVW piezometer borehole HP34 installed on the pavement of Bayswater Road, with a clearance of 5m from the wall of the LUL Central Line west-bound running tunnel. Figure 9 shows the steady-state pore water pressure profile measured by HP34. The "greenfield" profile of HP32 is also presented for comparison. It is observed that the steady-state pore water pressures measured in HP34 are 20kPa to 50kPa smaller than the "greenfield" values at corresponding depths, with the reduction larger at levels just above the LUL tunnel crown. It is believed that this deviation from the "greenfield" profile is related to the drainage effect of the segmental cast iron lined tunnel. While the shallower ground and the deeper ground are closer to and more quickly recharged by the upper and lower aquifers respectively, the groundwater around the depth of the existing tunnel is being drawn into the tunnel, creating a drop of the steady-state pore water pressure near the tunnel horizon.

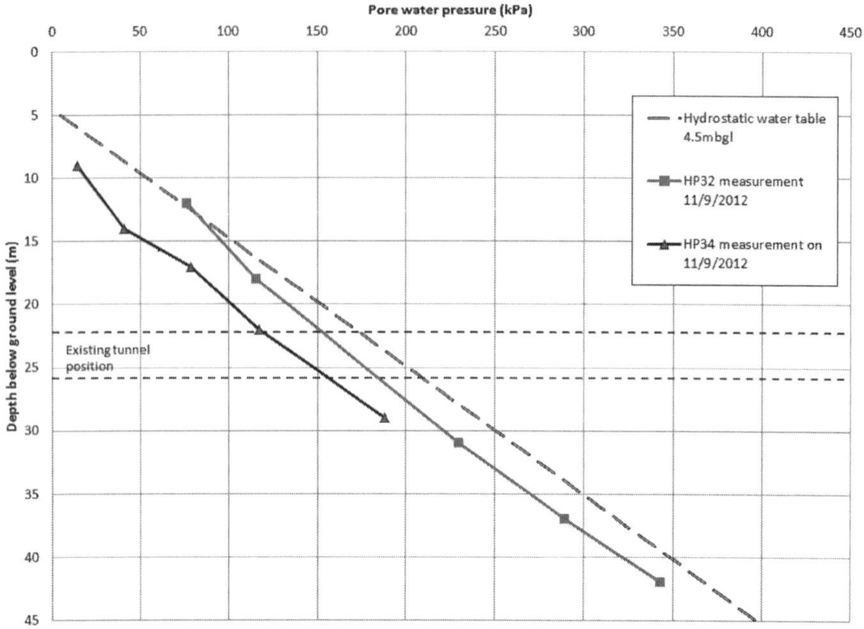

Figure 9 – Steady-state pore water pressures measured in the vicinity of the existing LUL running tunnel by VW piezometers (HP34)

Gourvenec et al (2005) describes a similar piezometer instrumentation scheme in the vicinity of an existing tunnel, situated at 20m below ground level in London Clay, with the same type of segmental cast iron lining in Kennington. They found that the steady-state pore water pressure measured in the ground as close as 1.5m (smaller than one tunnel radius) to the existing tunnel is no different to the far-field value, suggesting that there is no or little drainage effect of the existing tunnel at that site. This is attributed to the fact that the existing tunnel in Kennington is near the base of the London Clay formation (unit A2) where the permeability is among the highest in the London Clay formation (generally the coefficient of permeability in unit A2 is in the order of 10^{-9} m/s). It is argued that the segmental lined tunnel there is relatively impermeable and therefore not providing a drainage path for the groundwater. This is not the case for the Hyde Park site, where the existing segmental lined tunnel is fully within the relatively low permeability London Clay unit B2. The results from the falling head permeability test in borehole HP31A indicate a coefficient of permeability of 10^{-11} m/s for unit B2. This implies that the segmental lined Central Line tunnel close to the Hyde Park monitoring site is relatively more permeable than the surrounding ground, thus draining groundwater and distorting the steady-state pore water pressure profile near the tunnel, as measured by the piezometers in HP34.

Finite difference approximation

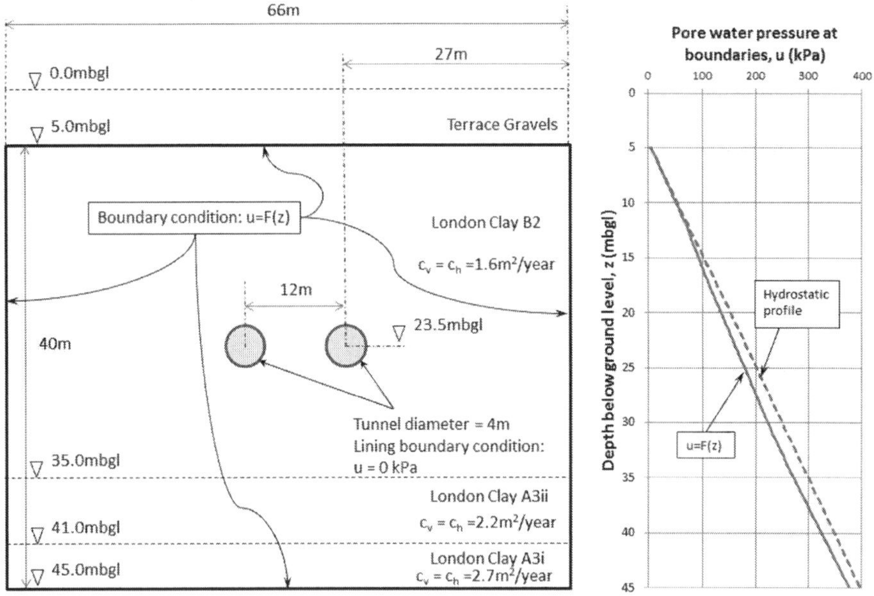

Figure 10 – 2-D finite difference model for simulation of drainage effect of existing segmental lined tunnels in London Clay

A simple 2-D finite difference analysis was performed to gain an insight into how the presence of the existing segmental lined tunnel affect the steady-state pore water pressure regime over a long period after the tunnel construction. Figure 10 shows the 2-D finite difference model with the boundary conditions shown. The 2-D mesh, with square elements sized 0.5m x 0.5m, extends from 5m below ground level (top of London Clay unit B2) to 45m below ground level (near the bottom of London Clay unit A3i) and is 66m wide with the existing tunnels at 23.5m below ground level. A far-field boundary condition of slightly under-drained pore water pressure profile as measured by the piezometers in HP32 is adopted along the four sides of the mesh. i.e. the top and bottom boundaries are constantly recharged by the shallow and deep aquifers respectively. The London Clay units were modelled as homogenous and isotropic with a coefficient of permeability of 5×10^{-11} m/s. However, the individual London Clay units were assigned different compressibility values. Depending on the relative permeability of the tunnel lining to the surrounding ground, the tunnel lining could be treated as a fully permeable drain, a fully impermeable boundary, or any state in between. For simplicity, the tunnel lining is modelled to be fully permeable and connected with atmospheric pressure (i.e. pore water pressure equals zero at all times). The finite difference formulations are based on the Terzaghi-Rendulic theory of consolidation. Consolidation was simulated for 100 years after the tunnel was constructed, as the length of the LUL Central Line tunnels beneath the site were built in the late 19th century.

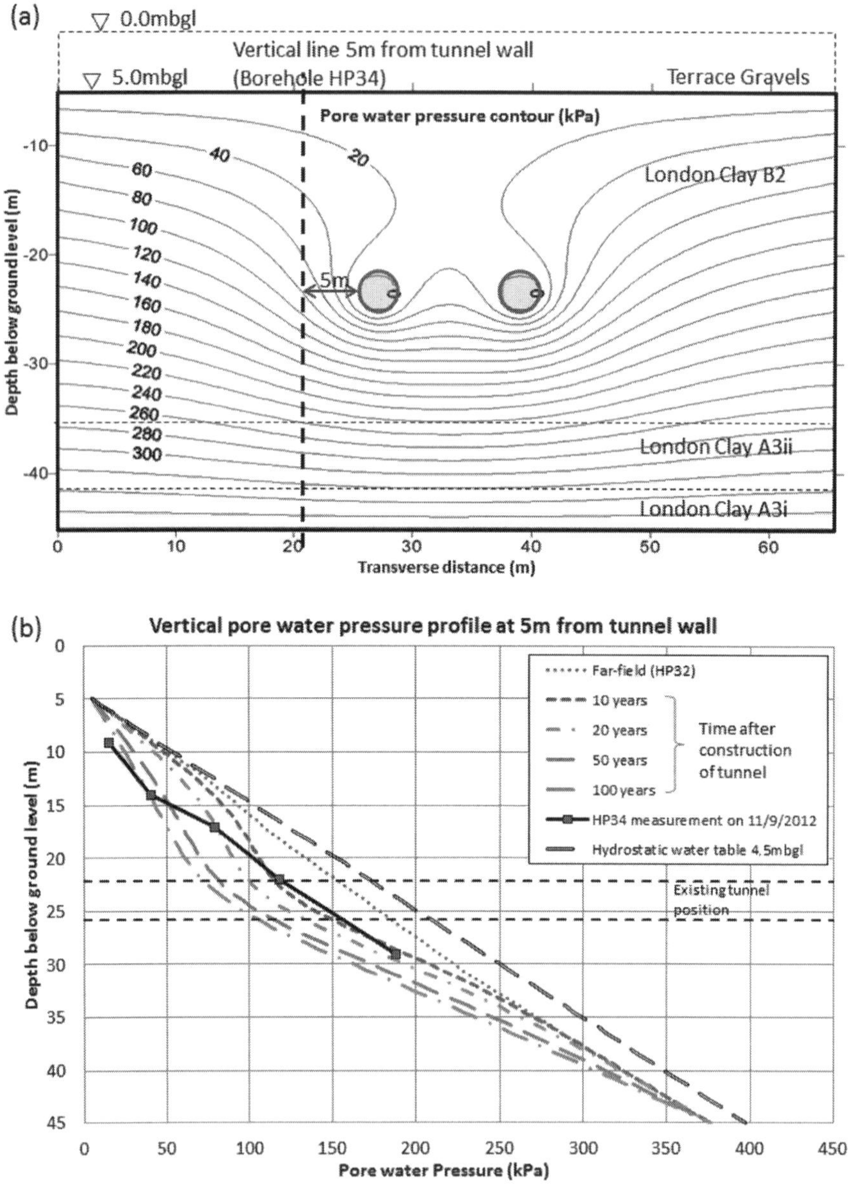

Figure 11 – Results of finite difference analysis (a) Pore water pressure contours (kPa) at 100 years after tunnel construction; (b) Pore water pressure profiles along a vertical line 5m from the tunnel wall for after 10, 20, 50 and 100 years of consolidation.

The simulation results of the finite difference analysis are shown in Figure 11. The pore water pressures in the near proximity to the existing tunnel drop as consolidation over 100 years takes place, as can be seen from the pore water pressure contours around the existing tunnel (Figure 11a). The pore water pressure profiles for after 10, 20, 50 and 100 years of consolidation along a vertical line 5m from the tunnel wall are shown in Figure 11b. The largest reduction in pore water pressure occurs at the tunnel axis level. The steady-state pore water pressures measured from the VW piezometer borehole HP34 (5m from the tunnel wall) are shown in the same figure, and fall within the range of the simulated profiles. Although the simplified 2-D finite difference model has a number of approximations and limitations, the analysis results clearly demonstrate the probable drainage effect of an existing tunnel in relatively impermeable ground, corroborating the field measurements from the MLVW piezometers.

Summary and conclusions

Subsurface instrumentation was installed at a research site to monitor the ground response from the Crossrail tunnel construction. Three types of piezometers were used to measure the pore water pressure around the new tunnel alignment before, during and after the construction. It has been demonstrated that the MLVW piezometers installed in fully grouted boreholes perform well in measuring the steady-state pore water pressures. The results were confirmed by nearby measurements made using conventional standpipe piezometers and the VW piezometers built into spade cells.

The field permeability values for units within the London Clay Formation were determined from falling head permeability tests carried out using three standpipe piezometers. The soil permeability at other depths has been qualitatively compared using readings from the equilibration of excess pore water pressure during the initial measurements of the MLVW piezometers. It was found that the London Clay unit B2, which is a very stiff silty clay, has a coefficient of horizontal permeability of the order of 10^{-11}m/s, whereas at the boundary between London Clay units B1 and A3ii, the latter being a hard clay characterised by the presence of fine sand partings and lenses, has a coefficient of horizontal permeability of the order of 10^{-9}m/s.

Baseline steady-state pore water pressure profiles with depth have been established from the measurements of the MLVW piezometers in both the “greenfield” ground and the ground in near vicinity of the existing LUL segmental cast iron lined tunnel. The drainage effect of the existing tunnel, characterised by a reduction in the steady-state pore water pressure near the tunnel, is indicated from the MLVW piezometer measurements near the tunnel and has also been confirmed by a simplified finite difference model.

The MLVW piezometers at the site have been proved to perform well. Using this technique greatly reduces the number of boreholes needed if conventional standpipe piezometers are used, for which practically a maximum of two response zones can be installed in a borehole. The results from the baseline monitoring period provide confidence in the MLVW piezometers in fully grouted boreholes for the subsequent measurements during and after the passage of the TBMs.

Acknowledgements
The authors wish to acknowledge Crossrail and the Engineering and Physical Sciences Research Council (EPSRC) who are major sponsors of the research project at Imperial College. Many thanks are due to the Imperial College research team, in particular Professors John Burland and David Potts, for their contributions in discussing the choice and layout of the instrumentation. A thoughtful review of the instrumentation plan from Mr. John Dunnicliff is also greatly appreciated. The support provided by the Royal Parks, London Underground Limited and Westminster Council during the installation work is gratefully acknowledged.

References
Contreras, I. A., Grosser, A. T. & Ver Strate, R. H. (2008). Geotechnical Instrumentation News – The use of the fully-grouted method for piezometer installation, Part 1 and Part 2. Geotechnical News, (June), 30-37.

Dunnicliff, J. (2008). Geotechnical Instrumentation News – Discussion of "The use of the fully-grouted method for piezometer installation" by Contreras et al (2008). Geotechnical News, (June), 38-40.

Gourvenec, S. M., Mair, R. J., Bolton, M. D., Soga, K. (2005). Ground conditions around an old tunnel in London Clay. *Proceedings of the Institution of Civil Engineers - Geotechnical Engineering* **158**, January 2005, 25-33.

Hight, D.W., Gasparre, A., Nishimura, S., Minh, N.A., Jardine, R.J. & Coop, M.R. (2007). Characteristics of the London Clay from the Terminal 5 site at Heathrow Airport. *Géotechnique* **57**, No. 1, 3-18.

Jones, M. A. (2007). Rising Groundwater in Central London. *Water and Sewerage Journal*, 4, 35-36.

King C. (1981), The stratigraphy of the London Basin and associated deposits. *Tertiary Research Special Paper*, Vol. **6**, Backhuys, Rotterdam.

Simpson, B., Blower, T., Craig, R.N. & Wilkinson B.R. (1989). The engineering implications of rising groundwater in the deep aquifer beneath London. *Special Publication 69*, CIRIA, London.

Standing, J. R. and Burland, J. B. (2006). Unexpected tunnelling volume losses in the Westminster area, London. *Géotechnique* **56**, No. 1, 11-26.

Wan, M. S. P. and Standing, J. R. (2014). Lessons learnt from installation of field instrumentation to monitor ground response to tunnelling. *Proceedings of the Institution of Civil Engineers - Geotechnical Engineering.* (Revised manuscript accepted November 2013).

Crossrail Project: Infrastructure design and construction
ISBN 978-0-7277-6078-4

doi: 10.1680/cpid.60784.539

Data management, analysis and visualisation on Crossrail Drive X Western Tunnels

Daniel Rendell, Civil Engineering Contracts Engineer, Getec UK
Paul Thurlow, General Manager, Getec UK

Abstract
Crossrail Drive X Western Tunnels extend from Westbourne Park to Farringdon Station. With the TBM and SCL contracts working under the most expensive real estate in the world, the safeguarding of assets is paramount for safety and stakeholder's assurance. The management of instrumentation data from geotechnical works, instrumentation and surveying methods has to be accurate, readily available and innovative. The challenges of providing a database and management for instrumentation and grouting systems that can be used by all engineers on the project and incorporate all the requirements of the specification have been great , but have led to innovation and methods that will change the way that data is reviewed on future monitoring projects. This paper describes methods used to provide data in formats that are used for review groups, tracking TBM and SCL face positions, compensation grouting works and value engineering.

Abbreviations
TBM - Tunnel Boring Machine
SCL – Sprayed Concrete Lining
I&M – Instrumentation and Monitoring
SRG – Shift Review Group
CTC - Contractors technical committee.
UCIMS - Underground Construction Information Monitoring Systems
AGS - Association of Geotechnical and Geoenvironmental Specialists
BIM – Building Information Modelling
FTP – File Transfer Protocol
TaMs – Tubes-à-Manchettes

Introduction
The Crossrail project is one of the largest construction sites in Europe. It also specified the largest monitoring package in modern times. To enable the installation of instrumentation, capture of data reporting and archiving required a considered specification to be issued for the project.

Crossrail has specified and let the 701 contract prior to main works tender. This tender allowed for instrumentation and monitoring of assets prior to main works as well as providing an overarching archive software called UCIMS (Underground Construction Information Monitoring Systems) that all contracts had to send processed data to and receive data from.

During the construction of the Jubilee Line Extension, instrumentation and monitoring was used extensively on different sites. There was very little co-ordination between sites and the data and results were largely left at site level. Attempts by academia to retrieve data and re-process the information was difficult due to different operating systems and data format.

The use of a common format for storing data was a logical step forward in monitoring data archiving. The AGS format is an established method of recording borehole data in the UK, but the use of this system for monitoring geotechnical and structural instrumentation data at such a scale was a first for instrumentation companies tendering on the project.

Monitoring software had to be re written to allow for automatic AGS export and import as many sections of the works overlapped between different contractors and their instrumentation suppliers. Data would be retrieved from UCIMS, but at critical phases of construction, the specified time frame in the works specification was not suitable, so instrumentation contractors took raw data from source and calculated results within their own software to facilitate construction progress. The use of an archive is commended, but for daily use on TBM and SCL works, co- ordination between data suppliers and their willingness to work together was key.

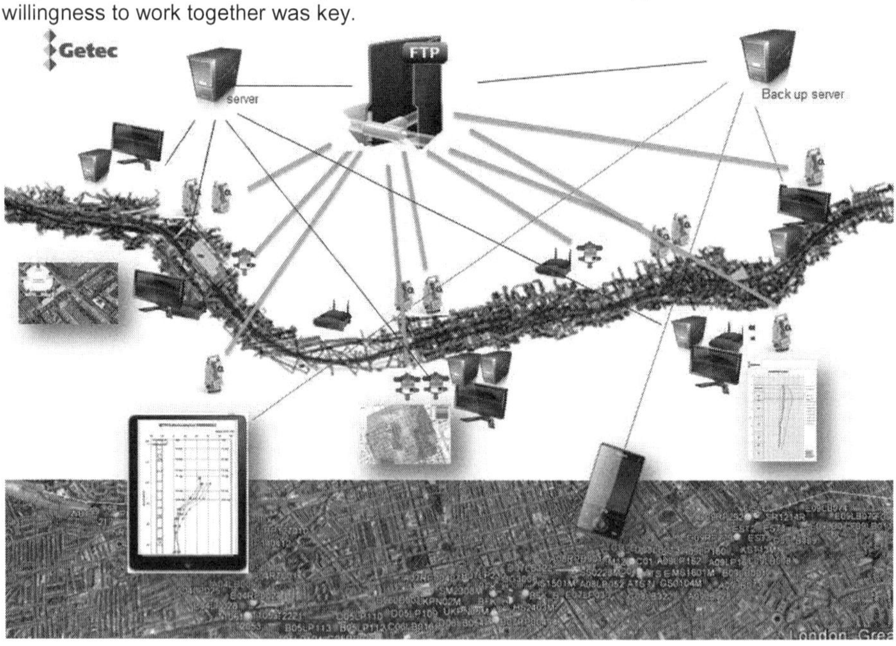

Figure 1: Schematic of data collection techniques.

The supply and visualisation of data is only useful if the software used on major infrastructure is able to expand as the project progresses and innovates. Contractors and Clients should invest in a database and monitoring system that is compatible with all end users and is able to expand as data load is increased. At the very least it should include the facility to track excavation progress and location real time and with absolute spatial position. It should also be compatible with mobile technology to enable Utility and Public stakeholder's access for particular activities.

Current and future monitoring software will be able to add the latest technology such as 3D models that are used in BIM, site reporting that is linked by date and time stamp to data and public information.

Monitoring software and its management of data points should also be linked to the method of use in the work place. The preparation of set views and graphs is essential for the daily review meetings as is management of new drawings, new instrumentation locations and software updates.

On a project of this scale data collection alone is not enough. There needs to be a way of storing the information in a secure way and also to view it from anywhere across the site at a speed that is acceptable to all end users. The software used for monitoring the I&M on the Crossrail Drive X Western Tunnel Drive was a specialist monitoring software called gtcVisual. gtcVisual is a software that was developed for the mining industry in the early eighties and was eventually used to also show grouting phases and results in a shared format. The software and its features have been upgraded extensively for Crossrail technical compliance and also to feature new instrumentation types such as fibre optics, shape accel array, wireless flat mesh tilt data and digital noise and vibration data.

The software has also been developed to provide fast and clear visualisation to the end user. The features far exceed the specification, but also provide user friendly input fields, data export for use in other geotechnical design software and clear journal features, that enable the history of instrumentation points on the project, but also the site activities.

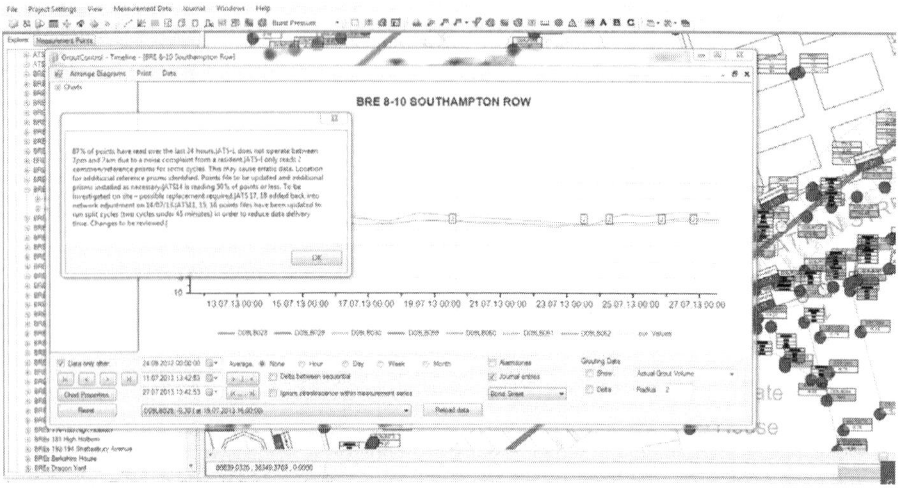

Figure 2: The journal feature in gtcVisual

Data Management

Managing and visualising data is key to delivering a successful project. The effectiveness of instrumentation is also a key consideration at the design stage. The instrumentation and monitoring industry is constantly changing with more accurate and functional products being released constantly. An adaptive software system must be in place which is able to adapt to these changes and ready to accept data in a wide variety of formats.

With all the different work packages on-going around the drive x route it was critical to ensure that the data collected was as up-to-date as possible so that on-site trends could be monitored closely and reviewed in a timely fashion and also at daily review meetings. To achieve this a reliable and robust system needed to be implemented which would automatically collect data from a variety of sources and then map that data to the correct measurement point ready for visualisation, manually downloading data was not an option. When considering the monitoring system there are generally four criteria which need to be considered:

Speed and capacity, security, Visualisation requirements and functionality for the end user

Figure 3: View of Tottenham Court Road Station showing HLC positions.

When considering the different delivery methods there are two options which can be used. That of a web-based service or a software based system. Generally data demands during a projects life cycle can reach high levels, coupled with a requirement to load a vast quantity of data through a connection shared with, in some cases, several thousand other users, this can impact on the performance of the system. It is for this reason that a software based system is much better suited for larger projects especially where users exceed ten.

Figure 4: Web based monitoring system used in parallel to main monitoring software

With software based systems some of the processing power is handled by the computer itself and the internet connection is only used for loading measurement data. This reduces the amount of traffic on a network and thus provides a better performance. Web-based systems are generally better suited for short time data access and where a limited number of users are involved.

The speed of the monitoring system is paramount when dealing with construction activities such as TBM passage and compensation grouting. As the majority of users were concentrating on viewing their particular work area only and not the whole project, provision was made to allow partial areas of the site to be available, making the software fit for purpose.

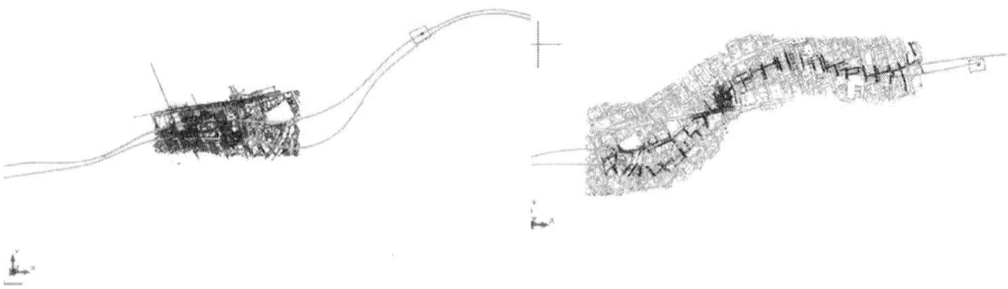

Figure 5: Expanded and local views of works sites.

The use of preset graphical presentations involving multiple points that were particular to specific buildings or assets were pre- set at the beginning of the project to ensure continuity within the site team. This also allowed for third parties to receive the same data in formats that were easy to read. This is a much more efficient way of analysing data instead of looking at each point in turn. From viewing the data in this way clear trends in deformations could be identified.

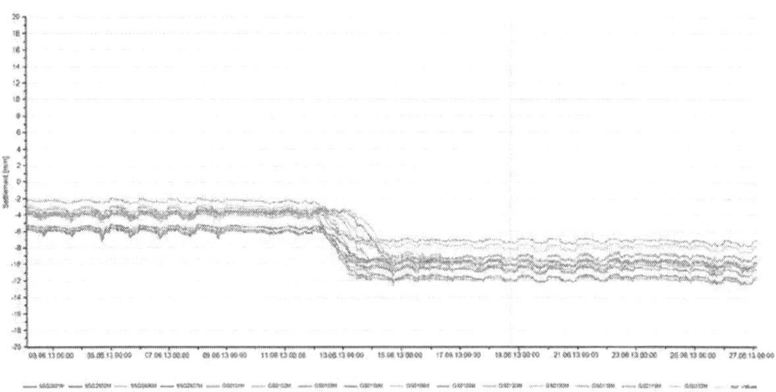

Figure 6: Trend analysis using multiple point data.

Loading data can slow down server resources and therefore consideration was given to provide facilities to show data from various time stamps or data formats on loading, to speed up operations. Averaging of the available data over a user defined time period is an example of this.

The storage capacity of servers and the backup procedures for data were also important factors to consider on this project. There needed to be sufficient capacity that expansion could easily take place if and when the works scope changed. Further to this there is also the consideration that measurement rates increase and decrease during a projects lifecycle according to the works taking place. For example during TBM passage the monitoring frequency is higher than that of when it has passed.

When data for automated instrumentation is measured it is then stored on a local logger or PC where it awaits uploading to a cloud storage facility such as an FTP site. Generally this will be to a specific folder on the FTP itself to improve functionality. Any new data files are then downloaded at regular intervals of 10 minutes. Once downloaded the new data files are directed to a specific location on the server where the file contents are read and recognised by the gtcVisual software and then added to the database. The data is then available to be viewed by anyone with access to the software. All of this means that the time from the data being measured to it arriving on a user's screen is less than 15 minutes. As the primary idea of the Observational method is to react upon observing deformations and recognising developing trends this quick turnaround is crucial. Combined with an easy to use export feature data can be downloaded for importing into specialist analytical software for further design checks and analysis, giving maximum flexibility to a user.

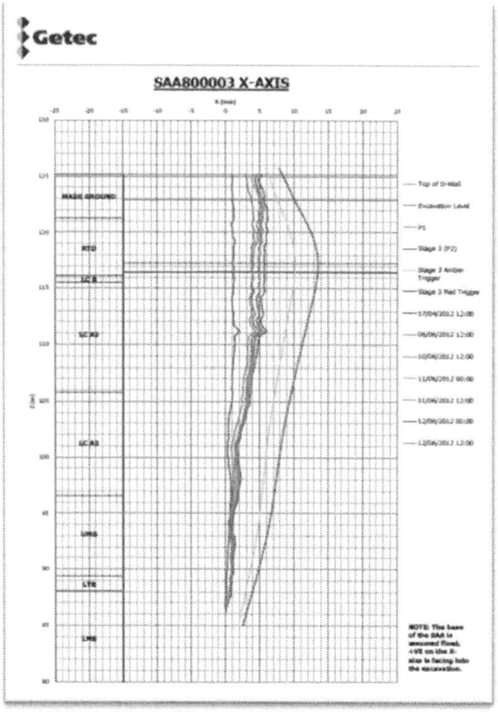

Figure 7: Inclinometer data with design lines, excavation and propping levels.

Further to the site loggers recording data there was also a large amount of data being received from other contractors along the route who were specifically involved with station boxes for example or LU running tunnels which the TBM would need to pass in close proximity to. This data was handled in much the same way in that a file format was discussed and agreed with the contractor and a download protocol set up on the FTP site and SQL server so that when a file was received it would be recognised and downloaded into the correct location on the server.

To ensure that all data is mapped correctly within the database each measurement sensor is assigned its own unique ID, its own unique point name, an alias name which in the case of Crossrail was its UCIMS name and its geodetic co-ordinates in Crossrail Grid. This information is used by the software to position the sensors on the user's screen in a precise way and also make each sensor identifiable for ease of use.

Analysis and Data Visualisation

Given the delicate nature of the works being carried out it was of critical importance to carefully monitor how deformations being measured on site compared to anticipated movements by assigning alerts and trigger values to each sensor, these were broken into three categories, Green, Amber and Red with a forth trigger threshold of Black being used in extreme circumstances. All charts which are plotted have the ability to also plot their associated trigger values so that appropriate safety checks can be carried out. When considering the method of presentation there are a few options which can be used;

- Individual or Collective measurement point charts
- 'Cut Profile' plots
- Contour plots

All of these could be plotted to not only show deformations since the instruments were first switched on but also from a specific moment in time, known as a time reference plot. This is especially useful when target grouting works are being carried out or the triggers assigned to a particular asset are only concerned with construction impacts and not previously observed movements.

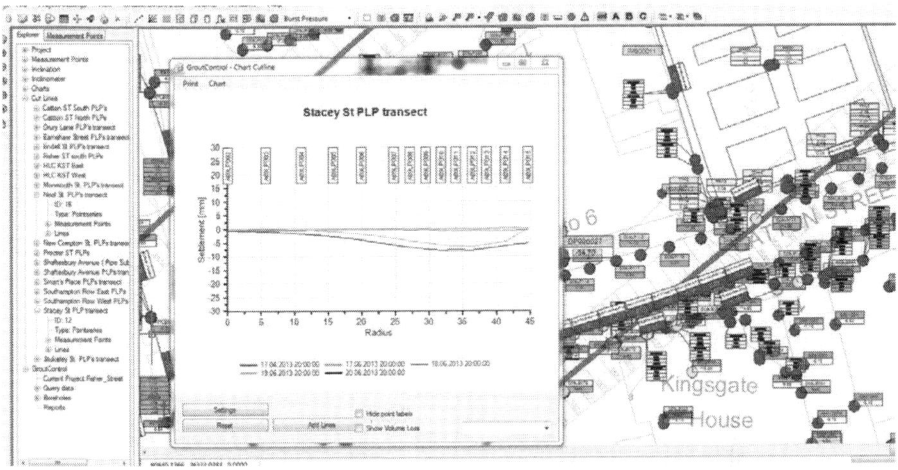

Figure 8: Visualisation of cut lines for precise level points.
Note the cut and slope have alarm triggers shown on screen.

Whilst viewing data for multiple points is a useful tool to have when monitoring it cannot always give a good indication as to what that means in relation to surrounding measurements and an overall view of what is happening across a site. In this instance virtual sensors are used to calculate parameters such as slopes and deflection ratios between select sensors either along a highway or public footpath. Track cant and twist were also calculated for rail specific assets were the condition of pairs of running rails was required to ensure the safety of the general public who would be travelling along these routes each day. In the event triggers were breached on these calculations the relevant parties would be notified and control measures put in place. However it never came to this as the combination of fast and reliable data ensured that construction methods during passage of the TBMs could be altered as they went before excessive deformations were caused.

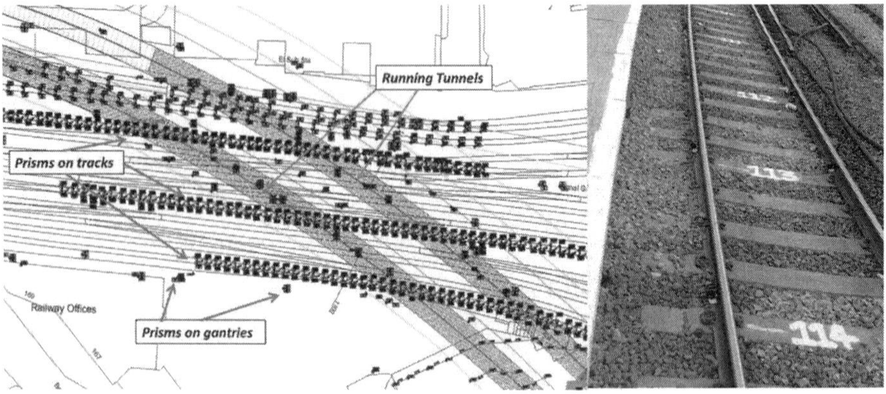

Figure 9 - Track Monitoring

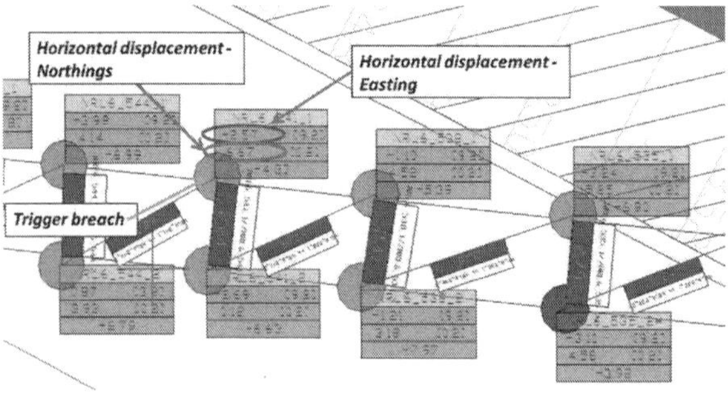

Prisms shown on existing tracks display the most current geospatial data which is then in-turn used to calculate the rail specific parameters. The result of which is displayed as a virtual sensor between the prisms.

Figure 10 - Twist and Cant between measurement Points

Figure 11 shows an example of the charts which were produced for the different site works. By viewing data in this collective manner it is easy to recognise that a common trend was occurring at this specific location, by checking the construction parameters adverse settlement was avoided and following TBM passage no further excessive settlement has been measured.

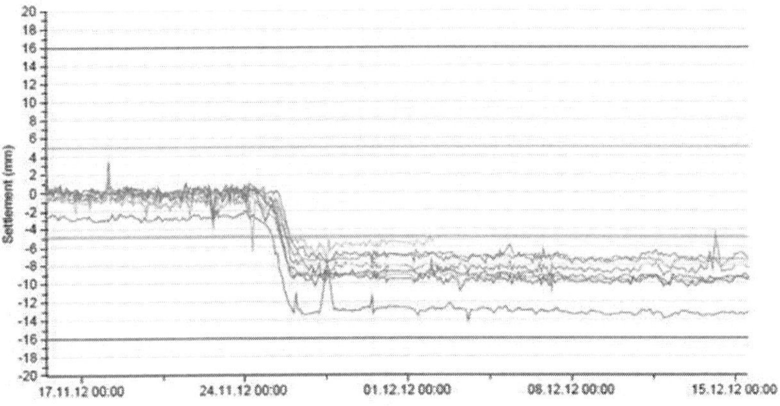

Figure 11: Prism data showing settlement as TBM crossed railway tracks.

By combining data within the monitoring database it was possible to interpolate between different measurement points and create contour mapping diagrams. This was then used to identify areas of heave and settlement which was then utilised by the site teams carrying out both the construction and the grouting works so that they could either alter their work methodology or focus their remediation techniques.

Figure 12: Interpolation of data to create a site contour plot

This information could then be used in conjunction with the "time reference" feature to see the same results but with regard to a specific date and time. This allows the user to then monitor construction induced settlements attributed to one specific section of works instead of the absolute settlement created during the life cycle of the project.

To ensure that the quality of data produced was to a high standard regular quality control checks were carried out on data on aspects such as the standard deviation which was checked over a defined time period. The risk in doing this was that the standard deviation equation does not distinguish between actual deformation and instrument noise or erroneous measurements. As the measurement points are nearly always moving it becomes difficult to determine a quality parameter. Therefore as deformations generally occurred gradually time periods of 24, 48 hours were chosen and the results plotted over a scatter graph to give an indication of data spread.

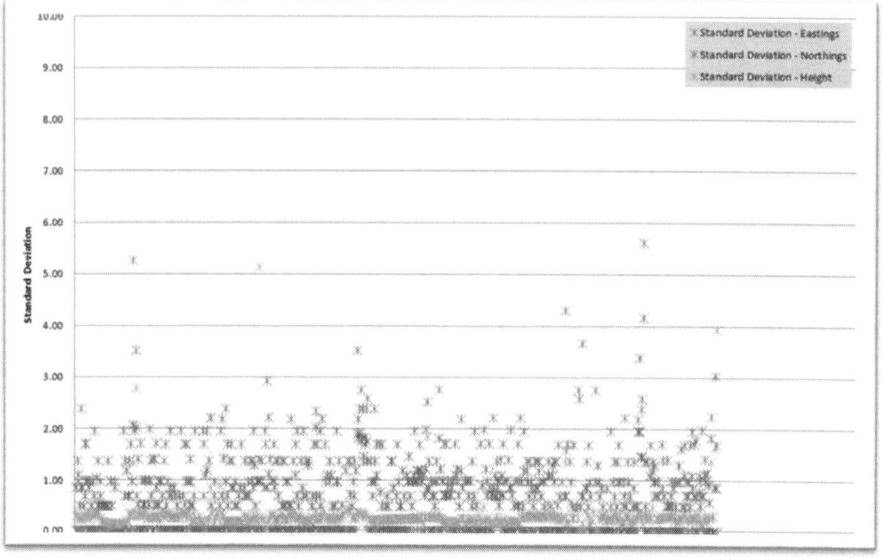

Figure 13: Standard Deviation Checks on Data

As construction works advance the influence zone generated by these works also advances and changes in size. It was especially important that during these works the exact location of any construction activities be known in as close to real time as possible. To this end the monitoring software used allows for multiple TBM and SCL faces to be tracked at any one time. Each of these uses a complex system of analytical queries to calculate the depth below ground level of an excavation and then from that extrapolate the zone of influence and visualise that on screen.

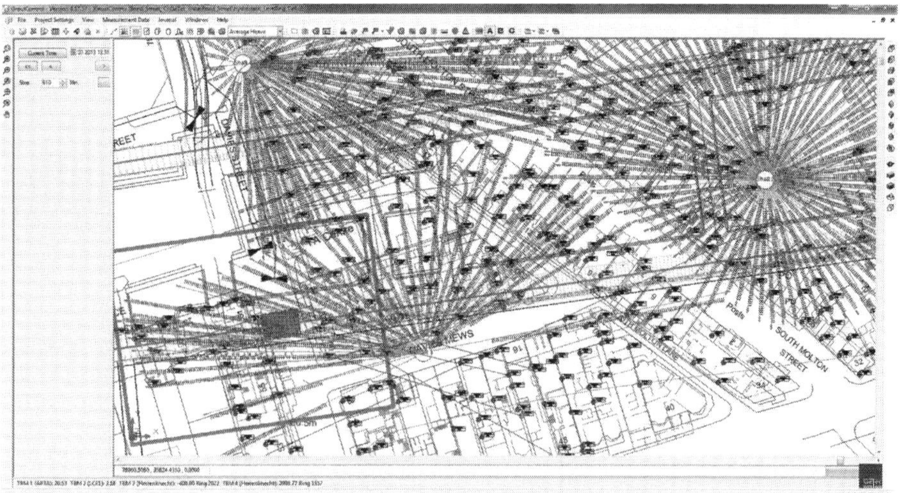

Figure 14: Visualisation of zone of influence marker for TBM with monitoring points and compensation grouting tubes.

The benefit of this is that the site engineers can then prioritise manual surveying teams to the areas of critical interest for a given shift. It also allows for on-site deformations to be correlated to site works, eliminating any question of erroneous data being measured. This process happens automatically each time a new construction advance takes place. Face loss at the tunnel head is also monitored through use of the volume loss calculation where predicted settlement troughs are compared against actual recorded settlements to determine if predicts behaviours are being observed.

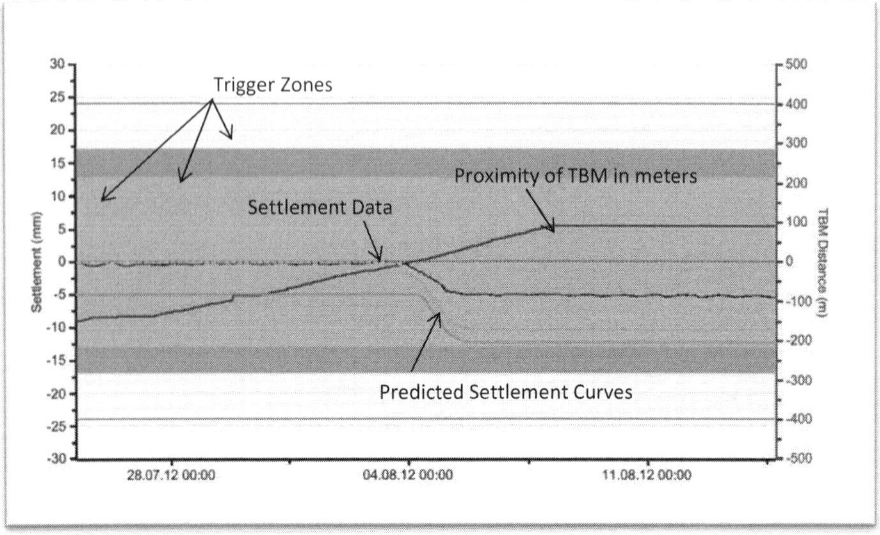

Figure 15: Volume loss curves.

The chart shown in figure 15 is an example of the volume loss curve which can be generated for a given point. The chart shows the predicted settlement curve which will result from construction works in relation to the horizontal and vertical distance the point is from the tunnel.

Asset Protection

With urban tunnelling projects there is nearly always a risk of interface between construction works and existing third party assets. These can be anything from buried water or telecommunication utilities to public transport links such as tube lines. In every case there will be an asset protection engineer (APE) somewhere who is interested in what monitoring is showing around and within that asset. Following the early success of the gtcVisual monitoring software many of these engineers bought into to the system giving them 24 hour access to data which they could then analyse and export to carry out further detailed calculations and checks.

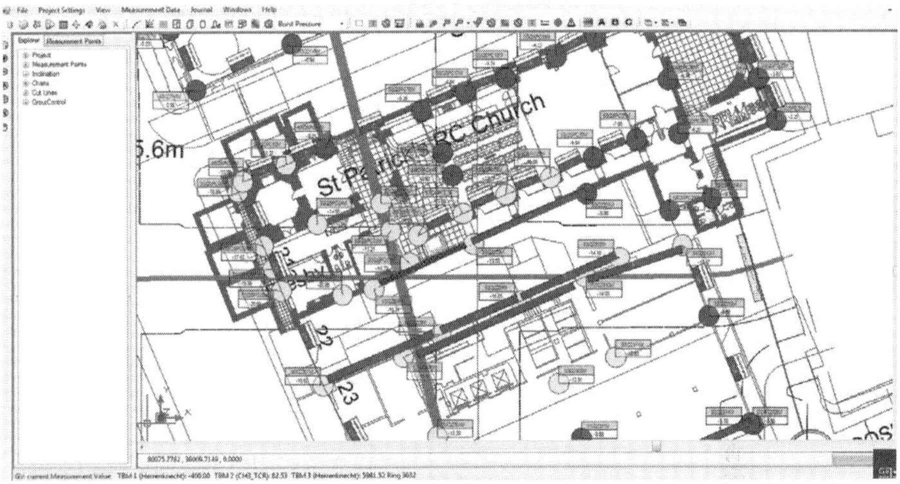

Figure 16; St Patricks Church View. Asset Protection Engineers were able to follow tilt and slope values easily.

This information was also used by stakeholders from whom permission was required before works could pass over or near to their assets. These LONOs (Letter of No Objection) were issued in a timely fashion because of the reliable and readily available data which could be accessed by the different engineers. Without this works would have progressed at a much slower rate and that was not acceptable.

Data on the move

There is a growing demand for data on the move. With the ability to view data on portable tablet devices engineers can reach new realms of control on site. This is especially critical where fast review and control is needed on site. Before the use of such applications the engineer would need to access data from an office away from the site and then spend time travelling between the two to investigate further. Now with the use of monitoring apps the latest data can be accessed and displayed in a clear and concise way whilst standing next to the asset .This not only improves the efficiency of those working on site but also improves response times if a problem is identified.

An example would be viewing inclinometer data which is being used for monitoring a station box excavation. Viewer applications can be fully customised to show excavation levels and different design curves all based around specific dates. The benefit with this being that the user can look back at historical data and the graphs will update to show the state of works at the time. This means that there is never a combination of new and old data being displayed at the same time, giving the user the most accurate display.

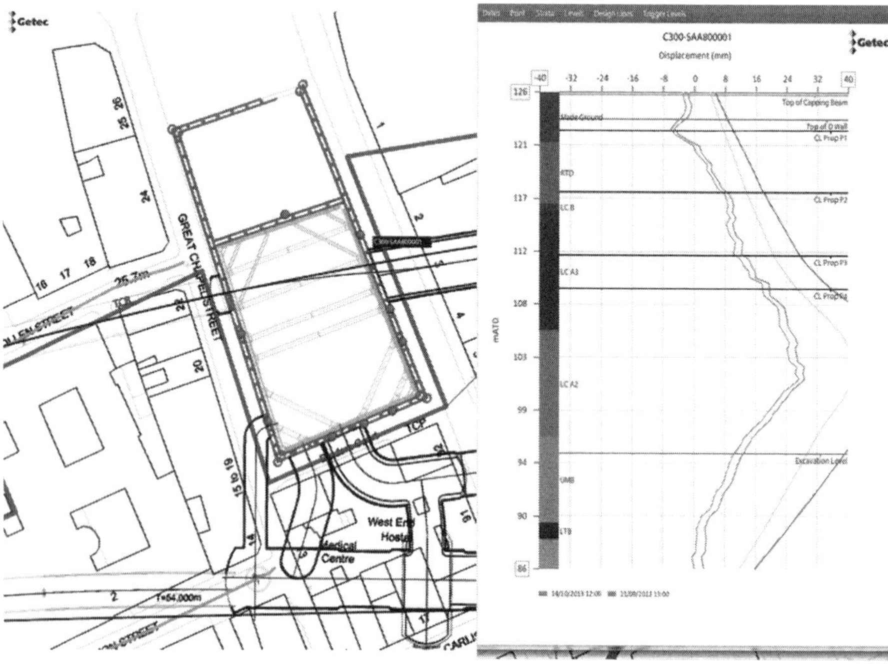

Figure 17: Data Views on Smart Devices

Another technology which is being utilised is the use of web-based applications. As the number of Another technology which is being utilised is the use of web-based applications. As the number of users is restricted and the pull on data targeted specifically to what is required the performance from such a system is not affected. Getec QuickView is the application which is being used to display simple data in the form of charts whilst also working in conjunction with a web based mapping tool. Providing detailed and reliable background mapping onto which monitoring data can be displayed. State of the art encryption and user logins ensures that all data is secure and only accessible by authorised persons.

The combination of technological advancements with hand held devices brings monitoring into a new field which is highly sought after, especially as more and more major construction projects are being carried out within the urban environment.

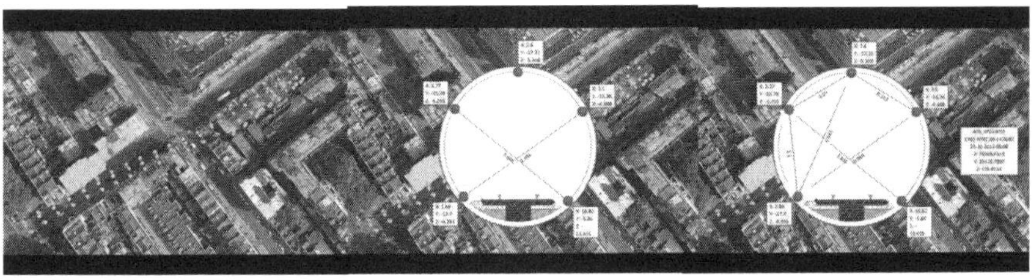

Figure 18 – Smart phone application showing tunnel ring deformation in true 3d using prism data

Monitoring compensation grouting

As well as monitoring the deformations created from constructing twin bore tunnels between Royal Oak Portal and Farringdon there are also station box excavations and sprayed concrete lining works which require mitigation measures such as compensation grouting. The compensation grouting specialist Keller Bam Richies used the gtcVisual software and one of its bolt on systems Grout Control to control the site works. The same database is used as for the monitoring data and the management of the data, drawings and measuring points is controlled by the I& M contractor. The grouting instructions are compiled on the grout control interface and sent directly to grouting containers on site. The operators sit in front of two screens, one of which is the gtcVisual project that is being used for the works.

Figure 19: Grouting container and gtcVisual display.

The operators select the shift information for the day and oversee the site operations. Details of the grout volume, pressure and flow are stored and sent back to the central database. The operator can view real time data and its change as a result of the injection.

The site engineer at the end of the shift can review visually the amount of grout injected, location and heave induced.

Figure 20: Visualisation of grouting activities over a series of days. Monitoring data, TBM and SCL face areas are also shown.

Figure 21: Heave values in mm for a grouting shift.

Conclusion

By maximising the available monitoring software technology and expanding it beyond its user end requirements it is not only possible to deliver a high class product which keeps both the client and site engineers well informed of all on-going geotechnical and structural deformations but also ensures that asset protection engineers and third party stake holders are provided with sufficient information to demonstrate the safety and security of their assets. By separating the available data into separate sites users are able to only view that which is important to them improving reaction times.

Integrating critical activities such as Compensation Grouting, which requires a high degree of control and data feedback provides the end user with confidence. The adaptability of software to meet the end user needs and provide more mobile data opportunities is crucial for future projects that contain Instrumentation and Monitoring.

Acknowledgements

The authors would like to acknowledge the support of the Crossrail Team and BFK the Main Contractor for the authors project that is described in the paper.

Crossrail Project: Infrastructure design and construction
ISBN 978-0-7277-6078-4

doi: 10.1680/cpid.60784.559

Crossrail Western Running Tunnels. Crossing of London Underground: Victoria Line and Bakerloo Line Assets

Gema Araque, MSc TBM Control Room Engineer, Team BFK, London, UK.
Ignacio Menoyo, I&M Engineer, Team BFK, London, UK.
Marios Vironos I&M Engineer, Team BFK, London, UK.
Sam Chernanko, BEng I&M Engineer, Team BFK, London, UK.
Daniele Fornelli, MSc I&M Engineer, Team BFK, London, UK.
Sergio Pellicer, BEng TBM Control Room Engineer, Team BFK, London, UK.

Abstract

The C300 running tunnels are of comprised a 6.2m internal diameter, 7.1m excavated diameter, and are to be driven by Earth Pressure Balance Machines (EPBMs) from West to East commencing at Royal Oak Portal and terminating at Farringdon Station via Paddington, Bond Street and Tottenham Court Road Stations.

Crossrail TBM (Tunnel Bored Machine) Drives cross different assets along their way. One of the most critical and interesting crossings is to the East of Bond Street Station, as a number of 3rd party assets are found within this area.

In less than 110 metres, the Westbound TBM (WB TBM) crosses under the Victoria Line running tunnels and the Bakerloo Line tunnels. The first asset is the Victoria Line Northbound running tunnel (VLNB) along with the Cross-passage connecting it to the Victoria Line Southbound running tunnel (VLSB), where the Westbound TBM passes just 3.095 metres under the first asset. Only 48 metres ahead, WB TBM will be vertically below the VLSB and Bakerloo Line Northbound and Southbound running tunnels (BKNB and BKSB) with a maximum clearance of 5.23 m.

This paper details the TBM's operating parameters during the construction of the Crossrail western running tunnels and analyses its effects in terms of induced settlements on the above assets.

Introduction

The C300 Crossrail Western Running Tunnels project, includes the excavation of new twin-bore tunnels 6.8 km in length under central London with two TBMs.

The TBM's were launched from the Royal Oak Portal, heading eastwards on an alignment that crosses through central London, via Paddington, Bond Street, Tottenham Court Road and Farringdon Stations. The crossing of the Victoria and Bakerloo Lines represented one of the more complex and challenging asset interfaces in the project.

The operation and control features of the TBM were designed to minimise sub-surface and ground level settlement at source. Ground loss arising from the TBM boring and tunnel construction operations was contractually limited to a maximum of 0.7% of excavated volume at the location of the Victroia and Bakerloo Line assets.

The TBM operating principles govern the use of the TBM in order to meet the specified volume loss requirements as well as asset protection displacement limits. This is achieved through close co-operation between the construction and monitoring teams, this is focused in the project control room. In the control room, trend analysis of the TBM parameters is undertaken 24/7 to anticipate any deviation from the specified parameter limits and targets to minimise settlements. An exhaustive monitoring regime of ground and asset movements allows the control room to monitor against the effects of the tunnel construction.

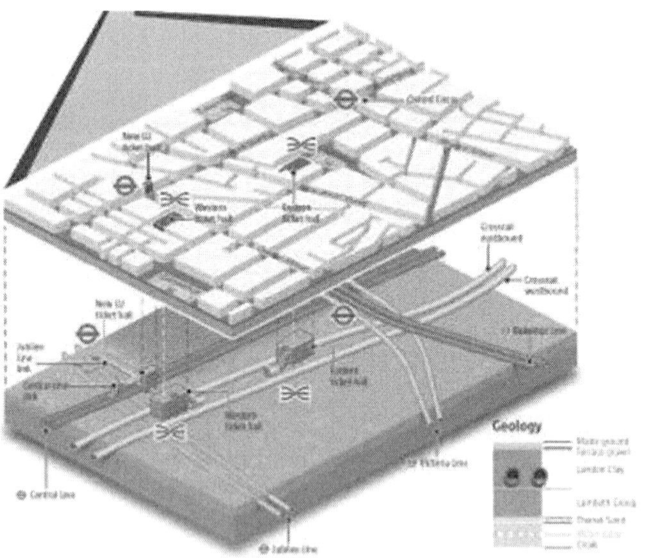

Figure 1. Location. Detail of Westbound and Eastbound Crossrail Tunnel crossing Bakerloo and Victoria Line.

The VL and BK Line assets are located between Bond Street and Tottenham Court Road these two stations (see figure 1). WB and EB Crossrail tunnels cross below VLNB (including cross-passage), Regent Street Sewer, BKNB and BKSB from west to east in 110m and a maximum clearance of 6m (please refer to Table 1).

The relevant information realting to the asset crossing dates, locations and clearances are shown below in Table 1.

In reference to Eastbound TBM, the second drive, it arrived to Victoria Line NB's Zone of Influence the 2nd of July 2013 advancing ring number 2370, going below Regent Street Sewer and Bakerloo Line NB in only 25 rings more and exiting from all the assets' Zone of Influence in ring number 2432 which was reached the 6th of July 2013. In Total 63 rings were built and 100.8 meters were advanced.

ASSETS		WESTBOUND TBM				EASTBOUND TBM			
		RING*	CHAINAGE*	DATE	Clearance (m)	RING*	CHAINAGE*	DATE	Clearance (m)
Victoria Line NB	Enter ZOI	2397	4156.94	04/05/2013	3.3	2370	4166.87	02/07/2013	4.7
	Exit ZOI	2424	4200.15	05/05/2013		2393	4203.68	03/07/2013	
Bakerloo Line NB, Victoria Line SB, Bakerloo Line SB	Enter ZOI	2431	4211.23	07/05/2013	4.9,9.6,4.8 (respectively)	2395	4206.95	04/07/2013	4.9, 9.9,5.0 (respectively)
	Exit ZOI	2465	4265.64	09/05/2013		2432	4266.05	06/07/2013	
* Ring advanced or Chainage Cutterhead Position ZOI = TBM Zone of influence									

Table 1. Assets Location and Dates reached. Summary table describing ring numbers, chainages and dates for each asset.

Assets under Study

Victoria Line south of Oxford Circus

The LU assets at this interface comprise the VLNB and VLSB running tunnels. A cross passage exists between the running tunnels which is located south of the CRL WB tunnel, displacements from this asset are not analysed in this paper.

The VLNB tunnel linings are mostly constructed from flexible cast iron segments with radial knuckle joints and the VLSB linings are mostly constructed from bolted cast iron segments. The crown of the VLNB and VLSB tunnels are approximately 20m and 16m below ground level and the tunnels are located within the London Clay. The interface is located south of Oxford Circus, east of Hanover Square. At the sub-surface level, the Victoria SB running tunnel crosses over the Bakerloo line running tunnels and passes below the Regent Street sewer. Figure 2 shows the position of the VL NB at the Crossrail interface

Bakerloo Line south of Oxford Circus

The LU assets at this interface comprise the BKNB and BKSB running tunnels. The location of the interface between the proposed Crossrail tunnels and the existing Bakerloo line is between

Oxford Circus station and Piccadilly Circus station beneath Regent Street.

The tunnel linings are bolted cast iron segments.

The crown of the BKNB and BKSB tunnels are approximately 24m below ground level and the tunnels are located within the London Clay. At the sub-surface level, about the WB TBM crossing, the BKNB and BKSB tunnels lie underneath the VLSB tunnel.

Figure 2 shows the positions of the assets in section.

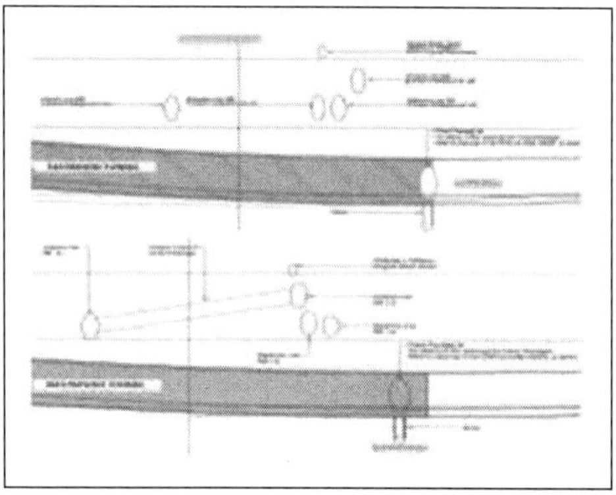

Figure 2. Section through LU assets showing vertical position relative to CRL western running tunnel.

TBM Parameters

TBM Characteristics & Parameters

The tunnel machines are an EPB type constructed and built by Herrenknecht AG, Schwanau, Germany. Earth Pressure Balance shields (EPB shield) are preferably used for cohesive soils with high clay, mud or silt contents which are characterised by low water permeability. The whole C300 route is covered by the London Clay formation. It is heavily over-consolidated and it decreases its thickness from west to east.

TBM Characteristics / TBM Parameters	
Tunnel maximun Slope	3.3 %
Tunnel minimun Radius	300 m
Excavation diameter	7100 m
Internal diameter	6200 m
Rings design segments	7+key (1.6 m)
Shield Diameter	7.06 m
Length shield + back-up	146 m
Weight	1100 t
Installed power	4000 KVA
Breakaway Torque	10179 KNm
Max. Thrust main cylinders	58540 KN
Rotation speed	0-3.2 rpm
Covering	< 50 m

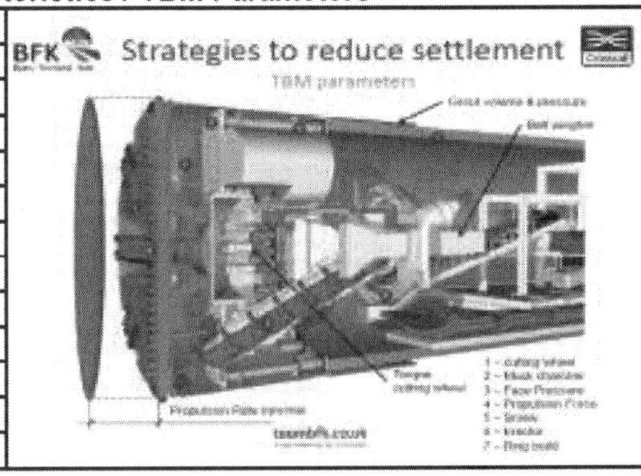

Figure 3. TBMs Characteristics/Parameters. Main characteristics of the TBMs.

The critical parameters that are required to control settlements at source during the excavation are: Pressure in the working chamber, grout injection (pressure and volume) and quantity of excavated material.

EPBface pressure

The soil excavated by the cutting wheel serves as the support of the tunnel face to avoid settlement, for that reason the ground excavated by the cutting head is mixed and accumulated under pressure in the cutting chamber. The ground is then extracted in a controlled manner with a screw conveyor. For controlling the stability of the tunnel face, there are installed earth pressure sensors which permit the face support pressure to be controlled precisely with continuous measurements. Shield advance speed and screw conveyor speed are the support pressure control variables (see appendix 7.2).

Grout Injection

The diameter of the cutter head is 7.1 m and the external diameter of the ring is 6.8 m. For that reason a gap is generated per advance at the back of the shield and control of surface settlement behind the TBM is also critical. The injection of grout is an effective tool in preventing potential surface settlement. Inadequate grouting of the void around the segments is a common cause of the settlements behind the TBM. Continuous grouting of tail void should be performed as TBM advances in order to compensate for volume loss.

For the C300 works, a two component cement grout (with additives) is used. They are mixed combining a retarded grout with an accelerator and injected by four grout lines (two on the crown and two on the invert), filling the gap between bored diameter and outer ring diameter. The procedure of the injection must be controlled taking into account the pressure, that depends on the pressure existing outside of the shield (see appendix 7.3) and the theoretical grout volume to fill the gap (4.8-5.2m^3).

Scales

Tunnel face support is achieved through balancing of the ground pressure in front of the TBM with the pressure inside the chamber. When the required chamber pressure is reached the excavation progresses maintaining the pressure and monitoring the volume of excavated material. The goal is to maintain the same volume of material extracted from the excavation chamber as the volume of the material excavated at the tunnel face. Two belt scales are installed on the belt conveyor to measure and monitor continuously, the volume of the excavated muck per advance to achieve a correct control of the weight of excavated material to avoid the cutter head over-excavation.

Other Parameters

Other mechanical parameters are involved for keeping the above mentioned parameters within designed limits/targets, these are: Propulsion Force (Thrust per advance) and Cutting wheel Torque (or moment of force of the cutting wheel per advance).

The TBM is equipped with 22 pairs of thrust cylinders divided into 4 groups, pushing the machine forward whilst reacting against the last ring built. In each group one cylinder is equipped with a measurement system to visualise the average length of this thrust cylinder group during the advance. The measured pressure and extension length of the cylinders will be recorded at the PLC system of the TBM. By regulating the pressure at the different cylinder groups the machine will move forward at different speeds. The maximum advance force of the main thrust cylinders is 58,540 kN at a hydraulic pressure of 350 bar. The double cylinders with a maximum stroke of 2,500mm are connected with steel plates at their free ends, which are covered with plastic plates to reduce damage on the surface of the built-in segments.

The machine is designed to work with a nominal torque of 7020 KNm at 2.4 rpm, a value which is related with the rotation speed and the ground (foam injection).

Excavated material should be a homogenous mix of coarse and fine particles, improving overall plasticity of the material to make it conveyable and avoiding the adhesion of clay particles into the metallic structure of the cutting wheel or its tools. Required physical properties of an excavated material are achieved through conditioning it with a foam solution which consists of water with the addition of a polymer in a percentage of 3 %, creating a material of paste-like consistency. Air bubbles are generated and the grain friction is reduced. Additional benefits of ground conditioning include reduced water permeability of conditioned ground as well as reduced stickiness of the materials, providing efficient mining.

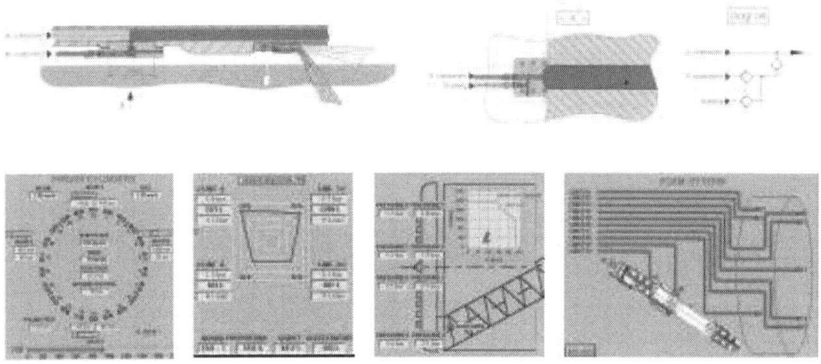

Figure 4. Principle of grout injection & PLC system to monitor TBM parameters.

Figure 4 shows the injection grouting in the gap generated by the excavation and a summary of the interface of the software installed to monitor the TBM parameters.

TBMs Parameters designed-observed

TBM monitoring data is analysed and during the daily Shift Review Group (SRG) the Drive Parameter sheet is provided to limit the values described before and summarised in the table below:

TBM PARAMETERS DESIGNED			ASSETS			
			Enter ZOI Victoria Line N/B	Vertically below Victoria Line N/B	Vertically below Bakerloo Line N/B & S/B- Regent ST. Sewer- Victoria	OUT ZOI LU ASSETS
Face Pressure		(TBM1)	1.55 bar	1.62 bar	1.70 bar	1.72 bar
		(TBM2)	1.66 bar	1.68 bar	1.72 bar	1.73 bar
Grout Pressure	Crown	(TBM1)	3.08 bar	3.11 bar	3.16 bar	3.17 bar
		(TBM2)	3.13 bar	3.14 bar	3.17 bar	3.18 bar
	Invert	(TBM1)	3.69 bar	3.72 bar	3.77 bar	3.78 bar
		(TBM2)	3.74 bar	3.75 bar	3.78 bar	3.79 bar
Grout Volume (TBM1 & TBM2)			4.8-5.2 m^3			
Propulsionforce (TBM1 & TBM2)			20,000 KN			
Cutterhead torque (TBM1 & TBM2)			4,800 KNm			
Mass Acumulator (TBM1 & TBM2)			127 t ± 10 t			

Table 2. TBM Parameters design. Limits for both TBMs.

As mentioned above, all these parameters are designed in relation with the ground the TBM is advancing through and all the assets involved. Under all the parameters described, the observed values are analysed in the graphs attached below.

In westbound TBM the EPB face pressure designed for this area is between 1.55 – 1.72 bar and 1.61 – 1.73 bar in case of eastbound TBM. It is necessary to understand that these values are the minimum pressures required to keep support at the face, in others words it is the minimum value to avoid any settlement. As the graphs are showing, the EPB face pressure value in westbound TBM is always above the target defined and around 0.2 bar plus. Sensors 1 and 8, locates in crown showed an EPB average value of 1.95 bar in eastbound TBM. This means the EPB face pressure was maintained correctly through the area under study.

The grout pressures values are similar to the EPB values. The target pressure assumed in the crown for grouting is between 3.08 – 3.17 bar in westbound TBM and 3.17 – 3.18 bar in eastbound TBM. In the invert, grout pressure for westbound TBM is between 3.69 – 3.78 bar and 3.78 – 3.79 bar for eastbound. The grout pressure graph is showing the values reached in the area studied and the grout pressure triggers values in the crown and the invert. These triggers values are the maximum and minimum values permitted to reach in the area established in the TBM operating procedure for EPB mode tunnelling 50% of the target pressure. This is achieved in the case of eastbound TBM with an average of 2.32 bar in crown and 2.55 bar in the invert.

Low grout pressure must also be avoided to avoid any settlement at the back of the shield. The lowest grout pressure values must be approximately 0.5 bar plus the EPB Face Pressure value - in the case of westbound TBM, this is approximately 1.7 bar. As the graph is shows there are some rings with less than 1.7 bar. The worst case, around ring number 2418, the reason of this low grout pressure is due to problems with the brushes in the bottom of the shield, and the grout and the pressure were lost by the invert. TBM was stopped to change the brushes and solved the problem as soon as possible.

In the grout volume graph, the observed volumes show all the rings are injected with the correct values. There are a few points in both TBMs with high and low values through the target. All these points are due to short or long advance and the grout volume injected is a little lower or higher than usual. This is confirmed by the average volume of 5.01m^3 per ring achieved in the eastbound TBM.

Is necessary to focus special attention in mass accumulator values as it's a parameter providing direct information of the material is extracting from the face. In the area the study is focused on, the material shoved is always the same, London Clay Formation. With a density of 2 t/m^3, the weight exported per 1.6 m of standard advance is around 117 (minimum target) and 137 tonnes (maximum target). In the graphs, all the values reached per ring are between the given triggers per standard advance, with a difference maximum of 2% between scales. Eastbound TBM is the machine that shows higher values with an average of 135 t.

In the rest of the parameters, the values designed are the same for both TBMs due to the same characteristics. Eastbound TBM achieved an average thrust of 17868 KN which is below the established limit of 20000KN. The same happens with the average torque of 3360 KNm which it is well within the limits of 4800 KNm. If we compare the thrust, torque and foam, we find that we have had an increase in the torque and the thrust if the foam injection decreased.

As shown by the graphs below, all observed parameters were within triggers.

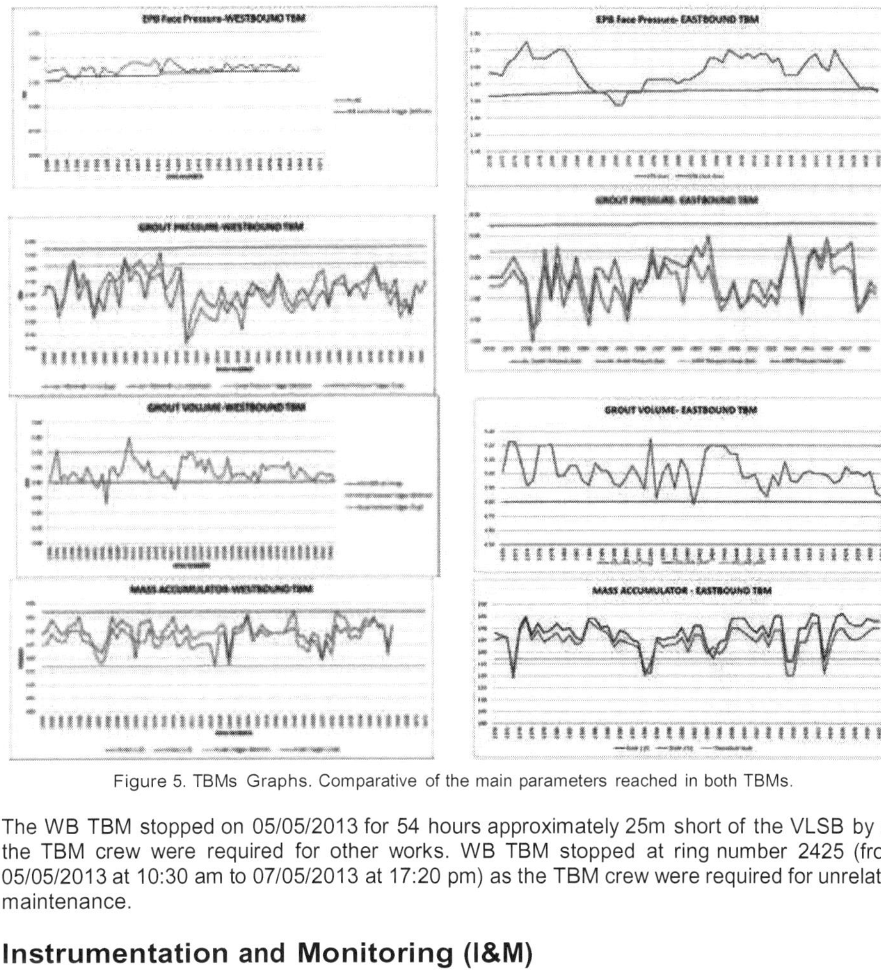

Figure 5. TBMs Graphs. Comparative of the main parameters reached in both TBMs.

The WB TBM stopped on 05/05/2013 for 54 hours approximately 25m short of the VLSB by as the TBM crew were required for other works. WB TBM stopped at ring number 2425 (from 05/05/2013 at 10:30 am to 07/05/2013 at 17:20 pm) as the TBM crew were required for unrelated maintenance.

Instrumentation and Monitoring (I&M)

Introduction to I&M

The monitoring installed inside the Victoria and Bakerloo lines comprise of Electro-level beams, PLP's (Precise Levelling Point), Tilt sensor arrays, Basset Convergence Systems, 3D

Geodetic Prisms, Manual Track Survey's and Retro Reflective Targets in both the North and Southbound running tunnels of the Victoria & Bakerloo lines south of Oxford Circus Station. This paper utilises data from the electro-levels for analysis of movements as these instruments provide the best accuracy and repeatability of all of the above instruments in these assets. Manual track surveys are also analyses to act as a confirmation of the electro-levels, these surveys are also required to check against the twist geometry of the tracks during passage.

An Electro-level Beam Sensor is attached to a rigid beam for installation in to the tunnel and measures rotation of structures in the direction perpendicular to the long axis of the beam. A beam provides a tilt reading, this tilt reading is then multiplied over the length of the beam providing differential vertical displacement for that beam. Both ends of the beam are fixed using anchor bolts, multiple beams were placed end to end along the tunnel, allowing a displacement profile of the structure to be calculated from anchor point to anchor point. This can be seen in figure 6 below which shows the typical installation of electro-level beams in the Victoria and Bakerloo Line tunnels. Settlement

A data logger installed in the asset to allow real-time reporting from the electro-levels, provided readings every 30 mins during the passage.

Manual track surveys were executed using a 3d geodetic prisms and total station, where a reading is taken on the top of each rail along the track. These readings were taken nightly during engineering hours.

Figure 6 – Examples of instrumentation and monitoring

Assets Displacements

Victoria Line Northbound

The chart shows the settlement's caused by the passage of both WB and EB TBM's separately and their combined effects as recorded by the electro-levels. Induced settlements in the Victoria Line NB (VNB) varied between drives with the electro-levels recording an initial maximum 13.2mm of settlement due to the WB passage and 7.8mm due to the EB passage. When considering the settlements at the locations of the TBM crossings due to both TBM's, the WB recorded 13.2mm and the eastbound 8.4mm. The difference can at least partly be attributed to the differences in minimum clearance between the WB and EB CRL tunnels and LU tunnel at crossing of 3.3m and 4.7m respectively. However, this amounts to 97% and 68% of predicted settlements for the WB and EB respectively at the locations of the observed maximum settlements on the electro-levels.

However, the maximum WB settlement does not occur at the predicted locations i.e. the location in where the CRL and LU tunnels intersect in plan – there is an offset of 4.4m. It was likely that there was an error in the calculation of the electro-level's as a correction by the monitoring sub-contractors can be observed after the crossing of the WB drive occurring along the entire string of electro-levels – this consists of correcting each electro-level by introducing an offset in the value in each electro-level between two readings depending on the magnitude of error in that beam. This appears to have corrected the values for the EB passage, whose settlements do occur at the predicted location at the EB crossing.

When comparing the electro-level values to the manual rail survey, the recorded displacements are comparable, this is the case for all of the LU tunnels. The rail survey has a lower accuracy and precision than the data provided by the electro-levels, however the manual rail survey is more reliable, partly because BFK do not calculate and report the electro-levels and partly due to the fact that it is a manual reading.

For the WB passage, the rail survey shows the location of the observed maxima better fits with the location of the predicted maxima, this further supports the error in the electro-levels as described above.

The rail settlements amount to 84% and 72% of predicted settlements for the WB and EB respectively at the locations of the observed maximum settlements on the. The EB, we can conclude that the displacement results are comparable given the manual precision of +/-1mm (difference in maximum settlements. If we consider that the WB maximum settlements recorded from the electro-levels were of the correct magnitudes but that the actual true location of this was over the predicted maximum, then we can conclude that 81% of predicted settlements due to the WB passage were recorded by the electro-levels – this is comparable to the manual rail survey.

However, one should consider that the manual rail survey reported a higher settlement than the electro-levels over the same period for the VNB, VSB and BKNB (exception was BKSB).

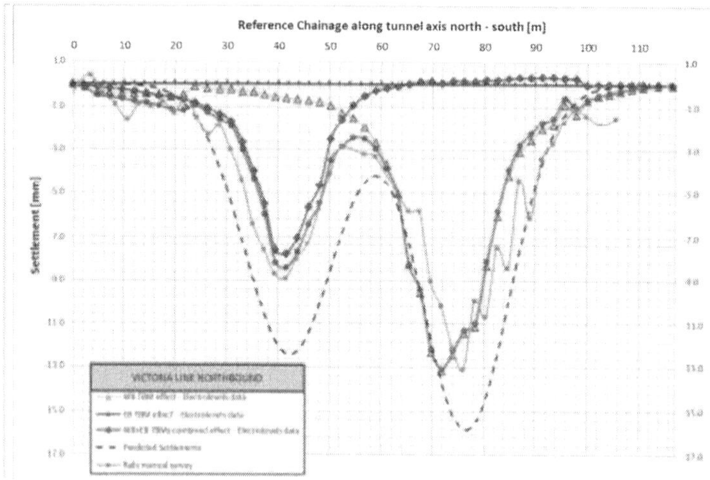

Figure 7 - TBM induced and predicted displacement profile in Victoria Line Northbound

Victoria Line Southbound

Induced settlements in the Victoria Line SB (VSB) varied between drives with the electro-levels recording a maximum 8.0mm of settlement due to the WB passage and 4.7mm due to the EB passage. When considering the settlements at the locations of the TBM crossings due to both TBM's, the WB recorded 8.7mm and the eastbound 6.0mm. Unlike the VLNB, there is very little difference between the minimum clearances between the WB and EB CRL tunnels and LU tunnel at crossing points 9.6m and 9.9m respectively.

Similar to the VNB, there is an offset of approximately 5.9m and 3.9m between the location of the observed maxima for the WB and EB respectively and the predicted maxima along the VLNB. Unlike in the VNB however, the rail survey doesn't disprove these values and the settlement troughs are comparable.

When considering the combined effect of the TBM's are recorded by the electro-levels, this amounts to 79% and 54% of predicted settlements for the WB and EB respectively at the locations of the observed maximum settlements on the electro-levels (which are offset from the predicted maxima).

This difference in behaviour with respect to the difference between WB and EB passage is difficult to explain. It cannot be attributed to TBM operating parameters which were very consistently within limits/targets between the two TBM's and indeed both TBM's achieved similar grout injection volumes, EPB and grout pressures – see Figure 5. The ground is considering to be homogeneous throughput the crossings and no reports to the contrary were given from the TBM crew.

The WB TBM stopped on 05/05/2013 for 54hours approximately 25m short of the VLSB by as the

TBM crew were required for other works. However no additional displacements were observed in the VLNB or VLSB due to this stoppage as the TBM was sufficiently far away from all the assets and EPB was maintained throughout the stoppage.

The section of the tunnel lining about the crossing with the WB TBM and north of this location is consistently bolted cast-iron type. Approximately 11m south of the crossing, where the tunnel intersects the cross-passage, the rings are constructed with a concrete surround, to provide additional support at this weak point. This might provide an explanation for the additional settlements, however, the construction of the Bakerloo Line tunnels are consistent throughout and the difference in settlements between WB and EB TBM passages also exists.

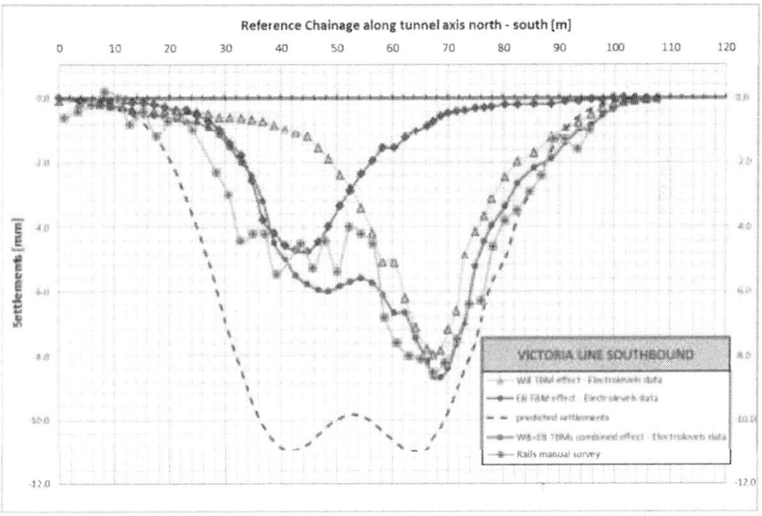

Figure 8 - TBM induced and predicted displacement profile in Victoria Line Southbound

Bakerloo Line Northbound

Induced settlements in the Bakerloo (BKNB) varied between drives. When considering the settlements at the locations of the TBM crossings due to both TBM's, the WB recorded 8.2mm and the eastbound 6.7mm. This amounts to 66% and 55% of predicted settlements for the WB and EB respectively at the locations of the observed maximum settlements on the electro-levels (which are offset from the predicted maxima).

The rail survey and the electro-levels are comparable, confirming the settlements on the electro-levels.

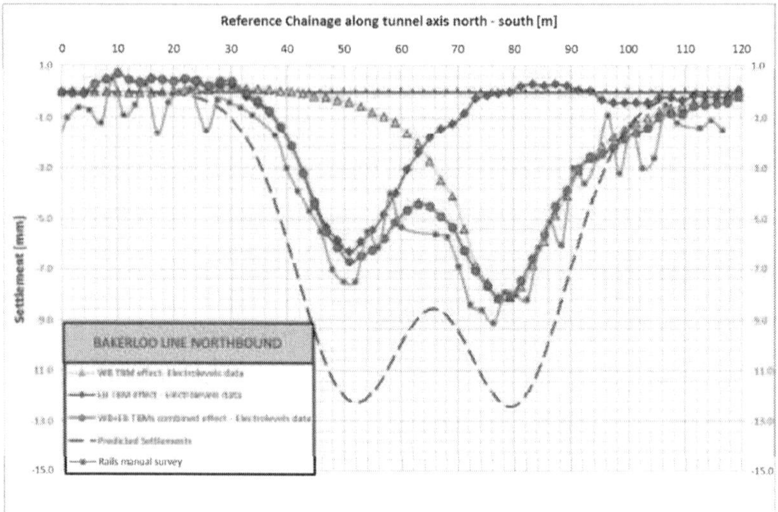

Figure 9 - TBM induced and predicted displacement profile in Bakerloo Line Northbound

Bakerloo Line Southbound

Induced settlements in the Bakerloo (BKSB) varied between drives. When considering the settlements at the locations of the TBM crossings due to both TBM's, the WB recorded 8.9mm and the eastbound mm. This amounts to 72% and 51% of predicted settlements for the WB and EB respectively at the locations of the observed maximum settlements on the electro-levels (which are offset from the predicted maxima).

The electro level readings between CH70-105 have been zeroed has the original values showed erroneous heave which was inconsistent with prediction and the manual monitoring.

The rail survey and the electro-levels are comparable, confirming the settlements on the electro-levels.

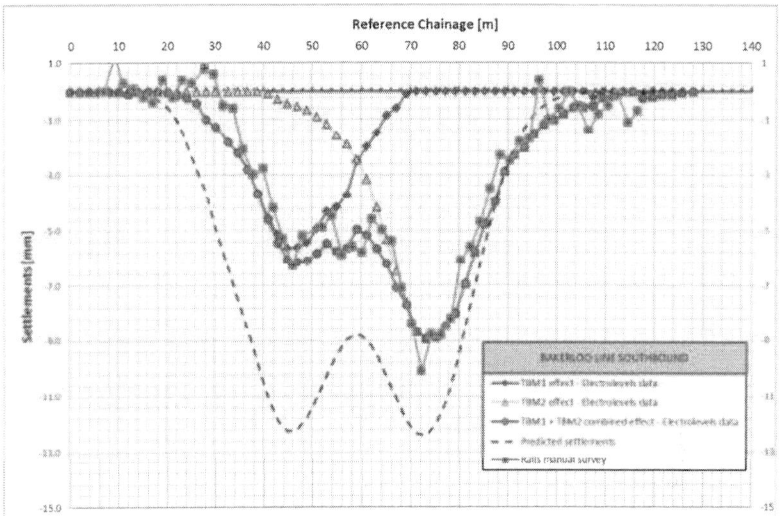

Figure 10 - TBM induced and predicted displacement profile in Bakerloo Line Southbound

Further Analysis of Displacements

Figures 11 and 12 presents a chart show displacements from the electro levels located closest to the respective CRL TBM crossings with respect to the distance relative to the position of the cutterhead.

For the WB TBM (Figure 11), the chart shows that the displacements during the approach of the TBM begin when the cutterhead is approximately 10m from the axis of the asset. The settlements during this period are a maximum of 3mm. The majority of the total settlements then occur over the length of the TBM shield.

Minimal settlements are induced behind the tail of the TBM (i.e. during grouting), this infers that the performance of the grouting operation was excellent. The exception to this is the VLNB, which continues to show settlements during and after grouting – approximately 50% of total settlements are induced in this period. After investigating the grouting volumes and pressures and EPB in the zone of the VLNB, all values are consistent and within limits/to targets, apart from localised exceptions which are highly unlikely to have caused this settlements, especially when considering the same behaviours occurs in the EB TBM/VLNB crossing where there was no localised deviations from the operational parameters.

Figure 12 shows that the same conclusions as the WB, with the majority of settlements occurring over the settlements occur over the length of the shield and very little settlements occurring after the passing of the tailskin. This once again confirms the excellent performance of the grouting operation. The exception again as stated above is the VLNB.

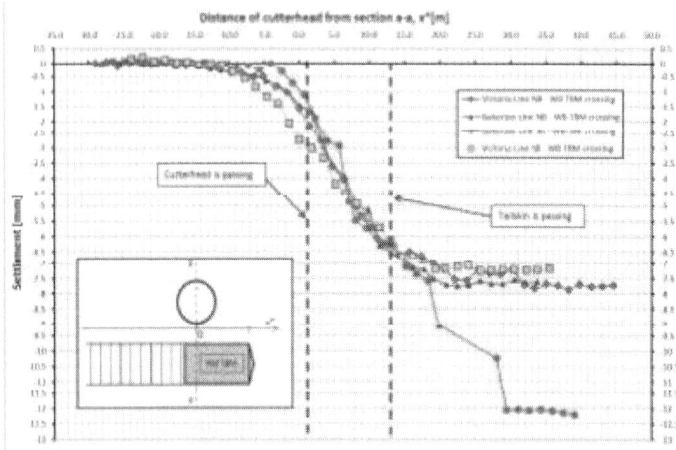

Figure 11 - WB TBM displacement with respect to distance from cutter-head

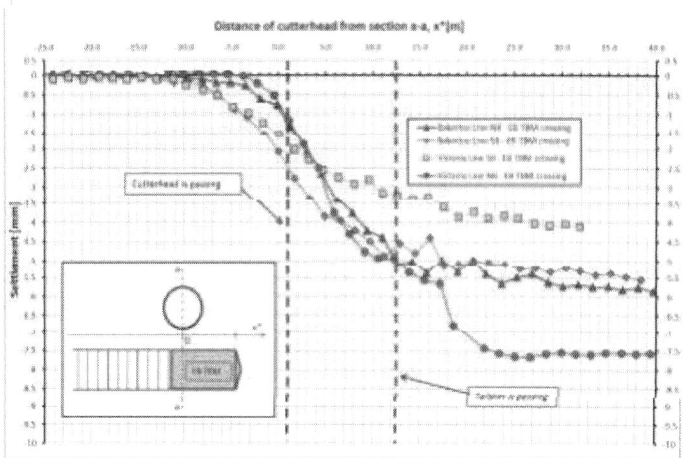

Figure 12 - EB TBM displacement with respect to distance from cutter-head

Conclusions

The analyses of the operating parameters and their comparison with the monitoring displacement results carried out in this paper, show that the general performance of the TBM's during the critical crossing of the LU Victoria and Bakerloo Line assets was extremely successful. This can be considered an achievement given the minimal clearances and the complexity of the crossing. Additionally, an unforeseen stoppage occurred between assets creating another challenge to the works, however this appears to have had little or no effect with respect to settlements given that the EPB was maintained correctly throughout and a significant distance between the TBM and tunnel linings exists.

The induced displacements didn't exceed predicted values and in almost all cases they were well within.

It is clear that the WB TBM induced more settlements than the EB TBM. In the case of the VLNB the settlements as a ratio to the predicted are comparable. For all the other assets, were the clearances were similar between both TBM's, a significant difference in displacements exists.

There are many possible causes for this that were investigated. The cross checking of the EPB, grout pressures and volumes and excavated volumes values show no significant differences between TBM's apart from a few localised values that are highly unlikely to affect displacements significantly.

Due to the uniformity of the foreseen geological conditions in this area and the relatively short distance between the LU tunnels the variability of the geotechnical parameters in the area should not be significant and it should not therefore significantly affect the TBM's induced settlements.

The construction of the LU tunnels varies significantly with flexible construction in the VLNB, part rigid-part flexible construction in the VLSB and fully rigid construction in the BKNB and BKSB. Given that all other possible causes for this differential settlement behaviour that were investigated were shown not to be the cause, it could be concluded with some confidence that the varying structure of the assets were likely to be at least partly responsible for this behaviour. As the differential did not occur in the VLNB it can be concluded that something specific to the VLNB that was not specific to the VLSB and BKNB and BKSB lines, did not produce this differential.

Further analyses of the specific kind of structure in the assets could show the answer to understand the behaviour of the Victoria and Bakerloo Lines due to the interaction with Crossrail Western Tunnel during the investigated crossing.

Appendices

London Underground Asset Triggers

London Underground								
Vertical Displacement of tunnel	LU08 Victoria Line Tunnel NB	+12	+15				Electro-levels	Unexpected movement of tunnel or other observed change considered to impact the safe operation of the railway shall constitute a Black trigger
Change in Tunnel Diameter (in any direction)		+/-10	+/-13	[illegible]			Basset Convergence System (primary) retro-flective targets from manual Survey (verification)	
Change to Longitudinal Gradient		1:1050	1:850	[illegible]			Electro-levels (along the length of the tunnel)	
Vertical Displacement of tunnel	LU08 Victoria Line Tunnel SB	+10	+13		Railway Safety		Electro-levels	Unexpected movement of tunnel or other observed change considered to impact the safe operation of the railway shall constitute a Black trigger
Change in Tunnel Diameter (in any direction)		+/-5	+/-8	[illegible]			Basset Convergence System (primary) retro-flective targets from manual Survey (verification)	
Change to Longitudinal Gradient		1:2000	1:2020	[illegible]			Electro-levels (along the length of the tunnel)	
Vertical Displacement of tunnel	LU08 Victoria Line Tunnel Cross passage	+12	+15				Electro-levels	Unexpected movement of tunnel or other observed change considered to impact the safe operation of the railway shall constitute a Black trigger
10m twist	LU 08 Victoria Line Tracks	+/-30	+/-37	[illegible]	+/-50		Manual track shoe survey spreadsheet	Unexpected movement of tunnel or other observed change considered to impact the safe operation of the railway shall constitute a Black trigger
2m twist		+/-14	+/-18	[illegible]	+/-25			
Vertical Displacement of tunnel	LU09 Bakerloo Line Tunnel	+12	+15				Electro-levels	Unexpected movement of tunnel or other observed change considered to impact the safe operation of the railway shall constitute a Black trigger
Change in Tunnel Diameter (in any direction)		+/-5	+/-8	[illegible]	Railway Safety		Basset Convergence System (primary) retro-flective targets from manual Survey (verification)	
Change to Longitudinal Gradient		1:1350	1:1070	[illegible]			Electro-levels (along the length of the tunnel)	
10m twist	LU 09 Bakerloo Line Tracks	+/-30	+/-37	[illegible]	+/-50		Manual track shoe survey spreadsheet	Unexpected movement of tunnel or other observed change considered to impact the safe operation of the railway shall constitute a Black trigger
2m twist		+/-14	+/-18	[illegible]	+/-25			

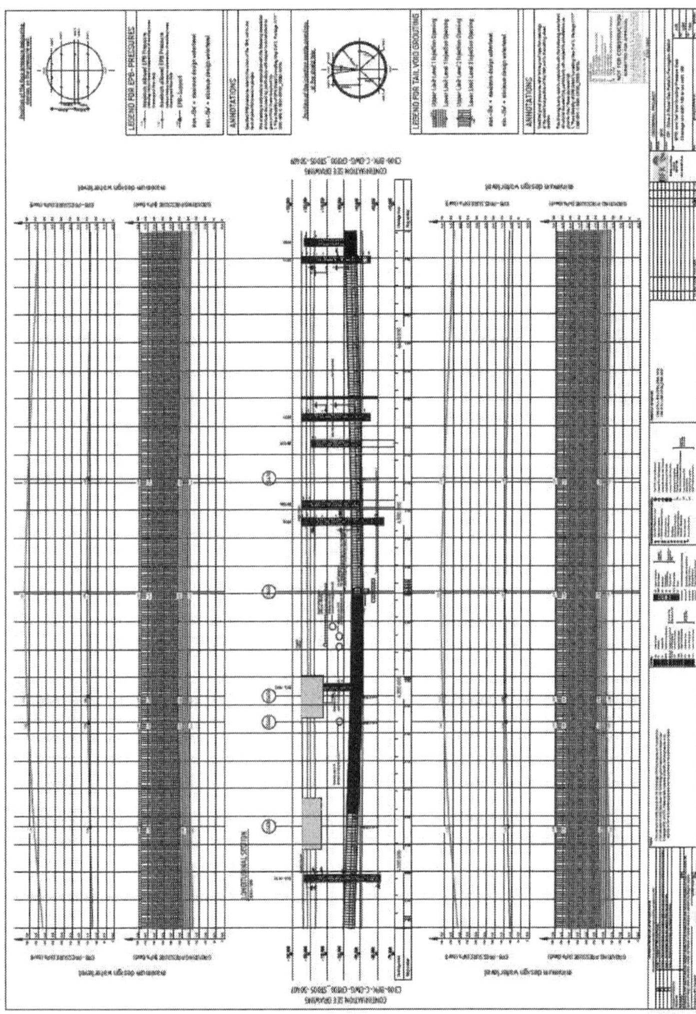

Overview pressures: EPB pressure crown, grout injection - TBM 2

BFK Bam Ferrovial Kier

REV 2: (EPB raised +0,2bar included in table)

Advance No.	Chainage Start cutting edge [m]	Chainage End cutting edge [m]	Pressure [bar] INPUT EPB crown (ring start)	Grout injection (Outlet/Lisene) related to confinement pressure Crown 1&4	Grout injection (Outlet/Lisene) related to confinement pressure Invert 2&3	Other
3570	5937.515	5939.115	1.31	2.97	3.58	
3571	5939.115	5940.715	1.31	2.97	3.58	
3572	5940.715	5942.315	1.31	2.97	3.58	
3573	5942.315	5943.915	1.31	2.97	3.58	
3574	5943.915	5945.515	1.31	2.97	3.58	
3575	5945.515	5947.115	1.32	2.98	3.59	
3576	5947.115	5948.715	1.32	2.98	3.59	
3577	5948.715	5950.315	1.32	2.98	3.59	
3578	5950.315	5951.915	1.32	2.98	3.59	
3579	5951.915	5953.515	1.32	2.98	3.59	
3580	5953.515	5955.115	1.33	2.98	3.59	
3581	5955.115	5956.715	1.33	2.98	3.59	
3582	5956.715	5958.315	1.33	2.98	3.59	
3583	5958.315	5959.915	1.33	2.98	3.59	
3584	5959.915	5961.515	1.33	2.98	3.59	
3585	5961.515	5963.115	1.34	2.99	3.60	
3586	5963.115	5964.715	1.34	2.99	3.60	
3587	5964.715	5966.315	1.34	2.99	3.60	
3588	5966.315	5967.915	1.34	2.99	3.60	
3589	5967.915	5969.515	1.35	2.99	3.60	
3590	5969.515	5971.115	1.35	2.99	3.60	
3591	5971.115	5972.715	1.35	2.99	3.60	
3592	5972.715	5974.315	1.35	2.99	3.60	
3593	5974.315	5975.915	1.35	2.99	3.60	
3594	5975.915	5977.515	1.36	2.99	3.60	
3595	5977.515	5979.115	1.36	2.99	3.60	
3596	5979.115	5980.715	1.36	3.00	3.61	
3597	5980.715	5982.315	1.36	3.00	3.61	
3598	5982.315	5983.915	1.36	3.00	3.61	
3599	5983.915	5985.515	1.37	3.00	3.61	
3600	5985.515	5987.115	1.37	3.00	3.61	

References

B. Maidl, M. Herrenknecht, U. Maidl and G. Wehrmeyer (2012). Mechanised Shield Tunnelling. Ernst & Sohn, Germany.

Crossrail (2013) C701-ITM-C-RGN-C125_WS084-50001 C701 Victoria and Bakerloo Line – Green Park to Oxford Circus Running Tunnels Installation Report, London

Crossrail (2012) C122-OVE-C2-RAN-C125-00011 Assessment of Ground Movements Effects: LU/08 Victoria Line South of Oxford, London

Crossrail (2012) C122-OVE-C2-RGN-CRG02-50002 Assessment of Ground Movement Effects LU09 Bakerloo Line South of Oxford Circus, London

Crossrail (2012) C122-OVE-C2-RSI-CR125-00002 Inspection for Analytical and Schedule of Condition: LU/08 Victoria Line South of Oxford Circus, London

Crossrail (2010) C300-PDP-C2-RGN-CRG03-00001. Geotechnical Baseline Report Contract C300. London.

Crossrail C300 (2012) Operating Manual S705/S706. Herrenknecht, Tunnelling Systems. Germany.

New Civil Engineer Magazine (2013) crossrail/crossrail-station-profile-bond-street/5208747.article (accessed 20/10/2013)

R. N. Craig and A. M. Muir Wood (Sir William Halcrow and Partners) (1978) A review of Tunnel Lining Practise in the United Kingdom. Transport and Road Research Laboratory. Department of the Environment Department of Transport. Crowthorne, UK, supplementary report 335.

Crossrail Project: Infrastructure design and construction
ISBN 978-0-7277-6078-4

doi: 10.1680/cpid.60784.581

The Protection of the 400kV Cables at Pudding Mill Lane

Alex Evans, BEng (Hons), IEng MICE. Atkins UK, Ground Engineering.

Abstract

The ground movement assessment and associated mitigation measures for the National Grid Electric 400kV power cables at Pudding Mill Lane, has been the subject of extensive debate and analysis since the commencement of the Crossrail project.

These power cables are very sensitive to ground movement and hence elaborate schemes were developed for asset protection. A refined ground movement assessment was undertaken to ascertain the risk to the cables. This assessment concludes that ground improvement works alone, would be sufficient to protect the cables. The settlement predictions where undertaken by standard Empirical assessment, modifications to the Empirical method (accounting for ground treatment stiffness) and Finite Element Modelling. The report details the good correlation of maximum settlement between the modified Gaussian method and the Finite Element Modelling, in the order of 25mm.

The observed movement shows that the most critical period for the cables were during the ground treatment phase, where a narrow heaved profile was observed. Analysis was undertaken at this point to ascertain the strain on the cables during depressurisation and tunnelling. This analysis summarised that the tunnel excavation would provide a relaxation in strain that could have built up in the cables during ground treatment.

The Eastbound TBM excavation was undertaken with minimal surface settlement (10mm), equating to a volume loss of 0.5%. The observed movement showed two distinct settlement profiles either side of the TBM, due to the extent of the ground treatment works.

Notations

UCS	is the unconfined compressive strength of the ground
CPT	is a cone penetration test
K_0	is the total stress coefficient of earth pressure
VL	is the volume loss from tunnelling
k	is the settlement trough width parameter
r	is the radius of excavation
a	is the distance from excavation centre to position of circumferential stress
H	is the depth to tunnel axis
ρ	is surface surcharge
MD	is Made Ground
AL	is Alluvium
RTD	is River Terrace Deposits

LC	is London Clay
TBM	is tunnel boring machine
E_u	is the Elastic Modulus of the ground
C_u	is the Compressive Strength of the ground
S_{max}	is the maximum surface settlement
S_1, S_2	is predicted sub-settlement beneath the ground treatment
RF	is the surface settlement reduction factor
FEM	is finite element modelling
TAM	is a tubular machete
σ_θ	is the circumferential stress around an excavation
EB	is the eastbound tunnel
WB	is the westbound tunnel

Introduction

The launch of tunnel boring machines from portals can be problematic, due to shallow cover which is usually accompanied with poor ground conditions. In the case of the Crossrail Drive Z bored tunnels this was further exasperated with the close proximity of six sensitive 400kV cables. These cables are a strategic power supply for Central London providing a key route for power throughout Greater London, protecting these cables was critical to the Crossrail project. The asset owners, National Grid, were consulted about the works that would be undertaken adjacent to their asset throughout the project.

Figure 1 – Location of the National Grid Power Cables.

The power cables are located in a rudimental cable trough positioned in the tow path of the River Lea. It is understood to be continuous and of the same construction along the whole length of the interface at Pudding Mill Lane. The oil cooled cables are highly sensitive to ground movements, any rupture of the outer sheaths would result in the oil draining, if unmanaged, the cables would overheat and subsequently this would be followed by ultimate state failure. The critical factor to protect these cables is at joint bays where an extension in length of the cable could result in the cables pulling apart. This creates a risk of large pull out forces that would hinder any reconnection.

During the Parliamentary Undertakings stage of the project a protection scheme was outlined and subsequently developed during the proceeding detailed design stage. This complex scheme involved physical intervention to the cables, to divert them onto a piled protection bridge. This intervention would expose the cables to a significant level of risk.

During the detail design stage, a further ground movement assessment was undertaken to establish the magnitude of movement to the cables to quantify appropriate mitigation. This paper will discuss the design of ground movement mitigation measures and will review the observed movement against that predicted.

Ground Treatment

The design philosophy of the ground treatment was to reduce the settlement of the cables through creating a block of improved ground. This stiffened ground would form an arch above the bored tunnels across the cone of settlement above the TBM, hence reducing the surface settlement. The analysis was undertaken as an arch, as opposed to a beam, due to the low tensile capacity of ground treatment, hence utilising the arch length when in compression.

The ground treatment specification, for asset protection purposes, required the ground treatment to provide a grouted block of 2m depth with a UCS of 300kPa. This was based on calculations that analysed the arch geometry (height and length) that could be formed within the grouted RTD above the tunnels.

Ground Conditions

The geology of the area to the west of Pudding Mill Lane portal comprises the Lambeth Group overlaid progressively by Harwich Formation, London Clay, River Terrace Deposits, Alluvium and Made Ground. Table 1 details the ground stratigraphy, taken from local borehole information.

Top of Stratum	Approximate Level (mATD)
Made Ground (Ground Level)	104.6
Alluvium	102.2
River Terrace Deposits	100.6
London Clay	98.5
Harwich Formation	94.6
Lambeth Group	92.7

Table 1 – Summary of ground stratigraphy.

The intent of the grouting was to modify the engineering properties of the River Terrace Deposits. In the terms of settlement predictions, the tunnel settlement would have a wider settlement trough. A gravel material would utilise a k value less than 0.5, with the intent of the grouting to increase this to very stiff clay, at least 0.5 or greater.

The arch length was utilised to establish a reduction factor for settlement. The grouted block was checked to ascertain the elastic-plastic deformation that would occur around the tunnel. This was undertaken using three approaches based on the theory of thick walled cylinders.

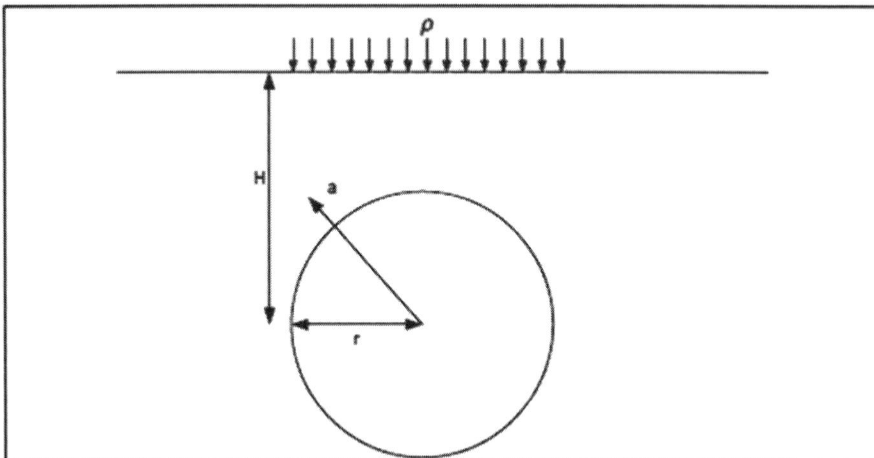

Kirsch	$\sigma_\theta = \frac{pH}{2}\left[(1+K)\left(1+\frac{a^2}{r^2}\right)+(1-K)\left(1+\frac{3a^2}{r^4}\right)\cos 2\theta\right]$
Lame	$\sigma_\theta = KpH\left(1+\frac{r^2}{a^2}\right)$
Roark	$\sigma_\theta = \frac{p3a^3}{2(a^3-r^3)}$

Table 2 – Formulae for stress around circular tunnels.

The Kirsch and Lame formulae provide an upper and lower bound envelope for stress around the tunnel respectively. This was used to ensure that the UCS specified was sufficient to ensure that the ground treatment would not be over stressed and deform into its plastic state. Therefore ensuring the ground treatment would provide suitable support to the increased stress around the tunnels and be able to reduce ground settlement.

Ground Treatment Verification

The design of the ground treatment to achieve the minimum requirements was undertaken by the Contractor and therefore this will only be briefly outlined in this paper.

The methodology treating the area beneath the 400kV cable involved installing horizontal TAMs, where the first two phases used a cementatous mix to provide an increased strength to the ground. This was then followed by two phases using a silicate mix to fill the remaining voids.

Insitu cone penetration tests (CPT) where undertaken in the ground treatment zone to verify the unconfined compressive strength of the ground treatment. The results demonstrate that the 300kPa UCS design requirement was generally achieved across the required ground treatment zone.

Ground Movement Analysis

The assessment was initially undertaken with a TBM volume loss of 0.5%, as the Drive Z tunnels where to be received at the portal. This was subsequently modified to 1% volume loss, as the TBMs would be launched from the portal.

The analysis undertaken assumes that no slip occurs between the ground, cable trough, cable surround and the cables. In reality there would be some slip between these surfaces, however it is difficult to quantify and would require the cables to be exposed. This level of analytical detail would not provide any substantial benefit to the protection of the cables. This paper assumes Greenfield settlement prediction and that the observed movement is also Greenfield, i.e. observed ground movement of the cable trough has been directly transpired to the cable. Reality is that the movement of the cables would be less than observed movement as they are protected from Greenfield movement by the concrete trough.

The empirical Gaussian method for ground movement prediction does not provide a sufficient mechanism to account for the improved characteristics of the ground. The ground movement assessment was undertaken in a phased approach, starting with the Gaussian method, then a modification factor and finally finite element analysis. This approach ensured a level of confidence in the prediction of settlement, due to the unique nature of the improved geology.

Standard Gaussian Method

The empirical surface settlement equation utilises a k factor to crudely represent the geology, influencing the shape of the settlement trough. Whilst this factor provides suitable predictions for Greenfield movement for well tunnelled materials in London, applying a k factor to the modified geology at the commencement of the Drive Z tunnels, would be too conservative and provide an unrealistic prediction.

For London Clay a k of 0.5 is usually the most suitable, for a gravel material like that above the tunnels at the

400kV cable (River Terrace Deposits), lower k values may be more realistic. This would provide a steeper trough and a greater magnitude of settlement. However in the case of the treated River Terrace Deposits they where thought to behave in a stiff manor, therefore the k value could be 0.5 or above. The minimum aim of the grouted River Terrace Deposits was to provide London Clay geology or better.

Modified Gaussian Method

A simplified approach was initially undertaken to quantify the reduction in settlement, from the ground improvement. As discussed previously the basis of the ground treatment design involved establishing the length of arch formed within the grouted block. For the minimum specified ground treatment parameters (2m deep with 300kPa UCS), an arch length of 6m was calculated. This was used to establish a reduction in surface settlement from a theoretical bridging of the subsurface settlement trough, at the underside of the ground treatment.

The arch was assumed to bridge 3m either side of the peak settlement (S_{max}) at the point of origin of the arch (S_1 and S_2) the settlement was obtained and used to establish a ratio against the peak surface settlement. This ratio was then used to reduce the surface settlement, accounting for the benefit from the ground treatment works.

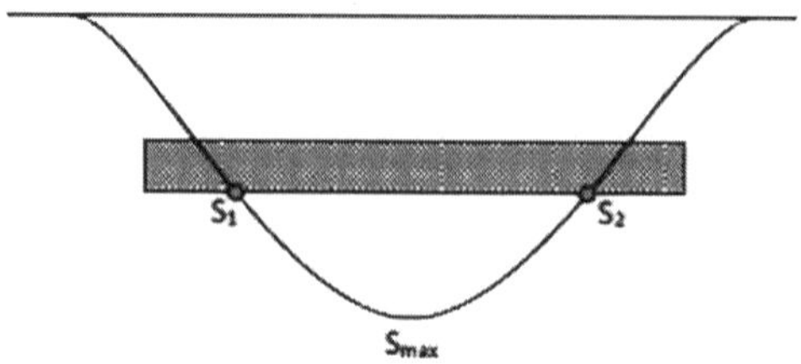

$$RF = 1 - \left[Smax - Median\,(S1, S2) / Smax \right]$$

Figure 2 – Reduction Factor Equation for Ground Treatment Arch.

During the ground treatment works the geological conditions where found to be slightly different. It showed that the RTD varied in depth (2m at portal to 1.5m beneath cables). Therefore the depth of grout block that could be successfully achieved was 1.5m under the cables, providing a ground treatment arch length of 5m. The analysis showed that a 1.5m deep grout block would provide a 16% settlement reduction in maximum settlement to that calculated with the Gaussian method.

Subsequently the design requirements, for asset protection, where changed so that the River Terrace Deposits alone would be treated. This decision was cognisant of the movement generated during the grouting works (Section 4.1).

Finite Element Analysis

A finite element analysis was undertaken to provide the client and asset owner with greater confidence in the predicted settlement performance of the ground treatment. In addition, it was used to model a cofferdam excavation to avoid the corner effect issue in software packages developed for settlement analysis. The modelling was undertaken in a staged approach, the ground treatment influence was only modelled for the TBM excavation stages.

The Made Ground, Alluvium and River Terrace Deposits were modelled using an elastic-plastic Mohr- Coulomb analysis. The London Clay and Lambeth Group where modelled using the BRICK model. The ground treatment was modelled using stiffness (E_u) to compressive strength (C_u) ratio of 1000:1. The FEM analysis, without the ground treatment, was calibrated against the Greenfield Gaussian curve.

The FEM results found that for a 2m deep ground treatment block the maximum final stage settlement was 22mm (with RF method S_{max} 24mm). The reduction factor method and FEM gave comparable results. Further FEM analysis was undertaken for the reduced depth of ground treatment (1.5m). Again there was excellent correlation of peak settlement, with both predicting a final stage S_{max} of around 25mm.

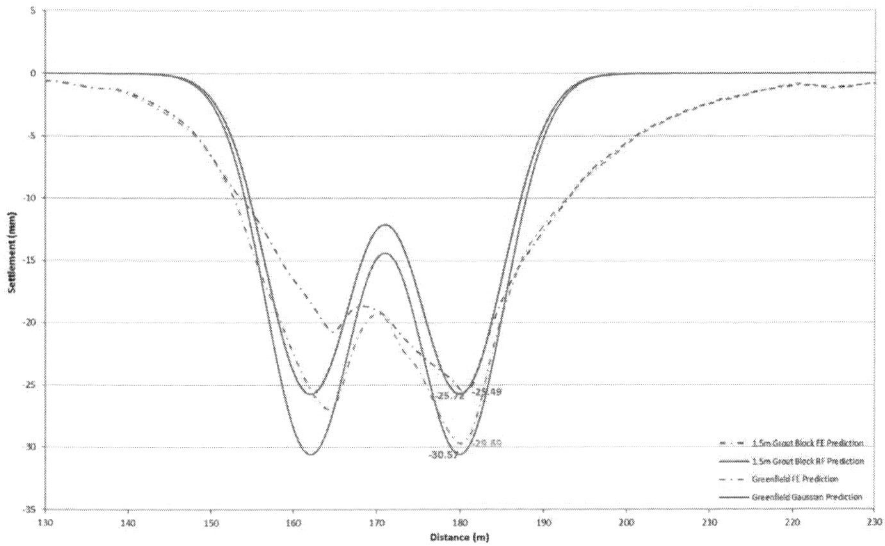

Figure 3 –Settlement Predictions for 1.5m Deep Grout Block.

The FEM analysis, taking account of the ground treatment, predicted a wider settlement trough than the Gaussian settlement. This prediction is inline with the previously discussed hypothesis that the grouting would act in a stiffer clay like manor, with a k value greater than 0.5.

Observed Ground Movement

To monitor the movement subjected to the cable precise ground levelling points where installed on the surface (studs on cable trough) and subsurface (1.5m deep studs) adjacent to the cable trough. The observed movement during the ground treatment phase showed heave, this was mainly attributed to the hydrofracture grouting in the Alluvium which was subsequently halted. The grouting parameters where also modified to control the heave. The most influential change to the works was the implementation of additional monitoring that involved reading levelling points, locally above the grouted TAM. Alert values were established to stop works, causing excessive movement to the cables. This provided real time control of heave during the ground treatment works.

Implications of Heaved Ground

The lifted profile of the ground would have theoretically caused strain in the cables, the profile that was observed (Figure 4) was of similar shape to an inverted Gaussian settlement trough. A back analysis of the data using the Gaussian method was undertaken to establish the theoretical strain in the cables.

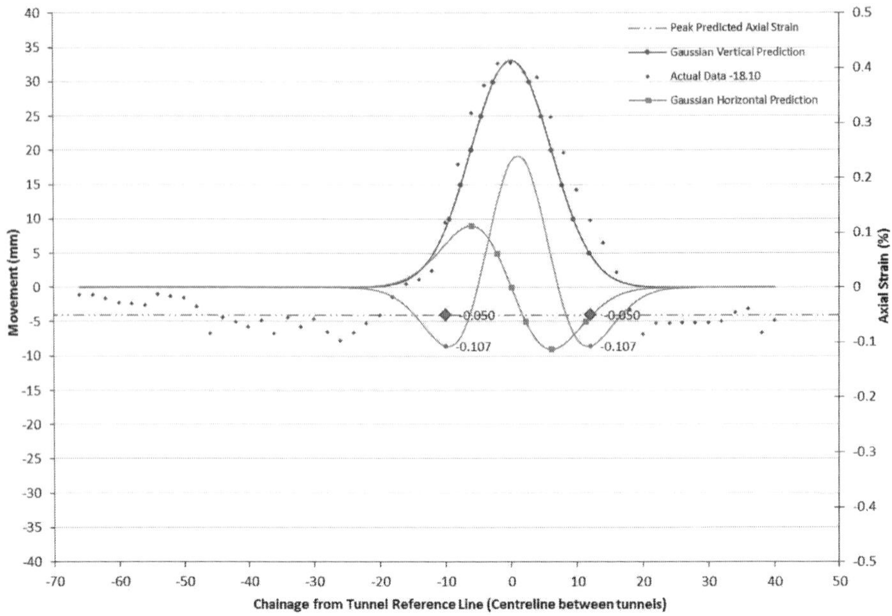

Figure 4 – Calculated Strain From Ground Treatment.

The tension and compression zones remain positioned either side of the point of inflection as described in the Gaussian function; the compression zone is the central area of the settlement trough. The rationale was based on the beam bending theory, using the extreme fibre at ground level replicating the theoretical state of strain (tension or compression) within the cable.

Whilst the magnitude of the movement was within the vertical trigger, the heaved shape was steeper than calculated for the TBM settlement where a k=0.5 and VL=1% where used. The regression analysis for the heave prolife found a k=0.45 and VL=1.28%. With these Gaussian curve parameters the calculated axial strain was greater than predicted. During this period the cables remained operational and critically the oil level remained stable.

Updated Assessment for Tunnelling

As the tunnels are excavated a Gaussian settlement trough would develop. However, the lifted profile caused a unique scenario where the post tunnelling settlement profile would not be in the Gaussian function.

To quantify the strain along the cables imposed during tunnelling, the horizontal movement prediction was superimposed over the theoretical horizontal displacement from observed movement (Figure 5). A direct axial strain calculation was then undertaken.

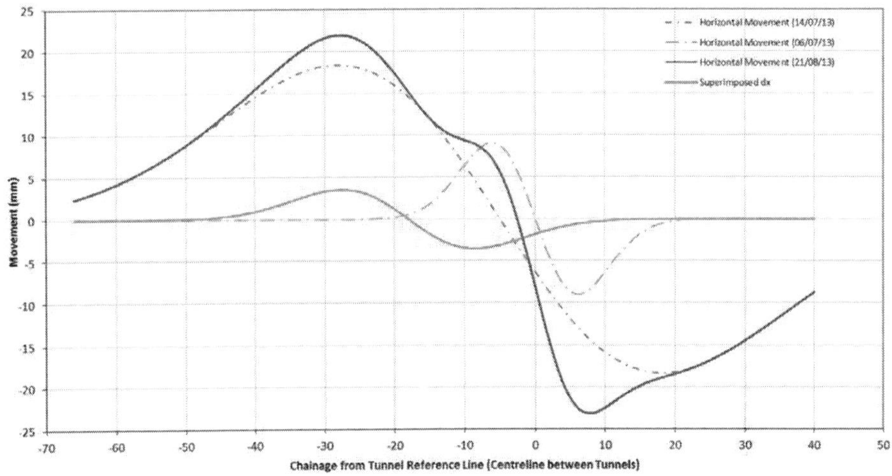

Figure 5 – Superimposition of Horizontal Movement

A horizontal superimposition method was undertaken, as opposed to vertical superimposition and calculating the horizontal movement. This was because the vertical profile created after the TBM excavation was not a Gaussian function and therefore, in this instance, the empirical horizontal formulae was not an appropriate methodology.

To ascertain the total predicted settlement on the asset, the vertical settlement prediction for each TBM, was superimposed against the observed movement. The settlement prediction for the eastbound tunnel was 18mm and for the westbound tunnel 20mm, with the predicted final stage TBM settlement of 25mm.

The associated theoretical strain was calculated for the passage for each TBM, incorporating strain from the observed ground treatment movement. The analysis showed that for the EB tunnel excavation, the theoretical average strain (0.048%) is comparable to the average axial strain originally calculated (0.041%). The calculated peak strain would slightly increase from its current value, to a peak strain of 0.114% during EB excavation. Immediately after the peak heave was observed, some consolidation was observed therefore the predicted increase in strain for the EB excavation, would not actually be evident this is discussed in greater detail in Section 4.3.

For the excavation of the WB tunnel, the TBM influence zone reduces the magnitude of change in settlement along the trough and hence reduces the axial strain (i.e. the TBM settlement trough smooth's out the settlement profile). The final stage average strain prediction is 0.032%, with the peak final stage strain predicted to reduce to 0.067%. The theoretical strain from the observed movement and predictions is summarised in Figure 6.

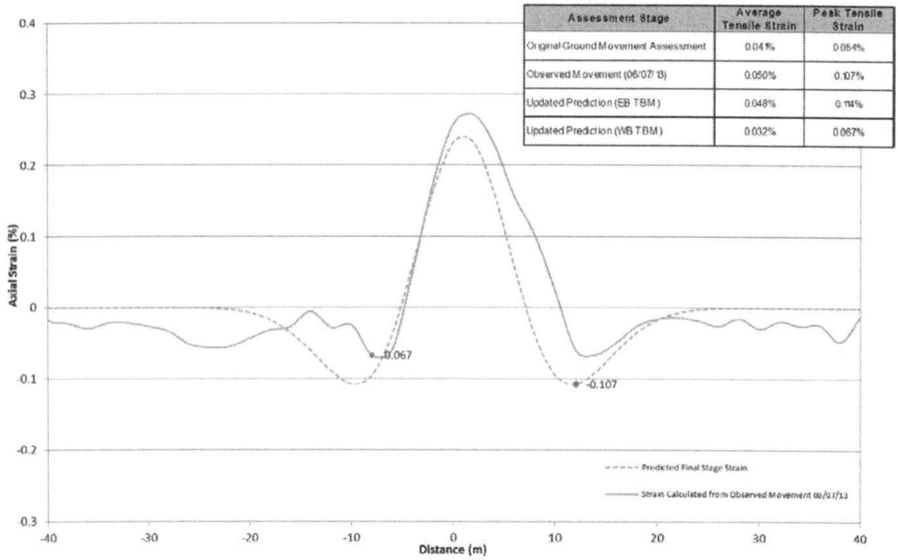

Figure 6 – Observed Strain (06/07/13) and Final Stage Prediction.

The lifted position of the cable provided a unique scenario. The mitigation intended to protect the cables during tunnel excavation actually required the TBM to provide strain mitigation.

Tunnelling Preparation

Following the observed peak heave, gradual consolidation settlement was noted. This is believed to be the increased pore water pressures, created during the ground treatment, naturally draining. This consolidation settlement was exasperated when the depressurisation works commenced. The depressurisation movement can be seen in Figure 7, from July 2013 to August 2013.

Although the depressurisation scheme was for the lower aquifer, it is believe that there was connectivity with the upper aquifer. This hypothesis has been drawn from the observed water level before depressurisation and the manner in which the cables settled post ground treatment. There are four clear trends in the movement of the cable trough;

1. Initially a gradual settlement (3mm / day) from the natural consolidation of the ground;
2. Followed by an increased settlement response (2.3mm / day) when the depressurisation commences,
3. Thirdly a slower settlement (0.6mm / day) once the depressurisation is in operation, and
4. Finally observed swelling (0.7mm / day) of the ground when the depressurisation is temporarily stopped.

Figure 7 – Vertical Movement of Cable Trough, range of observed data.

Tunnelling Settlement

At the time of writing this paper only the Eastbound TBM had passed beneath the 400kV power cables. The second machine will be launched at a later date.

A depressurisation scheme for the Harwich Formation was implemented to improve ground stability. The intent was to increase the proficiency of the TBM to balance face pressure, with the aim of improving surface settlement control. The FEM analysis was undertaken for undrained conditions, as the ground was now in a drained condition, the FEM assessment would theoretically predict less than 18mm for tunnelling. However, this would be offset by the additional settlement generated during depressurisation works. This sensitivity analysis was not undertaken, as it was deemed to be very convoluted and would have no material benefit to the ground movement assessment and impact to the 400kV cables.

The horizontal monitoring recorded negligible movement (less than 1mm extension) during depressurisation and tunnelling. When compared to the vertical movement during the same period (14mm), the horizontal stiffening influence of the cable trough can be clearly identified. The Gaussian horizontal movement for this magnitude of vertical movement predicts a horizontal extension in the order of 4mm. It was known that the cable trough would provide a stiffening effect, although it was thought that a greater change in horizontal movement would be observed.

This hypothesis also provides a view that the theoretical strain in the cable trough, during the ground treatment heave, may not have developed. The explanation for this is that slip has occurred between the cable trough and the surrounding ground. As the cable trough is very long, acting as a very deep beam in the horizontal plane, the stiffening effect is much larger than in the vertical plane. Therefore the cable trough has a greater tendency to follow the deformed ground profile in the vertical direction.

Observed Movement of Cable Trough

During the passage of the Eastbound TBM the settlement observed ahead of the TBM face was 10mm, slightly more than half the predicted value (18mm). This was reflected in the TBM excavated material volume balance, where there was very good correlation of observed excavated material volume to theoretical excavated volume.

There was minimal movement observed from convergence of the ground onto the TBM shield. This was attributed to the depressurised Harwich Formation below axis, the presence of London Clay above axis and bentonite support around the TBM shield.

Figure 8 – Regression Analysis for Observed Tunnel Movement.

During the build stage of the first ring beneath the cable (Ring #11) slight heave was observed (2mm). This was attributed to the tail skin grouting around the ring. In this location the cover is reduced, so it is believed that this caused heave of the cable trough. This hypothesis is enforced by the observations of heave during the grouting works, and more saliently, the heave during depressurisation recharge.

Following the passage of the TBM it was observed that for the second and third permanent rings built under the cable (Ring #12 and Ring #13) no heave was recorded. These rings were passed the River Lea retaining wall and therefore the influence of the grouting pressure is not evident at the surface, due to the sheet pile retaining wall providing protection.

A regression analysis for the movement of the EB tunnel shows that a volume loss of 0.5% was achieved with a k factor of 0.7. This is in agreement with the theory discussed earlier in the paper, where the intent of the ground treatment was to stiffen the ground and hence increase the k value.

The observed movement shows two distinct trends (Figure 9), a steeper trough for the monitoring points to the north of the tunnel centreline and a more graduated settlement profile to the south. This provides an indication of the influence of the grouting works. The ground treatment has a shorter extend to the north of the EB tunnel than to the south. If separate regression analysis is undertaken either side of the tunnel centreline, the k value and volume loss are significantly different.

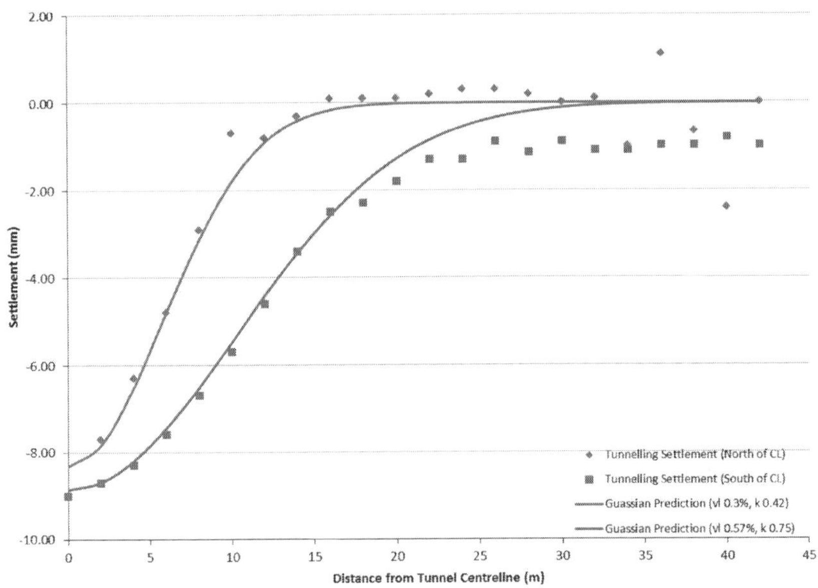

Figure 9 – Separate Regression Analysis either side of Tunnel.

The observed movement to the north of the tunnel (blue) shows a gravel ground movement response, hence a lower k (0.42). The results to the south of the tunnel (red) show a very stiff clay response with a higher k (0.75). The observations are in agreement with the hypothesis of how the ground would move and the intent of the ground treatment for asset protection.

To the south of the tunnel the ground treatment has a much larger extent and the grouting TAMs are at closer spacing. Therefore the ground treatment, along with the reinforcement influence of the steel TAMs, has resulted in a more global ground response. This is inline with the wider settlement trough predicted in the FEM analysis (Figure 3).

When comparing the observed movement to the FEM prediction, the prediction of the FEM was reduced to correlate with the observed data, so that a TBM face loss accounting for ground treatment could be ascertained. Based on this a TBM face loss of 0.60% was achieved (discounting heave during ring build), therefore a 20% settlement reduction could be ascertained from the benefit of the ground treatment and stiffening effect of the cable trough.

Observed Sub-Surface Movement

To further verify the benefit of the ground treatment, discounting the stiffening effect of the trough, the observed movement of the cable trough can be compared to the observed data of the adjacent 1.5m deep points. This monitoring array also recorded negligible movement from convergence onto the TBM shield. The settlement ahead of the TBM was recorded at around 15mm, using the Gaussian parameters used for the observed data of the cable trough (*k=0.7 VL=0.5*). The theoretical movement at 1.5m below ground is 12mm, this is less that the observed settlement (15mm).

To match the sub-surface settlement against the observed, a regression analysis used a volume loss of 0.60%, the k parameter remained unchanged. The difference between the observed movement of the 1.5m deep monitoring and the equivalent cable trough movement

(3mm) can be attributed to the stiffening influence of the cable trough.

TBM Performance

The ground settlement performance of the TBM must be cognisant of the settlement caused by depressurisation. The lower than predicted settlement must in part be attributed to the improved ground stability created by the depressurisation. It is not possible to accurately quantify how much benefit this provided. A comparison can be drawn from the observed movements on the Drive Y tunnels.

In similar geology on Drive Y, where neither depressurisation nor ground treatment was undertaken, the TBM performance was around 1% volume loss. In this location the TBM volume balance was similar to that for the Drive Z TBM. However the observed data on Drive Y shows a 20-80% distribution between settlement ahead of the TBM to that across the shield and ring annulus. Therefore it can be concluded the depressurisation provide significant benefit to prevent convergence of the ground onto the TBM shield and built ring.

Conclusion

The analysis and observations made in this paper provide an insight into the art of ground treatment. There are many variables and unknowns when undertaking ground improvement. The analysis works shows that ground settlement analysis using 3D FEM provides value in assessing the influence to third party assets. It can be further expanded to incorporate and quantify the reduction of ground movement that can be achieved from improvement techniques. To fully validate the reduction factor method further analysis work is required, with different types of ground treatment schemes.

Throughout the works to date the 400kV cables remained stable and where not adversely affected by the works. The most critical part is joint bays which were well outside the zone of influence of the Crossrail works. Should the joint bay within the influence zone of the Crossrail works, then it is very likely that a different protection scheme would be undertaken. Any future works around cable joint bays must be fully aware of their critical nature, which is vastly different to the nature of the cable discussed in this paper.

The ground surface movement during tunnelling are low for a TBM being launched at shallow cover. However the movement during ground treatment works was much larger than anticipated. The most efficient way to control the movement is to provide a thorough real time monitoring regime. Whilst the pressure and volumes for the ground treatment can be limited to reduce movements, monitoring was found to be the most successful method to reduce impact to the cables. As the ground treatment changes the characteristics of ground, particularly reducing the void ration of the RTD, the theoretical grouting pressure and volume take also reduce. At this point the rheological properties of the grout mix must be changed along with the grouting pressure and volume. If possible the movement should be calibrated against the grouting parameters and modified to limit heave.

Additional work should focus on the thick wall cylinder methodology to provide a prediction of radial displacements around the tunnel and surface settlement. These methodologies take account of the Young's Modulus and Poissions ratio of the ground treatment. The analysis should calibrate the maximum settlement for the Gaussian function against the radial displacement at surface. This can then be used to determine the reduce settlement, once the ground has been treated by modifying the Young's Modulus and Poissions ratio.

References

Mair, R J (1993) Subsurface settlement profiles above tunnels. *Geotechnique 43, No. 2*, 315 - 320.

Burland, J B (1997) Assessment of risk of damage to buildings due to tunnelling and excavation, *Earthquake Geotechnical Engineering 1997 Rotterdam* 1189-1201.

Brady, B H G & Brown, E T (2005) *Rock Mechanics For Underground Mining*, Springer Science, USA. Terzaghi, K (1943) *Theoretical Soil Mechanics* John Wiley and Sons, inc. USA.

Young W C, Budynas R G (2002) *Roark's Formulas for Stress and Strain McGraw-Hill*, USA.

Crossrail Project: Infrastructure design and construction
ISBN 978-0-7277-6078-4

doi: 10.1680/cpid.60784.597

Crossrail Sprayed Concrete Lining Depressurisation at Stepney Green Caverns

David Harris, BSc, MSc, DIC, CGeol, FGS Mott MacDonald Group Limited, Croydon, UK
Andrew Davis, BSc, MSc, CGeol, FGS Mott MacDonald Group Limited, Croydon, UK
Emilio Linde, BSc, MSc, DIC, EurGeol, FGS OTB Engineering Limited, London, UK

Abstract
Crossrail includes approximately 10km of Sprayed Concrete Lining (SCL) tunnels. One of the largest of these tunnels is being constructed at Stepney Green, East London to facilitate the construction of a sub-surface junction, allowing Crossrail trains from Maidenhead and the west to travel to Shenfield and Woolwich in the east. These caverns are the largest span underground excavations constructed below London to-date.

The east- and west-bound caverns lie approximately 20 m below existing ground level in a constrained urban site and are located within the London Clay and underlying Lambeth Group. The Lambeth Group is a variable series of clays, silts and sands and the higher permeability sands contain high pore pressures which are hazardous to SCL construction without depressurisation. The successful implementation of a practical depressurisation scheme was therefore crucial to safe construction and compliance with the zero harm Crossrail ethos.

This paper outlines the overall strategy adopted for investigating these water-bearing sands within the Lambeth Group and then focuses on the ground investigations and pumping tests which were required to determine typical material permeabilities and the response of the sands to pumping.

The paper reviews how that information was developed into the outline depressurisation scheme and how that was subsequently developed into the final contractor-designed depressurisation scheme that was implemented. Changes to the scheme during construction due to ground conditions are discussed together with the response of the ground during pumping. The safety critical nature of the scheme required close monitoring and implementation of a system of triggers and actions.

Introduction

Crossrail is a new railway currently under construction in London and when completed will significantly reduce journey times across the capital. It includes approximately 10 km of Sprayed Concrete Lining (SCL) tunnels at station, ventilation shaft and junction locations. One of the largest of these tunnels is being constructed at Stepney Green, East London, to facilitate construction of a sub-surface junction. This will allow Crossrail trains from Maidenhead and the west to travel to Shenfield and Woolwich in the east. The excavation and installation of primary sprayed concrete lining at Stepney Green was completed in August 2013.

The east- and west-bound cavern crowns lie 17 m and 21 m respectively below ground level in a constrained urban site and are situated within the London Clay and underlying Lambeth Group formations. The Lambeth Group is a variable series of clays, silts and sands and the higher permeability sands contain high pore pressures which are hazardous to SCL construction (CIRIA, 2004). Depressurisation of these high pore pressures was crucial to the safe construction of the caverns and compliance with the zero harm Crossrail ethos.

On the Crossrail project a distinction was made between use of the terms dewatering and depressurisation in order to facilitate discussions with third parties regarding the likely impact of any measures affecting groundwater. Dewatering was considered as applying to any pumping undertaken within the Lower Aquifer, with wider environmental implications, and depressurisation for pumping undertaken in the overlying Lambeth Group. For the purposes of consistency the term depressurisation has been used in this paper.

The SCL permanent works were designed by Mott MacDonald (Contract C121) with the construction undertaken by the contractor, Dragados Sisk Joint Venture (DSJV) under Contract C305. Temporary works design was provided to DSJV by OTB Engineering with the well systems being provided by WJ Groundwater.

Geology

The geology at Stepney Green is typical for that of east central London; comprising of a sequence of strata from Thanet Sand, the complex Lambeth Group strata, Harwich Formation and the lower A2 & A3 units of the London Clay. A summary is provided in Table 1 and a longitudinal section as Figure1.

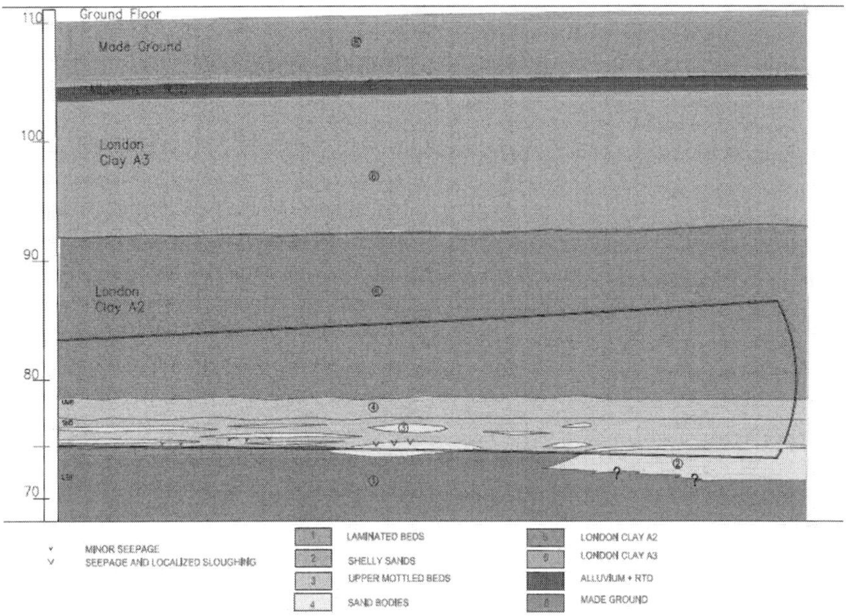

Figure 1 Schematic geological section along the westbound cavern

The eastbound SCL cavern is located predominantly within the London Clay A2 unit with Upper Mottled Beds at invert level. The westbound cavern is approximately 4 m deeper, causing the excavation to be within London Clay, Harwich Formation, with up to 4 m of Lambeth Group below axis level.

Strata		**Typical Thickness (m)**	**Typical Description**
Made Ground (MG)		2	Variable
River Terrace Deposits (RTD)		3	Medium dense orange brown fine to coarse SAND/GRAVEL and rare cobbles of flint.
London Clay (LC)		23	Firm to very stiff fissured CLAY with silt partings and laminations. Increase in silt partings and becoming slightly sandy in the A2 unit.
Harwich Formation	Swanscombe Member	1	Stiff to very stiff fissured sandy CLAY.
	Oldhaven Member		Medium dense slightly clayey fine to medium SAND.
Upper Lambeth Group (LG)	Upper Mottled Beds (UMB)	5	Very stiff to hard (friable microfissured) very closely to extremely closely fissured light grey to blue grey mottled light brown and red sandy to very sandy silty CLAY.
	Sand Channels (SC)	Variable up to 4m	Medium dense to very dense greenish grey fine to coarse SAND
	Laminated Beds (LTB)	3	Stiff to very stiff thinly laminated fissured grey to dark grey, sandy to slightly sandy, silty to very silty CLAY, with extremely to very closely spaced thin laminations of light

Strata		Typical Thickness (m)	Typical Description
			grey slightly sandy silt.
	Lower Shelly Clay (LSC)	1	Stiff to very stiff fissured thinly laminated to thinly bedded dark grey and grey, silty to locally very silty CLAY with occasional bands of light grey sandy silt. Frequent to shells and shell fragments.
	Lower Mottled Beds (LMB)	Variable up to 8m	Very Stiff to hard multi-coloured extremely closely fissured silty to very silty, slightly to very sandy CLAY.
	Upnor Formation (UF)	Variable up to 7m	Very stiff dark grey-green, brown very sandy silty CLAY, with occasional fine to coarse rounded flint gravel. Very dense black rounded medium and coarse flint GRAVEL Very dense brown or green to dark green speckled clayey to very clayey silty fine to coarse SAND, with a little to much fine and medium rounded flint gravel.
Thanet Sand (TS)		11	Very dense grey green silty fine and medium SAND.

Table 1: Summary strata at Stepney Green

Within the Lambeth Group there occur relict sand channel deposits related to the paleo-environment present during the geological period of deposition, representing a series of meandering channels in an estuarine type environment. These channels vary in frequency, width, depth, length and infill material and, as a result, are spatially unpredictable and therefore problematic to identify by Ground Investigation (GI) with any degree of certainty. There are examples of large extensive channels proven by Crossrail ground investigations, such as below Bond Street station where identification and extrapolation is more certain. The size of channels identified at Stepney Green however tend to be thinner, typically to a maximum thickness of 4 m, less extensive and encountered variably throughout the Lambeth Group. It was known from the GI that to the west of Stepney Green there was a thick persistent sand unit but towards the east, where the major caverns were to be excavated, the sand units became thinner and laterally impersistent.

Groundwater

The groundwater profile at Stepney Green (figure 2) is typical for the London area comprising of an upper aquifer located in the superficial deposits and a lower aquifer within the Chalk, Thanet Sand and Upnor Formation. The historical effects of groundwater abstraction from the Lower Aquifer has resulted in an underdrained profile with pore pressures within the Thanet Sand and Upnor Formation being much lower than those above in the London Clay and Upper Lambeth Group. The pore pressures within the SCL excavation depth range were variable and related to the underdrainage effects with maximum values of approximately 180 kPa at the base of the London Clay dropping to 120 kPa at the top of the Lambeth Group. The expected groundwater pressures to be encountered in any sand channels in the vicinity of the SCL works was therefore considered to be in the range 120 kPa to 150k Pa.This detailed profile was able to be determined due to the inclusion of several high quality multi-point monitoring systems included within the GI.

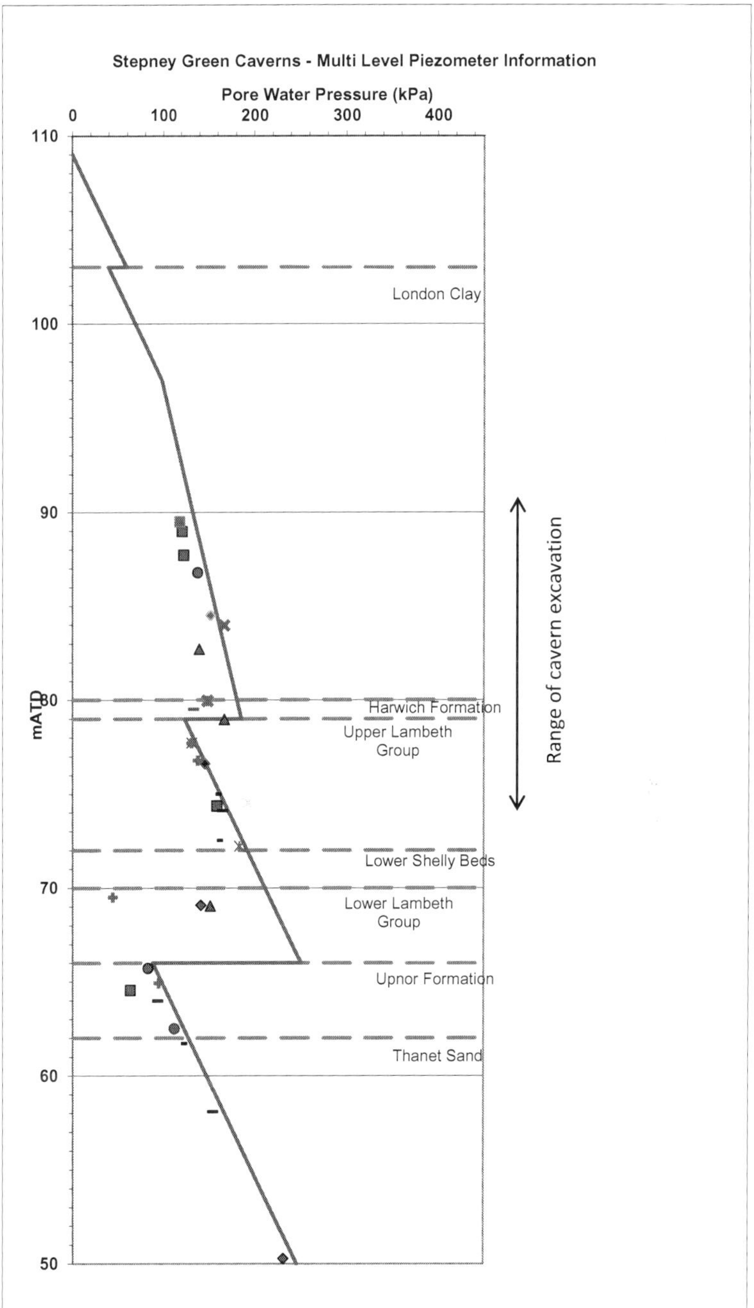

Figure 2 Muliti-point piezometer readings and adopted pore pressure profile.

Ground Investigation

Due to the relatively long planning stages of Crossrail, four phases of ground investigation were undertaken at Stepney Green over an eight year period to obtain geotechnical and groundwater information. The ground investigations included cable percussion and rotary boreholes, two in-situ pressuremeter testing boreholes and concluded in the final phase with a pumping test within the Lambeth Group. A total of 32 boreholes were drilled within 100 m of the proposed cavern locations.

Groundwater monitoring installations are summarised in Table 2 and consisted of conventional 25mm Casagrande piezometers, 50mm standpipes, vibrating wire piezometers, Continuous Multichannel Tubing (CMT) and a Multi-Port Sampling System (MPSS). Construction of groundwater monitoring installations was carefully supervised with none of the piezometers exhibiting erroneous hydraulic connectivity between aquifers. The multi-port systems proved particularly useful, as well as economic despite initially longer installation times.

Type	Total
19mm piezometer	22
50mm standpipe	5
150mm standpipe (well)	3
Vibrating Wire	14
Continuous Multichannel Tubing system	12
Multi Port Sampling System	6
Total	62

Table 2 Summary of groundwater installation details at Stepney Green

Pumping test

A pumping test was specified to determine the permeability of sand channels within the Lambeth Group and to provide data to allow design of the depressurisation requirements. The pumping test comprised three pumping wells with 7 days pumping and 7 days recovery monitoring. This was considered to more closely reflect the likely actual multiple pumping set up to be used during construction, rather than using a traditional single pumped well. The three pumping wells comprised 150mm slotted pipe constructed in 300 mm diameter boreholes. Each well had a screen length of between 8 m and 8.5 m length through the Upper Mottled Beds and Laminated Beds including sand channels. The wells terminated in the Lower Shelly Beds which were below the invert levels of the two caverns. The Harwich Formation was sealed off from the well screens to avoid potential misleading results.

Monitoring of piezometers, standpipes, CMT and VW piezometers was undertaken prior to and during the test at intervals in accordance with BS EN ISO 14686:2003. These were supplemented by manual monitoring of key installations, to ensure that in the event of a datalogger breakdown, manual data could be used to supplement the results.

A series of three step tests were undertaken to determine the maximum yield of each well. A constant discharge test was then undertaken using three submersible pumps. The headworks of the three wells were sealed and a vacuum pump used to improve yield. The test commenced with one well pumped at approximately 1litre/second with a vacuum applied. After one day a second well also commenced pumping under vacuum, increasing the combined flow rate to 1.3 l/s. Finally, the third well commenced pumping under vacuum, with a slight reduction in flow rate. Pumping stopped after 7 days at a final combined flow rate of 1.17 l/s. The maximum yields obtained during the test were from the three separate wells were:

First well	1.06 l/s with partial vacuum of -7.3 m H_2O,
Second well	0.58 l/s with partial vacuum of -8.1 m H_2O and,
Third well	0.45 l/s with partial vacuum of -6.0 m H_2O.

The maximum drawdown within the Upper Lambeth Group achieved at the end of the 7 day pumping period was typically 8m at a distance of 10m, and 3m at a larger distance of 85 m. There was insignificant drawdown in the underlying Lower Aquifer.

The wells took approximately 10 days to recover to pre-test levels, with the majority recovering within the first 7 days. The test showed that the sand channels generally had connectivity within the Lambeth Group, which provided confidence in the ability to successfully depressurise the Lambeth Group from surface wells. As expected there was insignificant drawdown in the overlying River Terrace Deposits or London Clay. Although the pumping test was not anticipated to produce surface settlement due to the limited duration and relatively small volumes of discharge, precise levelling was undertaken to demonstrate this, which could subsequently be used to assure adjacent stakeholders.

The pumping test results were analysed using a range of methods. The main method of analysis was the Cooper-Jacob Well Field Method (Kruseman and de Ridder, 1994) due to the pumping rate increasing in steps as the second and third wells were added. The conventional Theis curve fitting method was also used and the Theis' recovery and the steady state Theim's method were used (Kruseman and de Ridder, 1994). The results of the analysis indicated a transmissivity between 5 to 12 m^2/d (permeability of $6x10^{-5}$ to $1x10^{-4}$ m/s) and a storativity of $7x10^{-4}$ to $6x10^{-3}$.

SCL Ground Related Risks

SCL tunnel excavation is undertaken by spraying concrete directly onto exposed excavated ground, forming an immediate bond with the strata. The lining can follow complex profiles and layouts, in-fill over-break, and forms a strong, durable and watertight shell. As such it is an ideal method for the construction of complex geometrical layouts such as the Stepney Green caverns. The method does however require that the excavated ground remains stable in order that the sprayed concrete can be applied. Although the geology and groundwater conditions at Stepney Green are not unusual for east London, SCL introduces greater ground risk than conventional bored tunnelling. The need to control groundwater ingress and pressures where sandy materials occur in an open face or at invert is very important if stability, and safe conditions, are to be maintained prior to application of the sprayed concrete.

If the groundwater pressures at the face of the excavation are not reduced adequately they will cause instability of sand layers, resulting in the sand running into the excavation. If sand comprises a significant proportion of the face then direct face instability can occur. In situations where minor sand/silt strata occur (such as laminae) the wash in of soil may cause local undermining of the face and subsequent indirect instability. Any such loss of material from the face could trigger a fall of ground from the face, overbreak and cause additional settlement. Where the invert is within sandy material, the presence of elevated pore pressures can lead to the development of running sand conditions. In addition to stability concerns, groundwater ingress can result in poor working conditions which are unacceptable on health and safety grounds.

The extent of the risk and the form of the mitigation depend on a number of variables:

- The size of sand units that control the quantity of potentially stored groundwater;
- Their permeability which controls the rate of water inflow;
- Their ability to recharge which controls the time for which inflow may occur,
- The magnitude of the pore water pressures present which affects the amount of inflow
- The location of the sand channel in relation to the tunnel excavation profile

The risk of instability at Stepney Green was particularly hazardous due to the unusual size of the excavations required to construct the 17m diameter caverns.

Depressurisation Strategy and Design

Preliminary Design

Depressurisation works were considered as temporary works and the responsibility for the design and implementation lay with the contractor. However in order that the tender process could be assessed on an even basis, the designer derived an indicative preliminary design for pricing purposes. The successful contractor was then responsible for taking forward the design to suit their own assessment and programme of work to satisfy a performance specification.

The key aims of the preliminary depressurisation strategy were to address both the unpredictable nature and location of the water bearing strata, and the high pore pressures that would influence the practicality of maintaining safe excavations. To adequately reduce the level of risk there was a need to undertake SCL excavation with a degree of confidence that unexpected water bearing sandy materials were unlikely to be encountered. In order that the risk could be maintained As Low As Reasonably Practicable (ALARP) there was a requirement that construction investigation works were comprehensive.

To address that risk an extensive investigation and depressurisation scheme was developed combining both surface and in-tunnel works. The Stepney Green worksite is small and the adjacent third party land (including an urban farm, protected archaeological area and an all-weather sports facility) was not considered appropriate for undertaking construction based activities. The potential for surface depressurisation was optimised where possible by the inclusion of inclined wells drilled from within the worksite but terminating at tunnel level beyond the site. The potential to maximise use of surface well sites was therefore constrained and it was therefore considered that it was not possible to confidently identify and depressurise all water bearing materials by surface works alone.

Based on previous experience a maximum pore pressure of 100 kPa was adopted in the indicative designs as the limit at which sub-surface in-tunnel wells could practically be installed. This value was related to the difficulty of controlling groundwater ingress and bore stability during drilling, and the ability to subsequently install a well within the bore. Given the expected pore pressure values of up to 150 kPa it was considered that pore water pressures would need to be reduced, as a minimum, to less than 100 kPa by surface wells in advance of SCL tunnel excavation.

As it was considered that surface wells alone could not produce a dry excavation, it was envisaged that additional wells would be installed from within the shaft and/or tunnels. As the sand channels were not accurately known it was considered that a robust array of investigatory probing would be taken ahead of the advancing SCL face to identify any groundwater bearing strata. Wells would then target those locations where groundwater was identified. The number of wellpoints and wells required, and the time required to obtain the necessary depressurisation, would depend on the extent of the water bearing materials, their permeability, pore water pressure and amount of recharge potential. It was expected that different drilling patterns and numbers of probe holes would relate to the different excavation sequences.

The indicative tender design comprised a total of 25 surface wells and 140 in-tunnel gravity and vacuum wells. The surface well spacing was based on a typical 8 m drawdown cone being achieved around each well such that individual well spacing was 16 m. Surface wells were anticipated to comprise pumped and/or ejector wells, depending on the permeability of the strata to be depressurised. The combined depressurisation was considered to occur as a temporary "moving front" of depressurisation ahead of the excavation face, with the wells being redundant once the primary lining has been completed and gained adequate strength.

Contractor's Design

The contractor reviewed the preliminary design and revised the strategy to reduce the potential need for in-tunnel depressurisation measures by maximising the surface well layout. The intention was to reduce water levels to as low as possible from surface such that smaller quantities of in-tunnel measures were required. The increase in surface well coverage was possible due to the greater control that the Contractor had upon the location and timing of site access constraints. In-tunnel depressurisation mitigation measures were retained as part of the overall design to supplement the surface depressurisation measures. This revised approach was beneficial as it would reduce programme critical and expensive delays to tunnel excavation works by reducing the number of in-tunnel wells required.

An ejector spacing of approximately 8 m was implemented based on the pumping test data, experience of working in these conditions at other sites, the constraints imposed by surface access and the application of engineering judgement. This spacing was approximately half that provided in the preliminary design. Ejectors were located in two separate rings to surround the caverns where possible. The location of the ejectors depended upon the layout of the site and other existing surface constraints. Inclined wells (up to 30 degrees from vertical) were retained in the layout to achieve the desired coverage outside the boundaries of the site. Figure 3 illustrates the design layout of the scheme.

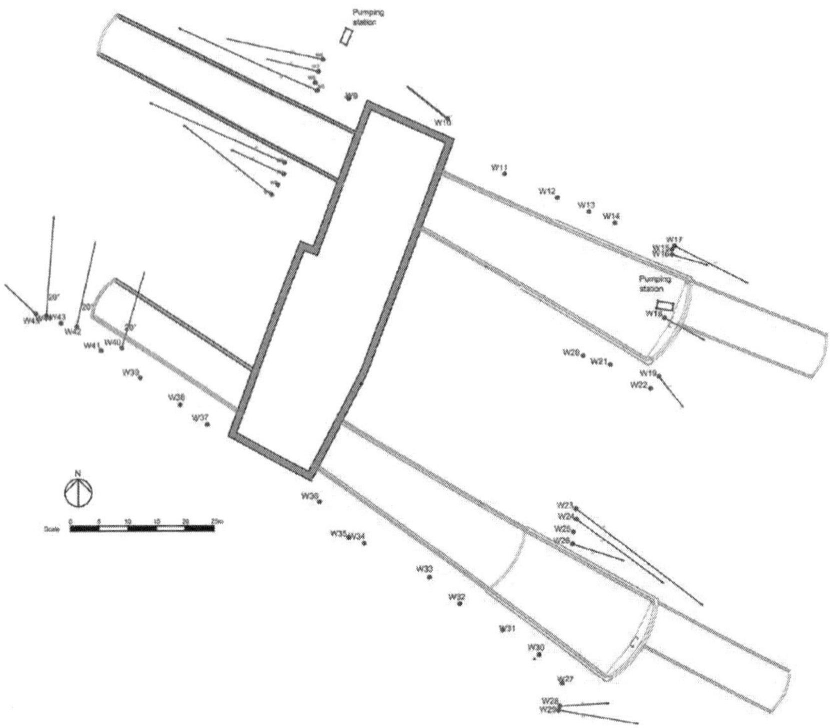

Figure 3 Surface well layout plan

The main features of the implemented depressurisation scheme were:

- No. of ejectors: 45 No.
- Depth: 40 m (110 to 70 mTD)
- Vertical installations: 27 No. Inclined at 10 deg: 7 No. Inclined at 20 deg: 8 No. Inclined at 30 deg: 3 No.
- Bore size: 200/250 mm nominal.
- Response zone: 70 to 83 mTD.
- Pumping Ejectors operated from a surface pumping station

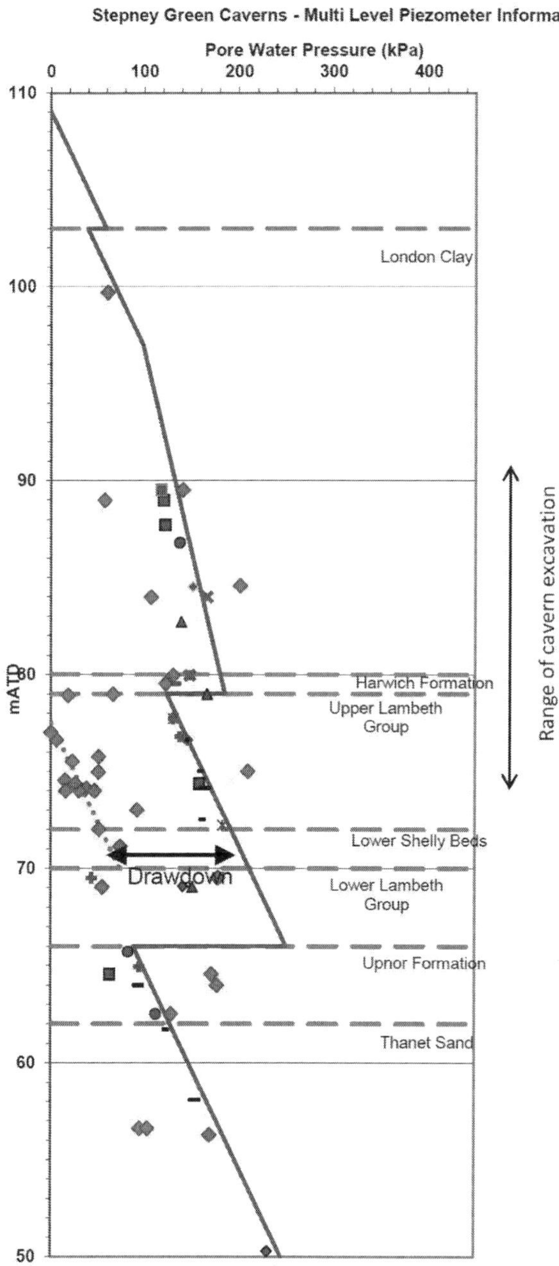

Figure 4 Pore pressure reduction achieved by surface wells

Following commissioning of the surface ejectors a drawdown equivalent to a pore pressure reduction of 140 kPa was observed in the piezometers as shown in figure 4. The monitoring indicated that:

- Most of the piezometers in the Upper Lambeth Group achieved an average drawdown of approximately 14 m (i.e. a pore pressure reduction of 140 kPa).
- Piezometers in the Upper Lambeth Group located close to the base of the London Clay were not significantly affected by the pumping. The likely cause of that was that those piezometers were installed in the Upper Mottled Beds, a cohesive material and therefore of low permeability.
- There was a certain degree of scatter in the drawdown data meaning that the effectiveness of the pumping was variable confirming the variability of the sandy layers within the Lambeth Group.

The average drawdown achieved was considered sufficient for the Eastbound Cavern to be constructed without subsequent in-tunnel depressurisation works, as that was to be constructed in London Clay. However, the drawdown was not sufficient for the deeper Westbound Cavern for which groundwater levels remained typically 1m above invert level. In order to achieve the desired drawdown in the Westbound cavern additional drawdown from within the tunnel was needed.

Initial trials were undertaken by the contractor using wells drilled from the shaft prior to SCL excavation and well layout proposals were derived as summarised in figure 5. Different well layout proposals were derived to suit the different excavation sequences required. Due to the Lambeth Group being only encountered below tunnel axis level, no measures were required for the excavation of the initial top heading, additional measures only being required for subsequent bench and invert excavations.

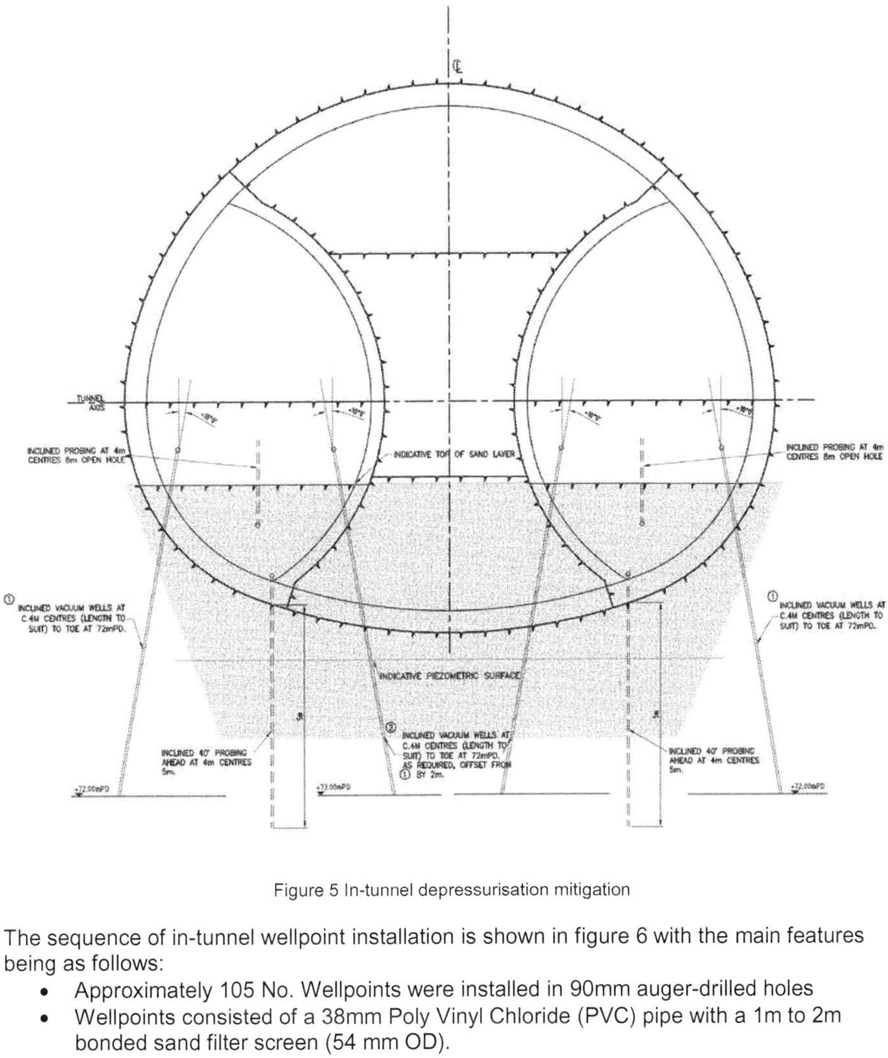

Figure 5 In-tunnel depressurisation mitigation

The sequence of in-tunnel wellpoint installation is shown in figure 6 with the main features being as follows:

- Approximately 105 No. Wellpoints were installed in 90mm auger-drilled holes
- Wellpoints consisted of a 38mm Poly Vinyl Chloride (PVC) pipe with a 1m to 2m bonded sand filter screen (54 mm OD).
- Wellpoints were of variable lengths depending on geology/elevation and angle.
- 25 mm PVC pipework was used to carry discharge water along the tunnel.
- A V-notch tank was used to measure the flow of the system.

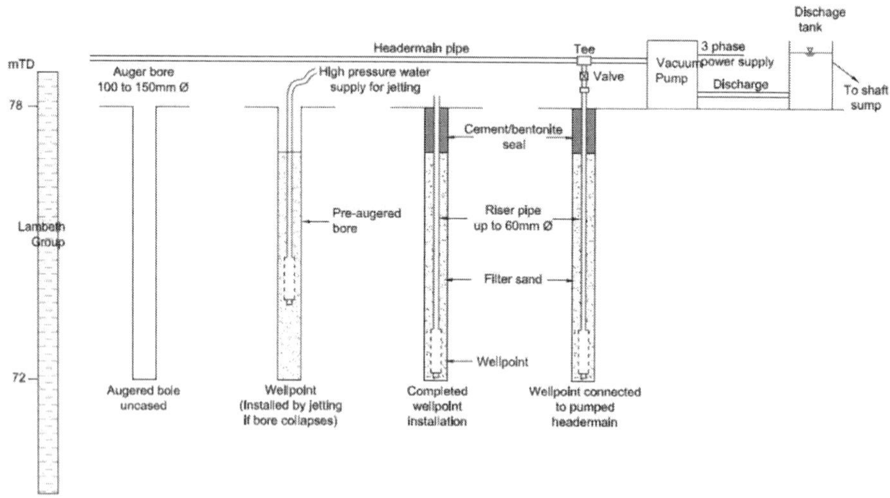

Figure 6 Typical in-tunnel well installation and set up

The inclusion of the in-tunnel wellpoints resulted in the depressurisation occurring as expected as a temporary "moving front" of depressurisation with the tunnel wellpoints being installed as the excavation progressed. Figure 7 shows an example of this front of reducing water level as the excavation approaches a piezometer at the end of the Westbound Cavern.

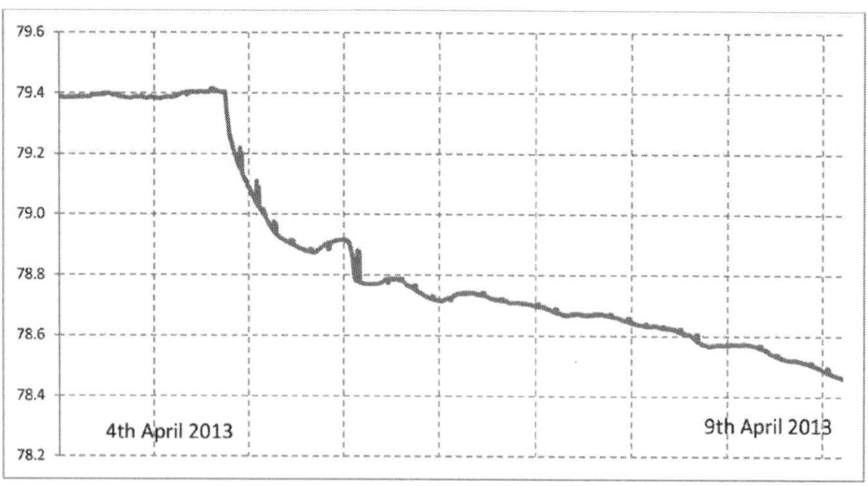

Figure 7 Example of pore pressure reduction achieved by additional in-tunnel wells

Managing, monitoring and performance during construction

Once installed and operational, a critical aspect of the depressurisation system was the continual monitoring of the piezometric levels, the maintenance of the system and management of the information. Separate strategies were developed for the surface and in-tunnel depressurisation measures.

Surface depressurisation system

Maintenance

The surface system consisted of 45 ejectors divided in two stations with three pumps in each: two duty pumps and a third one acting as a back-up in case of mechanical failure. In case of a power failure each station included stand-by generators. Daily maintenance was carried out by a resident operator with checks of the performance (flows, pressures, condition of the pipes etc.). To provide constant support to the 24/7 tunnelling operations an operator was on call to attend any failure in the system during night shifts.

Figure 8 Multiple surface well pump arrangement

An automated communication system was set up such that any loss of pressure or power failure would immediately be notified by text message to the dewatering operative and the SCL works "Person In Charge". Additionally a sound alarm would alert the rest of the site personnel to the situation. To ensure safety a "Safe stop" condition would be implemented in the SCL works, with removal of personnel, until remedial action had been taken to re-establish acceptable pore pressures.

Periodic maintenance activities included the removal of deposited iron crust caused by the presence of *Gallionella ferruginea* bacteria that affected the performance of the ejectors. The gradual reduction in ejector efficiency caused by these materials was noticeable from the piezometer records.

Monitoring & Management

For monitoring the drawdown 12 new piezometers were installed with piezometric level readings taken every two minutes and automatically sent to the Real Time Monitoring Centre (RTMC), where values were instantly visible. A back up unit guaranteed that the RTMC was always operational. Several gradational trigger levels were set to alert site personnel of unexpected changes of water levels that could affect the safety of SCL activities. If a piezometer reading was above a trigger level a text was sent to the dewatering operative and the SCL works "Person In Charge". Additionally a sound alarm would alert the rest of the site personnel. To ensure safety a "Safe stop" condition would be implemented in the SCL works, with removal of personnel, until remedial action had been taken to re-establish acceptable pore pressures.

Figure 9 Real Time Monitoring Centre

A daily review of the piezometric levels was carried out at the Shift Review Group meeting (SRG) in order to confirm the acceptability of continued excavation procedures.

In-tunnel depressurisation system

Maintenance

Similarly to the surface system, daily checks were carried out on the in-tunnel system. However, during excavation difficulties were encountered in achieving a completely dry excavation and a series of reviews and upgrades were undertaken to improve efficiency and performance.

First Upgrade

As a consequence of the presence of sandy material at the base of the invert together with water levels 500 mm above excavation level, some sloughing was observed during the excavation. However, where a sand/clay boundary was encountered in the face some residual water was typically encountered, even when the depressurisation measures appeared to have worked satisfactorily.

On occasions, contingency measures were required (sump pumps and caissons) which led to delays in construction progress. As invert level reduced as the westbound cavern excavation progressed away from the access shaft, there was an increasing risk of encountering "quick sand" phenomena. Due to that increasing risk the in-tunnel depressurisation system was reviewed in order to improve its performance. Several actions were identified and implemented:

- Lowering the vacuum pipework to reduce the suction head of the system.
- Increasing the diameter of the pipework to increase the flow. It was increased from PVC 25mm to PVC 50mm.
- Increasing the length of the porous filter in the well from 1m to 2m.
- New wells were installed with a layer of gravel to allow some 'breathing' that could help clean the screens and reduce clogging of the system.
- Back-flushing existing wellpoints to improve their operating efficiency.

When the above actions were completed, an immediate effect was observed in the piezometers resulting, as shown in figure 10, in an additional drawdown of 500mm.

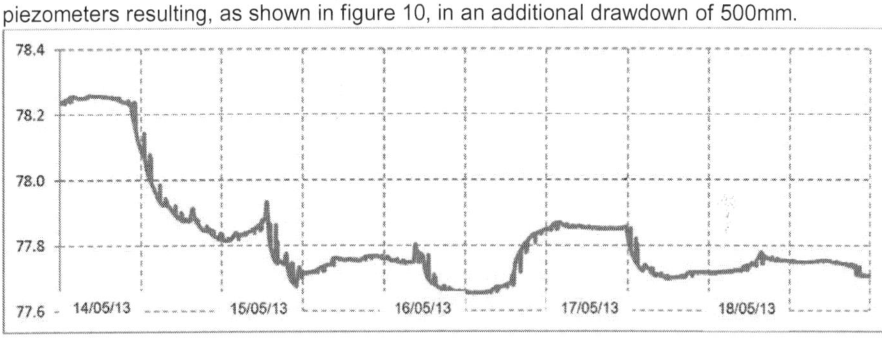

Figure 10 Pore pressure reduction achieved by additional in-tunnel measures (First Upgrade)

Second Upgrade

After completion of the first side drift of the westbound cavern, it was considered necessary to upgrade the depressurisation system once again to further improve performance and reduce groundwater level for the excavation of the second side drift and central pillar. Several actions were implemented as follows:

- Installation of new pipework and lowering of the pipework to reduce the suction head of the system.
- Upgrading the pipework from 50 mm PVC pipe to 100 mm steel pipe.
- New Geho piston pumps installed to replace the low flow DM-Vex vacuum pumps.
- The well filter specification in fine soils is always a compromise between meeting the conflicting requirement to keep out fines and let in water. As the amount of fines were not significant more permeable ones were specified for the wells. Figure 11 shows the revised wellpoint construction.

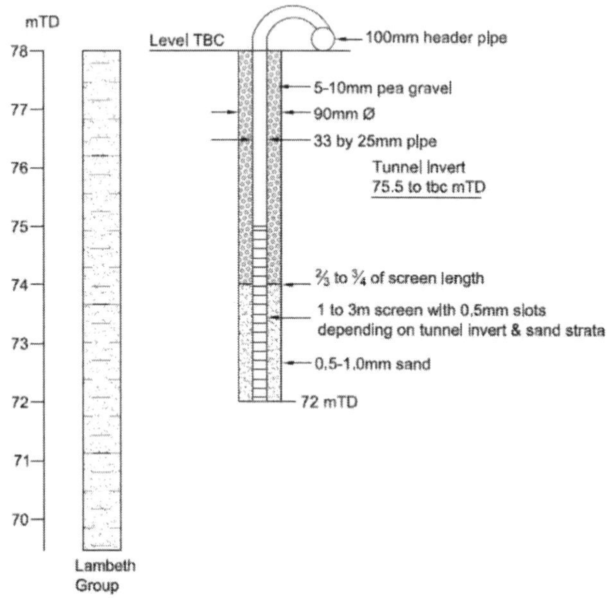

Figure 11 Revised wellpoint arrangement

These changes were initially tested in three trial wells the improved drawdown being demonstrated by the Figure 12. The changes ensured the remaining inverts of the westbound cavern were completed safely and without further requirements to reduce groundwater.

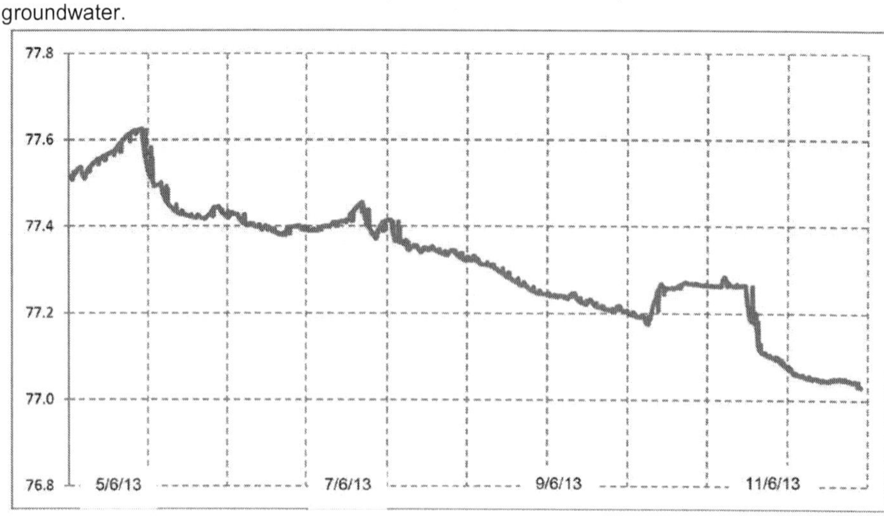

Figure 12 Pore pressure reduction achieved by additional in-tunnel measures (Second Upgrade)

Monitoring and management

The performance of the tunnel wellpoints was monitored by using un-pumped wellpoints as equivalent standpipe piezometers. This allowed the monitoring location and pumped wellpoints to be altered to suit the progress of the works. In addition, where required, dedicated piezometers were installed to 1 m below invert. Real time information from the piezometers was used to check trends and detect problems in any of the pumps. The invert excavation was the most critical part of the tunnel in terms of water bearing materials so a hold point was implemented within the excavation procedures effectively acting as a "permit to dig". The hold point required checking of water levels in the closest standpipe prior to the excavation, with the potential for revision of the depressurisation system and adoption of additional mitigation measures.

It was anticipated that, given the variability of the permeability and location of the sand channels, the design would have to be locally altered. Figure 13 illustrates an example where several contingency measures were implemented to aid in achieving a dry excavation. The need for such measures was infrequent and eliminated by the upgraded maintenance measures described above, improving both safety and efficency.

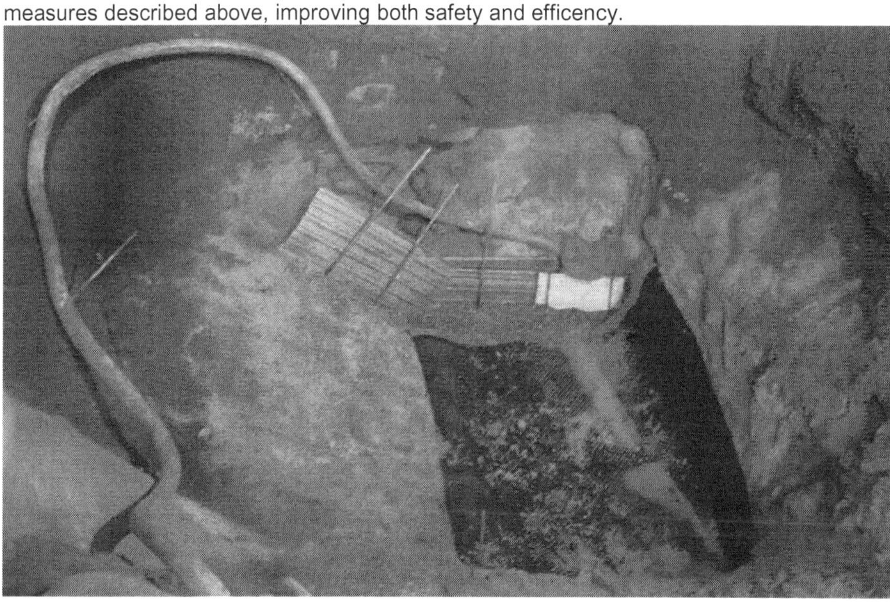

Figure 13 Additional contingency measures

Conclusions

The ground investigations, pumping test, designer's outline depressurisation design and final contractor's depressurisation design allowed the safe SCL construction of Stepney Green caverns in the Lambeth Group. An accurate ground model of the Stepney Green site was essential in understanding where to target investigations and to understand the variability of water bearing sandy materials in the Lambeth Group within which the deeper westbound cavern would be excavated. Close supervision and accurate measurement of pore pressures combined with use of multi-port piezometers ensured the groundwater data was reliable to inform the depressurisation plan.

Connectivity between Lambeth Group sand channels was demonstrated by both the pre-construction pumping test and by the use of surface ejectors and in-tunnel wellpoints. This was a key factor in being able to successfully use the contractors strategy of increased surface wells demonstrating the benefit of the GI work undertaken.

Key lessons learned included:

- Multi-port piezometers can be installed successfully allowing a reduction in the overall number of boreholes in a GI whilst providing reliable data for design purposes.
- Interconnectivity can exist between Lambeth Group sand channels, even if they are small and apparently separated by cohesive material.
- Adjustments and improvements to in-tunnel wellpoints such as increasing discharge pipe diameters can rapidly improve drawdown.
- Regular cleaning of ejectors rapidly improves drawdown.
- A certain residual amount of groundwater in Lambeth Group sandy materials can be expected, even when depressurisation appears to be successful.
- The drawdown response varied in relation to sand channel permeability, interconnectivity and morphology.
- The construction vindicated the requirement for a multi-level depressurisation approach using both surface and in-tunnel wells.

The use of automated instrumentation and communications in a real time context was successful in providing a robust safe action plan to forewarn of potential problems with the system.

The overall success was in part a result of the collaboration shown between the various parties in sharing knowledge of both the original design intent, the perception of the risks and the practical limits of implementing the systems. The knowledge sharing was aided by the ability to ensure continuity of personnel, with staff involved in the original GI's and pumping test being involved on site throughout construction. The significant improvements in depressurisation achieved with relatively minor changes to the system set-up also demonstrated the benefit of using experienced sub-contractors familiar with local conditions. The efficiency of the operation was such that the daily SCL advance rates improved from the programmed 0.75m/day to 1m/day.

Acknowledgements

The authors would like to acknowledge Crossrail, Dragados –Sisk Joint Venture, OTB Engineering, Mott MacDonald, W. J. Groundwater, ESG Ltd (formerly Soil Mechanics Ltd.), Geotechnical Consulting Group and Miguel Mucientes, formerly of Mott MacDonald, who undertook the pumping test analysis.

References

BS EN ISO 14686 (2003). Hydrometric determinations – Pumping tests for water wells – Considerations and guidelines for design, performance and use. BSI, London, UK.

CIRIA C583 (2004). Engineering in the Lambeth Group. CIRIA, London, UK.

Kruseman GP and de Ridder NA (1994) Analysis and evaluation of pumping test data. IILRI, Wageningen, Netherlands.

Crossrail Project: Infrastructure design and construction
ISBN 978-0-7277-6078-4

doi: 10.1680/cpid.60784.617

Index